IFMBE Proceedings

Volume 67

The International Federation for Medical and Biological Engineering, IFMBE, is a federation of national and transnational organizations representing internationally the interests of medical and biological engineering and sciences. The IFMBE is a non-profit organization fostering the creation, dissemination and application of medical and biological engineering knowledge and the management of technology for improved health and quality of life. Its activities include participation in the formulation of public policy and the dissemination of information through publications and forums. Within the field of medical, clinical, and biological engineering, IFMBE's aims are to encourage research and the application of knowledge, and to disseminate information and promote collaboration. The objectives of the IFMBE are scientific, technological, literary, and educational.

The IFMBE is a WHO accredited NGO covering the full range of biomedical and clinical engineering, healthcare, healthcare technology and management. It is representing through its 60 member societies some 120.000 professionals involved in the various issues of improved health and health care delivery.

More information about this series at http://www.springer.com/series/7403

Fatimah Ibrahim · Juliana Usman
Mohd Yazed Ahmad · Norhamizan Hamzah
Swe Jyan Teh
Editors

2nd International Conference for Innovation in Biomedical Engineering and Life Sciences

ICIBEL 2017 (in conjunction with APCMBE 2017), 10–13 December 2017, Penang, Malaysia

"Engineering Driven Innovation in Healthcare and Life Sciences Industries" 10th–13th December 2017, Golden Sands Resort Penang by Shangri-La, Penang, Malaysia

Springer

Editors
Fatimah Ibrahim
Department of Biomedical Engineering, Faculty of Engineering, Centre for Innovation in Medical Engineering
University of Malaya
Kuala Lumpur
Malaysia

Juliana Usman
Department of Biomedical Engineering, Faculty of Engineering
University of Malaya
Kuala Lumpur
Malaysia

Mohd Yazed Ahmad
Department of Biomedical Engineering, Faculty of Engineering, Centre for Innovation in Medical Engineering
University of Malaya
Kuala Lumpur
Malaysia

Norhamizan Hamzah
Department of Rehabilitation Medicine, Faculty of Medicine
University of Malaya
Kuala Lumpur
Malaysia

Swe Jyan Teh
Department of Biomedical Engineering, Faculty of Engineering, Centre for Innovation in Medical Engineering
University of Malaya
Kuala Lumpur
Malaysia

ISSN 1680-0737 ISSN 1433-9277 (electronic)
IFMBE Proceedings
ISBN 978-981-10-7553-7 ISBN 978-981-10-7554-4 (eBook)
https://doi.org/10.1007/978-981-10-7554-4

Library of Congress Control Number: 2017960293

Printed on acid-free paper

This Springer imprint is published by Springer Nature
The registered company is Springer Nature Singapore Pte Ltd.
The registered company address is: 152 Beach Road, #21-01/04 Gateway East, Singapore 189721, Singapore

Preface

The 2nd International Conference for Innovation in Biomedical Engineering and Life Sciences (ICIBEL 2017) in conjunction with the 10th Asia Pacific Conference on Medical and Biological Engineering (APCMBE 2017) was held in Penang, Malaysia, on December 10–13, 2017. ICIBEL was organized by the Centre for Innovation in Medical Engineering (CIME), Department of Biomedical Engineering, Faculty of Engineering, under the Innovative Technology Research Cluster, University of Malaya, and co-organized by the Society of Medical and Biological Engineering, Malaysia (MSMBE). This conference is endorsed by the International Federation for Medical and Biological Engineering (IFMBE), USA, and supported by the IEEE Engineering in Medicine and Biology Society (EMBS) Malaysia Chapter, Biomedical Engineering Society (BES), Singapore, IEEE University of Malaya Student Branch, Taiwanese Society of Biomedical Engineering, and Academy of Sciences Malaysia.

We are honored to accept 50 papers from Malaysia, Taiwan, Indonesia, Singapore, Iraq, Pakistan, India, Japan, Mexico and the USA after a peer-review process. This second proceeding volume provides an opportunity for readers to engage with a selection of refereed papers that were presented during the ICIBEL 2017 conference. The accepted papers will provide a platform to the state-of-the-art in research ideas, challenges, and findings to be shared with the academic world.

On behalf of the organizing committee, I would like to express our greatest appreciation to our reviewers, editors, keynote and invited speakers, paper presenters, participants, sponsors, exhibitors, supporters, organizing committee members, and advisory panels for ICIBEL 2017 for their support, commitment, and contributions that made this conference and proceedings successful.

Thank you.

Prof. Ir. Dr. Fatimah Ibrahim
Chairperson, ICIBEL 2017 Organizing Committee;
President, Society of Medical and Biological Engineering
Malaysia (MSMBE)
Kuala Lumpur, Malaysia

Conference Details

Name
2nd International Conference for Innovation in Biomedical Engineering and Life Sciences: ICIBEL 2017
In conjunction with
10th Asia Pacific Conference on Medical and Biological Engineering

Short Name
ICIBEL 2017 and APCMBE 2017

Venue
December 10–13, 2017
Golden Sands Resort Penang by Shangri-La, Penang, Malaysia

Proceedings Editors
Fatimah Ibrahim
Juliana Usman
Mohd Yazed Bin Ahmad
Norhamizan Hamzah
Swe Jyan Teh

Organized by
University of Malaya, Malaysia
Centre for Innovation in Medical Engineering (CIME)
Malaysia's Society of Medical and Biological Engineering (MSMBE)

Endorsed by
International Federation for Medical and Biological Engineering (IFMBE)

Supported by
IFMBE Asia Pacific Working Group
IEEE University of Malaya Student Branch
IEEE Engineering in Medicine and Biology Society (EMBS)
Biomedical Engineering Society, Singapore
Taiwanese Society of Biomedical Engineering
Academy of Sciences Malaysia

Organizing Committee

Chairperson
Fatimah Ibrahim

Co-chair
Prof. Dr. Tan Maw Pin

Publication
Dr. Fatimah Ibrahim
Juliana Usman

Secretary
Noraisyah Mohamed Shah

Treasurer
Mas Sahidayana Mohktar

Publicity/Logistic
Norhayati Soin

Technical
Mohd Yazed Ahmad
Norhamizan Hamzah
Swe Jyan Teh
Juliana Usman

Sponsorship/Exhibition
Siti Zawiah Binti Md Dawal
Wan Safwani Wan Kamarul Zaman

Secretariat Committee
Wan Safwani Wan Kamarul Zaman
Karunan Joseph
Nurhaslina binti Abd Rahman
Syarifah Aisyah Syed Ibrahim
Mohammad Faiz Zulkeflee
Yuslialif Yusup
Adhli Iskandar Putera Hamzah

International Advisory Board
Ahmad Sharif Hambali
Angela Ng Hwei Min
Anis Nurashikin
Badrul Hisham Yahya
Burhanuddin Yeop Majlis
Chandra Shekhar Sharma
James Goh Cho Hong
Marc Madou
Maziar Zarrehparvar
Nitish Thakor
Nur Aishah Mohd Taib
Sanjeev Saxena
Sasikala Thangavelu
Sergio O. Martínez
Toshiya Tamura
Zaidi Mohd Ripin
Zamri Radzi
Zamzuri Idris

About IFMBE

The International Federation for Medical and Biological Engineering (IFMBE) is primarily a federation of national and transnational societies. These professional organizations represent interests in medical and biological engineering. The IFMBE is also a non-governmental organization (NGO) for the United Nations and the World Health Organization (WHO), where we are uniquely positioned to influence the delivery of health care to the world through biomedical and clinical engineering.

The IFMBE's objectives are scientific and technological as well as educational and literary. Within the field of medical, biological, and clinical engineering, IFMBE's aims are to encourage research and application of knowledge and to disseminate information and promote collaboration. The ways in which we disseminate information include the following: organizing World Congresses and Regional Conferences, publishing our flagship journal Medical and Biological Engineering and Computing (MBEC), our Web-based newsletter—IFMBE News, our Congress and Conference Proceedings, and books. The ways in which we promote collaborations are through networking programs, workshops, and partnerships with other professional groups, e.g., Engineering World Health.

Mission

The mission of the IFMBE is to encourage, support, represent, and unify the worldwide Medical and Biological Engineering community in order to promote health and quality of life through advancement of research, development, application, and management of technology.

Objectives

The objectives of the International Federation for Medical and Biological Engineering shall be scientific, technological, literary, and educational. Within the field of medical, clinical, and biological engineering, its aims shall be to encourage research and the application of knowledge and to disseminate information and promote collaboration.

In pursuit of these aims, the Federation may, in relation to its specific field of interest, engage in any of the following activities: sponsorship of national and international meetings, publication of official journals, cooperation with other societies and organizations, appointment of commissions on special problems, awarding of prizes and distinctions, establishment of professional standards and ethics within the field, or in any other activities which in the opinion of the General Assembly or the Administrative Council would further the cause of medical, clinical, or biological engineering. It may promote the formation of regional, national, international, or specialized societies, groups or boards, the coordination of bibliographic or informational services, and the improvement of standards in terminology, equipment, methods and safety practices, and the delivery of health care.

In general, the Federation shall work to promote improved communication and understanding in the world community of engineering, medicine, and biology.

Contents

Part I
Telemedicine and Point of Care Healthcare Solutions

Pathological Hand Tremor Measurement—Challenges and Advances

Zaidi Mohd Ripin and Ping Yi Chan

Abstract

There are various wireless and wearable tremor assessment technologies available currently. In order to produce a reliable tremor assessment system to be used in practice, there is a need to improve the system in terms of processing time and quantification parameters. The future system should include an algorithm to separate postural and rest tremor and tremor stability index for better assessment of tremor in neurological disorders.

Keywords

Pathological tremor • Measurement technology

1 Introduction

Tremor is an involuntary rhythmical oscillatory muscle movement which can be categorized into various tremor types and syndromes. Tremors can be classified into two main types namely rest tremor and action tremor. Rest tremor involves muscles which are at full relaxation and supported against gravity. Action tremor occurs when there is voluntary muscle contraction. Action tremor can be furthered categorized into postural tremor, kinetic tremor, isometric tremor and task-specific tremor. Another major classification is the physiological and pathological tremor. Physiological tremor is related to normal functional tremble in maintaining the same posture, resting or moving, characterized by high frequency and low amplitude. Enhanced physiological tremor is tremor with large noticeable tremor amplitude. The frequency range for physiological tremor is wide, i.e. 3–30 Hz, whereas the frequency range for enhanced physiological tremor is 8–12 Hz [1, 2]. The common pathological tremors are essential and parkinsonian tremors. Essential tremor (ET) is the most common tremor disorder [3, 4] 20 times more prevalent than Parkinson's disease and usually exhibits postural and kinetic action tremors affecting mainly upper limbs (95%) and the head (35%), lower limbs (20%), voice (12%) and trunk (5%) [5]. It usually begins in the arm with extension-flexion of wrist or abduction-adduction of fingers [6] and spreads to other body parts for certain patients. The tremor frequency range is 4–12 Hz and the tremor progresses slowly over time and spread to other body parts [2]. It has been reported that 18.8% of patients in tertiary referral centre of USA have rest tremor [7].

Tremor is the early symptom for about 70% of the PD patients [8]. During the early onset of PD, the most predominant tremor is the rest tremor, which appears in 70% of the PD patients [9]. The typical frequency range involved is 3–10 Hz [10]. Some PD patients exhibit only tremor in one hand for many years [5]. Both ET and PD tremors hinder the patients' daily tasks like eating, writing, drinking liquid and dressing [1, 2].

2 Tremor Assessment Protocol

There are assessment tools and specific activities to be performed for a tremor type evaluation. The assessment tools used in clinical evaluation are tremor rating scales, whereas the assessment tools for biomechanical evaluation are measurement devices. The data obtained by the measurement

Z. M. Ripin (✉) · P. Y. Chan
TheVibrationLab, School of Mechanical Engineering, Engineering Campus, Universiti Sains Malaysia, 14000 Nibong Tebal, SPS, Pulau Pinang, Malaysia
e-mail: mezaidi@usm.my

F. Ibrahim et al. (eds.), *2nd International Conference for Innovation in Biomedical Engineering and Life Sciences*,
IFMBE Proceedings 67, https://doi.org/10.1007/978-981-10-7554-4_1

devices in biomechanical evaluation are processed before tremor quantification parameters are produced.

2.1 Types of Diagnostic Procedure for Tremor

A typical procedures in clinical diagnosis of tremor will begin with a complete and detailed *medical history* which includes age of onset, family history, usage of possible tremor stimulating drugs and mental condition [11]. *Functional neuroimaging* namely SPECT (Single photon emission computed tomography) and PET (Positron emission tomography) are used to diagnose tremor related disease and to discriminate PD and ET by detecting dopaminergic deficit in parkinsonian disorder [11]. In general, using the functional PET and SPECT can differentiate PD and ET but there are 15% of definite PD the absence of the dopaminergic deficit symptom [12] and requires further studies and confirmation [13]. Subsequently it is crucial to perform tremor assessment based on clinical or biomechanical evaluation.

2.1.1 Clinical Evaluation

Clinical evaluation of tremor focuses on understanding characteristics and evolution of tremor for the treatment of the associated disease. The evaluation is done through observational rating. Some ratings for assessing ET are Washington Height-Inwood Genetic Study of Essential Tremor (WHIGET) rating scale—wTRS [14], Essential Tremor Rating Assessment Scale—TETRAS and Fahn-Tolosa-Marin Tremor Rating Scale [15]. UPDRS and Hoehn and Yahr scales are the rating scales for Parkinson's disease. There is a reviewed version of UPDRS, namely MDS-UPDRS. All these rating scales are used to evaluate tremor at specific body segment and the severity of the tremor is given a score. These rating scores suffer from the subjectivity of a rater (a person who determines a rating).

2.1.2 Biomechanical Evaluation

Biomechanical evaluation involves the use of instruments to quantify tremor such as accelerometers [16], gyroscopes [17], inertial measurement unit, IMU [18], electrogoniometer [19], electromyography [20], spirography [10], active optical system and electromagnetic tracker system [11, 21].

Biaxial accelerometer was used in tremor pen stylus in quantifying the tremor symptom of Parkinson's disease [22]. Other example applications is the use of triaxial accelerometer in capturing the physiological tremor of normal persons and PD tremor [23]. Inertial measurement unit, IMU is an instrument which combines accelerometers and gyroscope, to measure linear and angular component of a motion. One example of the application of such measurement unit is the use of Xsens MTx in measuring the tremor frequency component of essential tremor patients [24].

2.2 Activities Perform During Upper Limb Tremor Assessment

The activities performed for assessment can be categorised according to the tremor type, i.e. rest tremor, postural tremor and kinetic tremor. In a study by Louis et al. [25] on assessing postural tremor of PD using WHIGET Tremor Rating Scale, the specific activities performed are holding arms in front of body and making wing beat posture. Stacy et al. [26] assessed postural tremor of ET using Fahn-Tolosa-Marin Tremor Rating Scale, whereby the subjects are asked to outstretch their arms with their wrist slightly extended and fingers are spread apart. In assessing postural tremor with The Essential Tremor Rating Assessment Scale (TETRAS), the posture assigned to subject is maintaining upper limb to be outstretched at 90° in front of body.

In general, the activities to be performed by subjects in kinetic tremor assessment are finger-to-finger, finger-to-nose, pronation-supination [27] and some specific tasks. Action tremor, including kinetic tremor of ET and PD can be rated using WHIGET Tremor Rating Scale which includes these activities: drinking water using a cup, drinking water using a spoon, finger-to-nose movement, pouring water between two cups and drawing Archimedes spirals. In the Fahn-Tolosa-Marin Tremor Rating Scale, the specific movements assigned to ET patients are finger-to-nose movement, writing and drawing.

Posture for assessing rest tremor has less variation. Arms are required to put on armrest based on Unified Parkinson's Disease Rating Scale (UPDRS) protocol for assessing PD tremor as in the study done by Louis et al. [25]. The protocol in assessing rest tremor using WHIGET Tremor Rating Scale requires subjects to put arms on laps instead [14].

2.3 Data Processing and Analysis

There are tremor quantification parameters developed to quantify tremor in way such that the severity of a tremor can be ranked [18, 22]. Direct analysis can be interpreted from the stated domains by evaluating the change in the magnitude over domain and also the distribution pattern. Examples of the parameters used in describing the characteristics of tremor are centre frequency (the median of frequency), the frequency distribution (standard deviation of centre frequency) [22, 23]. In addition, there are derivations of the measured data for instance expressing the measured signals in root mean square [22], power [23] and logarithm [18]. Some of the developed tremor quantification parameters are correlated with clinical evaluation.

3 Tremor Measurement Technology

Two commercially available tremor measurement technologies are discussed in detailed in this section. The first is Coordination ability test system, CATSYS which consists of a tremor pen, touch recording plate, reaction time handle and static posturography as shown in Fig. 1. Parameters used in tremor quantification with biaxial microaccelerometer of tremor pen stylus are tremor intensity, centre frequency, frequency dispersion and harmonic index [22]. The clinical trial tremor measurement result shows that tremor intensity is correlated with UPDRS (rest tremor of right and left hands are $\rho = 0.680$ and $\rho = 0.739$ respectively and postural tremor of right and left hands are $\rho = 0.803$ and $\rho = 0.793$ respectively).

Kinesia^{TM} Technology assesses PD tremor using wireless finger-worn motion sensors of triaxial gyroscopes and triaxial accelerometers. It has wrist-worn integrated data acquisition and wireless data telemetry also, as depicted in Fig. 2. The outputs of the motion sensors are angular velocity, angle and acceleration. The parameters which are used to be correlated with clinical scores are sensor output rms (root mean square), log of sensor output power and log of sensor output rms [18]. The quantification parameters correlated well with the average clinical scores in UPDRS in determining the severity of rest ($r^2 = 0.89$), postural ($r^2 = 0.90$) and kinetic tremors (0.69).

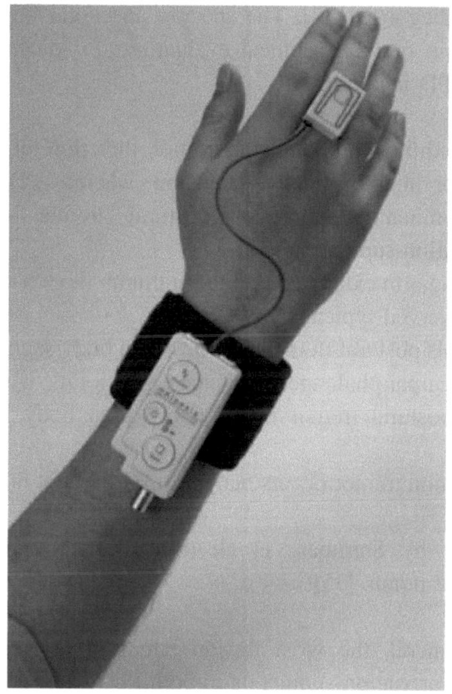

Fig. 2 Patient-worn Kinesia technology [18]

3.1 Challenges in Tremor Measurement

There are many tremor assessment systems developed [17, 19, 23, 28, 29], however these systems have limitations in providing a confirmed diagnosis of tremor associated disease especially Parkinson's disease and essential tremor. It was reported that there were as much as 10% of actual PD patients not diagnosed until the disease reached severe stage [30]. Besides, 25–50% of essential tremor patients were misdiagnosed to have Parkinson's disease [31, 32]. Co-occurrence of PD and ET adds complication into the

Fig. 1 A set of instruments in CATSYS system which consists of a data logger and four sensors, including a force plate for balancing measurement (static posturography), a reaction handle, a touch recording plate and a tremor pen [22]

clinical diagnosis [33]. The specific anecdotal impressions realized in previous clinical evaluations for distinguishing ET and PD include:

- ET exhibits predominant tremor direction of postural tremor in wrist flexion-extension, whereas PD exhibits predominant tremor of postural tremor in wrist pronation-supination [33].
- During arm extension, tremor in thumb flexion-extension is observed typically in PD [33].
- PD has postural tremor in more distal body segments, i.e. metacarpal-phalangeal and phalangeal joints; whereas ET has postural tremor in more proximal body segments [33].

  Intention tremor occurs mainly in ET and not in PD [34].

Study by Sternberg et al. (n = 100) showed several important points [35]:

- In general, the wrist flexion-extension is greater than wrist pronation-supination in both PD and ET. However, the predominance of wrist flexion-extension tremor over wrist pronation-supination in ET is greater than that of PD.
- Tremors in metacarpal-phalangeal and phalangeal joints are more intense than wrist joint in PD patients, whereas tremors in ET patients exhibit the opposite tremor characteristics.
- Intention tremor occurs in 28% of ET patients, whereas it occurs 4% in PD patients only.
- This result implied that detailed biomechanical measurement of the hand tremor can assist the clinicians in making more objective assessment of tremor.

4 Advances and Challenges in Hand Tremor Measurement

There have been many major development on hand tremor measurement, notably the improvement and the use of joint angle in the quantification of tremor [36, 37]. In this patented approach, the tremor of the hand and arm are measured based on the joint angle which can be related to the natural axis of the hand and can be used for better communication between physicians and also to communicate to the patients to describe the tremor direction. With the joint angle, the tremor can be described in degrees or degree/sec of the pronation-supination, flexion-extension or abduction-adduction of the hand and since the tremor can be a mixture of any of the joint angle, this has opened up a lot of opportunities for measurement of the hand tremor and for diagnostics based on the observations by Stenberg et al. [35]. TheVibrationLab of USM has also develop a two-degrees-of-freedom hand tremor simulator to study the accuracy and reliability of the joint angle tremor measurement system and the results have been submitted for consideration for publication [38].

4.1 Wearable Wireless Measurement System

One of the main problem with measurement of the hand tremor of the patients is the fact that tremor does not come at a specific time which meant that physicians may easily miss the observation of the hand tremor of the patient during consultation. A wearable measurement system that can measure the hand tremor anytime can provide ample data which can be used in diagnostics. The need for wireless system is evident due to the constraint of not obstructing the patients from their daily activities which led to the requirement of wireless hand tremor measurement system. A typical wireless system has been developed by TheVibrationLab USM which provides more freedom to the patient however due to reliant on the bluetooth technology the sensors must be within the proximity of the receiver. Another issue with the wireless sensor is the need to store the data and also the energy management of the system. A typical sampling of the patient hand tremor would be during the awake time and this entails a 16-h sampling and storage. A typical wireless sensor has a 4-h battery life. A practical solution is to have a large on-board battery and also computer to manage the energy and the memory.

Not all movement is tremor and the system should have the capability to filter out intended motion from the tremor which requires a relatively complex algorithm beyond the simple low pass filter.

4.2 An Algorithm to Separate Postural and Rest Tremor

The existing hand tremor measurement system that is available nowadays does not differentiate the postural from the rest tremor. The tendency is to accumulate the data and to statistically process the data by using the root mean square value. This made it difficult to correlate with the existing hand tremor assessment rating scale as the rating are based on a particular posture. What is required is for the measurement system to measure daily hand tremor and to relate with the rating scale. This is often not possible because of the required posture when rating is made. An intelligent system than can detect the posture, classify it and capture the data according to the posture would allow for better correlation with the clinical observations made using established protocols such as the UPDRS. Clearly more work is required in this front. The miniaturization of the computer has provided the opportunity to have a strap-on computer on the hand to network with all the sensors and to run a relatively complex algorithm for tremor measurement and classification. This feature is also more demanding as the use of more IMU sensors in order to identify the posture of the hand and also measure the tremor.

4.2.1 Tremor Stability Index

As the number of clinical research on hand tremor measurement system increases, there are now more detailed observations made that can improve the understanding of the tremor motion. One of the major landmark work is by di Biase et al. [39] which has shown that by calculating the tremor stability index based on the frequency shift of the tremor signal from a continuous time-series data of the hand tremor, the difference between ET and PD can be made with maximum sensitivity, specificity and accuracy were 95, 95 and 92%, respectively. This indicates that this new approach can be incorporated in the post processing analysis of the future measurement system and can greatly add value to the measurement system.

5 Conclusion

The various types of tremor, assessment protocol and the tremor measurement system have been presented. The challenges of the future measurement system are to ensure enough sampling time in a wireless and wearable device with processing capability to indicate whether the tremor measurement is made under rest or postural and to allow for comparison with existing tremor ranking scale such as UPDRS or WHIGET. The future system is also expected to be able to calculate the tremor stability index from the obtained data which would enable physicians to make a better diagnosis of the disease and its progression in the patient.

References

1. Orbeck, N., Denninger, T., Gewaltig, M.O., Leonid L.R., et al.: Scholarpedia **2**(10), 1379 (2007). https://doi.org/10.4249/scholarpedia.1379 (revision# 89108 [link to/cite this article])
2. Struble, L.M.: Tremors: learning to stop the shakes. Nurse Pract. Am. J. Prim. Health Care **35**, 18 (2010)
3. Louis, E.D., Ottman, R., Allen Hauser, W.: How common is the most common adult movement disorder? Estimates of the prevalence of essential tremor throughout the world. Mov. Dis. **13**, 5–10 (2004)
4. Rautakorpi, I., Takala, J., Marttila, R., Sievers, K., Rinne, U.: Essential tremor in a Finnish population. Acta Neurol. Scand. **66**, 58–67 (1982)
5. Jankovic, J.: Treatment of hyperkinetic movement disorders. Lancet Neurol. **8**, 844 (2009)
6. Louis, E.D.: Essential tremor. New Engl. J. Med. **345**, 887–891 (2001)
7. Louis, E.D.: Essential tremor. Lancet Neurol. **4**, 100–110 (2005)
8. WebMD: Parkinson's disease health center, symptoms (2010). Available: http://www.webmd.com/parkinsons-disease/guide/parkinsons-disease-symptoms. 29 Aug 2012
9. Jankovic, J.: Parkinson's disease: recent advances in therapy. South. Med. J. **81**, 1021–1027 (1988)
10. Mansur, P.H.G., Cury, L.K.P., Andrade, A.O., Pereira, A.A., Miotto, G.A.A., Soares, A.B., Naves, E.L.: A review on techniques for tremor recording and quantification. In: Critical Reviews™ in Biomedical Engineering, vol. 35, pp. 343–362 (2007)
11. Andrade, A.O., Pereira, A.A., Soares, M.F., de Almeida, G.L.C., Paixão, A.P.S., Fenelon, S.B., Dionisio, V.C.: Human tremor: origins, detection and quantification (2012)
12. Shahed, J., Jankovic, J.: Exploring the relationship between essential tremor and Parkinson's disease. Parkinson. Relat. Disord. **13**, 67–76 (2007)
13. Wang, J., Hoekstra, J.G., Zuo, C., Cook, T.J., Zhang, J.: Biomarkers of Parkinson's disease: current status and future perspectives. Drug Discovery Today **18**, 155–162 (2013)
14. Louis, E.D. Barnes, L., Wendt, K.J., Ford, B., Sangiorgio, M., Tabbal, S., Lewis, L., Kaufmann, P., Moskowitz, C., Comella, C. L.: A teaching videotape for the assessment of essential tremor. Mov. Disord. **16**, 89–93 (2001)
15. Stacy, M.A., Elble, R.J., Ondo, W.G., Wu, S.C., Hulihan, J.: Assessment of interrater and intrarater reliability of the Fahn–Tolosa–Marin tremor rating scale in essential tremor. Mov. Disord. **22**, 833–838 (2007)
16. Bankole, A., Anderson, M., Knight, A., Oh, K., Smith-Jackson, T., Hanson, M.A., Barth, A.T., Lach, J.: Continuous, non-invasive assessment of agitation in dementia using inertial body sensors. In: Proceedings of the 2nd Conference on Wireless Health, p. 1 (2011)
17. Salarian, A., Russmann, H., Vingerhoets, F.J., Dehollain, C., Blanc, Y., Burkhard, P.R., Aminian, K.: Gait assessment in Parkinson's disease: toward an ambulatory system for long-term monitoring. IEEE Trans. Biomed. Eng. **51**, 1434–1443 (2004)
18. Giuffrida, J.P., Riley, D.E., Maddux, B.N., Heldman, D.A.: Clinically deployable Kinesia™ technology for automated tremor assessment. Mov. Disord. **24**, 723–730 (2009)
19. Richards, C., Malouin, F., Bedard, P., Lemieux, R.: Characteristics of the electromyographic patterns of lower limb muscles during gait in patients with Parkinson's disease when OFF and ON L-Dopa treatment. Ital. J. Neurol. Sci. **18**, 195–208 (1997)
20. Morrison, S., Mills, P., Barrett, R.: Differences in multiple segment tremor dynamics between young and elderly persons. J. Gerontol. Ser. A: Biol. Sci. Med. Sci. **61**, 982–990 (2006)
21. Rajaraman, V., Jack, D., Adamovich, S., Hening, W., Sage, J., Poizner, H.: A novel quantitative method for 3D measurement of Parkinsonian tremor. Clin. Neurophysiol. **111**, 338–343 (2000)
22. Papapetropoulos, S., Katzen, H.L., Scanlon, B.K., Guevara, A., Singer, C., Levin, B.E.: Objective quantification of neuromotor symptoms in Parkinson's disease: implementation of a portable, computerized measurement tool. Parkinsons Dis. **2010** (2010)
23. Barroso, M.C., Esteves, G.P., Nunes, T.P., Silva, L.M., Faria, A. C., Melo, P.L.: A telemedicine instrument for remote evaluation of tremor: design and initial applications in fatigue and patients with Parkinson's Disease. Biomed. Eng. OnLine **10**, 14 (2011)
24. Uhríková, Z., Šprdlík, O., Hoskovcová, M., Komárek, A., Ulmanová, O., Hlaváč, V., Nugent, C.D., Růžička, E.: Validation of a new tool for automatic assessment of tremor frequency from video recordings. J. Neurosci. Methods **198**, 110–113 (2011)

25. Louis, E.D., Levy, G., Côte, L.J., Mejia, H., Fahn, S., Marder, K.: Clinical correlates of action tremor in Parkinson disease. Arch. Neurol. **58**, 1630 (2001)
26. Stacy, M.A., Elble, R.J., Ondo, W.G., Wu, S.C., Hulihan, J.: Assessment of interrater and intrarater reliability of the Fahn–Tolosa–Marin tremor rating scale in essential tremor. Mov. Disord. **22**, 833–838 (2007)
27. Rocon, E., Belda-Lois, J., Sanchez-Lacuesta, J., Pons, J.: Pathological tremor management: modelling, compensatory technology and evaluation. Technol. Disabil. **16**, 3–18 (2004)
28. Comby, B., Chevalier, G., Bouchoucha, M.: A new method for the measurement of tremor at rest. Arch. Physiol. Biochem. **100**, 73–78 (1992)
29. Hanson, M.A., Powell, H., Frysinger, R.C., Huss, D.S., Elias, W. J., Lach, J., Brown, C.: Teager energy assessment of tremor severity in clinical application of wearable inertial sensors. In: Life Science Systems and Applications Workshop, 2007. LISA 2007. IEEE/NIH, pp. 136–139 (2007)
30. Cummings, J.L.: Understanding Parkinson disease. JAMA J. Am. Med. Assoc, **281**, 376–378 (1999)
31. Bhagwath, G.: Tremors in elderly persons: clinical features and management. Hosp. Physician **37**, 31–49 (2001)
32. Benito-León, J., Louis, E.D.: Essential tremor: emerging views of a common disorder. Nat. Clin. Pract. Neurol. **2**, 666–678 (2006)
33. Thenganatt, M.A., Louis, E.D.: Distinguishing essential tremor from Parkinson's disease: bedside tests and laboratory evaluations. Exp. Rev. Neurother. **12**, 687–696 (2012)
34. Louis, E.D., Frucht, S.J., Rios, E.: Intention tremor in essential tremor: prevalence and association with disease duration. Mov. Disord. Off. J. Mov. Disord. Soc. **24**, 626 (2009)
35. Sternberg, E.J., Alcalay, R.N., Levy, O.A., Louis, E.D.: Postural and intention tremors: a detailed clinical study of essential tremor vs. Parkinson's disease. Front. Neurol. **4**, 1–8 (2013)
36. Ripin, Z.M., Chan, P.Y.: Pathological tremor measurement and reproduction. Presented at the international conference on engineering of Tarumanagara (ICET 2015), faculty of engineering, Tarumanagara University, Jakarta, Indonesia (2015)
37. Chan, P.Y., Ripin, Z.M.: Development of wearable inertial sensors for measurement of hand arm tremors. In: 2013 IEEE International Conference on Smart Instrumentation, Measurement and Applications (ICSIMA), pp. 1–6 (2013)
38. Chan, P.Y., Ripin, Z.M.: An in–laboratory validity and reliability tested system for quantifying hand–arm tremor in motions. IEEE Trans. Neural Syst. Rehabil. Eng. Paper submitted on 11th Apr 2017
39. di Biase, L., Brittain, J.S., Shah, S.A., Pedrosa, D.J., Cagnan, H., et al.: Tremor stability index: A new tool for differential diagnosis in tremor syndromes. Brain **140**, 1977–1986 (2017)

Non Invasive Cardiovascular Health Assessment in Post-adolescent Age Group Using Augmentation Index

Sai Karun, Suhan S. Nath, Kunal Bharathi, Manjunath Gaonkar, and B. Niranjana Krupa

Abstract

This paper presents a non-invasive method of classifying a subject's health into either one of two classes depending on the condition of the subjects' cardiovascular health. Novelty of the work lies in segregating the subjects who had good cardiovascular health and the subjects who were at risk. In the proposed work, VPG signals and other general information such as weight, height, BP were collected from 38 individuals in the post-adolescent age group. Furthermore, a signal from a person who was known to have good cardio vascular health was collected. This signal became the reference that was used to compare with other signals. After the initial pre-processing, the PPG and APG were obtained from the VPG. Then a total of 7 features of the wave contour from the APG and PPG signals were extracted. Based on the Augmentation Index the signals were classified into two classes using SVM and ELM classifiers. Where one class represented healthy individuals and the other class represents the individuals at risk of CVD. The average values of the extracted features were used and the final accuracy obtained was also an average value. The accuracy obtained using ELM classifier with K-fold cross validation was 77%. Whereas the efficiency achieved using SVM was 94.59%. Hence the proposed method can be used to assess the vascular health analysing PPG signals in the post-adolescent age group.

Keywords

Cardiovascular diseases • Photoplethysmogram (PPG) • Velocity photoplethysmogram (VPG) • Acceleration photoplethysmogram (APG) • Extreme learning machine (ELM) • Support vector machines (SVM)

1 Introduction

Cardiovascular diseases (CVD) are identified as one of the leading causes of death in the world today, accounting for 17.3 Million cases in 2013. CVDs are the diseases involving the heart or blood vessels. CVDs are the single largest cause of death worldwide [1]. The trends indicate the seriousness and danger that CVDs pose. The current tests that indicate whether a person has CVD is time consuming and very expensive. PPG is an optical method that is non-invasive and is used to detect flow of blood and changes in volume in peripheral vessels and smaller arteries at various locations in the body [2]. Light travelling through biological tissue can be absorbed by various substances. Almost all the changes in blood flow are mainly observed in the arteries and arterioles. The changes in the blood flow volume in the microvascular bed of tissue is detected using PPG sensor via reflection from or transmission through the tissue [3].

The peaks and valleys which are needed for analysis may not be very clear in a few PPG signals [4]. As a result, the Velocity Photo Plethysmogram (VPG), the first derivative of PPG is used. In the original PPG, sometimes there is a difficulty in detecting small changes in the phase of the

S. Karun · S. S. Nath · K. Bharathi · M. Gaonkar · B. N. Krupa (✉)
PES Institute of Technology, Bengaluru, 560085, India
e-mail: bnkrupa@pes.edu

F. Ibrahim et al. (eds.), *2nd International Conference for Innovation in Biomedical Engineering and Life Sciences*, IFMBE Proceedings 67, https://doi.org/10.1007/978-981-10-7554-4_2

inflections. So as an alternative the second derivative of the PPG signal is obtained which is also known as the Acceleration Photo Plethysmogram (APG). Even though the clinical significance of PPG has been documented there remains a lack of documentation on how healthy a person is in a young age group. Recent studies have found that 230,600 heart attack hospitalizations between 2001 and 2010 have occurred for patients between the ages 30 and 54 [5]. Another study also shows 10% of all heart attacks occur before the age of 45 [6]. So it becomes vital to identify people who are at risk of CVDs so as to prevent it. Post-adolescent age group consists of people in between 19 and 25 years. Thus, this investigation, is aimed at identifying people who are not healthy as others in the post-adolescent age group. Identification of healthy and unhealthy in the proposed work is done based on the Augmentation Index (AI) calculated from the APG signal [7].

An SVM model is a representation of the examples as points in space, mapped so that the examples of the separate categories are divided by a clear gap that is as wide as possible. New examples are then mapped into that same space and predicted to belong to a category based on which side of the gap they fall on. ELM theories show that the hidden neurons are important but can be randomly generated and independent from applications. ELMs, which can be biologically inspired, offer significant advantages such as fast learning speed, ease of implementation, and minimal human intervention.

In this study, algorithms have been developed to extract various features and using these features healthy and unhealthy classes in the post adolescent age group based on AI using ELM and SVM classifiers have been identified.

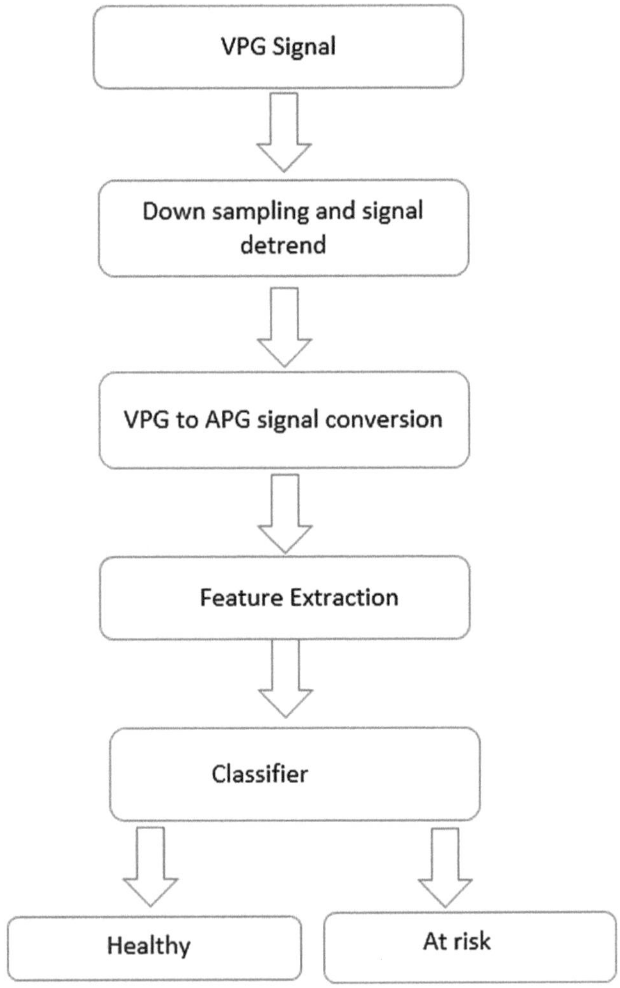

Fig. 1 Flow of the proposed work

2 Methodology

This section provides a detailed explanation of data acquisition, signal preprocessing, feature extraction and classification. Below illustrates the flow diagram of the work that was carried out (Fig. 1).

2.1 Dataset Overview and Acquisition

The dataset used in the current work consists of 36 VPG signals consisting of both genders were recorded using the pulse sensor [8]. In accordance with the declaration of Helsinki, informed consent was obtained from all the subjects. These signals were recorded for a duration of 2 min and sampled at a frequency of 50 Hz.

Before the signal was collected, it was verified whether the subjects satisfied a few inclusion criteria. Subjects were both non-diabetic and non-hypertensive, hadn't consumed food or caffeine based products six hours prior, alcohol products twenty-four hours, hadn't smoked two hours and hadn't exercised two hours prior to the recording [9].

As the procured signals had to be compared with a base/reference, a PPG signal was collected from an athlete who was known to have good cardiovascular health. The athlete belonged to the same post-adolescent age group. The Pulse Sensor was placed on the right hand index finger of the subject [10] while they were made to sit in upright position [11]. The VPG was obtained from the sensor.

Omron Blood Pressure Monitor, HEM-7120 was used to find the Blood pressure of the subject. This was noted before the VPG signals were recorded. The weight of each subject was measured using a digital weighing machine. Different demographic data such as height, weight, blood pressure, type of diet, etc. was collected. From their height and weight, Body Mass Index of each individual was calculated in kg/m^2. Information such as blood pressure, smoking status and alcohol consumption can give an indication of the cardiovascular health. A protocol was developed and followed during data acquisition to ensure that the data was uniform.

2.2 Preprocessing

In this section, unwanted artifacts, baseline wandering etc. were removed from the signal to make it suitable for feature extraction. The signal was first down sampled by a factor of 975, then, the signal was detrended to remove the trend from it. Next, the signal was split into 4 windows where each window consist of 1500 samples each which corresponds to 30 s, now one window was selected and the VPG signal was passed through a FIR bandpass filter to get the first derivative of VPG called the APG signal. A filter with Order 50, lower and upper cut-off frequencies of 0.1 and 10 Hz respectively. These filter specifications have been derived based on the power spectral density of VPG. Finally each cycle was extracted from the selected APG window.

2.3 Feature Extraction

A set of six features extracted from PPG, VPG and APG signals used in the work are presented below:

B/A Ratio. This ratio was determined from the Acceleration plethysmogram (second derivative of the PPG signal). The absolute values of the 'a' and 'b' waves of the APG were taken as 'A' and 'B' respectively [4].
P0/P1 Ratio. This ratio was determined from PPG and was calculated as the amplitude from the foot to the valley point (point where dichrotic notch is located) divided by the amplitude from the foot to the systolic peak [4].
ΔTBVP/T. ΔTBVP represents the absolute time delay between the diastolic and the systolic peaks and T represents the time period of the PPG signal [4]. The time factor to be multiplied to the samples was 0.02 s as the signal had a sampling frequency of 50 Hz.
Stiffness Index (S.I.). The assumption made here was that the path length was proportional to a person's natural height, the stiffness index has been defined as S.I. (m/s) = Body Length/ΔTBVP.
ΔTBVP was found in the same way as mentioned above. The data from the demographics was used along with ΔTBVP to find the S.I. [4]. Area under the systolic peak. The area under the systolic peak was calculated as the area lying under the systolic peak and the foot of the wave [4].
Augmentation Index (A.I.). Augmentation Index was defined as the ratio of the amplitude of late systolic peak(Y) to the amplitude of the early systolic peak (X) [12] (A.I. = Y/X).

Augmentation index has been shown to be a sensitive indicator of arterial status and known to be a predictor of adverse cardiovascular events in a diverse patient population [7]. The early systolic peak maps to the B wave in the APG, similarly, the late systolic peak maps to the C wave in the APG [12].

Detection of A wave was done by finding the maximum peak, B wave was determined by inverting the APG and finding the maximum peak and C wave was obtained by finding out the next maximum peak in the APG (see Fig. 2).

ZC1, ZC2, ZC3 and ZC4 represent the four zero crossings of the APG wave respectively (see Fig. 3).

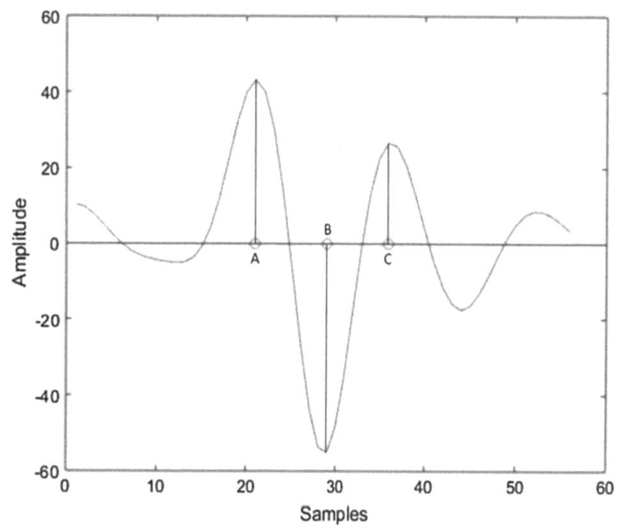

Fig. 2 A, B and C wave detection

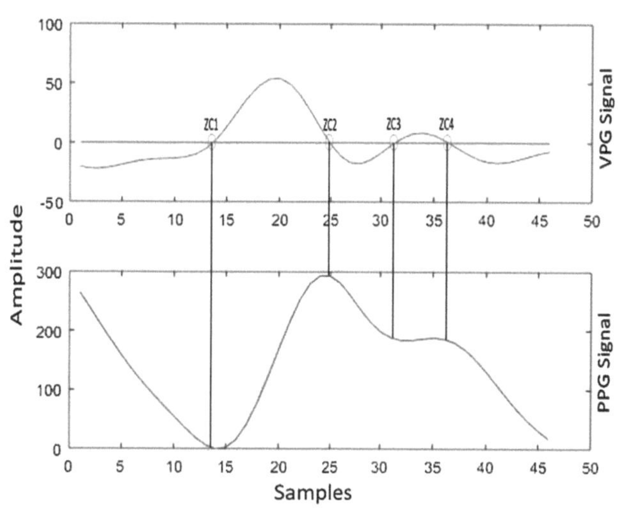

Fig. 3 Mapping of VPG on PPG

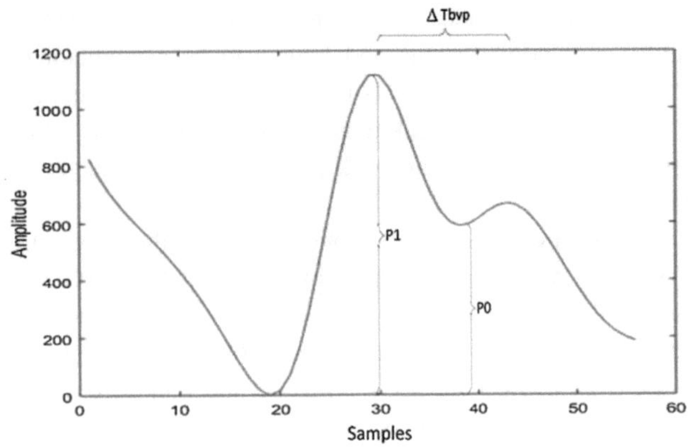

Fig. 4 Computation of ΔTBVP

P0 was obtained by determining the third zero crossing in the VPG signal and mapping it back to the PPG signal. P1 was obtained by calculating the Systolic peak amplitude in PPG. Diastolic peak required for the calculation of ΔTBVP was obtained by mapping the fourth zero crossing in VPG to PPG (see Fig. 4). Finally, for A.I. X and Y was obtained by adding the sample difference of A&B and sample difference of A&C to the starting point of systolic peak in the PPG signal, respectively.

Input feature set to classifier consist of following features, F = {B/A, P0/P1, ΔTBVP/T, S.I, Area under systolic curve, A.I.}.

2.4 Classification

The features extracted in the previous section was used to classify the samples into two classes. The dataset was divided into 2 classes based on the increase in A.I., one where the A.I. was between 0.70 and 0.919 and the other class where the A.I. was greater than 0.919. Classification has been performed using two machine learning algorithms; ELM and SVM.

Extreme Learning Machine (ELM). The ELM [13] for SLFNs indicates that hidden nodes can be randomly generated. The mapping of the input data to L-dimensional ELM random feature space, and the network output was

$$F_L(x) = \sum_{i=1}^{L} \beta_i h_i(x) = h(x)\beta \tag{1}$$

β = [β1, …, βL]′ represents the output weight matrix between the hidden nodes and the output nodes. For the better accuracy, K-fold cross validation was used and average efficiency was computed. Totally 25 samples were used for training and 13 for testing.

Support Vector Machines (SVM). SVM, a learning machine based on the principle of Structural Risk Minimization theory was used [14]. The SVM classification was performed with a Linear Kernel and four fold cross validation.

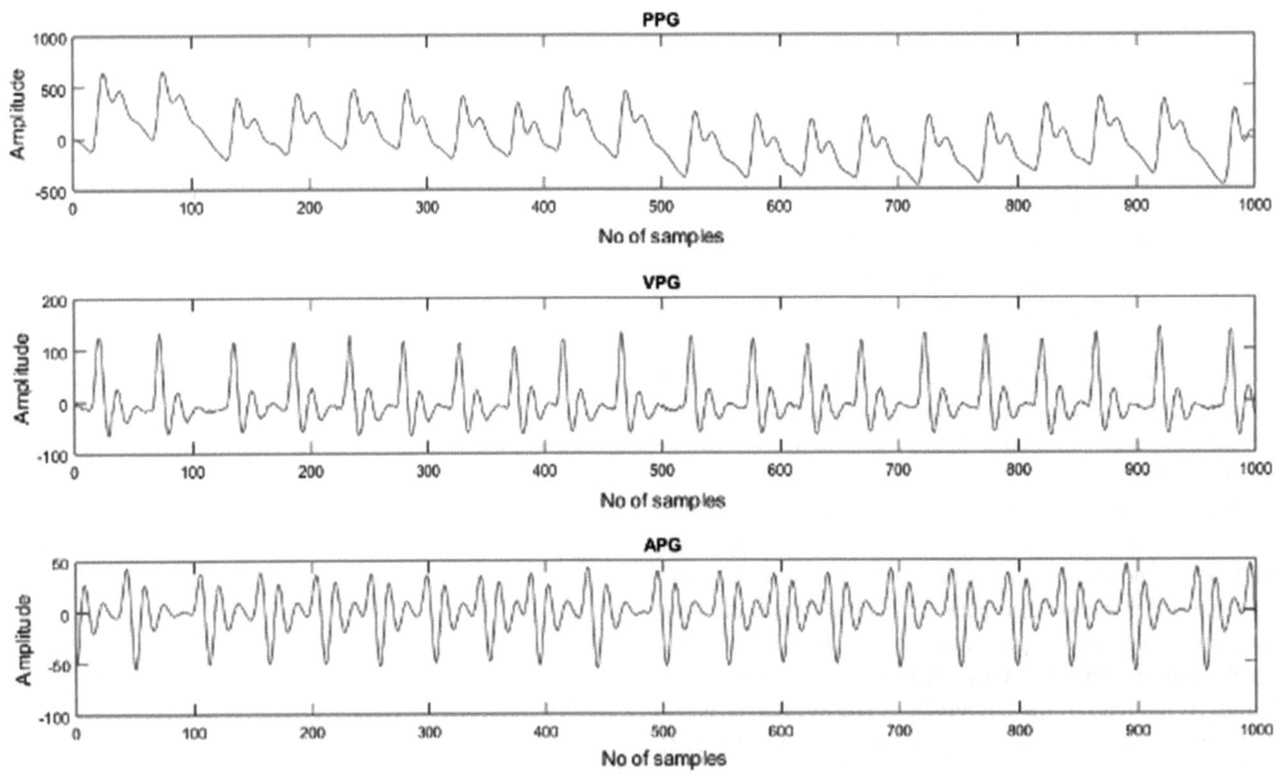

Fig. 5 Preprocessed PPG, VPG and APG signals

3 Results and Discussion

The detailed performance of the feature extraction and classification is discussed in this section. Also, the role of pre-processing is indicated by presenting the necessary plots.

Firstly, the signal was down sampled, after which the DC shift in the signal was removed. This signal is then divided into four windows, each of 30 s. Next, the signal is passed through a band pass filter so as to differentiate the input signal. Finally, the signal is segmented in a way that individual segments of a signal were obtained. The preprocessed PPG, VPG and APG are obtained respectively (see Fig. 5).

Using the formulae previously discussed, features for all 38 signals have been extracted and documented. A binary classification using Augmentation Index is performed on the set of features. Firstly, an ELM classifier with 23 neurons and a sigmoidal activation function is used for classification. This resulted in an average k-fold cross validation accuracy of 77%, for a *k*-value of four.

Later the same features set was given as input to an SVM classifier with linear kernel which yielded a 4-fold cross validation accuracy of 94.59%. It was observed that SVM was more successful in classifying the data than ELM.

4 Conclusion

In this paper, a novel method has been presented to identify unhealthy or subjects who are at risk of developing CVDs in a young age group. Augmentation index which is a major indicator of heart diseases was used to segregate data. Using popular machine learning algorithms the recorded data was classified and accuracies were compared. An accuracy of 77% was achieved using ELM with cross validation. Upon using SVM with cross validation an accuracy of 94.59% was achieved.

Both the algorithms effectively were able to distinguish unhealthy set of subjects and healthy set, but looking at the accuracy achieved it is evident that SVM performed better than ELM. By combining both signal processing and machine learning techniques, this paper proves that it is possible to effectively evaluate the cardiovascular health and is also able to distinguish unhealthy and healthy set of subjects from a young age group. However, it is necessary to test the proposed system on data belonging to subjects of other age groups and come up with an age index.

Acknowledgements The authors would like to thank the management of PES University for supporting the project.

References

1. Mendis, S., Puska, P., Norving, B.: World Health Organization, World Heart Federation World Stroke Organization: global atlas on cardiovascular disease prevention and control. In World Health Organization in Collaboration with the World Heart Federation and the World Stroke Organization, pp. 3–18. World Health Organization, Geneva (2011)
2. Pilt., K., Meigas, K., Ferenets, R., Temitski, K., Viigimaa, M.: Photoplethysmographic signal wave-form index for detection of increased arterial stiffness. Physiol. Meas. **35**(10), 2027 (2014)
3. Tamura, T., Maeda, Y., Sekine, M., Yoshida, M.: Wearable photoplethysmographic sensors—past and present. Electronics **3**, 282–302 (2014)
4. Jayasree, V.K., Radhakrishnan, P.: Selected cardiovascular studies based on photoplethysmography technique. Ph.D. thesis. Cochin University of Science and Technology, India, Jan 2009
5. News and Events Homepage: https://www.cardiosmart.org/News-and-Events/2014/07/Heart-Attack-Trends-Among-Young-Adults. Last accessed 29 July 2017
6. Harvard Men's Health Watch: http://www.health.harvard.edu/heart-health/premature-heart-disease. Last accessed 29 July 2017
7. Shimizu, M., Kario, K.: Role of the augmentation index in hypertension. Ther. Adv. Cardiovasc. Dis. **2**, 25–35 (2008)
8. Open Hardware: http://pulsesensor.com/pages/open-hardware. Last accessed 29 July 2017
9. Elgendi, M., Norton, I., Brearley, M., Abbott, D., Schuurmans, D.: Systolic peak detection in acceleration photoplethysmograms measured from emergency responders in tropical conditions. PLoS ONE **8**(10), 76585 (2013)
10. Elgendi, M., Norton, I., Brearley, M., Abbott, D., Schuurmans, D.: Detection of a and b waves in the acceleration photoplethysmogram. Biomed. Eng. Online **13** (2014)
11. Chellappan, K., Mohd Ali, M.A., Zahedi, E.: An age index for vascular system based on photoplethysmogram pulse contour analysis. In: 4th Kuala Lumpur International Conference on Biomedical Engineering, vol. 21, pp. 125–128. Springer, Berlin, Heidelberg (2008)
12. Elgendi, M.: On the analysis of fingertip photoplethysmogram signals. Curr. Cardiol. Rev. **8**, 14–25 (2012)
13. Huang, G.-B., Zhu, Q.-Y., Siew, C.-K.: Extreme learning machine: a new learning scheme of feedforward neural networks. In: IEEE International Joint Conference on Neural Networks, vol. 2, pp. 985–990 (2004)
14. Cortes, C., Vapnik, V.: Support-vector networks. Mach. Learn. **20**, 273–297 (1995)

Novel and Security Connections for Extending Mobile Audio Signal Delivery to Personal/Notebook Computer

Meng-Lun Hsueh, Huey-Dong Wu, and Bing-Yuh Lu

Abstract

This study proposes a direct wiring solution for connecting a mobile on the near side to a personal computer/notebook computer (PC/NB) for accessing audio signals from a distant mobile device to avoid any interferences and interrupts. Such signals can be communicated to the PC/NB. This supports achieving enriched resources of signal processing by both hardware and software in PC/NB. The results of a successful experiment are presented to validate the function of this wiring design. Briefly, it is ready for mobile communication with signal processing algorithms by connecting with the novel line without any additional processes in this study if the advanced algorithm is successfully tested in PC/NB. *In conclusion, a straightly wiring audio connecting line for mobile based communication to the PC/NB makes a better signal transfers that can be applied in auscultation much easily.*

Keywords

Earphone • Microphone • Mobile • Connecting line • Audio

1 Introduction

Wireless communications between mobile devices and personal computers/notebooks (PCs/NBs) can be implemented by Bluetooth, WiFi, Modem and other communication technologies. However, the settings and programming might be problematic for many users. Therefore, a novel wiring connection method was developed for achieving direct wired connections to deliver audio signals from mobile devices to the PC/NB.

M.-L. Hsueh
Department of Electronic Engineering, Hwa Hsa University of Technology, New Taipei City, Taiwan, Republic of China

H.-D. Wu
Department of Integrated Diagnostics and Therapeutics, National Taiwan University Hospital, Taipei City, Taiwan, Republic of China

B.-Y. Lu (✉)
Department of Electronic Engineering, Tungnan University, New Taipei City, Taiwan, Republic of China
e-mail: franklinlu888@outlook.com

Bluetooth is a wireless technology standard whose operating frequencies are between 2.4 and 2.485 GHz for exchanging data over short distances wirelessly. Ericsson, the telecom vendor, was originally employed it as a wireless alternative to RS-232 data cables in 1994 [1]. Some medical systems based on the Bluetooth Technology have been developed. For examples, Jung et al. [2] have implemented a ubiquitous Bluetooth based healthcare system which monitored for physiological signal in a body area network. Zhang and Xiao [3] have built a Bluetooth based system which remotely monitored the physiological signals in a sensor network. The main contribution of Ying and Xiao's study was the creation of a system infrastructure which reduced the cost for patients and increased the physiological data. The patients can be promoted the quality of medical care by this system. Wei et al. [4] have reported a portable system that measured ECG signal with the functions of monitor and analysis. The features of the system were portable (Bluetooth), easy to connect with computer, and real-time. Chien and Tai [5] have proposed a PDA system based on the wireless bluetooth technology to measure physiological signals.

F. Ibrahim et al. (eds.), *2nd International Conference for Innovation in Biomedical Engineering and Life Sciences*,
IFMBE Proceedings 67, https://doi.org/10.1007/978-981-10-7554-4_3

WiFi which based on the IEEE 802.11 standards is a technology for wireless local area network (WLAN). Wi-Fi compatible devices such as smart phones, PCs, Notebook PCs can connect to a wireless access point which has a range of about 20 m detecting distance. Chételat et al. [6] implemented a system for cardiorespiratory distant monitoring using biosensors and wearables. The system named "WELCOME" performed a precisely synchronous measurement of ECG. Chauhan et al. [7] proposed a system, "BreathPrint", which employed the audio signatures associated with the three individual gestures to process via the microphone sensor on smartphones and wearable devices.

In 2015, Lu and Wu [8] developed a real-time mobile-based auscultation (RMA) that can be used in distant health care systems to improve the convenience of performing real-time auscultation on patients. The advantages of RMA are easy to use, traditional performance of the telephone, and good to the elders. Especially, it is very easy to perform the distant auscultation of the experienced elder medical doctors. Based on RMA, a new design of the audio line was proposed to connect the PC based software for the scientific analysis of the auscultation in this study. *A better signal transfers that can be applied in auscultation much easily results from a straightly wiring audio connecting line for mobile based communication to the PC/NB.*

2 Methods

Modern PCs or NBs are equipped with connectors that can accommodate earphones and microphones. Figure 1 presents the proposed wiring design of pin connectors, where L denotes the left audio channel, R denotes the right audio channel, M denotes the microphone channel, and G denotes the ground. In this design, sound can be sent through the R and L channels in a mobile device to the microphone channel in a PC/NB for signal processing.

A mobile device is a basic unit in 3G and 4G mobile telecommunication systems. Therefore, when a distant mobile device calls up a mobile device on the near side that is connected to a PC/NB through the aforementioned line, sound can be sent from the distant source to the PC/NB on the near side. A diagram of this process is presented in Fig. 2. Similar applications have been widely discussed. Many researchers proposed advanced algorithm to enhance the region of interest (ROI) of investigating signals. In the mobile communication system, the algorithm should be

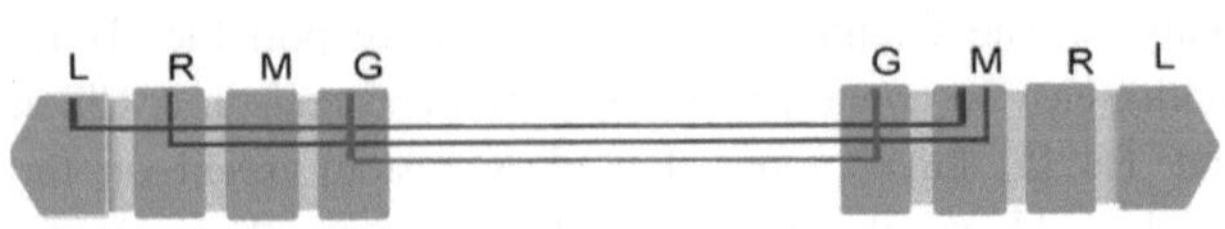

Fig. 1 Wiring design of the pin connectors

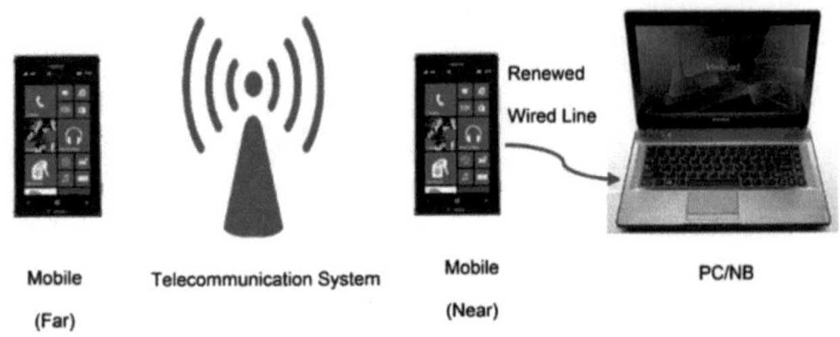

Fig. 2 An application of the wiring design of the pin connectors

implemented in the receiving mobile which is usually less resources in hardware and software then PC/NB. Therefore, the model in the Fig. 2 provides a direct solution to the implementation of the advanced algorithms much easily [9–11].

3 Results

A practical experiment that involved connecting a mobile device to a PC/NB through the prototype line was also implemented in Fig. 3. Mobile A on the far side was a Windows Phone 8X (hTC™) and Mobile B on the near side was an Amazing A6 (HwaWei™) which was connected to a PC/NB (Aspire V17, acer™) through the novel wired line shown in Fig. 3. Therefore, the sound from Mobile A was delivered to the speakers (Pebbles, JBL™). A simple program coded in Simulink (MathWork™) presented the sound signal from Mobile A in the oscilloscope display of the testing program. Therefore, the experiment was successful. All audio signals were sent through a telecommunication system. Briefly, the experimental design in Fig. 2 has been carried out by the system integration in Fig. 3. Furthermore, the novel wired design shown in Fig. 1 has been implemented, and pictures of the prototype of the audio line were

Fig. 3 The experiment of the diagram in Fig. 2

Fig. 4 The audio line: implementation of the wiring design of the pin connectors in Fig. 1

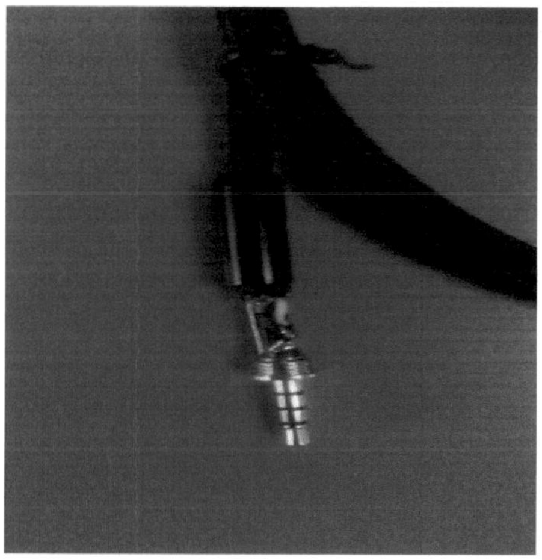

Fig. 5 The connector of audio line

depicted in Fig. 4. Furthermore, Fig. 5 presents the pin connector.

4 Discussion

In this novel wiring design, the M lines on both sides are not connected to avoid any unpleasant noise associated with insertion of the plug from the microphone of Mobile B. However, the communication of Mobile B depends on its microphone. Both M lines can be connected, if users need the function of communication. Reversely, the near side mobile and PC/NB can be put in the far side, and the wire can be employed backward. Therefore, many extended applications are feasible.

In 2015, Lu and Wu [8] developed a RMA system that can be used in distant health care systems to improve the convenience of performing real-time auscultation on patients. This novel wiring design can enable auscultation signals of distant patients to be sent as input to a PC/NB, in which they can be refined to enhance their quality and clarity by running various signal processing algorithms. Executing these processes on the PC/NB are easier than on Mobile B. *Compared with RS-232 or USB wired communication technologies, the connecting line that we proposed exhibits an easy to use real-time operation without any conversions and communication protocols. This straightly shows the benefits of this wired audio connecting line.*

However, the WiFi applications are hot now, but the security and privacy of the Internet communication is still a problem. Therefore, many research groups focus on the issue of the complex internet security. Kaiwartya et al. [12] reported an investigation to clear understand the security in wide area of network communications. This paper indicated that some unauthorized disclosure, unauthorized action, and data alteration are threating the privacy in the network. Certainly, this disclosed that the importance of protecting the privacy of personal medical data is a worldwide issue.

The RMA which based on the modern telecommunication can real-time transfer the respiration signals. The 3G or 4G mobile communication systems are commercial. Therefore, the respiration sounds can be transferred stably and safely. Through the audio line that this study proposed, the mobile and PC/NB can connect very easily. Furthermore, the PC/NB based analysis tools can process the acoustic signals from RMA without any modification of the codes in a ready program. *Therefore, the results supported many biomedical research groups to keep their PC/NB based algorithms and codes, but developed the distant heath care system as the APPs in mobile immediately. The "no harm" technology promotion disclosed the value and speed-up of the wiring method. The experienced experts will prefer the certainty of system upgrade. For examples, some filter technologies with space transformations usually need a huge computer resources, and the promoting version of APPs in smartphones may be completely new task of complex coding. Our method avoids the trouble shootings of the suffering programmers in many research groups related to the distant auscultation, but supplies a speedy and reliable solution.*

5 Conclusion

This study proposes a direct wiring solution for connecting a mobile on the near side to a PC/NB for accessing audio signals from a distant mobile device to avoid any interferences and interrupts. Such signals can be communicated to the PC/NB. This arrangement supports enriched signal processing in both hardware and software. Compared with the complex settings and programming requirements of wireless communication technologies such as Bluetooth and sockets, the wiring solution in this study provides a direct solution for successful operations of advanced algorithms. Briefly, it is ready for mobile communication by connecting with the novel line if the advanced algorithm is successfully tested in the PC/NB.

Acknowledgements The author thanks the reviewers for their valuable comments and the Ministry of Science and Technology, Taiwan, Republic of China, for the financial support through grants MOST103-2221-E-236-001 and MOST103-2627-E-002-005.

References

1. Wikipedia. https://en.wikipedia.org/wiki/Bluetooth. Last accessed 19 Aug 2017
2. Jung, J., Ha, K., Lee, J., Kim, Y., Kim, D.: Wireless body area network in a ubiquitous healthcare system for physiological signal monitoring and health consulting. Int. J. Signal Process. Image Process. Pattern Recogn. **1**(1), 47–54 (2008)
3. Zhang, Y., Xiao, H.: Bluetooth-based sensor networks for remotely monitoring the physiological signals of a patient. IEEE Trans. Infor. Tech. Biomed. **13**(6), 1040–1048 (2009)
4. Wei, Y.C., Lee, Y.H., Young, M.S.: A portable ECG signal monitor and analyzer. In: The Proceedings of the 2nd International Conference on Bioinformatics and Biomedical Engineering, pp. 1336–1338, IEEE, Shanghai, China (2008)
5. Chien, J.R.C., Tai, C.C.: A new wireless-type physiological signal measuring system using a PDA and the bluetooth technology. Biomed. Eng. Appl. Basis Comm. **17**(5), 229–235 (2005)
6. Chételat, O., Wacker, J., Rapin, M., Porchet, J.A., Meier, C., Fahli, A., Haenni, E., Caldani, L., Mancuso, C., Paradiso, R., Arneth, L.: New biosensors and wearables for cardiorespiratory telemonitoring. In: The Proceedings of the IEEE-EMBS International Conference on Biomedical and Health Informatics (BHI), pp. 481–484. IEEE, Las Vegas, USA (2016)
7. Chauhan, J., Hu, Y., Seneviratne, S., Misra, A., Seneviratne, A., Lee, Y. BreathPrint: Breathing acoustics-based user authentication. In: Proceedings of the 15th Annual International on Conference on Mobile Systems, Applications, and Services, pp. 278–291. ACM, NY, USA (2017)
8. Lu, B.Y., Wu, H.D.: Auscultation using modern mobile communication. Acous. Austra. **43**, 303–309 (2015)
9. Palaniappan, R., Sundaraj, K., Ahamed, N.U.: Machine learning in lung sound analysis: a systematic review. Biocyber. Biomed. Eng. **33**, 129–135 (2013)
10. Patell, S.B., Wodicka, G.R., Callahan, M.G.: Active noise reduction stethoscope for lung sounds measurement in loud environments. J. Acous. Soc. Am. **99**, 2477–2500 (1996)
11. Shengkun, X., Jin, F., Krishnan, S., Sattar, F.: Signal feature extraction by multi-scale PCA and its application to respiratory sound classification. Med. Biol. Eng. Comput. **50**, 759–768 (2012)
12. Kaiwartya, O., Prasad, M., Prakash, S., Samadhiya, D., Abdullah, A.H., Rahman, S.O.A.: An investigation on biometric internet security. Int. J. Netw. Secur. **19**(2), 167–176 (2017)

Sensitivity Improvement of Respiratory Movement Detection in Non-contact In-Bed Cardiopulmonary Measurements During Sleep

Mayuko Takano and Akinori Ueno

Abstract

Unobtrusive measurement methods for electrocardiograms and respiratory movements (RMs) have the potential to become key home monitoring technologies. This study reports on an improvement in the RM signals obtained from subjects when a non-contact in-bed cardiopulmonary measurement system was used; our aim was to improve the detection sensitivity (p_{sens}) of RMs that reflected chest and abdominal motions (RM_{chest} and RM_{abd}, respectively). A suitable capacitance value for a virtual midpoint circuit (C_{VMC}) in the system was determined for stabilization during prolonged measurements. The system was then modified by using a suitable C_{VMC} value in conjunction with a modified electrode structure, a revised front-end voltage follower, and a readjusted passband for filtering. The performance of the modified system was evaluated via overnight experiments conducted using seven healthy volunteers. The results revealed that p_{sens} for both the RM_{chest} and RM_{abd} signals of the modified system were better than those found for the conventional system tested; on average, p_{sens} for RM_{chest} and RM_{abd} were 15.8 and 13.5%, respectively, greater than those found for the conventional system. Although the modified system still needs to be improved with regard to p_{sens} for both RM_{chest} and RM_{abd}, it seems to be a promising development as far as daily cardiopulmonary monitoring during sleep is concerned.

Keywords

Respiratory movement • Virtual midpoint circuit • Capacitive sensing

1 Introduction

The number of patients with chronic cardiorespiratory disorders is increasing, and for this reason home monitoring approaches are expected to play a key role in the treatment of such patients in the future [1]. Clinical trials have demonstrated that management programs that include home monitoring improve clinical outcomes, particularly for patients with heart failure hospitalization [2]. Because sleep-disordered breathing is associated with acute unfavorable effects on cardiovascular physiology [3], home monitoring should concern itself not only with electrocardiograms (ECGs), but also with respiratory movement (RM) during sleep. Between July 2003 and December 2007, the US Veterans Health Administration (VHA) introduced a national home telehealth program called Care Coordination/Home Telehealth; this program identified the care needs for both chronic obstructive airways disease and chronic heart failure [4]. The VHA also reported that large-scale enterprise-level home telehealth implementation was an appropriate and cost-effective way to manage chronic care patients in both urban and rural settings [4].

In home monitoring scenarios, ambient and unobtrusive methods are required for the measurement of patients' vitals so that the status of patients can be monitored without impairing their everyday lives. Consequently, a number of

M. Takano (✉) · A. Ueno
Tokyo Denki University, Tokyo, 120-8551, Japan
e-mail: 16kmj27@ms.dendai.ac.jp

F. Ibrahim et al. (eds.), *2nd International Conference for Innovation in Biomedical Engineering and Life Sciences*, IFMBE Proceedings 67, https://doi.org/10.1007/978-981-10-7554-4_4

research activities in the past decade have explored different ways in which this can be done [1, 5, 6]. These studies indicated that capacitive sensing is a promising method for making cardiopulmonary measurements. Yama et al. proposed a non-contact and unobtrusive method for measuring narrow-band ECG (10–40 Hz) and RM signals; their method used capacitively coupled electrodes, and they used sleeping babies as the subjects in their study [7]. However, the narrow-band ECG can provide less information than conventional (i.e., 0.5–100 Hz band) ECGs due to the lack of low-frequency component (i.e., 0.5–10 Hz) including T wave and P wave of ECG. Moreover, the method reported in the study conducted by Yama et al. was only able to measure a single-channel RM that reflected abdominal movement. In obstructive sleep apnea, chest and abdominal movements exist even when there is no airflow through the airways, and these movements are known to be reversed phase [8]. It is therefore desirable for the RM signals of these two areas to be measured independently of one another; one way in which this could be done is through the introduction of a virtual midpoint circuit (VMC) into the measurement system [9]. However, it is still unknown what effect capacitance values in VMCs have on the stability of RM measurements. In this paper, we have explored what the optimal values are for this capacitance, and we have introduced it into a modified measurement system. Following this, we evaluated the performance of the modified system via overnight measurements.

2 Measurement System

2.1 Multilayered Cloth Electrode

Figure 1a shows the positions of the electrodes and block diagram of the cardiopulmonary measurement system used in this study. The ECG signal was measured as the potential difference between electrodes (1) and (3), which were placed under a bed sheet at the positions of subjects' upper backs and buttocks, respectively. RM measurements reflecting chest motion (RM_{chest}) were also derived from this potential difference through the use of a separation filter and amplifiers. In order to measure the RM associated with the movement of the abdomen (RM_{abd}) independently of that of RM_{chest}, electrode (2) was placed at the position of subjects' waists. RM_{abd} measurement was derived from the potential difference between electrodes (2) and (3). The electric potentials at electrode (1), (2) and (3) were also used to synthesize the circuit ground potential via a VMC.

Figure 1b shows a cross-sectional diagram of a conventional cloth electrode and its connection to a typical front-end voltage follower. The top-most layer is a sensing layer, and the output from the front-end voltage follower is fed into the third layer in order to configure a driven shield. Figure 1c shows a cross-sectional diagram of an electrode and its connection to a front-end voltage follower those are modified in this study. As shown in the Fig. 1c, we adopted a doubly shielded structure for the electrode by introducing the fifth layer as a ground plane [10] in the hope of increasing the modified electrodes tolerances against electrostatic discharges and movement artifacts [11].

2.2 Front-End Voltage Follower

The purpose of the front-end voltage follower was to suppress voltage losses at the capacitive coupling between the surface of a subject's body and the electrodes; it did this by having an ultra-high input impedance and by leading the potential of the electrode to the subsequent circuitry with a low output impedance. In the connection that used a conventional electrode, a standard buffer was used for the voltage follower, as shown in Fig. 1b. In the modified electrode connection, however, a bootstrapped buffer was used (see Fig. 1c). The bootstrapped buffer provided low input impedance for the direct current component of the input electric potential, whereas it provided an ultra-high input impedance for the alternating current component of the input electric potential. In non-contact ECG measurements, the bootstrapped buffer has been shown to contribute to the shortening of baseline restoration times after body movements [12].

2.3 Signal Detection Circuitry

In order to detect RM_{abd} and RM_{chest} signals independently of one another, we introduced the VMC [9] into the signal detection circuity, and we grounded the circuit (GND) using three electric potentials measured by electrodes (1), (2), and (3). We set a common capacitance (C_{VMC}) in the VMC switchable; this was used so that adequate values could be used for RM signal detection. The ECG and RM detecting devices used consisted of differential separation filters, instrumentation amplifiers, and other filters and amplifiers with a gain of 60 dB in the same manner as in the report of [9]. The frequency band used to detect the ECG signal was set to 0.5–40 Hz, and a band elimination filter of 50 Hz was used to suppress commercial power-supply noise. For the detection of the RM signals, a bandpass filter of 0.10–0.50 Hz was used in the conventional system and 0.15–0.50 Hz was used in the modified system.

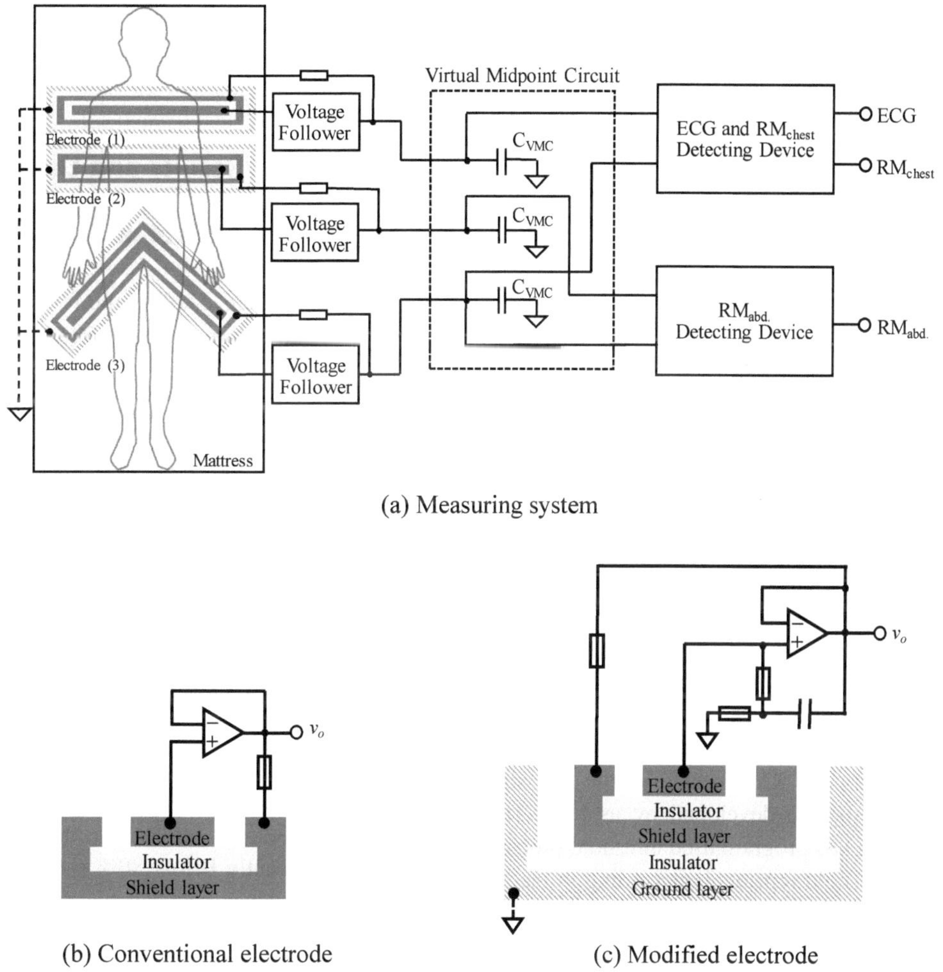

Fig. 1 Schematics showing the conventional and modified measurement systems. **a** Positions of the electrodes and block diagram of the conventional measurement system along with the additions for the modified system (which are indicated by the shaded areas and dashed lines) are shown. The cross-sectional diagram of the electrode and the connection to the front-end voltage follower in the **b** conventional and **c** modified measurement systems. The output voltage of the voltage follower is fed back into the third layer in both systems

3 Method

All of the experimental procedures were approved by the Human Life Ethics Committee of Tokyo Denki University. All of the subjects provided informed consent prior to their participation in our experiments.

3.1 Evaluation of the Capacitance in the VMC

CVMC can affect stability when the electric potential is measured at the VMC's ground; this can affect the stability of the signal baselines of both RMchest and RMabd. We examined this effect by changing CVMC during tests in which subjects slept for either 60 min or 6 h. One male subject, who wore commercially available nightwear made out of cotton that was 0.36 mm thick, participated in the experiment. The conventional electrodes depicted in Fig. 1b were placed on a mattress according to the layout shown in Fig. 1a and were covered by a cotton bed sheet that was 0.33 mm thick. Each electrode was connected to a bootstrapped voltage follower, and the signal detection circuitry was used to measure ECG, RMchest, and RMabd simultaneously.

In the 60 min sleep test, C_{VMC} was changed every 20 min in the order of 1, 10, and then 100 μF; during this time, the subject lay on the bed in a supine position. A signal was also measured using a band-type respiration transducer (BN-RESP-XDCR, BIOPAC Systems) on the abdomen and a breathing amplifier (RSP100C, BIOPAC Systems); this was used as a RM reference signal (RM_{ref}). In the 6 h sleep test, C_{VMC} was changed every 2 h in the order of 1, 10, and then 100 μF; during this time, the subject slept on the bed in a supine position. RM_{ref} was measured in the same way as in the 60 min sleep test.

3.2 Overnight Measurements

We assessed the conventional and modified measurement systems by twice conducting overnight experiments in different nights while each of the seven healthy subjects that participated slept; during one sleep cycle, a CVMC value of 1 pF, the conventional electrode structure, and the conventional voltage follower were all used, while in the other, CVMC evaluated by the experiment in Sect. 3.1, the modified electrode structure, and the modified voltage follower were all used (see Table 1). The physical information of all of the subjects is shown in Table 2. The subjects were asked to lie on the bed in a supine position when the measurements were started and to keep lying on their beds for at least 6 h; however, they were free to change their body positions. The reference signal used for RM was the same as in Sect. 3.1, and so too was the clothing used. Initial 6 h measurements of RMchest, RMabd and RMref were analyzed as described in the following Sect. 3.3.

3.3 Analysis of the RM Signals

In order to pre-process the signals for analysis, we applied a digital bandpass filter (IIR, 0.15–0.5 Hz, Q = 0.707) twice to the RM signals of RM_{ref}, RM_{chest}, and RM_{abd}. In the 60 min sleep test, we computed the frequency spectrum of the RM signals for each capacitance condition by using fast Fourier transform.

For the 6 h sleep test and overnight measurements, we detected each RM event from the RM signals with a dedicated program developed by us [13]. The detection algorithm in this program used threshold processing and renewed the threshold repeatedly according to the mean amplitude of the three most recently elapsed RM events, including the newly detected one. The algorithm regarded all events that had amplitudes more than 1.8 times the threshold as body movements, and so excluded them from analysis. If the starting point of one cycle of an individual subject's RM event was included in the interval of the corresponding cycle of a reference RM event, then the RM detected by our system was deemed to have been detected correctly.

The detection sensitivity (p_{sens}) was calculated using Eq. (1) [14]. In this equation, N_{TP} represents the number of RM events correctly detected in each output signal of the conventional or modified systems, and N_{FN} is the number of undetected RM events in the output signals of the conventional or modified systems:

$$p_{\text{sens}} = \frac{N_{TP}}{N_{TP} + N_{FN}} \times 100 \quad (1)$$

4 Results and Discussions

4.1 Effect of C_{VMC} on the RM Signal Quality

Figure 2 shows the frequency spectra of the RM signals for durations of 20 min for different C_{VMC} values. In the spectra for RM_{ref}, both the center frequency (around 0.3 Hz) and its power are independent of C_{VMC}. However, in the spectra for RM_{chest} and RM_{abd}, the power increased as C_{VMC} increased, and the shape of the spectra as well as the center frequency of that was not completely unrelated to C_{VMC} value. While 100-μF C_{VMC} produced the largest power at the center frequency, the shape of the spectra of the 10 μF C_{VMC} meant that this value was the most advantageous; this was because of the bunched spectral to some specific frequencies at this capacitance.

According to the results obtained for the 6 h sleep test that are shown in Table 3, p_{sens} at 10 μF C_{VMC} exceeded 80% on average for both RM_{chest} and RM_{abd}. Although p_{sens} at 100 μF C_{VMC} was found to have the best value (i.e., 83.5%) for RM_{abd}, its p_{sens} value for RM_{chest} was significantly below 80% (i.e., 70.6%). A comparison of the p_{sens} values between the first and second 1 h sections also indicated that the use of the 10 μF C_{VMC} was the most suitable for the stable detection of RM signals, because the sensitivities found were not very different, and they both exceeded 80% for both RM_{chest} and RM_{abd}.

4.2 RM Sensitivities Obtained by the Conventional and Modified Systems During the Overnight Experiments

Figure 3a and b show examples of the ECG and RM signals obtained by the conventional and modified systems,

Table 1 Comparison of the measurement conditions between the conventional and modified systems

Item	Conventional system	Modified system
Electrodes	Three-layer	Five-layer
Front-end circuit	Voltage follower	Voltage follower with bootstrap
C_{VMC}	1 pF	1/10/100 μF
Frequency band in RM detecting device	0.1–0.5 Hz	0.15–0.5 Hz

Table 2 Age and physical constitutions of the subjects that participated in the overnight experiments

Subject	Age (years)	Height (m)	Weight (kg)	BMI (kg/m^2)
#1	22	1.73	73.4	24.5
#2	23	1.75	65.0	21.2
#3	21	1.87	73.0	20.9
#4	20	1.73	62.0	20.7
#5	20	1.60	55.0	21.5
#6	22	1.67	70.0	25.1
#7	19	1.72	78.8	26.6

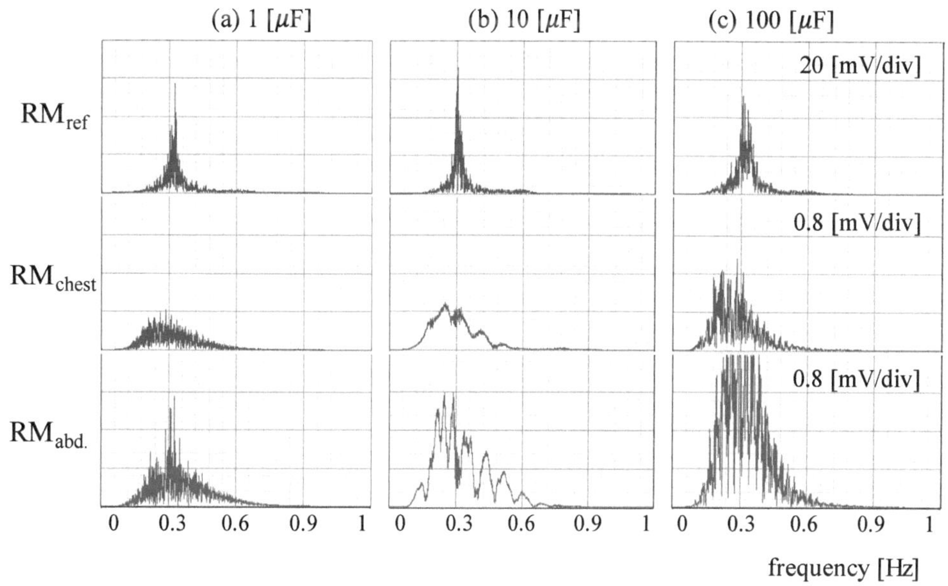

Fig. 2 Frequency spectra of 20-min RM recordings obtained with various VMC capacitances; **a** 1 μF, **b** 10 μF, and **c** 100 μF. The temperature and volumetric humidity used in these readings were 26.8 °C and 11.6 g/m^3, respectively

Table 3 Sensitivity (%) of the RM signal measurements for different C_{VMC} values

Time section (min)	Chest (μF)			Abdomen (μF)		
	1	10	100	1	10	100
0–60	66.6	83.1	64.1	75.8	81.0	75.8
60–120	67.0	80.1	70.3	68.1	81.1	86.2
Average	66.8	82.3	70.6	72.5	81.1	83.5

The temperature and volumetric humidity measured during these readings were 25.8 °C and 15.9 g/m^3, respectively

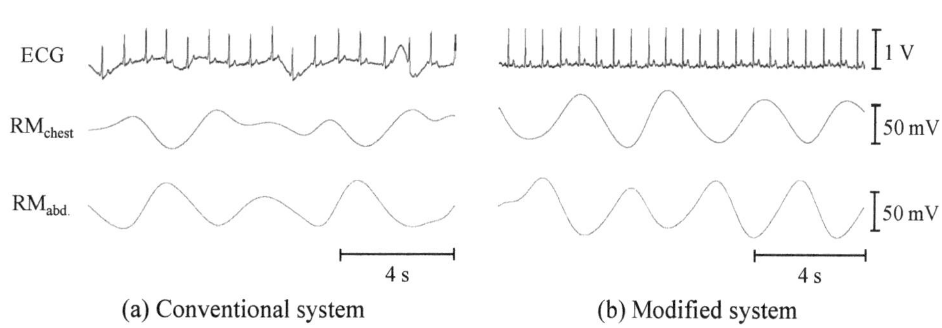

Fig. 3 Recordings of the ECG and RM signals from Subject #1 in the overnight measurements as obtained by the **a** conventional and **b** modified systems

respectively, during the sleep cycles of subject #1. Both have the same time elapsed from the onset of the measurement. The baseline of the ECG signal in Fig. 3a was less stable than that in Fig. 3b. This instability often developed after body movement was recorded by the conventional system. The RM amplitude in Fig. 3a was more variable than it was in Fig. 3b, irrespective of whether it was RM_{chest} or RM_{abd}. The poor repeatability of the RM amplitudes arose more frequently in the recordings measured by the conventional system. A comparison between RM_{chest} and RM_{abd} in Fig. 3a revealed that the RM_{chest} measurements were inferior to those of RM_{abd} in the conventional system. In contrast to the conventional system, both RM_{chest} and RM_{abd} measurements in the modified system were found to have more stable amplitudes for all of the subjects.

Figure 4 summarizes p_{sens} for the 6 h measurements of RM_{chest} and RM_{abd} made for each subject during the each overnight experiment. As can be seen in Fig. 4, p_{sens} increased in the modified system for both RM_{chest} and RM_{abd} for all of the subjects; the mean increases were 15.8 and 13.5%, respectively. In the measurement for Subject #6, the increase in the mean p_{sens} of RM_{chest} was subtle (1.1%), and the standard deviation (SD) of this p_{sens} was larger than that found for the conventional system. Because an advanced analysis of mean p_{sens} for every 1 h segment revealed a smaller each SD of RM_{chest} measured by the modified system than that measured by the conventional system, the larger value of SD for the 6 h period can be attributed to the occasional larger variation of the respiration rate in the measurement using the modified system. These results indicate that the modified system is more suitable for prolonged RM measurements than the conventional system. However, the mean p_{sens} found for RM_{chest} and RM_{abd} were 67.6 and 72.2%, respectively; these values are below 80%. As such, the measurement system needs to be improved further. We found that this was particularly the case for low-body-mass-index (low-BMI) subjects, because Subjects #2, #3, and #4, all of whom had lower BMI values than the other subjects, all showed low p_{sens} for both RM_{chest} and RM_{abd}.

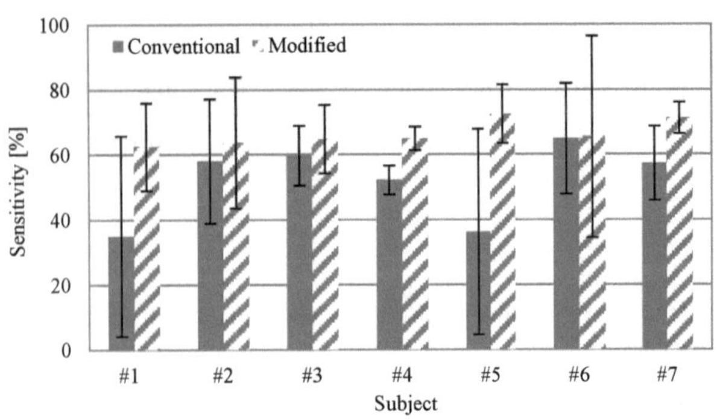

(a) Chest

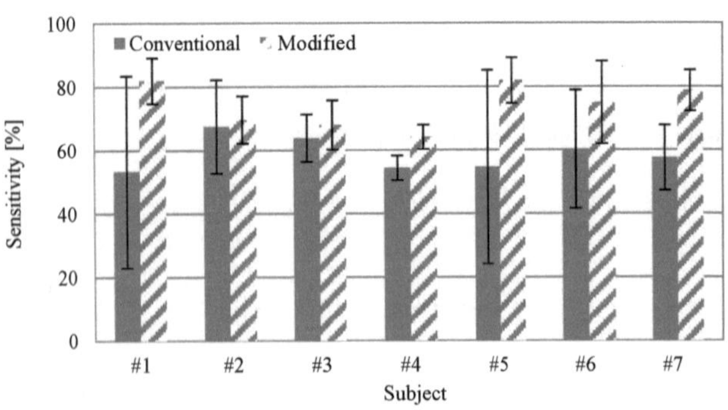

(b) Abdomen

Fig. 4 Comparison of the RM sensitivities (%) at the **a** chest and **b** abdomen positions between the conventional and modified systems in the overnight measurements

5 Conclusion

In this study, we examined suitable values of CVMC in a non-contact in-bed cardiopulmonary measurement system so that we could improve psens of RMchest and RMabd during sleep. 10 μF was identified as the most suitable value for this parameter, and we used it in our modified system along with the modified electrode structure and front-end voltage follower we proposed. We evaluated the performance of the modified system in overnight experiments, and we found that it demonstrated improved signal quality and therefore produced better psens for the modified system than for the conventional system. Although the modified system still needs to be improved in order for better psens to be obtained for RMchest and RMabd, it seems to be a promising development as far as daily cardiopulmonary monitoring during sleep is concerned.

Acknowledgements This research was supported by the Strategic Research Foundation Grant-aided Project for Private Universities from Ministry of Education, Culture, Sport, Science and Technology, Japan, 2015–2017 (S1512001).

References

1. Brüser, C., Antink, C.H., Wartzek, T., Walter, M., Leonhardt, S.: Ambient and unobtrusive cardiorespiratory monitoring techniques. IEEE Rev. Biomed. Eng. **8**, 30–43 (2015)
2. Konstam, M.A.: Home monitoring should be the central element in an effective program of heart failure disease management. Circulation **125**(6), 820–827 (2012)
3. Shahar, E., Whitney, C.W., Redline, S., Lee, E.T., Newman, A.B., Nieto, F.J., O'Connor, G.T., Boland, L.L., Schwartz, J.E., Samet, J.M.: Sleep-disordered breathing and cardiovascular disease:

cross-sectional results of the sleep heart health study. Am. J. Resp. Crit. Care Med. **163**(1), 19–25 (2001)
4. Darkins, A., Ryan, P., Kobb, R., Foster, L., Edmonson, E., Wakefield, B., Lancaster, A.E.: Care coordination/home telehealth; the systematic implementation of health informatics, home telehealth, and disease management to support the care of veteran patients with chronic conditions. Telemed. E-Health **14**(10), 1118–1126 (2008)
5. Folke, M., Cernerud, L., Ekström, M., Hök, B.: Critical review of non-invasive respiratory monitoring in medical care. Med. Biol. Eng. Comput. **41**(4), 377–383 (2003)
6. Lim, Y., Hong, K., Kim, K., Shin, J., Lee, S., Chung, G., Baek, H., Jeong, D., Park, K.: Monitoring physiological signals using nonintrusive sensors installed in daily life equipment. Biomed. Eng. Lett. **1**(1), 11–20 (2011)
7. Yama, Y., Ueno, A.: Unrestrained facile measurement of narrow-band ECG and respiratory variation in infants with a capacitive sheet-type sensor. Trans. Jpn. Soc. Med. Biol. Eng. **47** (1), 42–50 (2009). (in Japanese)
8. Murata, A.: Diagnosis and treatment of sleep apnea syndrome: sleep apnea syndrome triggers life-threatening disorders during sleep. Nihon Ika Daigaku Igakkai Zasshi **3**(2), 96–101 (2007). (in Japanese)
9. Ueno, A.: Development of measurement technique for simultaneous detection of two electrocardiograms and two breathing effort signals using sheeted fabric electrode. In: NTS (ed.) Personal Healthcare—Cutting-Edge Electronics for Realizing Ubiquitous and Wearable Medicine—. Passus 2, Chapter 2, pp. 49–56. NTS Press, Tokyo (2013) (in Japanese)
10. Fukuyama, Y., Suzuki, R., Takayama, S., Ueno, A.: Multi-layered fabric electrode for movement artifact reduction in capacitive ECG measurement. In: Proceedings of the 35th Annual International Conference of the IEEE Engineering in Medicine and Biology Society (EMBS), pp. 555–558. IEEE, Osaka (2013)
11. Kato, Y., Motohara, S., Omuta, T., Azran, A., Ueno, A.: Artifact tolerance test for capacitive wearable chest-belt electrocardiograph —effect of electrode configuration -. IEEJ Trans. Electron. Inf. Syst. **137**(4), 607–615 (2017)
12. Takano, M., Yamagishi, S., Ohmuta, T., Fukuoka, Y., Ueno, A.: Non-contact simultaneous measurements of electrocardiogram and respiratory movements using capacitive sheet electrodes. Adv. Biomed. Eng. **6**, 26–38 (2017)
13. Sakai, H., Matsumoto, S., Ueno, A.: A study of vital care monitor using bed-type respirometer. In: 35th Annual International Conference of the IEEE Engineering in Medicine and Biology Society (EMBS), Late Breaking Poster #3388, Osaka (2013)
14. Lee, J.S., Heo, J., Lee, W.K., Lim, Y.G., Kim, Y.H., Park, K.S.: Flexible capacitive electrodes for minimizing motion artifacts in ambulatory electrocardiograms. Sensors **14**(8), 14732–14743 (2014)

Model Iterative Airway Pressure Reconstruction During Mechanical Ventilation Asynchrony: Shapes and Sizes of Reconstruction

Chee Pin Tan, Yeong Shiong Chiew, J. Geoffrey Chase, Yeong Woei Chiew, Christopher Pretty, Thomas Desaive, Azrina Md Ralib, and Mohd Basri Mat

Abstract

Model-based methods estimating patient-specific respiratory mechanics may help intensive care clinicians in setting optimal ventilation parameters. However, these methods rely heavily on the quality of measured airway pressure and flow profiles for reliable respiratory mechanics estimation. Thus, asynchronous and/or spontaneous breathing cycles that do not follow a typical passive airway profile affect the performance and reliability of model-based methods. In this study, a model iterative airway pressure reconstruction method is presented. It aims to reconstruct a measured airway pressure affected by asynchronous breathing iteratively, trying to match the profile of passive breaths with no asynchrony or spontaneous breathing effort. Thus, reducing the variability of identified respiratory mechanics over short time periods where changes would be due only to asynchrony or spontaneous artefacts. A total of 2000 breathing cycles from mechanically ventilated patients with known asynchronous breathing were analyzed. It was found that this method is capable of reconstructing an airway pressure free from asynchronous or spontaneous breathing effort. This work focuses on several cases, detailing how iterative pressure reconstruction method performs under different cases, as well as its limitation.

Keywords

Mechanical ventilation • Spontaneous breathing • Asynchrony • Airway pressure reconstruction

1 Introduction

Application of model-based methods in the intensive care environment has become increasingly popular [1, 2]. It aims to provide personalized care based on patient-specific condition and patient-specific response to treatment [3–5]. Similar trends have also emerged in mechanical ventilation (MV) treatment for respiratory failure patients. These model-based methods use mathematical models to describe and estimate patient-specific respiratory conditions in real-time, as identified from clinical data, and are able to provide useful new insight to attending clinicians in order to devise suitable, more patient-specific treatment.

C. P. Tan · Y. S. Chiew (✉)
School of Engineering and Advanced Engineering Platform Health Cluster, Monash University Malaysia, Selangor, Malaysia
e-mail: chiew.yeong.shiong@monash.edu

J. Geoffrey Chase · C. Pretty
University of Canterbury, Christchurch, New Zealand

Y. W. Chiew
Lam Wah Ee Hospital, George Town, Pulau Pinang, Malaysia

T. Desaive
University of Liege, Liege, Belgium

A. M. Ralib · M. B. Mat
International Islamic University Malaysia, Kuantan, Pahang, Malaysia

F. Ibrahim et al. (eds.), *2nd International Conference for Innovation in Biomedical Engineering and Life Sciences*, IFMBE Proceedings 67, https://doi.org/10.1007/978-981-10-7554-4_5

In particular, these model-based methods are able to estimate respiratory system parameters such as respiratory system elastance (1/compliance), airway resistance, impedance, reactance, time constant decay, alveoli threshold opening pressures, viscoelasticity, alveoli recruitment, perfusion etc.; these information can then be used by clinicians to optimize MV settings. However, currently, accurate estimation of patient-specific respiratory system parameters relies on specialized clinical protocols that require patients to be sedated and paralyzed during data acquisition. Some patients are required to undergo protocols such as staircase recruitment maneuvers [6], super syringe method [7], and/or low flow maneuver [8], all of which are added invasive, care interrupting protocols that also consume nursing time and effort. They are thus not ideal.

Sedation and paralysis during MV is given partly to reduce patient's discomfort, as well as to reduce asynchronous events [9]. When the patient is synchronized with the ventilator, data acquisition is more consistent, the airway pressure and flow are uninterrupted, and, the respiratory system parameters can be accurately estimated [10, 11]. Without patient-ventilator synchronization, the data are altered by the patient's spontaneous and unpredictable breathing effort, leading to inaccurate parameter identification [12]. In addition, frequent use of sedation and paralysis drugs may also affect patient recovery time and increase ventilator dependency, worsening patient outcomes [13]. Thus, there is an increasing need for research to model and/or to account for the effects of asynchronous breathing or spontaneous breathing (SB) to enhance model-based respiratory mechanics estimation [10, 14, 15].

In this study, the authors investigate an airway pressure reconstruction method for airway pressure profiles altered by asynchronous or SB. This reconstruction method, model iterative pressure reconstruction (MIPR), attempts to model the asynchronous breathing cycle to a non-asynchronous cycle using a single compartment lung model. In particular, this study presents the airway pressure reconstruction algorithm, demonstrating how pressure reconstruction can be performed on several irregular airway pressures and analyses their performance and effect on respiratory mechanics estimation.

2 Methods

2.1 Patient Data

Retrospective data from respiratory failure patients from the CURE pilot study were used [16]. Patients were ventilated using Puritan Bennet 840 ventilator, using synchronous intermittent mandatory ventilation (SIMV) volume controlled mode (tidal volume = 6–8 ml/kg). Patients with clinically observed ventilation asynchrony were selected. As patients do not experience asynchronous breathing deliberately, these datasets provide a wide range of asynchronous breaths in shape and size of asynchrony to test the performance of the pressure reconstruction method. All data were sampled at 50 Hz and processed using MATLAB (R2014b, The Mathworks, Natick, MA, USA).

2.2 Model Iterative Pressure Reconstruction (MIPR)

A 3-step iterative pressure reconstruction (MIPR) method is developed, incorporating methods proposed by Damanhuri et al. [17] and Newberry et al. [18], where the maximum original airway pressure and model fit pressure are used to improve airway pressure reconstruction. Figure 1 shows the sequence of how airway pressure of an asynchronous breathing cycle is reconstructed iteratively to create an unaffected 'non-asynchronous' cycle using MIPR.

Step 1: Model fit to original inspiratory airway pressure:

The inspiratory airway pressure of an asynchronous breathing cycle fitted using a single compartment linear lung model [19, 20]. The model is defined as an elastic compartment with a connecting airway as shown in Eq. (1):

$$P_{aw}(t) = E_{rs}V(t) + R_{rs}Q(t) + P0 \tag{1}$$

P_{aw} is the airway pressure, t is time, E_{rs} is the respiratory system elastance, V is the air volume, R_{rs} is the respiratory system airway resistance, Q is the flow and $P0$ is the offset pressure. Figure 1a shows the model fit with asynchronous data.

Step 2: New maximum airway pressure:

The reconstruction of the new airway pressure is performed using the intersection between the original airway pressure and model fitted pressure. This reconstructed airway pressure begins to 'fill' the empty space caused by asynchronous breathing as shown in Fig. 1b. This reconstructed airway pressure is then used as a new airway pressure for subsequent model fitting as shown in Fig. 1c.

Step 3: Iterative pressure reconstruction:

Pressure reconstruction is performed iteratively following Fig. 1d–g, where Fig. 1g shows all iterations together, illustrating how the pressure profile becomes 'normal' and unaffected by asynchrony. The pressure reconstruction process stops after 20 iterations. Figure 1g shows all 20 iterations with the final reconstructed airway pressure located at the top-most of the pressure reconstruction curves. Figure 1h shows a typical normal airway pressure not affected by asynchrony.

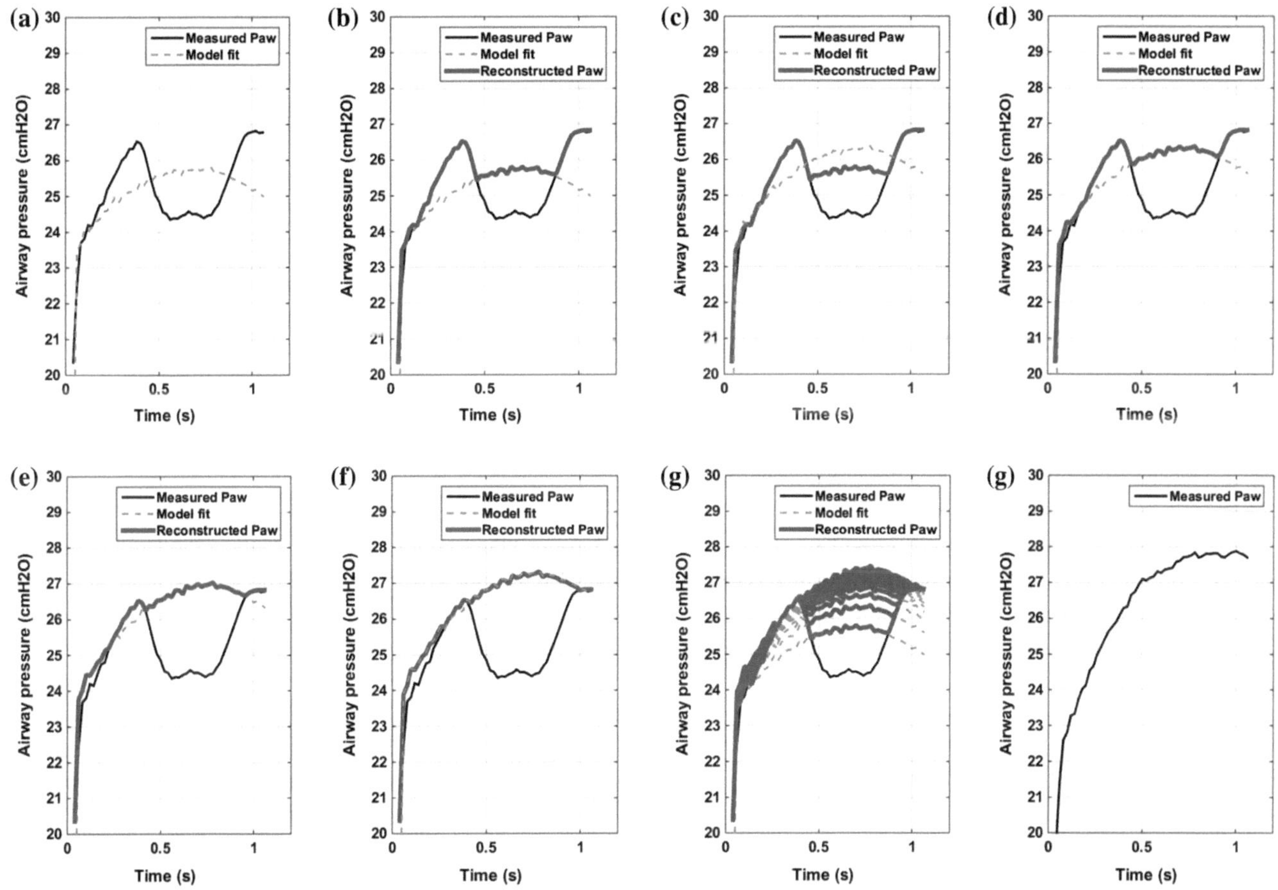

Fig. 1 A figure caption is always placed below the illustration. Short captions are centered, while long ones are justified. The macro button chooses the correct format automatically

In this study, several different shapes and magnitude of asynchronous airway pressure profiles are reconstructed to highlight the details and performance of the MIPR.

2.3 Data Analysis

2.3.1 Respiratory Mechanics

The airway pressure generated by MIPR and the original airway pressure are used for respiratory mechanics estimation, yielding parameters E_{rs} and R_{rs}. Estimated parameters are compared between the original and reconstructed profiles. A student t-test is used to test the difference of E_{rs} distribution, with p-value < 0.05 considered significant. This analysis determines if the MIPR method provides a measurable change in the identified respiratory properties.

2.3.2 Coefficient of Variation and Spread Analysis

The MIPR performance in estimating respiratory system parameters is assessed using variation and spread analysis. The coefficient of variation (CV) and Range90 are calculated for the estimated parameters using the reconstructed and the original pressure profiles. CV is calculated as the mean absolute difference (MAD) of the parameters divided by the mean. Range90 is calculated as the difference between the 95th percentile and the 5th percentile of the parameter distribution [21]. Higher CV and Range90 values indicate higher variability, and lower values indicate less variability which is desirable over short periods given no other changes in care of patient condition. Thus, lower or reduced values indicate the success of the method in accounting for asynchrony.

3 Results and Discussion

3.1 Case 1: A Typical Pressure Reconstruction

Figure 2 shows a typical successful pressure reconstruction. The observed shape and magnitude of asynchronous breathing is higher compared to Fig. 1, and the MIPR method is capable of reconstructing the airway pressure to an airway pressure free of asynchronous breathing. In each iteration, the intersection and overlaying of model fit and the

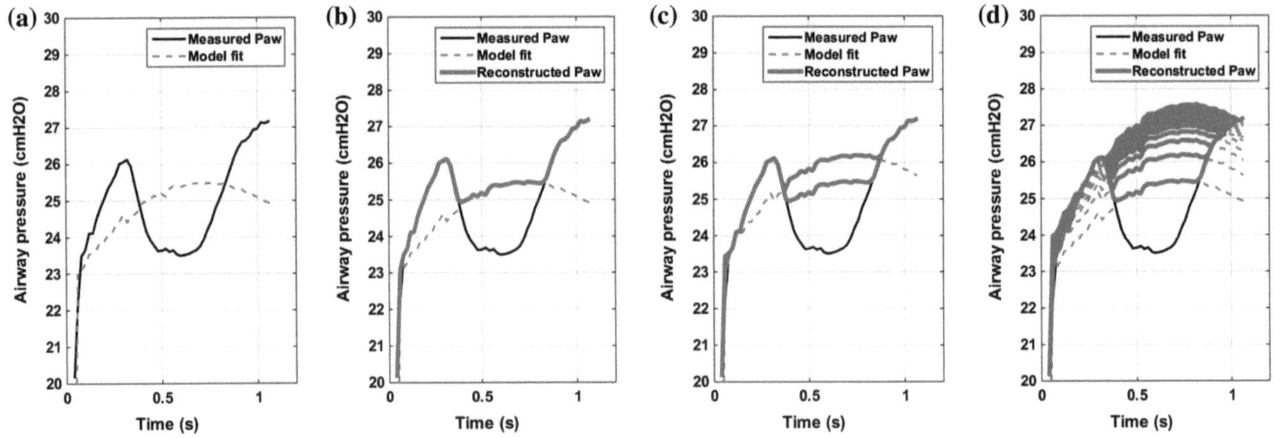

Fig. 2 Pressure reconstruction performed iteratively, resulting in a non-asynchronous airway pressure

original airway pressure (solid line) provides a new airway pressure profile for model identification. As it reconstructs iteratively, the error between the model fit and reconstructed pressure profile decreases, resulting in a non-asynchronous airway pressure profile.

3.2 Case 2: Late Asynchronies

In some cases, patient induced asynchronous breathing occurs at the end of the breath, which cannot be reconstructed using the typical pressure reconstruction method. In particular, the MIPR method fails due to the amount of 'missing data' at the end of inspiration. Figures 3a–d shows a sample of failed pressure reconstruction due to late asynchronies. The reconstruction process depends on the quality of model fitting using linear regression where it attempts to best fit a passive model to data. However, if most data are 'incorrect', the first step of pressure reconstruction will not provide a 'correct trend' for the iterative reconstruction process.

In these cases of late asynchronies, modification of the MIPR algorithm is required. The MIPR method is thus modified to include a pressure filling method as proposed by Damanhuri, Chiew [17], where the end of inspiratory pressure should at least be similar to the maximum observed airway pressure. Similarly, any airway pressure points that is in between the end of inspiratory pressure and the observed peak will be 'filled' with an estimated maximum airway pressure as shown in Fig. 3e–h. From this modified starting point, the MIPR method is used to obtain an unaffected final airway pressure.

3.3 Case 3: Early Asynchrony

Another potential anomaly in inspiratory airway pressure is the early patient breathing asynchrony or spontaneous breath triggering. This condition occurs when the initial pressure step increase due to the step or ramp airway flow increase is not observed. The increase in airway pressure is an important feature during parameter identification. If it is missing, model fitting fails. During model fitting of the original airway pressure, this lack of the pressure increase results in non-physiological respiratory mechanics parameters, such as negative elastance or resistance, which are impossible.

In some instances, this negative value captures patient triggering effort as an active external source, resulting in negative elastance or resistance [22]. However, that is not the case in this paper, where the airway pressure reconstruction has to assume a positive value. Hence, the reconstruction algorithm will force a positive respiratory elastance and airway resistance for model fit. The minimum parameter used in this study were $E_{rs} = 5$ cm H_2O/l and $R_{rs} = 5$ cm H_2Os/l. These minimal respiratory parameter values give a baseline feature of the modelled airway pressure, resulting in better reconstruction. Figure 4 shows an example of early asynchrony, and how the iterative pressure reconstruction caters for this form of asynchrony or spontaneous breath.

3.4 Respiratory Mechanics and Variation Analysis

In this study, a total of 2000 breathing cycles were analyzed. The values of E_{rs} and R_{rs} estimated from the original airway pressure are 24.0 cm H_2O/l [interquartile range (IQR): 20.5–28.5] and 10.9 cm H_2Os/l [IQR: 10.1–11.8]. The estimated E_{rs} and R_{rs} after pressure reconstruction are 28.6 cm H_2O/l [IQR: 26.6–32.1] and 10.9 cm H_2Os/l [IQR: 10.4–11.5]. Notice that the E_{rs} estimated from reconstructed airway pressure is significantly higher than the original ($p < 0.05$); this result is expected, where, the MIPR reconstruct the asynchronous airway pressure to a higher value, thus,

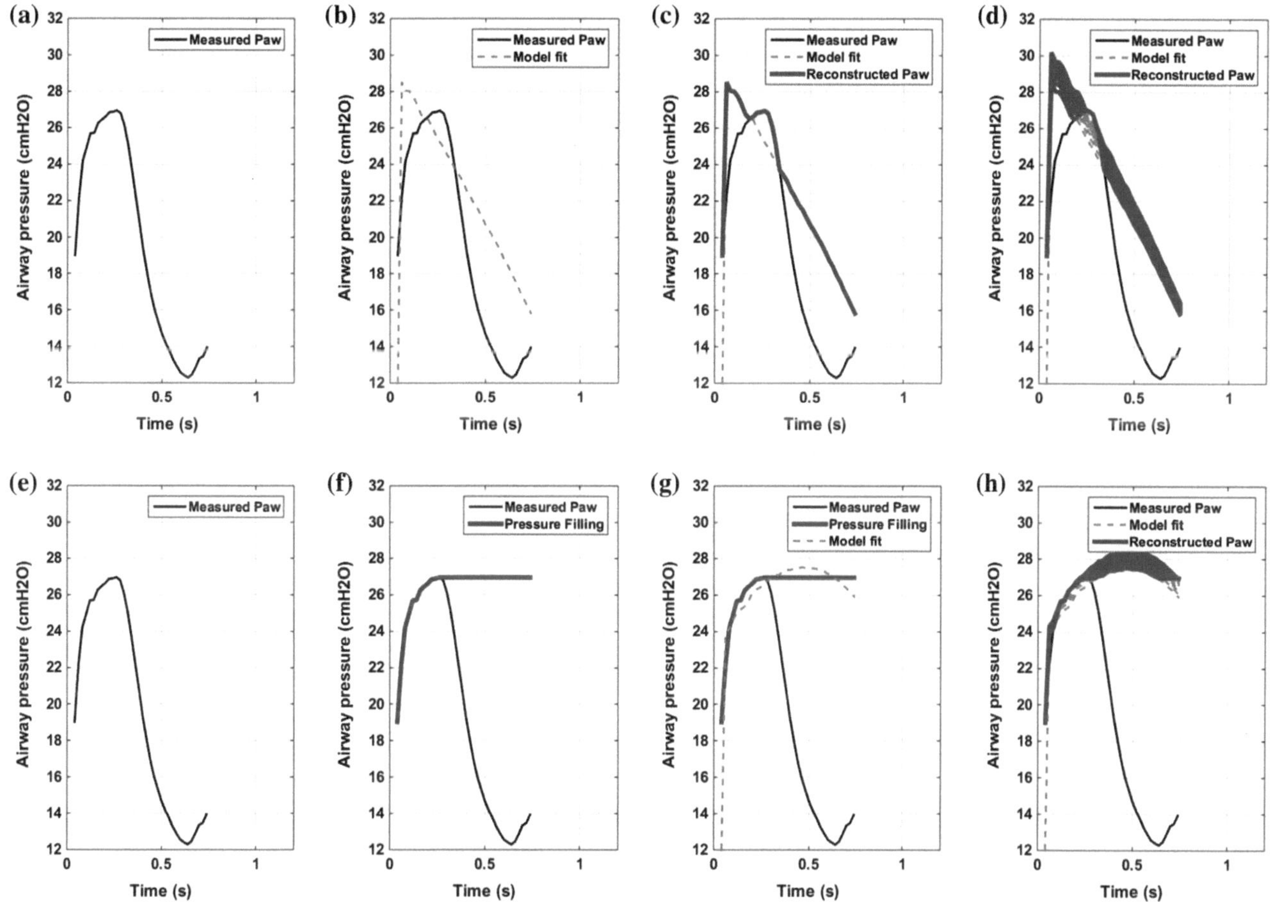

Fig. 3 **a–d** Failed pressure reconstruction due to missing of end of inspiratory pressure data. **e–h** Modified pressure reconstruction with pressure filling

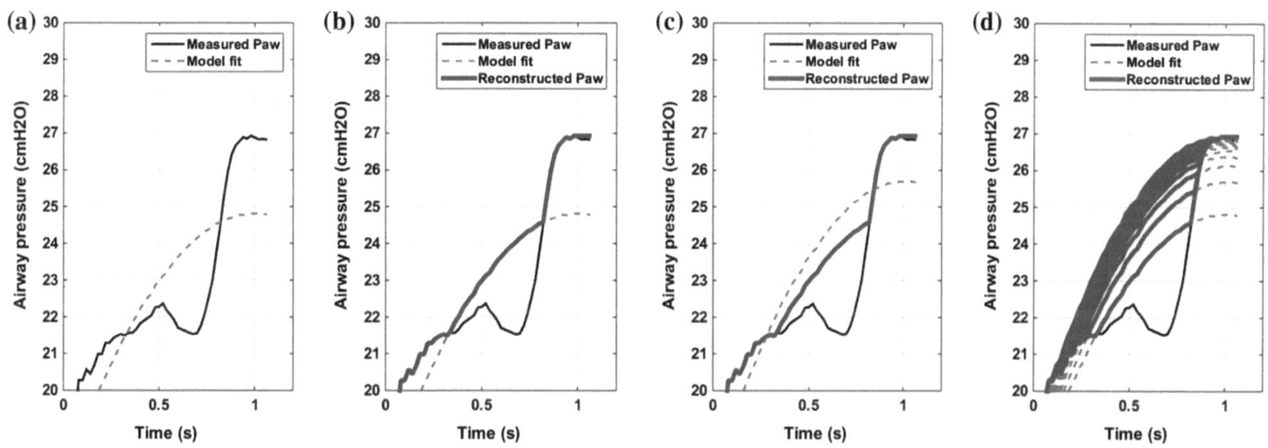

Fig. 4 Early asynchrony pressure reconstruction. **a** The airway pressure does not show pressure increase as start of breathing cycle due to early asynchrony. **b** Pressure reconstruction for the first iteration. **c** New model fit using reconstructed pressure. **d** Complete pressure reconstruction

resulting in higher E_{rs} given the same tidal volume. No significant difference was found in the estimated R_{rs}.

CV and Range90 for E_{rs} after reconstruction fell from 0.3022 and 41.5 cm H_2O/l to 0.1698 and 23.5 cm H_2O/l. Similarly, the CV and Range90 for R_{rs}, fell from 0.1567 and 8.7 cm H_2Os/l to 0.0949 and 4.8 cm H_2Os/l. These results show MIPR is able to provide a more consistent respiratory mechanics estimation, reducing the variability caused by spontaneous and asynchronous breathing, matching expectations of minimal change over the periods analyzed.

4 Discussion

The MIPR iteratively generates a maximal simulated airway pressure that is free from the patient-specific breathing effort. This simulated airway pressure follows the single compartment linear lung model, and is able to model the likely, underlying passive patient pressure profile, resulting in the identification of respiratory mechanics as if there is no asynchronous and/or spontaneous breathing effort. However, it is important to note that this method may not necessarily reconstruct the airway pressure profile to its original, unaffected state; this limitation is due to the fact that the asynchronous affected pressure may have its underlying pressure profile altered. As seen in Fig. 1g and h, the reconstructed pressure may have lower peak airway pressure compared to other normal breathing cycles, and thus may not necessarily be perfect, although there is no way to be certain if the peak pressures that results are within natural breath-to-breath variability, as seen here.

In addition, there is no means to obtain a direct measurement to compare the reconstructed pressure. Thus, there is no way of knowing the exact airway pressure profile without a time machine. Invasive measuring tools or specific clinical protocols that requires sedation and paralysis may provide an insight to the original airway pressure. However, as noted in Redmond et al. [23] and Bibiano et al. [24] patient-specific respiratory mechanics are protocol dependent, affecting the absolute values of the respiratory mechanics. Thus, only a general trend can be derived from these values and there may be no 'absolute value' to be compared with.

In this study, the MIPR estimated airway pressure is compared with a normal breathing cycle. This normal unaffected breathing cycle, selected when there is no apparent anomaly in the pressure/flow profile. It occurs within a short time frame before or after the MIPR breathing cycle. It is assumed that within a short time frame, if there is no significant changes in clinical protocol or MV settings, the patient-specific respiratory mechanics should be within a specific variation [25]. It thus provides a good comparison between the respiratory mechanics of MIPR to a normal breathing cycle, as long as this assumption holds.

Comparing the results with and without MIPR, the coefficient of variation for MIPR is lower than without. A higher variability during volume controlled ventilation indicates asynchronous breathing, resulting in misidentification of respiratory mechanics. This result also suggests that MIPR is also able to estimate the expected, consistent respiratory mechanics in the presence of asynchrony. The CV and Range90 results reflect this outcome.

In this study, the MIPR is stopped after 20 iterations of pressure reconstruction; this number is chosen arbitrarily based on experience with the algorithm on this dataset. In future, a convergence criteria could be determined using the sum of square error (SSE) between the reconstructed airway pressures between each successive iterations of model fitted airway pressure. If the SSE between the model fitting and the reconstructed pressure, does not vary by more than a threshold between successive iterations for 5 consecutive times, convergence is deemed to be met.

5 Conclusion

The iterative pressure reconstruction (MIPR) method is shown to be effective and robust for the wide variety of cases seen in this study. It was found that there is significant changes in the model-estimated E_{rs}, and the estimated parameters are more consistent with lower variability. Thus, this algorithm offers potential improvements for real-time respiratory mechanics estimation. It is however, with limitations. Thus, pressure reconstruction should be monitored carefully, and the application and its performance warrants further investigations.

Acknowledgements The authors would like to thank the Health Research Council of New Zealand (HRC) (Grant Number: 13/213), the Ministry of Higher Education Malaysia (MOHE) Fundamental research grant scheme (FRGS) (Ref: FRGS/1/2016/TK03/MUSM/03/2), and the Advanced Engineering Platform Health Cluster of Monash University Malaysia for funding this research.

References

1. Kretschmer, J., Wahl, A., Moller, K.: Dynamically generated models for medical decision support systems. Comput. Biol. Med. **41**(10), 899–907 (2011)
2. Penning, S., et al.: First pilot trial of the STAR-Liege protocol for tight glycemic control in critically ill patients. Comput. Methods Programs Biomed. **2012**(0) (2012)
3. Tawhai, M.H., Hoffman, E.A., Lin, C.-L.: The lung physiome: merging imaging-based measures with predictive computational models. Wiley Interdisc. Rev. Syst. Biol. Med. **1**(1), 61–72 (2009)
4. Chiew, Y., et al.: Feasibility of titrating PEEP to minimum elastance for mechanically ventilated patients. Pilot Feasib. Stud. **1**(1), 9 (2015)
5. Rees, S., et al.: Using physiological models and decision theory for selecting appropriate ventilator settings. J. Clin. Monit. Comput. **20**(6), 421–429 (2006)
6. Hodgson, C., et al.: A randomised controlled trial of an open lung strategy with staircase recruitment, titrated PEEP and targeted low airway pressures in patients with acute respiratory distress syndrome. Crit. Care **15**(3), R133 (2011)

7. Albaiceta, G.M., et al.: Application of continuous positive airway pressure to trace static pressure-volume curves of the respiratory system. Crit. Care Med. **31**(10), 2514–2519 (2003)
8. Stahl, C.A., et al.: Dynamic versus static respiratory mechanics in acute lung injury and acute respiratory distress syndrome. Crit. Care Med. **34**, 2090–2098 (2006)
9. de Wit, M., et al.: Observational study of patient-ventilator asynchrony and relationship to sedation level. J. Crit. Care **24**(1), 74–80 (2009)
10. Major, V., et al.: Respiratory mechanics assessment for reverse-triggered breathing cycles using pressure reconstruction. Biomed. Signal Process. Control **23**, 1–9 (2016)
11. Szlavecz, A., et al.: The clinical utilisation of respiratory elastance software (CURE soft): a bedside software for real-time respiratory mechanics monitoring and mechanical ventilation management. Biomed. Eng. OnLine **13**(1), 140 (2014)
12. Langdon, R., et al.: Implementation of a non-linear autoregressive model with modified Gauss-Newton parameter identification to determine pulmonary mechanics of respiratory patients that are intermittently resisting ventilator flow patterns. IFAC-PapersOnLine **48**(20), 354–359 (2015)
13. Strøm, T., Martinussen, T., Toft, P.: A protocol of no sedation for critically ill patients receiving mechanical ventilation: a randomised trial. Lancet **375**(9713), 475–480 (2010)
14. Kannangara, D.O., et al.: Estimating the true respiratory mechanics during asynchronous pressure controlled ventilation. Biomed. Signal Process. Control **30**, 70–78 (2016)
15. Vicario, F., et al.: Noninvasive estimation of respiratory mechanics in spontaneously breathing ventilated patients: a constrained optimization approach. IEEE Trans. Biomed. Eng. **PP**(99), 1–1 (2015)
16. Major, V., et al.: Assessing respiratory mechanics of reverse-triggered breathing cycles—case study of two mechanically ventilated patients. IFAC-PapersOnLine **48**(20), 505–510 (2015)
17. Damanhuri, N.S., et al.: Assessing respiratory mechanics using pressure reconstruction method in mechanically ventilated spontaneous breathing patient. Comput. Methods Programs Biomed. **130**, 175–185 (2016)
18. Newberry, F., et al.: Iterative interpolative pressure reconstruction for improved respiratory mechanics estimation during asynchronous volume controlled ventilation. In: Ibrahim, F., et al. (eds.) International Conference for Innovation in Biomedical Engineering and Life Sciences: ICIBEL2015, 6–8 Dec 2015, Putrajaya, Malaysia, pp. 133–139. Springer Singapore, Singapore (2016)
19. Chiew, Y.S., et al.: Model-based PEEP optimisation in mechanical ventilation. Biomed. Eng. OnLine **10**(1), 111 (2011)
20. van Drunen, E., et al.: Visualisation of time-varying respiratory system elastance in experimental ARDS animal models. BMC Pulm. Med. **14**(1), 33 (2014)
21. Moorhead, K., et al.: NAVA enhances tidal volume and diaphragmatic electro-myographic activity matching: a Range90 analysis of supply and demand. J. Clin. Monit. Comput. **27**(1), 61–70 (2013)
22. Chiew, Y.S., et al.: Time-varying respiratory system elastance: a physiological model for patients who are spontaneously breathing. PLoS ONE **10**(1), e0114847 (2015)
23. Redmond, D.P., Chiew, Y.S., Chase, J.G.: The effect of respiratory manoeuvres for patient-specific respiratory mechanics monitoring**The study acknowledges funding support from the EU FP7 IRSES Marie Curie action and royal society of New Zealand. IFAC-PapersOnLine **48**(20), 135–140 (2015)
24. Bibiano, C., et al.: Effects of different models and different respiratory manoeuvres in respiratory mechanics estimation. In: Kyriacou, E., Christofides, S., Pattichis, C.S. (eds.) XIV Mediterranean Conference on Medical and Biological Engineering and Computing 2016: MEDICON 2016, March 31st–April 2nd 2016, Paphos, Cyprus, pp. 50–55. Springer International Publishing, Cham (2016)
25. Kim, K., et al.: Breath-to-breath respiratory mechanics variation: how much variation should we expect? Crit. Care **19**(Suppl 1), P260 (2015)

Smart Balance Board to Improve Balance and Reduce Fall Risk: Pilot Study

Kang Xiang Khor, Phei Ming Chern, Che Fai Yeong, Eileen Lee Ming Su, Muhammad Farhan Mustar, Najib Bin Abdullah, Kang Xian Khor, and Hadafi Fitri Mohd Latip

Abstract

Falls in the elderly population is a significant public health concern, and there are up to 37.3 million people experiencing severe falls each year. Balance board is commonly used for balance training, however, the device is not widely used at home due to lack of interactive element and training guide. A new, effective, and widely applicable strategy to prevent falls is urgently needed due to the increasing number of falls in elderly worldwide. This work demonstrates the feasibility of FIBOD, as a smart balance board that is able to provide interactive balance training and objective assessment for balance. Two elderly subjects performed 24 sessions of training within four weeks. Results show that the training using FIBOD could improve balancing capability, reduce falling risk, and increase the motivation to do balancing training. This study shows the feasibility of FIBOD as a balancing training device to improve balance skill and reduce the occurrence of fall.

Keywords

Balance board • Balance training • Interactive • Virtual reality

K. X. Khor
Malaysia-Japan International Institute of Technology, Universiti Teknologi Malaysia (UTM), 54100 Kuala Lumpur, Malaysia

K. X. Khor · M. F. Mustar · N. B. Abdullah
Techcare Innovation Sdn. Bhd, 81310 Skudai, Johor, Malaysia

P. M. Chern
Clinical Research Centre, Cheras Rehabilitation Hospital, 56000 Kuala Lumpur, Malaysia

C. F. Yeong
Center of Artificial Intelligence and Robotics, UTM, 54100 Kuala Lumpur, Malaysia

C. F. Yeong (✉) · E. L. M. Su
Faculty of Electrical Engineering, UTM, 81310 Skudai, Johor, Malaysia
e-mail: cfyeong@utm.my

K. X. Khor
Malacca State Department of Health, Ministry of Health, 75450 Melaka, Malaysia

H. F. M. Latip
Sports Innovation and Technology Centre, UTM, 81310 Skudai, Johor, Malaysia

1 Introduction

Falls in the elderly is a significant public health concern. Statistic shows that 37.3 million people experienced severe falls each year [1]. Complication from falls, including hip fractures, fear, decreased quality of life from self-imposed activity restriction, social isolation and depressive symptoms [2]. The average cost of hospitalization for fall related injuries are projected to increase to US$240 billion by year 2040 [3]. Worldwide, elderly people who are over 60 years are growing faster than any other 18 age group. The number persons who are over 60 years were projected to grow to almost two billions by 2050. The increasing number of injuries may induce a huge burden to hospital, therefore, a new, effective, and widely applicable strategies to prevent hip fractures are urgently needed [3]. Balance board is recommended to improve proprioceptive and reactive capabilities,

F. Ibrahim et al. (eds.), *2nd International Conference for Innovation in Biomedical Engineering and Life Sciences*, IFMBE Proceedings 67, https://doi.org/10.1007/978-981-10-7554-4_6

which may reduce the likelihood of fall injuries [4]. Many studies have investigated the clinical efficacy for fall injury and ankle sprain prevention and treatment, however the training is not widely use due to lack of motivation [5] and feedback which is important in motor relearning [6]. The goal of this study is to evaluate the functionality and feasibility of a smart fitness balance board, FIBOD (formerly known as Innovaboard) that able to provide structure training program with visual feedback on smart device for two elderly people to be used at home.

2 Methods

2.1 Subjects

This case study comprised of one male and female elderly subjects. The male subject was 67-year-old, right handed, 66 kg in weight and 162 cm in height. The female elderly subject was a 61-year-old, right handed, 57 kg in weight and 160 cm in height. Inclusion criteria were (1) sufficient cognitive and language abilities to understand the training instructions, (2) did not have any injuries within the past six months prior to the study and (3) able to stand on the used equipment safely, without reporting any pain. Both subjects signed the informed consent form to take part in the study.

2.2 FIBOD—Smart Fitness Balance Board

FIBOD was used in this study. It is a smart fitness balance board that allows users to train and access their balance skill with virtual reality games and objective assessment program in smart device. The board is 15.5 inches in diameter and weighted 1.3 kg. It can achieve surface tilt up to 20° in anterior-posterior and medial-lateral directions. The sensor has a resolution of 0.01°, accuracy of ±0.04% and repeatability of ±0.06°. FIBOD software consisted of fall risk assessment program, comprising three stability indices, which are medial-lateral stability index (MLSI), anterior-posterior stability index (APSI), and overall stability index (OSI) [7]. The software consisted of a virtual reality training program involving combination of six training configurations for anterior-posterior and medial-lateral directions with three levels of difficulties (easy, medium, hard).

2.3 Intervention

Each subject received a total of 24 training sessions over a period of four weeks. Each session, the training during was about 5–10 min and subject was asked to stand on both legs on the FIBOD. Subject can train more than one times within a day as long as they completed 24 sessions within a period of four weeks. During training, they can hold the handle bar when they are unbalance. Both subjects received instructions on how to train with the device safely and the correct way to interact with it before the training. Subjects could rest or stop in the middle of the training if they were exhausted or encountered any other personal issues.

In the beginning of each training session, subject will perform FIBOD fall risk assessment and then proceed for training program. The training program consisted of three level; easy, medium, hard. Each level of training program comprised of different combination of training direction. The performance of each training program are analyzed and presented with a score at the end of the training. The score calculation is mainly based on the time used to complete the task. All subjects are required to start from easy level and the subject is only allowed to proceed to next level if they are able to hit 100% in that particular level. Figure 1 shows the experimental setup during the study and the software interface for assessment and training program.

2.4 Outcome Measures

The outcome measures were used to identify the changes in balance skill and falling risk. The primary outcome measure is Berg Balance Scale (BBS), secondary measures include FIBOD fall risk assessment scale (FFRS), eye-opened and closed Single-Leg-Stance Test (SLS-O and SLS-C). During FFRS, subject will perform with eyes open, stand unassisted on both legs and they are required to maintain the FIBOD position as flat as possible, within 10 s. The assessment data was stored in the cloud automatically after test for further progress analysis by the doctor. The assessments were performed in the first session and the last session of the training. During SLS-O and SLS-C, subjects are required to stand on dominant leg with eyes open and eye close. Their arms will be placed on the hips, the subjects must stand unassisted on one leg and is timed in seconds from the time one foot is flexed off the floor to the time when it touches the ground or the standing leg or an arm leaves the hips. The maximum duration for SLS is 45 s. Participants unable to perform the one-leg stand for at least 5 s are at increased risk for injurious fall [8]. The four assessments were performed in the first session and the last session of the training. All assessment values are average of three repeated test except BBS.

3 Results

Both subjects completed the 24 sessions of training. Figure 2 shows the result of the primary and secondary outcome measures compared with the average benchmark score obtained from three healthy subjects.

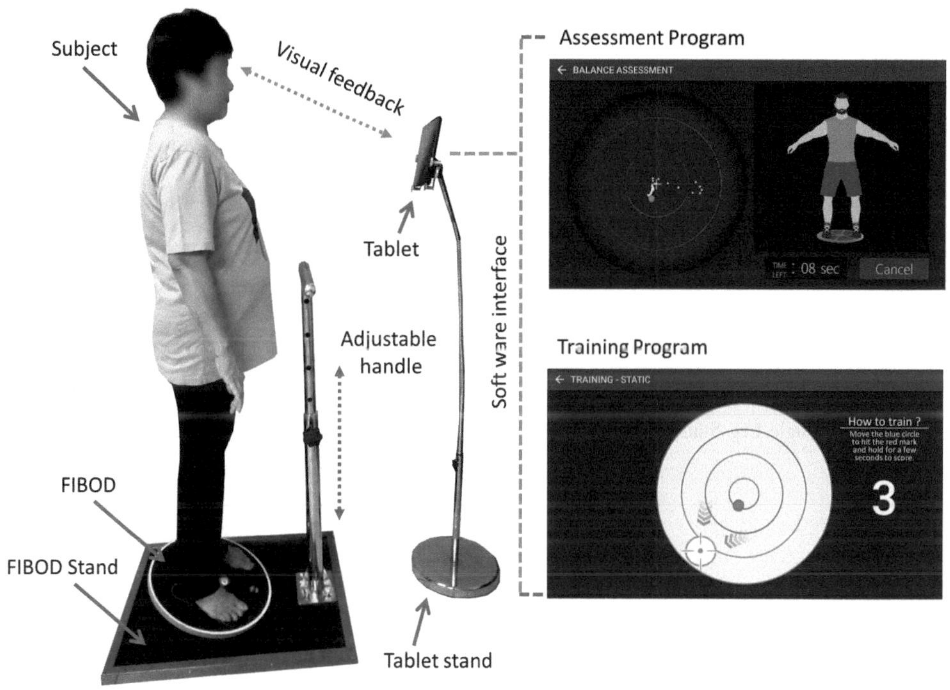

Fig. 1 Experimental setup of the subject and software interface illustration. Subject stood on the FIBOD, which was placed on the FIBOD stand. The stand provided an adjustable handle for subject to hold if they feel unstable during the training (left figure). Software interface for assessment and one of the training modules (right figures)

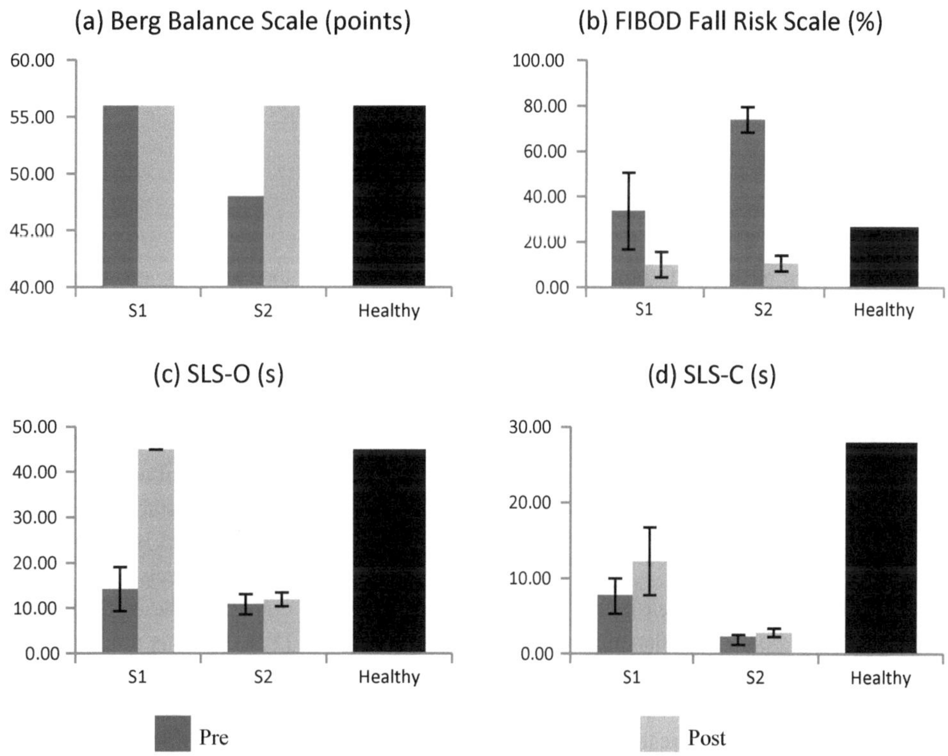

Fig. 2 Comparison of primary and secondary outcome measures before and after the training: **a** Berg-Balance Scale (BBS), **b** FIBOD Fall Risk Scale (FFRS), **c** Eye-opened Single Leg Stance Test (SLS-O) and **d** Eye-closed Single Leg Stance Test (SLS-O)

Both subjects improved in all the assessment tests. S1 able to reach the hard level, while S2 able to reach medium level of the training program. S1 recorded improvement in FFRS by 23.67% (70.3%), SLS-O by 30.82 s (217.35%), SLS-C by 4.46 s (57.25). S2 recorded an improvement of 7 points achieving full mark in BBS (14.6%), FFRS by 63.33% (85.58%), SLS-O by 1.03 s (9.5%), and SLS-C by 0.53 s (23.25%). The subjects reported no discomfort or pain during or after training with FIBOD. Based on the feedback from the interview after the training session, both subjects

Table 1 Demographics and outcome measure of subjects before and after intervention

Subject	S1	S2
Gender	Male	Female
Age	68	61
Weight (kg)	57	65.6
Height (cm)	162	160
Berg Balance Scale (points)		
Week 1	56	48
Week 4	56	55
Changes	0	7
Changes (%)	0	14.6
Eye-opened Single Leg Stance Test (s)		
Week 1	14.18 ± 4.85	10.84 ± 2.25
Week 4	45 ± 0	11.87 ± 1.55
Changes	30.82	1.03
Changes (%)	217.35	9.50
Eye-closed Single Leg Stance Test (s)		
Week 1	7.79 ± 2.18	2.28 ± 0.22
Week 4	12.25 ± 2.53	2.81 ± 1.08
Changes	4.46	0.53
Changes (%)	57.25	23.25
FIBOD Fall Risk Scale (%)		
Week 1	33.67 ± 16.92	74 ± 5.57
Week 4	10 ± 5.57	10.67 ± 3.51
Changes	−23.67	−63.33
Changes (%)	−70.30	−85.58

were motivated to do the training due to the visual feedback and increment of challenges. They believed the training therapy would help them improve their balance skills, and they were satisfied as users of the device. Table 1 shows the demographics and detail outcome measure of subjects before and after intervention.

4 Discussion

This study explores the feasibility of using the FIBOD, a smart balance board to balancing training and assessment. Study shows that the training program with objective assessment improve balancing capability of both subject. The interactive virtual reality training program increase motivation for the subject to perform more training. The result further strengthens the evidence of balance board in improving balance capability and reducing falling risk. This improvement in balance may be associated with improved joint mechanoreceptor functioning and lower extremity muscles strength [9]. The limitations of this study included the subject group's heterogeneity, the relatively small number of subjects, and the lack of comparison groups to differentiate the benefit from the training.

Accurate assessment of risk is important in predicting falling risk and designing interventions to prevent falls for patients. Based on the result, BBS has a ceiling effect and thus may not adequately assess balance in the early stages of postural instability in individuals or healthy person. This ceiling effect is also addressed by other researchers [10, 11]. Ceiling effects limit the usefulness of an evaluation, as it constraint the comparison among better functioning subjects [12]. There is a need for fall risk assessment scale that have minimal or no ceiling effect, so that wider population can be used to gauge the falling risk. Based on the data of S1, the FFBS is still able to determine the fall risk percentage, even BBS show the maximum score. Therefore, FIBOD fall risk assessment scale can potentially be used as a quick assessment scale for fall risk. Future study will include the evaluation of FIBOD training program and fall risk assessment scale with bigger number of samples.

5 Conclusion

This work demonstrates the feasibility of FIBOD, as a smart balance training and assessment device. Results show that the training using FIBOD could increase balancing capability, reduce falling risk, and increase the subjects' motivation to do balancing training.

Acknowledgements This work is supported by the Universiti Teknologi Malaysia, Clinical Research Center Cheras Rehabilitation Hospital, Lab2Market commercialization grant [08906] and Collaborative Research in Engineering, Science and Teknologi Center (CREST) R&D grant [P37C2-13].

References

1. World Health Organization: WHO global report on falls prevention in older age (2007)
2. Bartimole, L., Fristad, M.A.: Taiji (Tai Chi) for fall prevention in the elderly: training the trainers evaluation project. Explor. J. Sci. Heal (2017)
3. Cummings, S.R., Rubin, S.M., Black, D.: The future of hip fractures in the United States. Numbers, costs, and potential effects of postmenopausal estrogen. Clin. Orthop. Relat. Res. **252**, 163–166 (1990)
4. Willardson, J.M.: Core stability training: applications to sports conditioning programs. J. Strength Cond. Res. **21**(3), 979 (2007)
5. Fitzgerald, D., Trakarnratanakul, N., Smyth, B., Caulfield, B.: Effects of a wobble board-based therapeutic exergaming system for balance training on dynamic postural stability and intrinsic motivation levels. J. Orthop. Sports Phys. Ther. **40**(1), 11–19 (2010)
6. Molier, B.I., Prange, G.B., Krabben, T., Stienen, A.H., van der Kooij, H., Buurke, J.H., Jannink, M.J., Hermens, H.J.: Effect of position feedback during task-oriented upper-limb training after stroke: five-case pilot study. J. Rehabil. Res. Dev. **48**(9), 1109 (2011)
7. Arnold, B.L., Schmitz, R.J.: Examination of balance measures produced by the biodex stability system. J. Athl. Train. **33**(4), 323–327 (1998)
8. Springer, B.A., Marin, R., Cyhan, T., Roberts, H., Gill, N.W.: Normative values for the unipedal stance test with eyes open and closed. J. Geriatr. Phys. Ther. **30**(1), 8–15 (2007)
9. Balogun, J.A., Adesinasi, C.O., Marzouk, D.K.: The effects of a wobble board exercise training program on static balance performance and strength of lower extremity muscles. Physiother. Can. **44**(4), 23–30 (1992)
10. Leddy, A.L., Crowner, B.E., Earhart, G.M.: Functional gait assessment and balance evaluation system test: reliability, validity, sensitivity, and specificity for identifying individuals with parkinson disease who fall. Phys. Ther. **91**(1), 102–113 (2011)
11. Lemay, J.-F., Nadeau, S.: Standing balance assessment in ASIA D paraplegic and tetraplegic participants: concurrent validity of the Berg Balance Scale. Spinal Cord **48**(3), 245–250 (2010)
12. Mao, H.-F., Hsueh, I., Tang, P.-F., Sheu, C.-F., Hsieh, C.-L.: Analysis and comparison of the psychometric properties of three balance measures for stroke patients. Stroke **33**(4), 1022–1027 (2002)

Part II

Biomechanics, Ergonomics, and Rehabilitation

Pushing a Manual Wheelchair Requires More Muscular Force Than Pulling

Hamza Toor, Saad Jawaid Khan, Hashir Iqbal, Maria Tahir, Bareera Amjad, Mafia Afzal, Muhammad Shafique, and Juliana Usman

Abstract

One of the most challenging problems faced by regular manual wheelchair users is the increased physical demand while navigating it and the resulting shoulder, wrist and hand injuries. These problems call for a device that reduces injury risk factor owing to a prolonged wheelchair use while simultaneously increasing user independence and mobility. The study suggests a modified wheelchair design that operates by means of a pulling motion as opposed to pushing. This motion offers a more efficient solution to the problems of shoulder, elbow, wrist injuries and user fatigue, since a pulling motion requires the use of a stronger and more capable group of muscles relative to those involved in standard pushing mechanism.

Keywords

Wheelchair • Biomechanics • Mechanical advantage

1 Introduction

Manual wheelchairs are the preferred mode of transportation for those with physical limitations, due to their low cost and low maintenance requirements. Manual propulsion is also preferred whenever possible as a practical and accessible form of physical exercise that is compatible with the individual's life style for long-term exercise sustainability. Pushing the wheel-chair for a longer period of time lead to upper limb injury (shoulders) [1–3]. Studies have shown atleast three biomechanical parameter are related to upper arm pathology for wheelchair users: high force requirement, repetitive motion and extreme joint postures [4, 5]. The whole upper extremity undergo large range of motion during propulsion and required to generate substantial force at joint angles near physiological limits [6]. Injury to any muscles leads to a severe pain. It is important to overcome all these injuries caused by pushing the wheelchair and it is essential to do for the wheel-chair users who are at the high risk of secondary health problem.

The way of pushing a wheelchair doesn't make use of the strongest muscle groups. Worse, the apparently unnatural motion of constantly pushing a wheel may cause repetitive stress injuries [7]. Pulling forward better mimics rowing where larger muscles like the upper back and biceps come into play.

Manual wheelchair users will develop upper extremity overuse injuries or pain during their lifespan, in part due to the high physical demand [8]. The correct functionality of the upper limbs is an essential condition for the autonomy of people with disabilities. The push phase muscle (Triceps, Teres major) and the pull phase muscles (bicep brachia, Trapezius, Anterior deltoid, latissimus dorsi muscles) undergo large ranges of motion during propulsion and, in some cases, are required to generate significant forces at

H. Toor · S. J. Khan · H. Iqbal · M. Tahir · B. Amjad · M. Afzal · M. Shafique
Department of Biomedical Engineering, Riphah International University, Islamabad, Pakistan

S. J. Khan (✉) · J. Usman
Department of Biomedical Engineering, University of Malaya, Kuala Lumpur, Malaysia
e-mail: saadjawaid.khan@riphah.edu.pk

J. Usman
Centre for Applied Biomechanics, University of Malaya, Kuala Lumpur, Malaysia

F. Ibrahim et al. (eds.), *2nd International Conference for Innovation in Biomedical Engineering and Life Sciences*,
IFMBE Proceedings 67, https://doi.org/10.1007/978-981-10-7554-4_7

joint near the physiological limit [9, 10]. High forces are generated by muscles and applied to the hand rim to propel the wheelchair in the forward and reverse direction of translatory motion. Two major muscles involve in the push phase of propulsion are the Triceps and Teres major. The triceps is an extensor muscle of the elbow joint and an antagonist (to biceps) it can fixate the elbow joint when the forearm and hand are used for fine movements of wheel chair. The Teres major muscle is the essential muscle for conducting the push phase this muscle is involved in the extension of the forearm due to its insertion on humerus.

Muscles convoluted in the pull phase are four in numbers, the bicep muscle influence the movement of both shoulder and elbow joints. During propulsion, the shoulder is maintained at 70° of abduction. Bicep crosses the shoulder acted to transfer the power to the hand rim. Trapezius, in pull phase, during propulsion of wheelchair this muscle is active and stabilizes the shoulder blades. It creates high forces and apply to the hand rim to propel the wheelchair during its pull phase. The lateral fibers of deltoid abduct the arm by pulling the humerus toward the acromion. Abduction of the arm results in the arm moving away from the body. Contraction of the posterior fibers extends and laterally rotates the arm by pulling the humerus toward the spine of the scapula. This helps in pulling of wheel chair as the extension and lateral rotation moves the arm posteriorly. During the pull phase, the scapular part of the muscle latissimus dorsi supports the movement of shoulder. Due to the crossing over of the muscle fibers at the insertion point (posterior side of humerus), the muscle's contraction leads to an inward rotation of the humerus, which helps to pull the wheel chair. Therefore, the purpose of the study was to find the mechanical efficiency of the muscles involved in the push and pull phases of manual wheelchair propulsion. We hypothesize that pull phase consume lesser energy than push phase during wheelchair propulsion.

2 Methodology

2.1 Research Participants

The study involved 6 able-bodied adults. The mean age of the participants were 21 years (±3) in a good health and have no previous record of pathologies of upper extremities.

Experimental Setup:

The complete setup based on the Bio Pac MP 3.7.7 EMG conducting electrodes are used (2.5 cm conducting area). Each channel of Biopac further consists of three leads. One of the reference lead was placed on the origin of the Muscle, Active lead was placed on the Muscles belly, whereas the ground lead was grounded on the bony area. Adhesive tape was used to fix the electrodes on the skin in order to limit the effect of interference due to sliding of the electrodes on the skin (motion artifact of soft tissues).

2.2 Test Protocol

We distinguish two phases in the propulsion: push phase and the pull phase. During propulsion, the shoulder is maintained at approximately 70° of abduction. At the onset of the propulsive phase of motion, the shoulder is extended and internally rotated and subsequently ends up flexed and externally rotated at the onset of the recovery phase. During propulsion, the area covered by the subject was 135 in. and each trial consisted 7 power strokes with time duration of 6.5 s. Subject was seated on the manual wheel chair, the electrodes were placed on the Triceps and Teres muscles during the Push phase, the active lead was placed on the muscle belly, the reference lead was placed on the muscle origin and one of the lead was ground. During the push phase the hands partially follow the circular motion and the wheel can be moved imparting a tangential force upward with the elbow flexed (between the beginning and the middle of the push) and one downward with the elbow extended (from the middle to the end of the movement) The elbow is flexed throughout the pushing phase, starting from an angle of about 60°, gradually increasing until the hand assume the most distant position from the ground and subsequently decreasing until to reach the minimum distance from the ground near by the recovery phase. In the pull phase, the EMG was recorded for Triceps, Deltoid, Trapezius and Latissimus dorsi muscles. The pull phase tracked the same mechanism but the wheel chair moves in the backward direction instead of forward direction.

2.3 Statistics

Shapiro-Wilk test was applied to the data to assess normality. Paired t-test was used to find paired differences at $\alpha = 5\%$ using IBM SPSS version 21 (SPSS Inc., USA).

3 Results

3.1 Normality Test

Shapiro-Wilk test showed that the data for both push and pull phases are normally distributed with p-values of 0.196 and 0.570 respectively.

3.2 Paired t-Test

A paired t-test was run on a sample of 06 participants to determine whether there was a statistically significant mean difference between the push phase compared to the pull phase. Participants spent more force in push phase (1135.29 ± 276.00 mV) as compared to the pull phase (676.94 ± 222.32 mV), $t(05) = 3.098$, $p = 0.027$.

3.3 Calculations

EMG calculations are based on the final EMG signal peak stroke values of each push phase trail and pull phase trial during propulsive phase. In push phase, the major muscle involve for the conduction of push was the Triceps muscle and the maximum peaks value during one stroke for the triceps muscle is 1075.50 mV during the trial of 6.5 s. Table 1 and Fig. 1 represent the forces of triceps and teres major in push phase of wheelchair propulsion.

When the wheel chair was moved in the forward direction (push), the triceps muscle and Teres muscle was involved in dragging the wheel chair, the average force triceps muscle delivered was 853.88, while the average force contributed by the Teres Major is 480.99. Sum of both muscle forces results in the movement of wheel chair and that average force was 1135.33.

During Pull phase i.e. moving the wheel chair in the backward direction total load was distributed on four muscles the average force delivered by the bicep muscle, Trapezius muscle, Deltoid muscle and latissimus dorsi was 202.33, 116.51, 269.84 and 199.88 Respectively. Table 2 and Fig. 2 represent the forces generated by biceps, trapezius, deltoid and latissimus dorsi during pull phase of wheelchair propulsion. Sum of all four muscles forces results in the movement of wheel chair and that average force was 676.91

The comparison relates that more force is required to move the wheel chair in the forward direction then the movement on wheel chair in backward direction. Table 3 and Fig. 3 represent the comparison between the total sum of forces between push and pull phase of wheelchair propulsion.

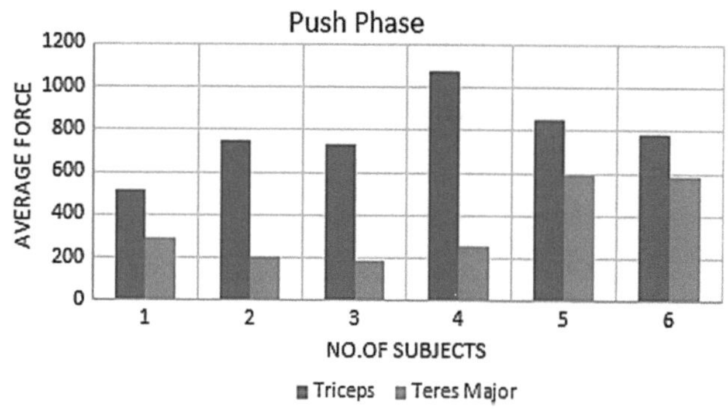

Fig. 1 Graphical representation of push phase average force. EMG amplitude is calculated in milli-volts (mV) as a surrogate measure of muscle force

4 Discussion

The purpose of this study was to evaluate the difference between push and pull phases of propelling a manual wheelchair through electromyography. The strength of the study is that it provides useful information to the clinicians and manual wheelchair manufacturers. The manufacturers can modify wheelchair design based on this study allowing the patients to move the wheelchair forward using a pulling mechanism rather than a pushing mechanism. Having stated the strengths, our study also has certain limitations. Our sample size is small and based on able bodied participants. Therefore, we are not able to generalize our results to persons with disabilities.

Our study showed that manual wheelchair used mechanism which is unnatural in any way. During pushing mechanism (wheelchair moving in forward direction), only triceps and teres major come into play. Triceps is not so strong while teres major doesn't contribute much during this propulsion. On the other hand, during pulling mechanism, the number of muscles is increased and consists of relatively stronger muscles like biceps and deltoid. After having

Table 1 Trend of push muscles and total force required to drag wheel chair during the push phase

Subjects	Push phase		
	Triceps (mV)	Teres major (mV)	Sum (mV)
1	513.78	289.26	803.05
2	746.04	203.42	949.47
3	730.60	185.49	916.09
4	1075.50	255.51	1331.01
5	853.88	590.95	1444.83
6	782.02	585.30	1367.32

Table 2 Trend of pull muscles and total force required to drag wheel chair during the pull phase

Subjects	Pull phase				
	Biceps (mV)	Trapezius (mV)	Deltoid (mV)	Latissimus Dorsi (mV)	Sum (mV)
1	128.12	45.88	288.52	184.71	647.25
2	151.65	64.12	291.18	176.32	683.28
3	164.50	67.95	268.37	150.64	651.48
4	305.84	117.35	357.61	215.90	996.71
5	310.58	171.63	310.58	206.28	310.58
6	153.28	232.13	121.47	265.42	772.32

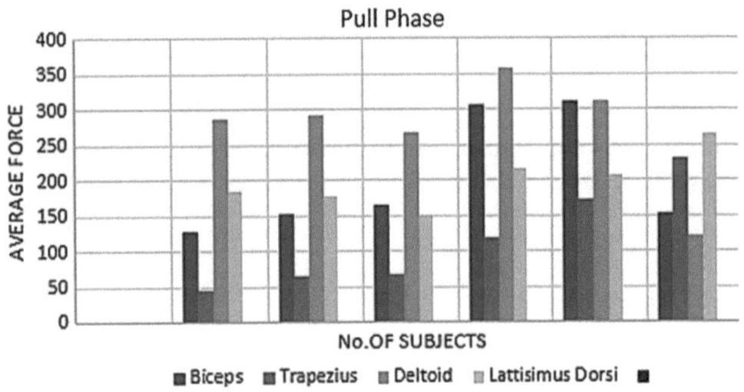

Fig. 2 Graphical representation of pull phase average force. EMG amplitude is calculated in milli-volts (mV) as a surrogate measure of muscle force

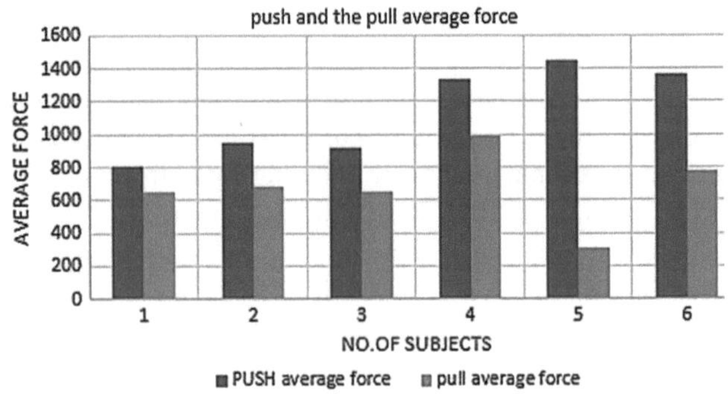

Fig. 3 Comparison between the total sum of push and pull phase. EMG amplitude is calculated in milli-volts (mV) as a surrogate measure of muscle force

Table 3 Comparison of the sum of push and pull phase muscles

Subjects	Propulsive phase	
	Push phase (sum in mV)	Pull phase (sum in mV)
1	803.05	647.25
2	949.47	683.28
3	916.09	651.48
4	1331.01	996.71
5	1444.84	310.58
6	1367.32	772.32
Mean	1135.33	676.92

electromyographic comparison between push and pull mechanisms, it is clear that pulling requires lesser energy and force than pushing. Through this study we are proposing that wheelchair manufacturers should use some mechanical tool like gears etc. so that when the user pulls the wheelchair, they move in the forward direction. In this way, the user will engage more and stronger muscles in forwarding the wheelchair which require less force and will consume less energy. From the results, we can also deduce that the major cause of energy expenditure is the main contribution of triceps muscle in pushing phase. People should be trained to use teres major muscle, while pushing the wheelchair, rather than triceps for lesser energy consumption. Teres major is a large muscle as compared to triceps. Therefore, teres major can generate required amount of force with lesser consumption of energy as compared to the triceps.

5 Conclusion

Average muscle force required to push the wheel chair in forward direction is 40% greater than the force required to pull the wheel chair in the backward direction.

Acknowledgements The authors would like to thank Dr. Imran Amjad, Assistant Professor & Assistant Dean of Riphah Academy of Research and education, Riphah College of Rehabilitation Sciences, Riphah International University. His guidance in muscle palpation and placing the electrodes correctly made the data collection possible.

References

1. Rankin, J.W., et al.: The influence of wheelchair propulsion technique on upper extremity muscle demand: a simulation study. Clin. Biomech. (Bristol, Avon) **27**(9), 879–886 (2012)
2. Finley, M.A., et al.: The biomechanics of wheelchair propulsion in individuals with and without upper-limb impairment. J. Rehabil. Res. Dev. **41**(3B), 385 (2004)

3. Sie, I.H., et al.: Upper extremity pain in the postrehabilitation spinal cord injured patient. Arch. Phys. Med. Rehabil. **73**(1), 44–48 (1992)
4. Fay, B.T., et al.: Wrist kinematics and indicators of carpal tunnel syndrome during manual wheelchair propulsion. In: Engineering in Medicine and Biology. 21st Annual Conference and the 1999 Annual Fall Meeting of the Biomedical Engineering Society. BMES/EMBS Conference. Proceedings of the First Joint. IEEE
5. Mercer, J.L., et al.: Shoulder joint kinetics and pathology in manual wheelchair users. Clin. Biomech. **21**(8), 781–789 (2006)
6. Rao, S., et al.: Three dimensional kinematics of wheelchair propulsion. Gait Posture **2**(1), 57 (1994)
7. Jayaraman, C., Moon, Y., Sosnoff, J.J.: Shoulder pain and time dependent structure in wheelchair propulsion variability. Med. Eng. Phys. **38**(7), 648–655 (2016)
8. Flemmer, C.L., Flemmer, R.C.: A review of manual wheelchairs. Disabil. Rehabil. Assist. Technol. **11**(3), 177–187 (2016)
9. Sabick, M.B., Kotajarvi, B.R., An, K.-N.: A new method to quantify demand on the upper extremity during manual wheelchair propulsion. Arch. Phys. Med. Rehabil. **85**(7), 1151–1159 (2004)
10. Lin, H.-T., et al.: Muscle forces analysis in the shoulder mechanism during wheelchair propulsion. Proc. Inst. Mech. Eng. H J. Eng. Med. **218**(4), 213–221 (2004)

Wearable Instrument Applications for Indonesian's Worker Clothing

Melani Sukirman, Pringgo Widyo Laksono, and Ilham Priadythama

Abstract

UU No I 1970 tells about the assurance for workers' health conditions. Current years, there is no monitoring aids that can help supervisors to monitor the workers condition. A standard worker's clothing can protect from harm situation but it cannot deliver the actual physical condition of workers. If there is something happen dealing with physical condition of the workers it will not only interfere the operation and production line but also lead deathly risk. Monitoring of labor conditions is easily done in a laboratory but doing it directly in the production floor still face many challenge. Therefore, it is necessary to develop a device to monitor physiological conditions that can be used on the site. Wearable instrumentation is a solution to face these problems. This research aims to develop a physiological detector based on wearable instrumentation that able to indicate physical conditions of workers in real time. From these results, it is known that the TMP 36 sensor has a fairly accurate measurement values for different only 0.1–0.2 °C from the actual temperature. This instrumentation system accuracy was verified by observing physiological condition of 30 subjects. Further we are going to attach the system to a worker's clothing which was specifically designed for simple and comfort usage.

Keywords

Physiology detector • Wearable instrumentation • Sensors

1 Introduction

Occupational health and safety insurance is a right of every employee. It became a company responsibility since UU No I 1970 [1] published. It is a safety and health regulation, mainly for worker, and additionally for society and environment closed to the plant and its working place [2]. In industries, some tasks are dealing with high load or risk so that they require physical condition or excellent concentration. Whereas physical problem may be occurred without any signed. It can be happen at any time and place. Especially for high risk task, losing concentration in working may lead to the deathly risk. There is no ideal way to guarantee worker health and safety condition without a real time physiological monitoring. The supervisors would face difficulties to monitor the worker's condition every time if they are utilizing manual devices, accordingly it could develop interruption for them and making the production slowing down or stop. Recently, monitoring process has just performed at the laboratory and could not observe all the time.

Three main parameters to determine the workload are energy expenditure, working heart rate, and body temperature [3]. In the wearable instrumentation, body temperature and heart rate become parameters which are widely use to indicate the condition and work load. Using heart rate to calculate working load intensity level has some advantages. It is easy, fast, not necessary to utilize a high cost

M. Sukirman · P. W. Laksono · I. Priadythama (✉)
Sebelas Maret University, Surakarta, Indonesia
e-mail: priadythama@gmail.com

P. W. Laksono
Center Study of Technology Development and Industrial Collaboration, Surakarta, Indonesia

F. Ibrahim et al. (eds.), *2nd International Conference for Innovation in Biomedical Engineering and Life Sciences*,
IFMBE Proceedings 67, https://doi.org/10.1007/978-981-10-7554-4_8

instrumentation, and harmless to the subject [4]. The body temperature difference is influenced by many factors, either internals or externals. Internally, it has a high correlation with excessive thermal production, maximum thermal production, and extreme thermal emission. The external factors are dealing with working environment. Working in too cold or hot environment may be reduced the working performance. Generally, in a high temperature and humidity condition would require a high physical work. Therefore a severe exhausted feeling would be occurred.

One of the equipment which supports the working condition that regulated by occupational health and safety industrial standard is worker's clothing [5]. Generally this kind of apparel is provided by the company which has been designed sufficiently safe and comfortable according to the job. Some of industrial worker's outfits were made from high quality fabric, which is strong and is ready for storing some kind of hand tools. The material is not easily ripped, even though some heavy objects are carried on them. However, it can't inform actual worker's physical condition. Wearable instrumentation system is not a heavy thing. It can consist of several devices but the size is relatively small and easy to attach. Therefore, installing it in the worker's clothing is a good opportunity.

This paper was not the first time which utilized the idea of installing the wearable system in an industrial's clothing equipment. A safety vest design has been created by employing several sensors which are able to send acquired physiological data via Bluetooth [6]. It automatically sends a message using GPS coordinate from user to supervisor or control headquarters instantly when a danger condition is detected. However, its module dimension was relatively sizeable so that was not appropriate for wear pack. The study was not also included ergonomics consideration of devices placement, whereas it became necessary in a design which was dealing with comfort usage. The aim of this research was designing an industrial worker's outfits which were equipped with system that can detect physiological user condition in a real time based on wearable instrumentation.

2 Method

This paper provided five stages to install wearable instrumentation of the worker clothing. Each step results are reported on the next section.

1. Component selection: this stage was conducted in order to select components or devices which were suitable to its requirements design.
2. Connection setup: this stage was developed to create a hook up connection between components or devices to create a complete instrumentation system.
3. Instrument test: This stage was operated to perform tests in order to ensure the system works well before it would be installed to the cloth.
4. Clothing design: This was the final stage of this study in this paper. The clothing design was manufactured with the detail placement of every instrumentation components or devices.
5. Clothing prototype testing: It was an analysis of the respondent physiological condition when they were sawing a square of aluminum and iron wearing the work outfit which wearable instrumentations have been installed on. The test was performed by repeated data retrieval to avoid refraction.

3 Result and Discussion

3.1 Component Selection

The components which used for making a prototype instrument circuit system were Arduino Lilypad, breadboard, FTDI, thermal sensor, heart rate sensor, cable, USB cable, LCD, and LEDs. The Lilypad has a tiny sized computer system which can be programmed for variety commands or input and output, and easy for creating and developing a programming language [7]. It was chosen due to its unique shape and the number of pins (pin I/O) therefore it can distribute several functions [8]. Sensor itself is defined by a tool which transduced a physical quantity such as pressure or acceleration to produce signal (usually electric) which be used as an input for control system [9]. Several criteria such as identification of physic which was measured, determining stimulus, accuracy, precision, and linearity were three characteristics for selecting TMP 36 as a temperature sensor. While XD-58C was preferred for heart rate sensor due to its accuracy, precision, and dead band. Pulse heart rate sensor XD-58C employed LED and photo resistor to detect heart rate.

3.2 Connection Setup

Components were arranged by single copper cable. First and foremost, LCD was plugged into breadboard. Secondly, its pins were associated to digital Lilypad pin out. Next, a sensor data processing pin of body temperature and heart

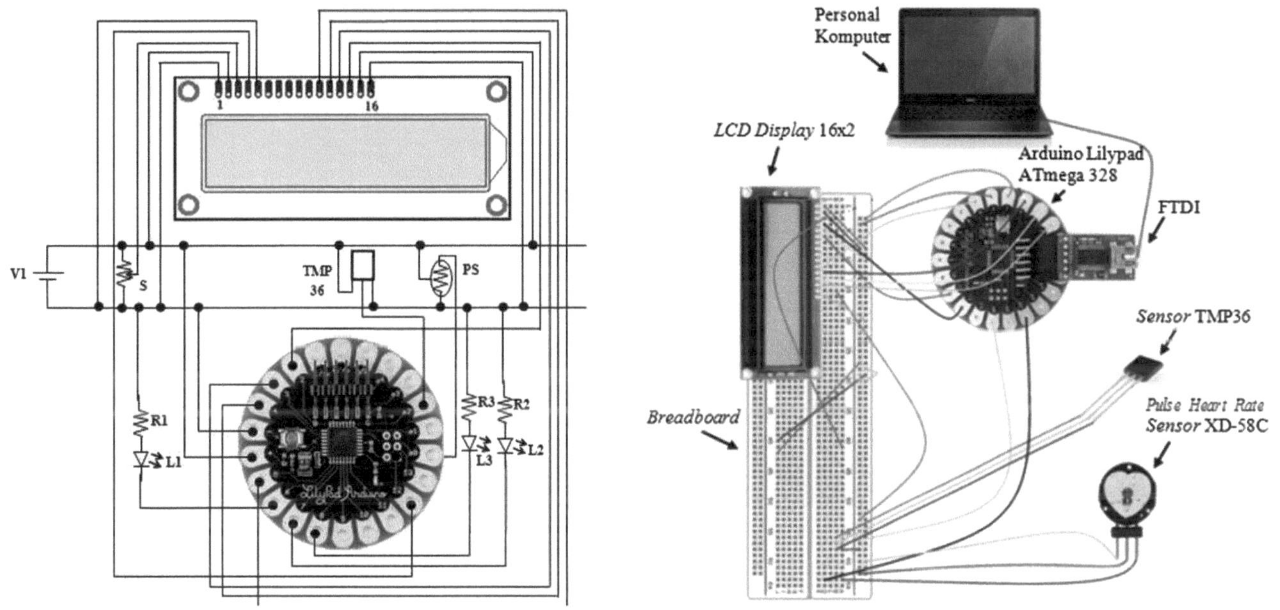

Fig. 1 Wiring diagram of Lilypad complete circuit system

rate were attached to Lilypad analog pin out. Thereafter, mini USB cable was allied via FTDI board to personal computer, where a programming language can be transferred or downloaded using Arduino IDE software. Finally, the data was processed on microcontroller, and the result was sent to LCD, where body temperature and heart rate level were clearly shown on it. Instrument circuit system scheme explanation can be clearly seen on Fig. 1.

The instruments circuit system testing was operated to ensure the system works properly before being attached into the working apparel by composing programming logic on the Arduino IDE software initially. Programming language which performed to show the heart rate measurement was processing data that provided by pulse sensor. It has been paired on the circuit system.

Based on the programming algorithm that has been constructed, the LCD screen can display the temperature value and the worker's heartbeat based on body temperature sensor TMP36 and heart rate sensor XD-58C that was mounted on the instruments circuit system which can be seen in Fig. 2.

3.3 Instrument Test

An initial test of instrument circuit system aimed to verify the wearable system which been made before installment to worker's clothing. We compare detected pulse and temperature data from sensors with digital heart rate monitor and digital body thermometer. The testing was conducted many

Fig. 2 Illustration of thermal and heart rate monitoring circuit system

Fig. 3 Illustration of thermal and heart rate monitoring circuit system which was installed on the clothing

times to increase its validity. However, the heart rate detection by using XD-58C sensor appeared some noises since due to its high sensitivity. The calibration was accomplished on 30 respondents with 10 data repetition of each respondent. The measurement difference average was 0.166 °C for the temperature and for the heart rate, the average distinct was 1 beat/min. From these results, it was noticed that TMP 36 sensor has a fairly accurate measurement value because it was only 0.1–0.2 °C of temperature different from the actual. For overall, the circuit system calibration was still within the tolerance limits (allowance). It means that the measurement deviation was not too far from the real condition and the temperature and heart rate monitoring tool were feasible to be installed.

3.4 Clothing Design

The components which stitched on the working apparel were Arduino Lilypad, LCD (Liquid Crystal Display), 3 LEDs (Light Emitting Diode), coin cell battery, TMP36 temperature sensor, XD-58C heart rate sensor. The installation on it was done by sewing each component using conductive thread. String connections were soldered to enhance the contact. The selected conductive threads are able to manipulate an electrical current which serves as a substitute for connecting cables between components. It was sewn on the working clothings which convenient to the design and an instruments circuit system that already has an ability to portray the temperature and heart rate levels. It can be clearly seen in Fig. 3.

The physiological parameter accuracy has necessary been considered for the sensor installation. In the other hand, the LCD and LEDs must be easily seen by worker or his worker supervisor. Moreover, all connectors must not be too long to minimize noise or signal loss. Therefore, a certain places must be determined for placing the sensors, Lilypad, LCD, and LEDs. Basically, mouth is the best place to attach temperature sensor. However, considering worker's convenience, neck was more tolerable. Consequently, due to lower temperature on the neck, the instrument must be adjusted to a higher level −40 to +125 °C. For the heart sensor, the best place for detection was at fingertip. However, considering the hand is commonly utilized by workers when doing a task, we prefer to set the heart rate sensor on ear. These placements will be affecting to the conductive thread sewing path design.

There were 3 LEDs indicators. The indication of the proper LEDs installation system must be placed on front side the cloth, because it should be easily noticed by the worker when the device was still not properly installed. The other two LEDs have to be easily observed by worker's supervisor. Two LEDs were placed on the back for reducing movement interference and easy to be monitored. These LEDs were going to blink when the physiological condition became abnormal. The shoulder positioned nearby can be sufficient for the distance visibility.

LCD was placed below collar or installed on backside of the neck to make it accessible for supervisor when detailed workers condition is necessary to be checked. For LCD display, to show the detail information of the temperature and heart rate, supervisor will come close to the worker after he got an abnormal physiological signal (blinking LEDs). Thermal sensor was attached on the neck by utilizing thermal conductor tape to keep the sensor sensitivity. The neck installment can shorten the overhang cable which was come out from collar.

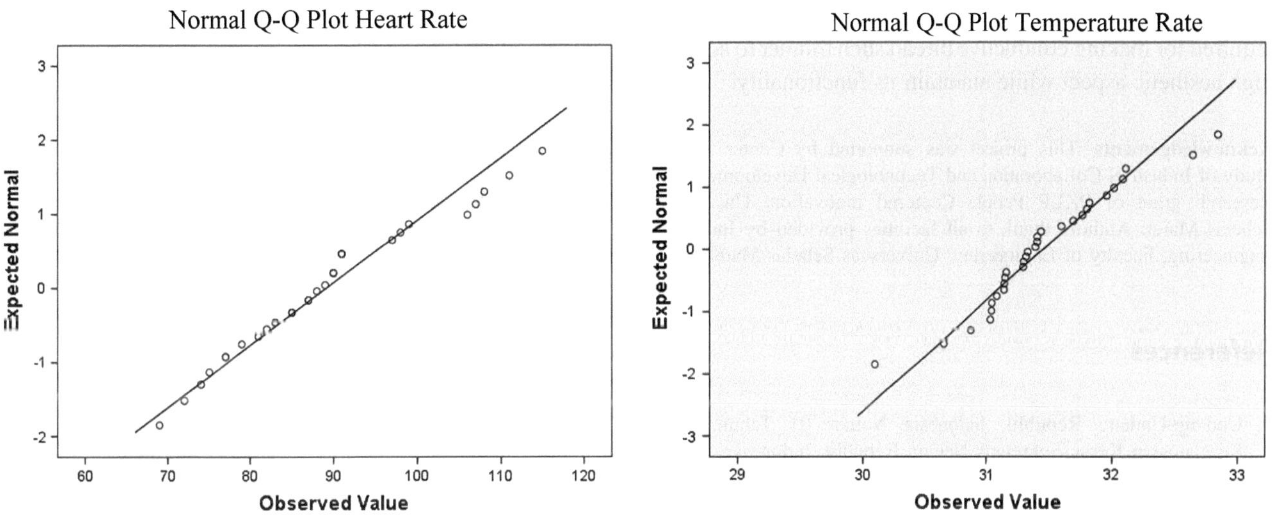

Fig. 4 Normal Q-Q plot Heart Rate and Temperature

The Worker's Clothing Prototype Test

The tests were conducted on students who work at a workshop. Data retrieval was repeated 10 times. The subjects had to be 47–86 kg of weight and 158–186 cm of height range. They were asked to saw a square of aluminum and iron pipe by working in a standing position. The work piece was clamped using a vise and the it mounted on a table, so the ideal posture can be achieved.

The both physiological data show good normality which was based on Shapiro Wilk test with $p \geq 0.05$. Its testing results can be seen by using Normal Q-Q Plot diagram that can be observed in Fig. 4. Both the temperature and heart rate data, seems fit to the lines. It means that the data were normally distributed.

Based on the subjective observation towards clothing prototype convenience, it was comfortable to wear for working activities. It can reduce heat since it uses a premium class material. After several minute of hard activity, the LED upper boundary was flashing, it portrayed the worker physiological response had already exceeded a limit. The LCD applications also did not overload while executing and successfully shows the detailed information of worker physiological condition. In the other hand, the circuit system also did not interfered by worker movement. The heart rate noise issue detection was not anymore occurred since we have installed a dark tape to isolate the external light. Over all, the system was running very well. The only drawback was the conductive thread installment. Even though, it designed for wearable instrument, it was too thick and its flexibility was poor compared to common thread. Due to its metallic color, the stich looks coarse and cannot blend well with the clothing design. A clamp terminal was needed when the thread want to be soldered with pins or other components. Otherwise, it would result in poor contact and easily detach. Unfortunately, the existence of clamp terminal will reduce its aesthetics aspect since it difficult to be tidily maintained.

4 Conclusion

In brief, the result portrays the wearable instrumentation circuit system design can be installed on the clothes and was working well. Based on the physiological condition, it can show through LEDs and LCD whether the employees were in good or bad circumstance. The design was comfort and easy to wear. It was attached on the uniform which was able to display the workers temperature and heart rate directly at the LCD (Liquid Crystal Display) in real time. TMP36 and XD-58C sensor have a fairly accurate measurement value. The median measurement results of simulated temperature sensor TMP 36 and heart rate sensor XD-58C was 0.166 °C and 1 beat/min. It has a very good accuracy with 0.1–0.2 °C of temperature range and below 60 beats/min of heart rate. Therefore, it can be concluded that those sensors can be applied and recommended for the apparatus to measure the worker fatigue level. However, sensors have to be equipped with a protecting device to maintain its sensitivity. Further study will be required for making conductive thread stich thinner to achieve high aesthetic aspect while maintain its functionality.

Acknowledgements This project was supported by Center for the Study of Industrial Collaboration and Technological Development and Research grant of PKLP People Centered Innovation, Universitas Sebelas Maret. Authors thank to all facilities provided by Industrial Engineering, Faculty of Engineering, Universitas Sebelas Maret.

References

1. Undang-Undang Republik Indonesia Nomor 01 Tahun 1970. Keselamatan Kerja. Sekretaris Negara Republik Indonesia, Jakarta (1970)
2. Ridley, J.: Keselamatan dan Kesehatan Kerja Edisi K3. Erlangga, Jakarta (2009)
3. Adiputra, N.I.: Beban Kerja Tari Bali Baris Modern dan Pengaruhnya Terhadap Parameter Fisiologis Tubuh. [Disertasi], Universitas Airlangga, Surabaya (1992)
4. Widodo, S.: Penentuan Lama Waktu Istirahat Berdasarkan Beban Kerja Dengan Menggunakan Pendekatan Fisiologis. [Tugas Akhir], Universitas Muhammadiyah Surakarta (2008)
5. Mutiara Homepage. http://maydamutiara.blogspot.co.id/2012/01/manfaat-fungsi-pakaian-seragam-kerja.html. Last accessed 16 Apr 2016
6. Markus, A., Marko, L.: Design implementation of a wearable safety suit. IT Converg. Pract. (INPRA) **1**(1), 28–36 (2013)
7. Monk, S.: Arduino Project for the Evil Genius. The McGraw-Hill Companies, New York (2010)
8. Arduino: Lilypad Homepage. http://www.arduino.cc/en/Main/ArduinoBoardLilypad. Last accessed 25 May 2016
9. Feming, W.J.: Overview of automotive sensors. IEEE Sens. J. **1**(4) (2001)

Soft Tactile Sensors for Rehabilitation Robotic Hand with 3D Printed Folds

Kirthika Senthil Kumar, Hongliang Ren, and Yun Hol Chan

Abstract

The tactile sensing technology has now progressed to the new developments in the field of robotics. In order to keep up with the trends, in this paper we have developed a soft fabric based tactile sensor which is incorporated with a 3D printed robotic hand for rehabilitation uses. Various types of tactile sensors were made with differences in the conductive material used. Conductive thread and conductive fabric based sensors were fabricated with the piezoresistive principle. On comparison with the conventional types of rigid tactile sensors, these newly fabricated sensors are advantageous in many ways in terms of their uncomplicated design, high sensitivity, lightweight and flexible nature. The 3D printed robotic hand has its inspiration from origami and its potential application in the rehabilitation process for stroke patients is validated in this paper.

Keywords

Tactile sensor • Textile sensor • 3D printed robotic hand

1 Introduction

Tactile information present on a human hand and fingers give precise information and good resolution [1]. This information helps us to understand the interaction between us and the environment around us. As technology develops, robotics is entering the dynamic environment. One main requirement for a robotic machine nowadays is for it to be aware of its surrounding in terms of the shape, size, texture and other physical properties of the object it meets. This sense of touch will assist its interactions with the real-world objects. Especially in the field of biomedical, tactile sensors have been used in various devices in minimally invasive surgeries [2], rehabilitation treatments [3], motion analysis [4]. In this scenario, the performance of the conventional type of rigid tactile sensors are limited as most of the real-world objects do not come in definite shape and size. Therefore, a flexible and soft tactile sensor is needed to assist in delicate handling of objects by the robot. To realize this a fabric sensor technology is employed [5]. Soft robotics is another major field where it is utilized in many areas such as construction, product design and in various biomedical applications [6]. In biomedical, it has been used as a replacement for rigid mechanics [7]. Hence, in this paper, soft robotics is utilized in creating a rehabilitative device for stroke patients as it provides much more flexible maneuverability. We have also aimed to design and fabricate a flexible textile based soft tactile sensor to be implemented on the robotic hand and validate its performance.

2 Sensor Design and Fabrication

2.1 Sensor Structure

The main component of the sensors are made up of fabrics. We have used the stretchable piezoresistive fabric, stretchable conductive fabric and conductive thread. This sensor works on the piezoresistive principle such that the piezoresistive fabric material is placed in between conductive

K. Senthil Kumar · H. Ren (✉) · Y. H. Chan
Department of Biomedical Engineering, National University of Singapore, Singapore, Singapore
e-mail: ren@nus.edu.sg

F. Ibrahim et al. (eds.), *2nd International Conference for Innovation in Biomedical Engineering and Life Sciences*,
IFMBE Proceedings 67, https://doi.org/10.1007/978-981-10-7554-4_9

material [8]. A non-conductive layer is used to enclose the grid forming a protective outer covering. The working principle of this sensor is piezoresistive type. When an external force is applied, a compression is produced onto the piezoresistive fabric layer which results in decrease in resistive at that point. This reduction in the resistance in turn affects the output voltage measured between the top and bottom layers of the sensor.

The outer layer of the sensor consists of a non-conductive iron-on vinyl fabric which is fusible on one side, HeatnBond (Therm O Web). It helps to hold the other sensor components together in place on top of each other in layers. The middle layer of the sensor consist of the resistive layer, which is the stretchable piezoresistive fabric (LTT-SLPA from Eeonyx), it consists of 72% Nylon and 28% Elastane. It is also coated with doped polypyrrole polymer which determines the material's resistance. Its approximate thickness is of 0.38 mm.

The conductive fabric used is made up of 76% Nylon and 24% elastic fiber, plated with medical grade silver. It can stretch in both directions. Its approximate thickness is 0.40 mm and resistivity less than 0.5 Ω per square making it easier for conduction. The conductive thread (Shieldex 117/17 dtex Z-turns HC+B, STATEX Produktions + Vertriebs GmbH) has its resistivity about 500 Ω per meter.

The piezoresistive material has a varied resistance in the range of MΩ to kΩ. In this case in the sensor it behaves as a variable resistor. The conductive layers are connected to an external voltage source. When there is a mechanical deformation applied to the sensor, there is a change in the resistance. This change reflects as a change in the potential difference across the conductive layers. These layers are connected to the analog input of a microcontroller.

The area of overlapping between the two conductive layers form a taxel. Each taxel is connected to an analog pin of the microcontroller. In case of multiple taxel sensors multiplexing is done to overcome the shortage of analog pins in microcontrollers. A low pass filter is designed to minimize the noise from the sensor output. In its undisturbed state, the sensor has high resistance, which reduces the current flow confirming minimum energy loss. This benefits in the longer run of a battery powered portable system.

2.2 Sensor Fabrication Procedure

The textile based tactile sensors have a very simple and fast fabrication process. The sensors fabricated in this paper has a sensing area of dimension 2 × 2 cm^2. A single taxel sensor is made by using the different layers mentioned in the section above. The size of the conductive layers is made slightly lesser in dimensions than the resistive layer of the sensor. This is done to avoid cross contact or short circuit between the two conductive layers. The outer layers of the sensor are made larger than the sensing element to protect and secure the inner layers in place. The layers are arranged as shown in Table 1 for each sensor type. For extending the electrical connections from the conductive layers, thin strips of non-stretchable conductive fabric were inserted between the piezoresistive layer and each of the conductive layers. Then soldering of connecting wires is done.

2.3 Types of Sensors

Various patterns were designed for the sensors consisting of multiple layers of conductive materials and by varying the number of piezoresistive fabric layer. A comparative study was done between the different layered sensors. The various sensor patterns are shown in Table 1 [9].

SPF and SPT—Single Piezo layered sensors

The single piezo layered sensor consists of 5 layers of materials. The outermost layers (layer 1&5) consist of 4 × 4 cm^2 non-conductive iron-on vinyl fabric. The middle layer (layer 3) has the piezoresistive fabric which is cut to the dimensions 2 × 2 cm^2. The layers 2 and 4 consists of the conductive material of area slightly lesser than 2 × 2 cm^2. Single Piezo fabric sensor (SPF) is a conductive fabric type sensor. In this sensor, the conductive material used is Stretch Conductive Fabric (LessEMF Inc.). The Single piezo thread sensors (SPT) differ from the SPF sensors in the 2nd and 4th layers where a conductive thread is used instead. It is sown following a pattern onto the iron-on vinyl fabric. The two patterns of thread in the conductive layers are orthogonal to each other to ensure higher number of overlaid points hence higher conductivity.

MPF and MPT—Multiple Piezo layered sensors

The construction of the MPF and MPT sensors are different from the previous types in terms of the number of resistive layers. In order to increase the overall sensitivity of these sensors by increasing the resistance, multiple piezoresistive fabric layers are added. This type of sensor is similar to the single layered piezoresistive sensors with additional layers of the piezoresistive fabric in the middle layers. The multiple piezo fabric (MPF) based sensor consist of the stretchable conductive fabric as the conductive layer whereas in the multiple piezo thread sensor (MPT) the conductive material used is the silver coated conductive thread. With every additional layer of the piezoresistive fabric, the thickness of the sensor increases by 0.38 mm approximately.

Table 1 Various layers of the sensors

S P F	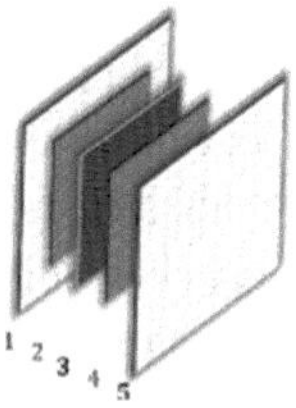	*Layers 1,5* Non conductive fusible interface (iron-on vinyl sheet) *Layers 2,4* Conductive fabric *Layer 3* Piezoresistive fabric	
S P T	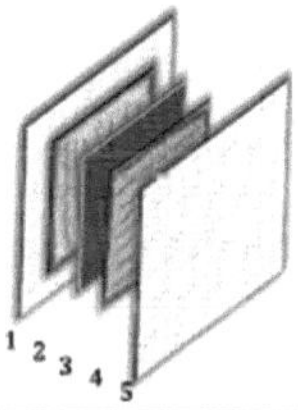	*Layers 1,5* Non conductive fusible interface (iron-on vinyl sheet) *Layers 2,4* Conductive Thread (horizontally and vertically sewn) *Layer 3* Piezoresistive fabric	
M P F	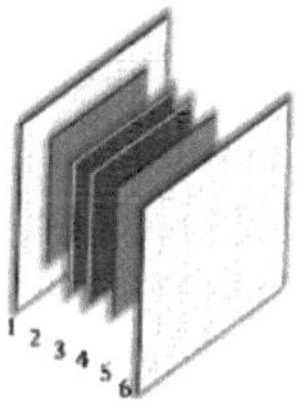	*Layers 1,6* Non conductive fusible interface (iron-on vinyl sheet) *Layers 2,5* Conductive fabric *Layer 3,4* Piezoresistive fabric	
M P T	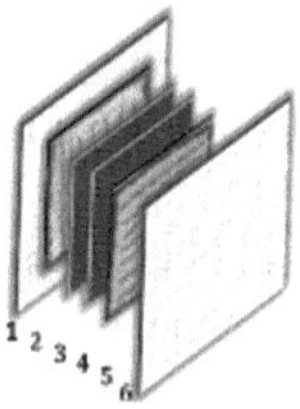	*Layers 1,6* Non conductive fusible interface (iron-on vinyl sheet) *Layers 2,5* Conductive thread (horizontally and vertically sewn) *Layer 3,4* Piezoresistive fabric	
F T F		*Layers 1,7* Non conductive fusible interface (iron-on vinyl sheet) *Layers 2,6* Conductive Fabric *Layer 3,5* Piezoresistive fabric *Layer 4* Conductive thread sewn orthogonal to each other on both sides	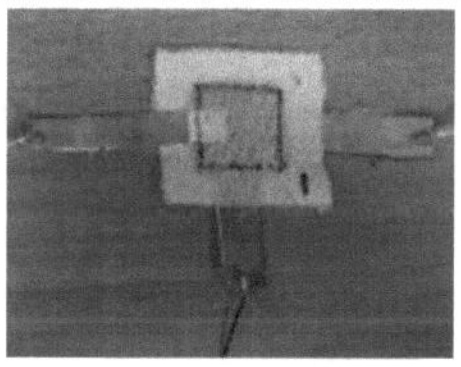
T F T	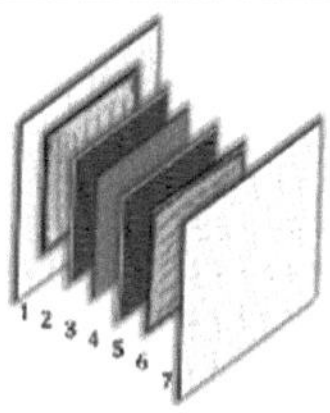	*Layers 1,7* Non conductive fusible interface (iron-on vinyl sheet) *Layers 2,6* Conductive thread (horizontally and vertically sewn) *Layer 3,5* Piezoresistive fabric *Layer 4* Conductive thread sewn orthogonal to each other on both sides	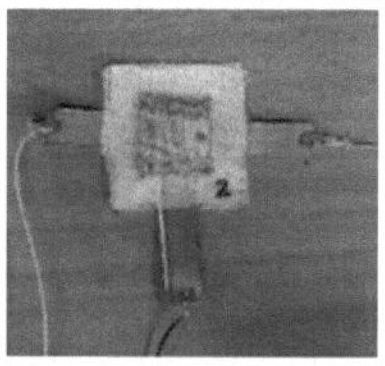

FTF—Fabric Thread Fabric sensor

This sensor is a combination of both the conductive materials. This sensor is designed as a seven-layered sensor, with alternate layers of piezoresistive fabric and conductive material. The top and bottommost layer consists of the iron-on vinyl fabric material which is used to secure the sensor materials together. The second and sixth layer contains the conductive fabric. A single layer of conductive thread is sewn unto both sides of the iron-on vinyl fabric of which its dimensions are lesser than that of the piezoresistive fabric layer. The thread is patterned to be orthogonal to both sides, crossing each other at multiple points, ensuring higher number of intersection points. This forms the middle layer and is placed between the third and fifth layers of piezoresistive fabric.

TFT—Thread Fabric Thread sensor

This TFT sensor is designed identical as that of the FTF sensor. One major difference as compared to the FTF mentioned above lies in the middle layer. A conductive fabric used instead of the conductive thread. This is sandwiched between two layers of piezoresistive material. Conductive thread is sewn unto the first and last layer of iron-on vinyl fabric, in an orthogonal pattern. Both FTF and TFT sensors have two channels of signals representing the potential difference between the conductive layers.

3 Soft and Flexible Anthropomorphic Robotic Hand

Origami method were used to design the folding of the fingers [10]. They were used at the bending parts of the finger in the mechanical system. With the basic design, each finger was customized based on the physical dimensions and movability. The index finger was the base design for the other fingers. Whereas modifications were made for the thumb as it requires 2 degrees of freedom to perform adduction and abduction. A customized hand support system was designed as a part of the mechanical system.

The finger and hand design were 3D printed using Ninjaflex (Fenner Drives, Inc.) composed of thermoplastic and rubber as shown in Fig. 1. The flexibility of this thermoplastic elastomer material provided a force that is equal and opposite in direction to the force exerted by the drive mechanism. The wire thread was chosen as the best material to be used as the wire mechanism for the hand exoskeletal system. Wire mechanism has been utilized by numerous soft robotic exoskeletal system. This wire mechanism was used to drive the mechanical system using four TowerPro MG995 servomotors. Two servomotors will be providing 2 DOF of adduction/abduction and flexion/extension individually for the thumb mechanism. One servomotor will be providing 1 DOF of flexion/extension for index finger mechanism. The last servomotor will provide 1 DOF of flexion/ extension for subsequent three fingers that will be connected to a single wire. Together this system aims to provide strength for stroke patients during the rehabilitative process.

4 Experiments and Results

4.1 Tests and Comparative Results of Sensors

To characterize the sensors, the Instron Universal Testing machine, was used. A cylindrical probe was used to apply a constant compressive force, perpendicular to the surface of the sensor. The compressive force range was varied for each type of sensor. A maximum load (L_{max}) was decided for each sensor. The sensors were loaded from idle state to L_{max} and then unloaded to bring them back to idle state at a rate of 2.5 mm/N. The sampling rate of the machine and the sensor reading was set to 100 ms. Drift test was also performed by loading the sensor with a constant force of 5 N for a duration of 10 min.

The results are illustrated in Fig. 2. The output of the sensors is linear in their optimum force range (OFR). This OFR varies for each type of sensor depending on the type of conductive material and the number of resistive layers introduced. By observing the results, we can see that the OFR for SPF sensors is 0–5 N, this is suitable for application with lower range of force. The additional layers of the piezoresistive material has increases the OFR up to 10 N in the MPF sensors. The enhancement in the property of the conductive thread type sensors had be shown by resulting in a wider force range. In SPT sensor the OFR is 0.5–10 N and in MPT the sensor force range has been

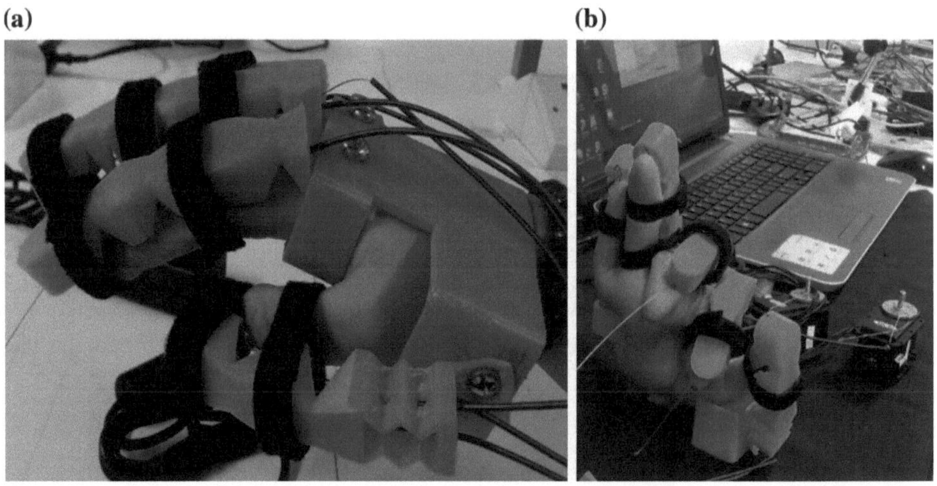

Fig. 1 **a** The 3D printed hand exoskeleton with the mechanical and actuating system. **b** The actuating hand with the thumb, index finger and middle finger in action

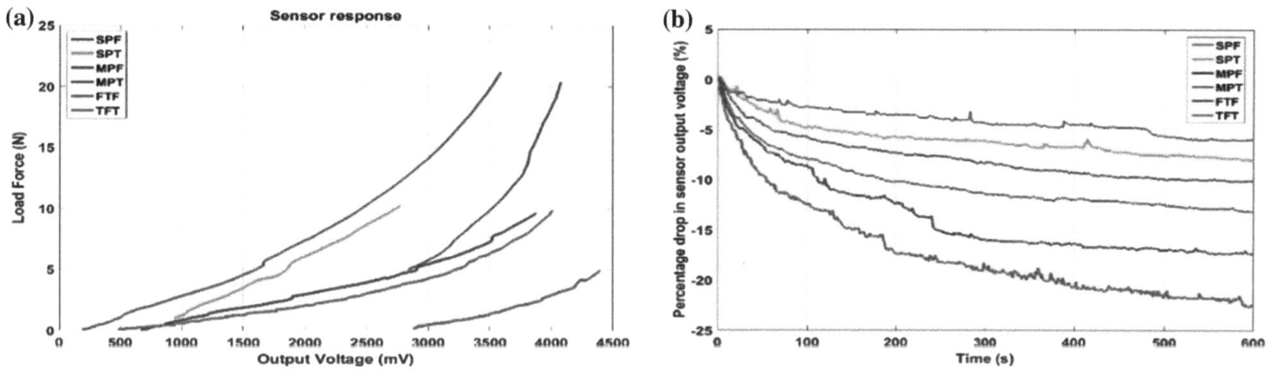

Fig. 2 a Response of the sensor in it optimum force range. **b** Drift response for a constant load for 10 min

Table 2 Force exerted by the 3D printed hand on the sensors

Voltage (V)	Force (N)
Force exerted by index finger mechanism	
3.27	32.7
Force exerted by middle finger mechanism	
3.71	37.1
Force exerted by thumb mechanism	
2.93	29.3

escalated up to 20 N. The FTF sensors exhibits a high sensitivity in the lower force range whereas the TFT sensors work well upto a range of 20 N with stable output.

4.2 3D Printed Hand Exoskeleton

Experiments were performed to measure the force needed for maximum flexions and maximum angle of flexion of finger mechanism. Alongside the fabricated sensors were attached to the interior surface of the 3D printed hand. The force exerted by the hand is measured with the usage of these sensors. The raw readings of the sensor are first filtered through a low pass filter realized digitally in MATLAB. The results obtained for each finger are shown in Table 2.

5 Conclusion

Flexible and soft textile based tactile sensors can be utilized as a sensor for robotic hands. Especially in the field of soft robotics. In this paper, we have aimed to develop a suitable soft sensor for rehabilitative procedures. We have also established the performance of the sensors through realistic experiments and shown its advantages over the other types of materials. The origami inspired exoskeleton has proven to be feasible for its application in the field of robotics and rehabilitation. The experiments performed has shown that it has the potential to provide adequate strength for force support in users, during their rehabilitation process and daily life activities. For future works the ability of the sensor can be used for other common applications such as slip detection, grip force measurements and so on.

Acknowledgements This work is supported by the Singapore Academic Research Fund under Grant R-397-000-227-112, NUSRI China Jiangsu Provincial Grant BK20150386 & BE2016077 awarded to Dr. Hongliang Ren.

References

1. Lee, M.H., Nicholls, H.R.: Tactile sensing for mechatronics—a state of the art survey. Mechatronics **9**, pp. 1–31 (1999). Last accessed 21 Nov 2016
2. Ahmadi, R., Packirisamy, M., Dargahi, J., Cecere, R.: Discretely loaded beam-type optical fiber tactile sensor for tissue manipulation and palpation in minimally invasive robotic surgery. IEEE Sens. J. **12**(1), 22–32 (2012)
3. McKinney, Z., Heberer, K., Nowroozi, B.N., Greenberg, M., Fowler, E., Grundfest, W.: Pilot evaluation of wearable tactile biofeedback system for gait rehabilitation in peripheral neuropathy. In: 2014 IEEE Haptics Symposium (HAPTICS), Houston, TX, pp. 135–140 (2014)
4. Yang, C.-M., Lin, Z.-S., Hu, C.-L., Chen, Y.-S., Ke, L.-Y., Chen, Y.-R.: A novel dynamic sensing of wearable digital textile sensor with body motion analysis. In: Proceedings of the Annual International Conference of the IEEE Engineering in Medicine and Biology Society (EMBC), pp. 4898–4901 (2010)
5. Castano, L., Flatau, A.B.: Smart fabric sensors and e-textile technologies: a review. IOP Publ. Smart Mater. Struct. **23**(5) (2014)
6. Kang, B.B., Lee, H., In, H., Jeong, U., Chung, J., Cho, K.J.: Development of a polymer-based tendon-driven wearable robotic hand. In: 2016 IEEE International Conference on Robotics and Automation (ICRA), pp. 3750–3755. IEEE (2016)
7. Polygerinos, P., Wang, Z., Galloway, K.C., Wood, R.J., Walsh, C. J.: Soft robotic glove for combined assistance and at-home

rehabilitation. Robot. Auton. Syst. **73**, 135–143 (2015). https://doi.org/10.1016/j.robot.2014.08.014

8. Lee, W.W., Kukreja, S.L., Thakor, N.V.: A kilohertz kilotaxel tactile sensor array for investigating spatiotemporal features in neuromorphic touch. In: Biomedical Circuits and Systems Conference (BioCAS), pp. 1–4. IEEE (2015)
9. Kirthika, S.K., Vedhagiri, G.P., Ren, H.: Fabrication and comparative study on sensing characteristics of soft textile-layered tactile sensors. IEEE Sens. Lett. (2017)
10. Johnson, M., Chen, Y., Hovet, S., Xu, S., Wood, B., Ren, H., et al.: Fabricating biomedical origami: a state-of-the-art review. Int. J. Comput. Assist. Radiol. Surg. 1–10 (2017)

The Effect of Different Size of Slicing Spheres of Polyaxial Screw Head on Static Compression Mechanical Test

Po-Yi Liu, Yu-Tzu Wang, Hsien-Wen Wang, Shao-Fu Huang, Yang-Sung Lin, Po-Liang Lai, and Chun-Li Lin

Abstract

Background The pedicle screw fixation system is the gold standard treated method in clinical for the unstable disease. Generally, pedicle screw can be classified into two categories included monoaxial and polyaxial constructions. Although the polyaxial screws were common used owing to it can provide more degree of freedom on the screw-to-rod connection to facilitate rod seating. However, the complexity of structural design may reduce their mechanical strength. The size of slicing spheres of polyaxial screw head was found as the important issue to influence the mechanical strength of the pedicle screw system. Therefore, this study investigated the effects of different size of slicing spheres of polyaxial screw head on the vertebrectomy model under compression test. *Material and Method* The different size (60, 75 and 90% diameter) of slicing spheres of polyaxial screw head were assembled for destructive mechanical testing. All pedicle screw systems assemblies were tested in compression test according to FDA regulation testing method of ASTM F1717. The stiffness and ultimate load were determined for different slicing spheres of screws. *Result and discussion* The results of testing found that stiffness were 10.91 ± 2.03, 19.34 ± 1.21, and 23.39 ± 0.85 N/mm and ultimate loads were 185.43 ± 14.57, 264.44 ± 21.68, and 364.54 ± 10.43 N for 60, 75, and 90% diameter, respectively. Significant differences ($p < 0.05$) were found among these three groups. *Conclusion* The mechanical strength presented by stiffness and ultimate load increased as the size of slicing spheres of polyaxial screw head increased.

Keywords

Polyaxial pedicle screw • Nut tightening torque • Size of slicing spheres

1 Introduction

Treatment of posterior spinal fusion with pedicle screw is the current gold standard treated method in clinical for the unstable disease and progressive scoliosis [1–5]. The clinical benefits of pedicle screws and rods are improved spinal stabilization and increased rates of fusion, particularly for unstable disease of the lumbar spine. Generally, pedicle screw base on posterior spinal instrumentation can be classified into two categories included monoaxial and polyaxial

P.-Y. Liu · Y.-T. Wang · H.-W. Wang · S.-F. Huang · Y.-S. Lin · C.-L. Lin
Department of Biomedical Engineering, National Yang-Ming University, Taipei, Taiwan

P.-Y. Liu · P.-L. Lai (✉)
Department of Orthopedic Surgery, Chang Gung Memorial Hospital at Linkou, Bone and Joint Research Center, College of Medicine, Chang Gung University, Taoyuan, Taiwan
e-mail: polianglai@gmail.com

F. Ibrahim et al. (eds.), *2nd International Conference for Innovation in Biomedical Engineering and Life Sciences*,
IFMBE Proceedings 67, https://doi.org/10.1007/978-981-10-7554-4_10

constructions. Monoaxial screws are immobile at the screw head and are thus superior for performing the vertebral correction. However, monoaxial screws have a potential disadvantage regarding the difficulty of seating the rod into the screw head [6, 7].

In contract, polyaxial pedicle screws are a ball-and-socket designs in pedicle screw junction. This allow deviation of the screw away from the perpendicular to the longitudinal rod, which facilitates application of a screw–rod system into the curved spine [8, 9]. Therefore, polyaxial screws are commonly used in clinical applications. The design changes have improved surgical versatility, but they have introduced complex locking devices between the screw head and the rod. This raises questions over the mechanical strength and stiffness of the polyaxial designs. The size of slicing spheres of polyaxial screw head and the nut tightening torque are found as the important issue to influence the mechanical strength of the pedicle screw system. The slicing spheres of polyaxial screw head is the opening of the polyaxial pedicle tulip head to connect screw body. There are not yet any published mechanical data with regard to these design, with particular reference to the size of slicing spheres of polyaxial screw head and the nut tightening torque.

The rates of spinal fusion and the deformed correction have high relationship with mechanical strength and mechanical stiffness of pedicle screw fixation system. In previous, the ultra high molecular weight polyethylene (UHMWPE) was used as dummy vertebrae to investigate the mechanical strength of pedicle screw fixation system. Besides, the testing method of ASTM F1717-15 was published by American Society for Testing Materials (ASTM). It was a reference of spinal implants regulation functional testing when the spinal implants pass the FDA approved. Therefore, this study used the method of ASM F1717 to investigate the effects of different size of slicing spheres of polyaxial pedicle screw head and nut tightening torque on the vertebrectomy model under compression mechanical test.

2 Material and Method

2.1 Pedicle Screw Design and Manufacture

Three different size of slicing spheres of polyaxial pedicle screw tulip head (90, 75, and 60% of diameter of screw spherical head), each measuring 4.5 mm in diameter and 35 mm in length, were used in this study. The polyaxial pedicle screws were designed with a ball-and-socket joint connecting the screw head to the screw body. Three different size of slicing spheres of polyaxial pedicle screws are shown in Fig. 1. The polyaxial pedicle screws were made of Ti6Al4 V by a manufacturer with good manufacturing

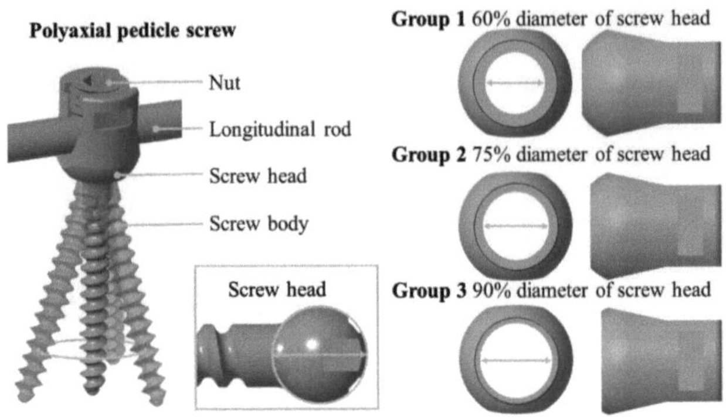

Fig. 1 The different size of slicing spheres of polyaxial screw tulip head

practices and International Organization for Standardization 13,485 quality management systems (Huang-Liang Co, Ltd, Kaohsiung, Taiwan).

2.2 Specimen Preparation and Biomechanical Testing

All pedicle screw fixation systems assemblies were tested in static compression bending test according to FDA regulation testing method of ASTM F1717. Four same size of slicing spheres of polyaxial pedicle screws, four nuts, two rods, and two UHMWPE blocks were assembled a test specimen. The two polyaxial pedicle screws were inserted into the UHMWPE block. the rod was set into same side poly axial pedicle screw tulip head. Nut tightening torque was driven with a push pull force gages (ALGOL-AK-10, ALGOL INSTRUMENT CO., LTD., Taoyuan, Taiwan). The torque was generated by using a hex wrench as moment arm to connect the force gages. Two different nut tightening torque (8 Nm and 12 Nm) were applied to screw in the nut. The test setting of compression bending test are shown in Fig. 2. The depth of screw insertion was standardized to position the center of rotation of the polyaxial link 10 mm above the surface of the UHMWPE blocks for three different size of slicing spheres of polyaxial pedicle screws. This gave a total lever arm from the center of rotation of the link to the axis of loading of 50 mm. the distance between two axis of pedicle screw body (Active length) was 76 mm. UHMWPE blocks were used to simulate the vertebral bodies as stated in ASTM F-1717–15. A pair of aluminum yokes with pins supported the UHMWPE blocks and permitted free rotation of the blocks in the sagittal plane during deformation of the assembly. The upper yoke was mounted on a ball-race that permitted free rotation of the assembly around the longitudinal axis under compressive load. All mechanical testing was performed on a material testing system under software control using a 5-kN load cell (ElectroPuls™ E3000;

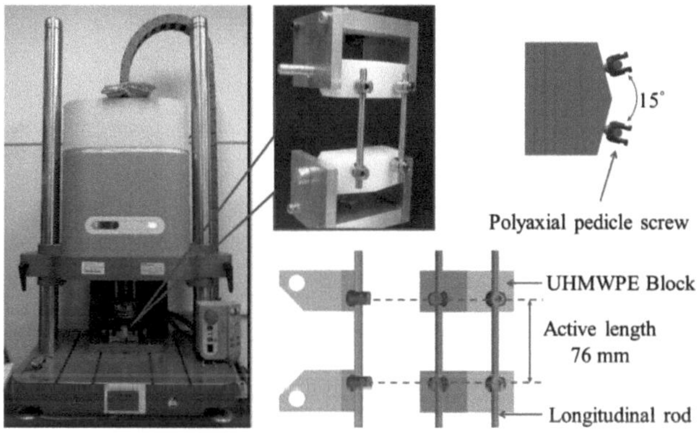

Fig. 2 The experience setup of compression test

Instron, High Wycombe, UK). Static compression bending test was performed in air under ambient conditions. Compression bending test as conducted in displacement control at a rate of 25 mm/min as per ASTM F1717–15. Load-displacement curves were collected for each assembly tested. From the curves, and stiffness and ultimate load were measured. Stiffness is defined in ASTM F1717–15. Ultimate load was taken as the peak load before mechanical failure (sudden decrease in measured load) or the load at 25 mm displacement if the assembly did not fail (no decrease in measured load).

2.3 Statistical Analysis

The curve of the force versus displacement were analyzed. The stiffness and ultimate load in each group were determined and then averaged. A one-way analysis of variance and post hoc Tukey tests were used to detect a significant difference in the ultimate load in three groups. The student T-test were used to detect a statistical difference in the ultimate load when using different nuts tightening torque to assemble the pedicle screw fixation system.

3 Result

The typical force versus the displacement curves for using different nuts tightening torque in three different size of slicing spheres designs are shown in Fig. 3. The measured mechanical properties (stiffness, and ultimate load) in static compressive test are shown in Fig. 4. The stiffness, and ultimate load of using 12 Nm nut tightening torque was statistically larger than using 8 Nm nut tightening torque in each groups (60, 75, and 90% of diameter of screw spherical head) ($p < 0.05$). And the stiffness, and maximum load were significant increase as the size of slicing spheres increase ($p < 0.05$).

4 Discussion

As spinal surgery and minimally invasive techniques advance, polyaxial pedicle screws are becoming more commonly utilized. The ease of rod reduction makes polyaxial screws more favorable when compared to traditional monoaxial screws. However, previous studies have shown that the heads of polyaxial screws fail at a lower value than traditional monoaxial screws and that polyaxial screws typically slip at the screw-head interface [8, 10, 11]. Although polyaxial screw designs can include angle modifications, with which there is a greater capacity of angular motion through the screw-tulip universal joint in a specific plane, the biomechanical properties of failure through this friction-limited joint are not improved. There is still a tendency for slippage through such a joint, when comparing it to monoaxial screw designs [12]. The loss of segmental lordosis from such polyaxial screw tulip head slippage could ultimately contribute to sagittal malalignment. The possibility of a resulting sagittal imbalance has been associated with pain, pseudoarthrosis, and adjacent segment degeneration [13–15]. Many studies have been conducted about the factors which affect the stability of polyaxial pedicle screws.

Fig. 3 The force versus displacement curve

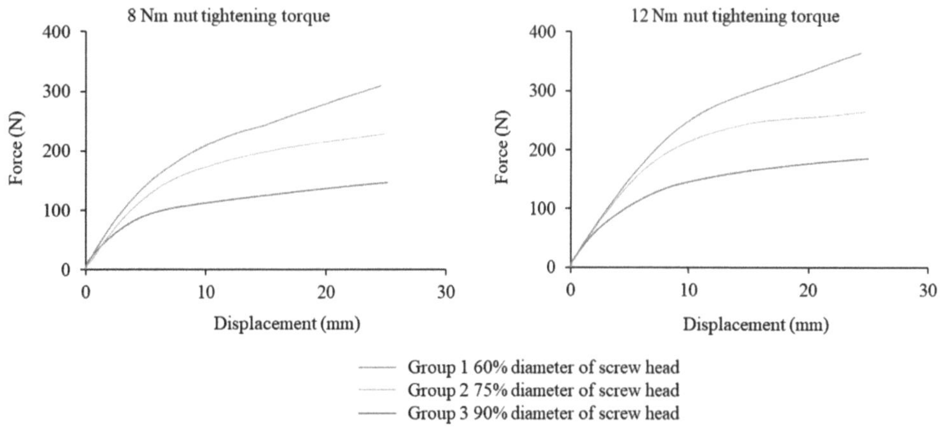

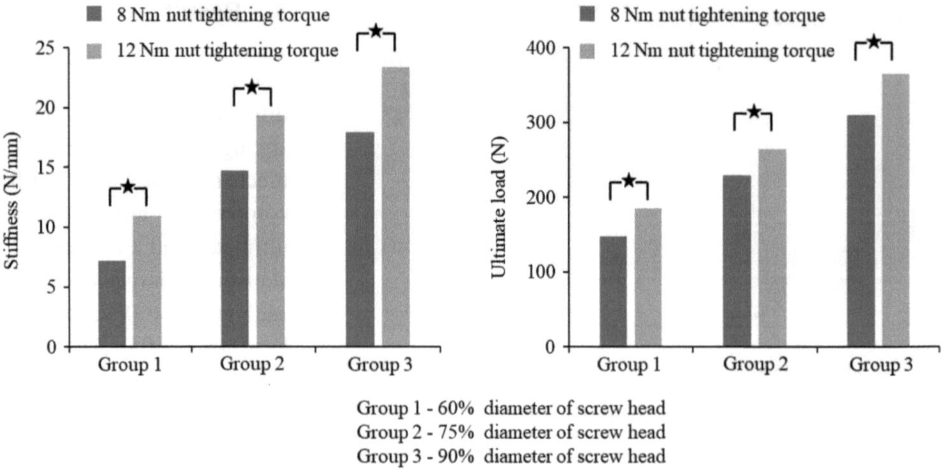

Fig. 4 The Bar chat of stiffness and ultimate load

Screw head design, nut tightening torque, and insertion techniques were directly related to pedicle screw fixation system stiffness and mechanical strength. Nut tightening torque and the size of slicing spheres were the most important feature of the screw controlling mechanical strength.

In previous study, the different nut design was presented to investigate the pedicle screw fixation system stabilization. The outside nut, pin-nut, helical or dovetail wedges, and thicker walled polyaxial head are designed to prevent the head from deforming. The single outside nut–locking mechanism was statistically weaker than any other design [10]. However, no study investigated the influence of the size of polyaxial pedicle screw head tulip. In this present study, three different size of slicing spheres of pedicle screw head were used to compare the mechanical strength. The results shown the group 3 (The size of slicing spheres is 90% screw head diameter) of had statistically higher stiffness, and ultimate load than other groups (The size of slicing spheres is 75 and 60% screw head diameter). The ultimate load was increase as the size of slicing spheres increase. However, the risk of pedicle screw head cut off could increase as the size of slicing spheres increase. In this present study, it didn't happen to the pedicle screw tulip head cut off. Therefore, the size of slicing spheres is 90% screw head diameter was a feasible size than the lower size. In nut tightening torque, some guidance of pedicle screw fixation system indicated that nut tightening torque was not recommended to exceed 12 Nm during final tightening. When the nut tightening torque exceed 12 Nm, the longitudinal rod or nut may deformation or failure. Nevertheless, the lowest nut tightening torque was not recommended. In this present study, the 12 Nm (largest recommended nut tightening torque) and 8 Nm (66% largest recommended nut tightening torque) were used to assemble the pedicle screw fixation system. The results shown the group of 12 Nm inserted torque have statistically higher stiffness, and ultimate load than the group of 8 Nm inserted torque ($p < 0.05$). Therefore, the larger nut tightening torque could enhance the mechanical strength.

According to the ASTM testing standard (ASTM F1717–15) for vertebral fixation, we adopted UHMWPE to simulate standardized vertebral bodies to test the biomechanical properties of polyaxial screw fixation system. UHMWPE blocks were instrumented with polyaxial pedicle screws and the rod was the only connection between blocks. This represented the worst condition for implants when the vertebrectomy was performed. The UHMWPE block simulated vertebral segment ensured the same condition for pedicle screw fixation and eliminated the influence of variability of vertebra and morphologies often associated with cadaveric tissues. Thus, the test in this study is more precise and reproducible than that in cadaver models. However, the test results cannot predict the actual situation of the in vivo application of the three different size of slicing spheres of polyaxial pedicle screws.

This study had several limitations. First, we utilized only one design of polyaxial pedicle screw to investigate the influence of size of slicing spheres. We also had a small sample size of screws (n = 3); while the sample size was adequate to detect important differences, and some indeed were detected, smaller (but still potentially clinically important) differences might have appeared as no-difference findings in this report. Transverse connectors were not used in this study because we aimed to assess the rationality of the size of slicing spheres of polyaxial pedicle screws without interference of transverse connectors. Previous studies demonstrated that transverse connectors do not influence the results of compression bending mechanical tests but significantly improve the torsional stiffness according to their sites, numbers and shapes.

5 Conclusion

The mechanical strength presented by stiffness, and ultimate load increased as the size of slicing spheres of polyaxial screw head increased and the failure occurred at the multiaxial link in static compression bending test. The nut tightening torque play an important role for pedicle screw fixation system. the mechanical strength increased as the nut tightening torque increased.

Acknowledgements This work was kindly supported by grants CRRPG3E0141 and CRRPG3E0142 from Chang Gung Memorial Hospital, and grant MOST 105-2622-8-010-001-TE4 from Ministry of Science and Technology, Taiwan.

References

1. Kuklo, T.R., Potter, B.K., Polly Jr., D.W., Lenke, L.G.: Monaxial versus multiaxial thoracic pedicle screws in the correction of adolescent idiopathic scoliosis. Spine (Phila Pa 1976) **30**, 2113–2120 (2005)
2. Vora, V., Crawford, A., Babekhir, N., Boachie-Adjei, O., Lenke, L., Peskin, M., Charles, G., Kim, Y.: A pedicle screw construct gives an enhanced posterior correction of adolescent idiopathic scoliosis when compared with other constructs: myth or reality. Spine (Phila Pa 1976) **32**, 1869–1874 (2007)
3. Rose, P.S., Lenke, L.G., Bridwell, K.H., Mulconrey, D.S., Cronen, G.A., Buchowski, J.M., Schwend, R.M., Sides, B.A.: Pedicle screw instrumentation for adult idiopathic scoliosis: an improvement over hook/hybrid fixation. Spine (Phila Pa 1976) **34**, 852–857 (discussion 858) (2009)
4. Cetin, E., Ozkaya, M., Guler, U.O., Acaroglu, E., Demir, T.: Evaluation of the effect of fixation angle between polyaxial pedicle screw head and rod on the failure of screw-rod connection. Appl. Bion. Biomech. **2015**, 150649 (2015)
5. Wang, H., Li, C., Liu, T., Zhao, W.D., Zhou, Y.: Biomechanical efficacy of monoaxial or polyaxial pedicle screw and additional screw insertion at the level of fracture, in lumbar burst fracture: an experimental study. Indian J. Orthop. **46**, 395–401 (2012)
6. Voleti, P.B., Shen, F.H., Arlet, V.: Failure of monoaxial pedicle screws at the distal end of scoliosis constructs: a case series. Spine Deform. **2**, 110–121 (2014)
7. Wang, X., Aubin, C.E., Crandall, D., Parent, S., Labelle, H.: Biomechanical analysis of 4 types of pedicle screws for scoliotic spine instrumentation. Spine (Phila Pa 1976) **37**, E823–835 (2012)
8. Serhan, H., Hammerberg, K., O'Neil, M., Sturm, P., Mardjetko, S., Crawford, A.: Intraoperative techniques to reduce the potential of set-screw loosening in long spinal constructs: a static and fatigue biomechanical investigation. J. Spinal Disord. Tech. **23**, e31–e36 (2010)
9. Essig, D.A., Miller, C.P., Xiao, M., Ivancic, P., Jegede, K., Badrinath, R., Smith, B.G., Grauer, J.N.: Biomechanical comparison of endplate forces generated by uniaxial screws and monoaxial pedicle screws. Orthopedics **35**, e1528–e1532 (2012)
10. Fogel, G.R., Reitman, C.A., Liu, W., Esses, S.I.: Physical characteristics of polyaxial-headed pedicle screws and biomechanical comparison of load with their failure. Spine (Phila Pa 1976) **28**, 470–473 (2003)
11. Schroerlucke, S.R., Steklov, N., Mundis Jr., G.M., Marino, J.F., Akbarnia, B.A., Eastlack, R.K.: How does a novel monoplanar pedicle screw perform biomechanically relative to monoaxial and polyaxial designs? Clin. Orthop. Relat. Res. **472**, 2826–2832 (2014)
12. Stanford, R.E., Loefler, A.H., Stanford, P.M., Walsh, W.R.: Multiaxial pedicle screw designs: static and dynamic mechanical testing. Spine (Phila Pa 1976) **29**, 367–375 (2004)
13. Glassman, S.D., Berven, S., Bridwell, K., Horton, W., Dimar, J.R.: Correlation of radiographic parameters and clinical symptoms in adult scoliosis. Spine (Phila Pa 1976) **30**, 682–688 (2005)
14. Lazennec, J.Y., Ramare, S., Arafati, N., Laudet, C.G., Gorin, M., Roger, B., Hansen, S., Saillant, G., Maurs, L., Trabelsi, R.: Sagittal alignment in lumbosacral fusion: relations between radiological paramcters and pain. Eur. Spinc J. **9**, 47–55 (2000)
15. Park, P., Garton, H.J., Gala, V.C., Hoff, J.T., McGillicuddy, J.E.: Adjacent segment disease after lumbar or lumbosacral fusion: review of the literature. Spine (Phila Pa 1976) **29**, 1938–1944 (2004)

FES Standing: The Effect of Arm Support on Stability and Fatigue During Sit-to-Stand Manoeuvres in SCI Individuals

Musfirah Abd Aziz and Nur Azah Hamzaid

Abstract

Functional Electrical Stimulation (FES) has been widely used as part of physiotherapy for spinal cord injury (SCI) patients. The ability to do sit to stand (STS) manoeuvre is an important and practical indicator of functional independence in SCI individuals. One of the factors that contributes to STS movement is arm support. The objective of this study is to instrument an arm support with pressure sensors, and to analyse the impact of using standing frame during FES-assisted STS movement to the SCI patients' stability and rate of fatigue throughout multiple STS movements. FlexiForce sensors were used to analyse the force exerted on the frame's handle. Experiments on STS activity with two SCI subjects were completed in two consecutive days (with and without assistance of FES) in a motion analysis laboratory. The instrumented standing frame (SF) was calibrated via a series of hanging test with ten healthy subjects with different body weights to provide an insight on the weight distribution along the SF. This test demonstrates the instrumented standing frame's ability to measure the force exerted on the frame with minimum accuracy of 85% to total body weight. Both SCI subjects showed shorter time taken to complete a STS cycle without the assistance of FES. They showed early stage of fatigue with assistance of FES thus longer time taken recorded in performing STS activity. SCI subjects' centre of force slightly inclined to the right side of the standing frame in both sessions without and with FES to compensate several conditions of lower limb joint contractures.

Keywords

Spinal cord injury (SCI) • Functional electrical stimulation (FES) • Sit-to-Stand (STS)

1 Introduction

Spinal cord injury (SCI) has two major categories which are complete and incomplete injury. While there is permanent loss of sensory and motor function below the spinal cord lesion in patients with complete SCI, incomplete injury otherwise refers to spinal cord injury without any sensory and/or motor function below the neurological level including S4–S5 sacral spinal nerve [1]. Based on the American Spinal Injury Association (ASIA) Impairment Scale (AIS) classification, the target population for this Functional Electrical Stimulation (FES) standing research is AIS C patients, i.e. those with motor incomplete spinal cord injury. Their motor function is preserved below the neurological level, and more than half of the key muscle functions below the single neurological level of injury have a muscle grade less than Grade 3 on MRC scale [1].

Sit to stand (STS) movement is one of the most basic yet important routine manoeuvres for human. Hence, the ability to perform STS movement is a very practical indicator to determine functional independence of the SCI patient [2]. Accomplishment of STS activity in SCI patients results in physiological postural transformation, from a balanced position to a vulnerable position with lower muscle limbs

M. Abd Aziz · N. A. Hamzaid (✉)
Department of Biomedical Engineering, University of Malaya, Kuala Lumpur, Malaysia
e-mail: azah.hamzaid@um.edu.my

F. Ibrahim et al. (eds.), *2nd International Conference for Innovation in Biomedical Engineering and Life Sciences*, IFMBE Proceedings 67, https://doi.org/10.1007/978-981-10-7554-4_11

weakness and postural instability during STS. With muscle mass deterioration and weakness, instability and fatigue become a concern to patients. As patients' stability and adapted movement are highly dependent on users' personal method, these aspects are then determined by their own conduct during STS movement and quiet standing.

The ability to perform functionally-independent STS movement is influenced by factors which are chair seat height, feet position, and use of armrests [3]. The presence of arm support reduces the moment needed at the hip probably without altering the joints range of motion. Scientific evidence on the importance of arm support during STS for SCI individuals is insufficient and inconclusive. Hence, a simple instrumentation of standing frame (SF) is designed suitably to quantify the force exerted at the upper limb during STS.

2 Methodology

2.1 Sensor Calibration

Calibration is an important procedure to use sensors with reliable accuracy and consistency. During calibration test, the force value was established by measuring its voltage. Thus, a sensor needs to be calibrated by set a known force to the sensor using dead weight method and measuring its conductance. Four FlexiForce sensors were calibrated individually. A load with known weight, starting from 50 N was placed on top of each sensor. The weight was left for 5 s (predicted time for SCI patient to perform a cycle of STS activity with standing frame). This helped to minimize the drift error. The weight of load was increased gradually until 400 N.

2.2 Instrumentation of the Standing Frame

The standing frame used in this study was instrumented to identify the characteristics of FlexiForce sensor acting upon the standing frame. FlexiForce sensors were placed at the bottom of the four legs of the standing frame as shown in Fig. 1. Ten subjects with different body masses were instructed to lift their body with their arms holding onto the arm support of the standing frame. This step allows their body weight to be distributed among the four sensors at the base of the walking frame.

2.3 Subject Selection

Three male subjects with SCI AIS C took part in this study as shown in Table 1. The mean age was 40 years (SD,

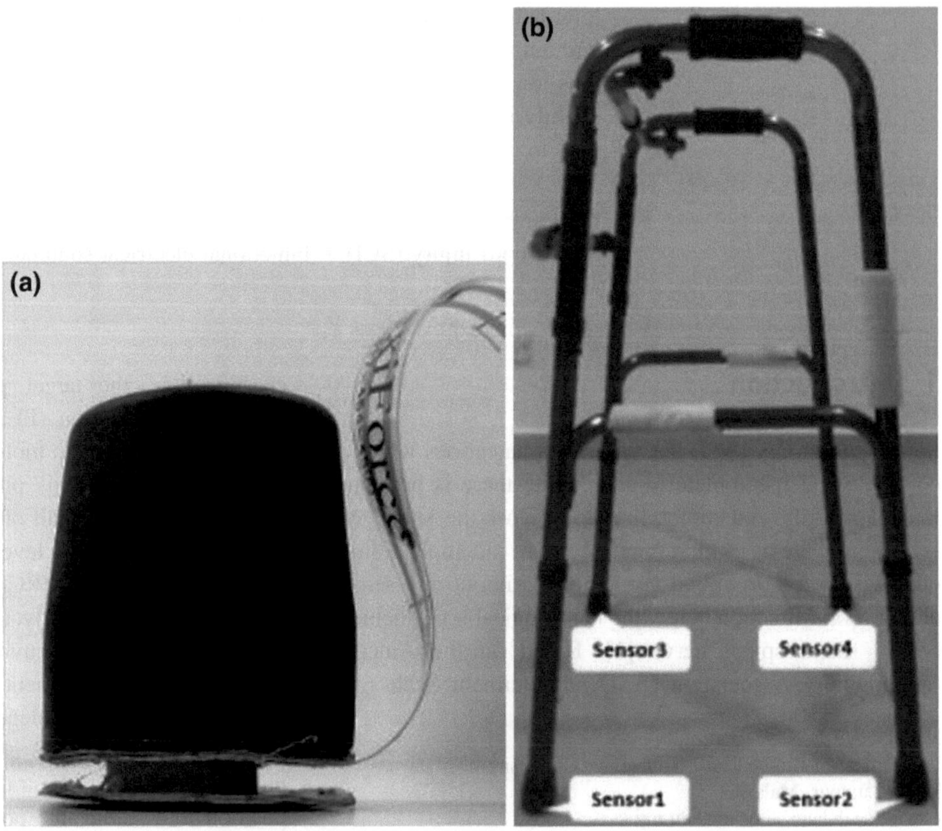

Fig. 1 **a** Placement of FlexiForce sensor **b** Sensors' placement on the standing frame's feet

Table 1 Subjects' profile

Participant	Level of injury	Mechanism of injury	Response to FES	Maximum tolerable current for FES (mA)	
				Quadriceps	Gluteal maximus
Subject 1	T12–L2	Fall from tree	No	–	–
Subject 2	T11–T12	Construction site accident	Yes	58	68
Subject 3	L1–L2	Accident during work	Yes	32	40

6.1 year, range 33–43 years); average mass was 69.67 kg (SD, 20.6 kg, range 54–93 kg) and average height was 165.8 cm (SD 5.3 cm, range 162.5–172 cm). Informed consent was sought from all subjects for this study. All of them were able to stand up from a chair independently upon verbal instruction without using their hands. However, Subject 1 was unable to complete the research protocol due to neuropathic pain on his back and he is FES non-responsive at his left quadriceps muscles.

2.4 Experimental Protocol

The subjects were positioned on armless chair with standardised height of 45 cm. Subjects' knees were flexed at 90° and both feet were positioned on the force plate. On Day 1, subjects were tested on their regular method of transition from a seated position to standing without the assistance of FES. Subject was directed to attempt stand up from the seated position at the end of a countdown by an investigator. The subject was monitored by a physiotherapist and researchers throughout this test. Three seconds after the subject achieves full upright standing (full knee extension), the subject was directed to sit down. Then, the subject was allowed to take 5-minute rest between tests. The tests were repeated up to 10 times or until the subject was tired. At the end of the first session, FES stimulation was determined for the subject for the next experiment session. The subject was asked for his feedback on his tolerance with higher stimulation intensity. 48-hour rest was given to the subject before the next test (Day 2).

In Day 2, subjects were tested on their ability to perform a transition from a seated position to a standing one with the assistance of FES. In addition, the FES electrodes were placed by a physiotherapist and researcher on gluteal maximus and quadriceps femoris muscles bilaterally. The protocol for Day 2 was similar with Day 1.

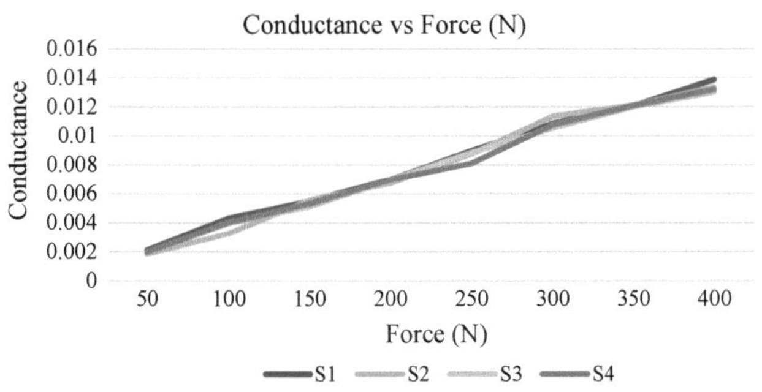

Fig. 2 FlexiForce sensor calibration

3 Results and Discussion

3.1 Calibration of FlexiForce Sensors

Four FlexiForce sensors was calibrated individually by putting a load with known weight at 50 N intervals. The graph of conductance versus force was plotted as shown in Fig. 2. A line of best fit was drawn to find a linear equation of each sensor.

Body hang test with ten normal subjects were done to validate the distribution of weight acting upon the instrumented standing frame. The characteristic of net body weight of each subject at standing frame is determined with this formula:

$$\text{Net body weight} = \text{ValueSensor1} + (2 * \text{ValueSensor2}) + \text{ValueSensor3} + (2 * \text{ValueSensor4}) \quad (1)$$

The percentage of net body weight is calculated using the formula:

$$\text{Percentage of net body weight} = (\text{Net body weight}/\text{Actual weight}) * 100 \quad (2)$$

The hanging test shows the instrumented standing frame setup has minimum 85% accuracy with comparison the total body weight of the subject. Several identifiable random errors has reduced the results' accuracy but these errors remained consistent throughout the experiment. These errors were difficult to be precisely factored in most experiments, but repetitions of experiments have been done to minimise effect of the errors on result's reliability. Thus, any reduction

Table 2 Average time recorded to do a cycle of STS

Participant	Time taken for STS (s)	
	No FES	FES
Subject 2	1.11 ± 0.1277	2.55 ± 0.2625
Subject 3	1.07 ± 0.1677	1.36 ± 0.0951

in accuracy did not affect the comparison between sit to stand manoeuvre with and without standing frame.

3.2 Experimental Results

The average time taken for each SCI subject to perform one full cycle of STS were tabulated in Table 2. Based on the average time recorded, both of them showed longer time to complete a cycle of STS when using FES. Both SCI subject felt more confident and comfortable to stand without using FES. Subject 2 took the longest time to perform a complete STS cycle with assistance of FES due to joint contracture at his left ankle. This contracture has affected his foot movement contacting with the force plate. In addition they also first time experience with FES in STS activity.

Next, Fig. 3 displayed the translocation of the centre of net force for upper limb to the right side of the standing frame in both sessions. It happened because Subject 2 used more strength at his right upper limb to move his trunk forward for standing. In addition, he also has left ankle and right knee joint contracture thus making him harder to place his left forefoot at the force plate and extend the right leg during standing activity. Subject 2 showed smaller variation of coordinate's x-axis and y-axis with standard deviation of less than ±0.03. Hence in order to preserve upper limb stability, Subject 2 displayed greater net force on right side of standing frame to compensate current given to the lower limb with several joint contractures during STS activity.

For Subject 3, the net force distribution of the upper limbs during STS in both session followed the translocation pattern of Subject 2. The centre of net force has inclined to the right side of the standing frame as illustrated in Fig. 4. He practiced more strength at his right upper limb to move his trunk forward as to compensate his weakness on lower limb joint. Furthermore, Subject 3 was used to wearing bilateral solid ankle foot orthosis (AFO) during his outside ambulation activity.

Figures 5 and 6 manifested the rate of fatigue for two SCI subjects during experiment without and with FES. For every trial, the value of force exerted by subjects' arms on the frame and the force exerted by subjects' feet on force plate while the subject at the highest knee moment during STS were taken to identify the rate of fatigue for this experiment.

Based on Fig. 5 for STS with assistance of FES, Subject 2 presented early fatigue in lower limb at trial 6 (t6), proven by the rise in force produced by the upper limb. He

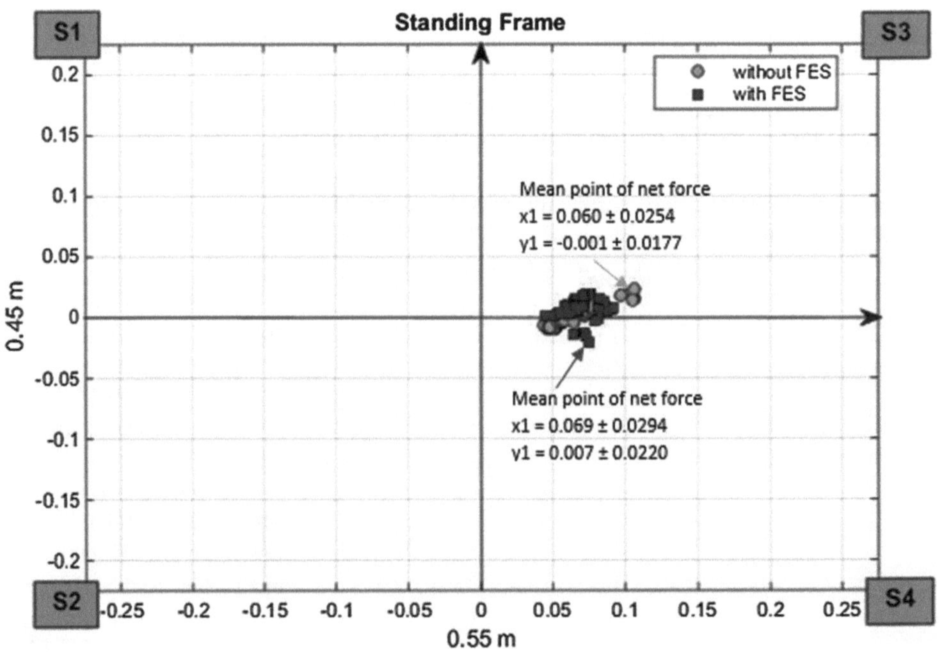

Fig. 3 Net force distribution detected with four sensors at standing frame during STS for Subject 2

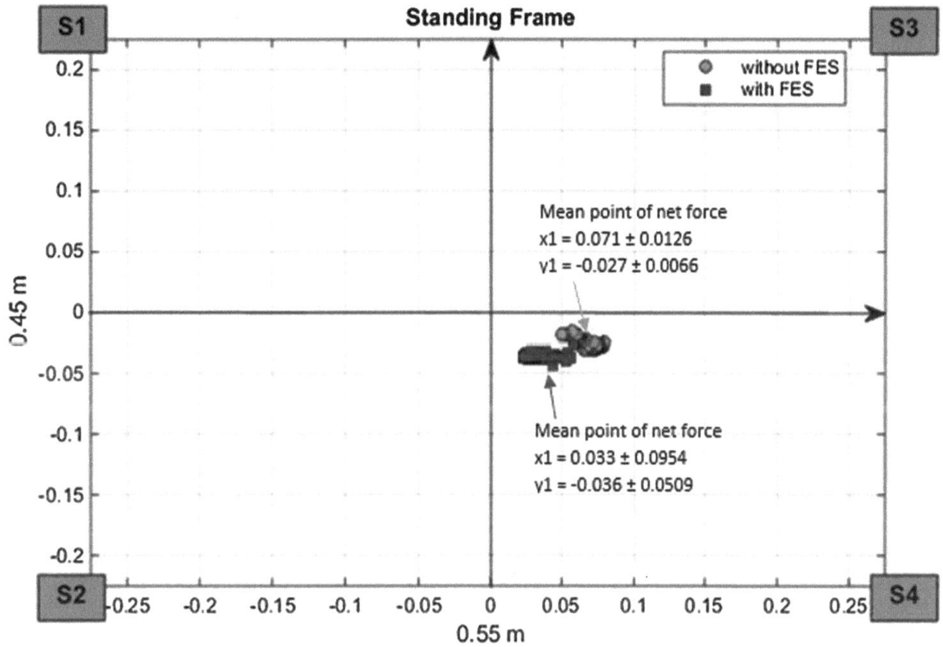

Fig. 4 Net force distribution detected with four sensors at standing frame during STS for Subject 3

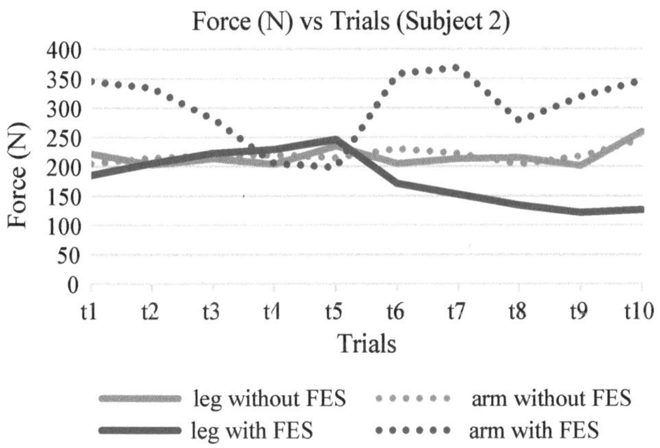

Fig. 5 Force exerted at arm and feet versus trials during Sit-to-Stand using FES and without FES for Subject 2

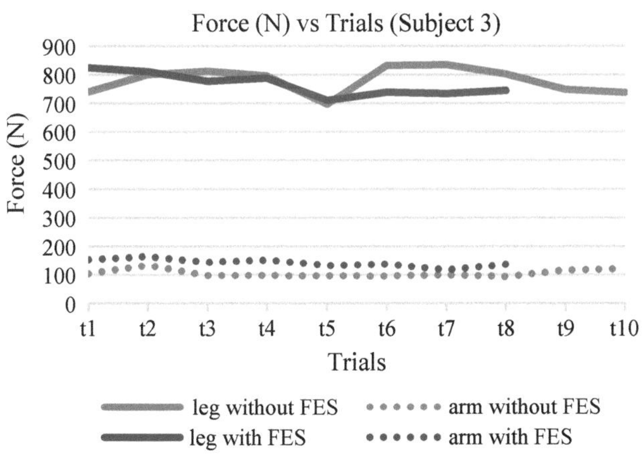

Fig. 6 Force exerted at arm and feet versus trials during Sit-to-Stand using FES and without FES for Subject 3

displayed evidence of early fatigue at trial 9 when performing STS without FES.

Meanwhile in Fig. 6, Subject 3 displayed evidence of fatigue on every session of experiment. With FES assistance on trial 5, subject 3 showed high reduction of force at leg and the value maintained for next trial until trial 8. However, without FES, Subject 3 showed early fatigue at trial 5 and this situation continued at trial 8 onwards.

4 Conclusion

The sensors on the instrumented standing frame were calibrated. At least 85% accuracy of body weight acting upon standing frame was obtained when normal subjects bear all their body weight to the frame. According to the average time taken in performing a STS cycle, both subjects presented shorter time without using FES as they felt more confident and comfortable to stand. They also achieved early stage of fatigue during using FES thus longer time needed to do a STS cycle. However, SCI subjects displayed translocation of the centre of net force to right side of the standing frame to achieve stability in both sessions, as they used more strength on the right upper limb to move their trunk forward for standing. Besides, they have several conditions of lower limb joint contracture.

Acknowledgements The authors appreciate significant contribution and continuous support by Assoc Prof. Dr. Nazirah Hasnan from Department of Rehabilitation Medicine, University of Malaya. This research was fully supported by the Ministry of Higher Education, Malaysia and University of Malaya through Fundamental Research Grant Scheme (FRGS) Grant No. FP027-2015A.

References

1. Kirshblum, S.C., et al.: International standards for neurological classification of spinal cord injury (Revised 2011). J. Spinal Cord Med. **34**(6), 535–546 (2011)
2. Lee, M.Y., Lee, H.Y.: Analysis for Sit-to-Stand performance according to the angle of knee flexion in individuals with Hemiparesis. J Phys Ther Sci **25**(12), 1583–1585 (2013)
3. Janssen, W.G., et al.: Determinants of the Sit-to-Stand movement: a review. Phys. Ther. **82**(9), 866–879 (2002)

A Low Cost Street Crossing Electronic Aid for the Deaf and Blind

Nur Hasnifa Hasan Baseri, Ee Yeng Ng, Alireza Safdari, Mahmoud Moghavvemi, and Noraisyah Mohamed Shah

Abstract

This paper describes the design and implementation of a low cost electronic device to aid the visually impaired and deaf-blind to cross the road safely using an Arduino controlled transmitter and receiver. The prototype consists of two parts; the transmitter is placed on the green traffic light and the receiver which is held by the user. The transmitter sends the traffic light signal status and information on wait duration by broadcasting them out to the receiver using XBee. The visually impaired or deaf-blind person holding the receiver will be notified once he/she is within 10 meters from the traffic light, and guides the person to cross the road via sound and vibration. The prototype is able to function successfully, with few limitations and recommendations highlighted for future improvements.

Keywords

Visual impairments • Navigation • Electronic devices

1 Introduction

Traditionally, the visually impaired uses walking sticks or guide dogs to help them move around but this is efficient only in familiar areas. Walking stick or white cane is used as sensors to detect the presents of obstacles in their path and to alert passerby of their presence. However white canes are only efficient over a short distance [1].

A lot of technological advancement in recent years to have been made to aid the visually impaired in their movement. For example, smartphone apps that uses image processing to extract road information such as the location of crosswalk, and current status of the traffic light can be found in [2–5], with variations in the algorithm used and the solution it provides.

Many works also focus on improving the user-interface for the blind, such as Google maps and Waze. Pocket Navigator [6] is a simple map-based navigation system for smart phones with Android operating system. Once the destination is setup, the visually-impaired user can be guided by the vibration of the smart phone. Another project called NAVIC uses virtual augmented reality together with image processing and satellite navigation [7] to recognise a location via the camera mounted on the blind and guide them to the desired location. It functions similar to a vehicle GPS; a user will command the device with buttons and voice control, while the device feed the user with information through earphones. However due to low accuracy of the GPS system, location-based services for the blind are instead supported by RFID technology [8–11]. This technique involve the use of multiple RFID tags placed on the floor to be detected by the user's white cane. An external server will then process the input data and inform the exact location of the user. However, to implement the system outdoor, multiple RFID tags have to be placed on the street and the device has to be online 24/7 to communicate with the server database.

Few method of improving the infrastructure of the pedestrian streets to be more disabled-friendly are also introduced [12]. This involved placing physical materials that can be detected by the blind such as slopes, guide strips, perpendicular bar tiles and waring bar tiles, to indicate the

N. H. Hasan Baseri · E. Y. Ng · A. Safdari · M. Moghavvemi · N. M. Shah (✉)
Department of Electrical Engineering, Faculty of Engineering, University of Malaya, Kuala Lumpur, Malaysia
e-mail: noraisyah@um.edu.my

F. Ibrahim et al. (eds.), *2nd International Conference for Innovation in Biomedical Engineering and Life Sciences*, IFMBE Proceedings 67, https://doi.org/10.1007/978-981-10-7554-4_12

and align the blind to a pedestrian crossing. Another method is beaconing Accessible Pedestrian Signals (APS), which is basically a device which makes noise on the opposite end of the crossing when it is time to cross [13].

1.1 Motivation

This paper describes a device that enables the blind to cross a road at a traffic light without assistance from other individual(s). The prototype of the device is able to guide the visually impaired user to a traffic light and convey information on its current status. It uses ZigBee technology and Arduino controllers, which are less costly. By setting up the transmitter unit prototype in a semi open area and switching on the receiver unit prototype at 10 meters apart, the receiver unit instantly picks up signals from the transmitter unit and guide the user to the transmitter using beeping audio via the earpiece. As the user arrive within 1 m from the transmitter, the receiver gives voice indication/vibration if the state of the traffic and the wait duration to safely cross the road

2 Methodology

The goal of this task is divided into two: to guide the visually impaired user to the traffic light, and to convey information of a traffic light to the user, to determine when to safely cross the street. This can be identified in the following three points:

1. To guide the blind to a pedestrian crossing:

A hand-held device that is capable to know if the pedestrian is going towards or away from the pedestrian crossing.

2. To send reliable traffic light information to the blind's hand-held device.

There will be 2 information that will be send to the blind:

- The current state of the traffic
- The period to next road busy state.

3. To convert data received by the hand-held device into suitable interfaces for the user. The interfaces should suit both kinds of user:

- Visually impaired user
- Visually and hearing impaired user.

Figure 1 gives a basic structure of the communication device. It is a one way communication from the transmitter to the receiver. The wireless communication technologies considered for this device were Bluetooth classic, Bluetooth Low Energy, ZigBee, ANT, Sensium and Radio Frequency Identification (RFID). Based on the operating space, the ZigBee technology would be an ideal choice. The low data rate of ZigBee did not affect the performance of the device due to the low data requirement. A module called XBee by Digi International [14] that uses ZigBee technology was adopted. It is capable to operate up to 90 m outdoor with line-of-sight. The module cost is higher but has the advantage of a wider functionality such as error correction and the extraction of the received signal strength indication (RSSI) values. XBee is a wireless communication module that utilizes the IEEE 802.15.4 protocol. This protocol is known as Low-Rate, Wireless Personal Area Network (LR-WPAN). It provides 250 kbps data transfer between nodes on a CSMA/CA network. The XBee-PRO 802.15.4 is used as the transmitter and XBee 802.15.4 as the receiver, both in AT modes for serial communication.

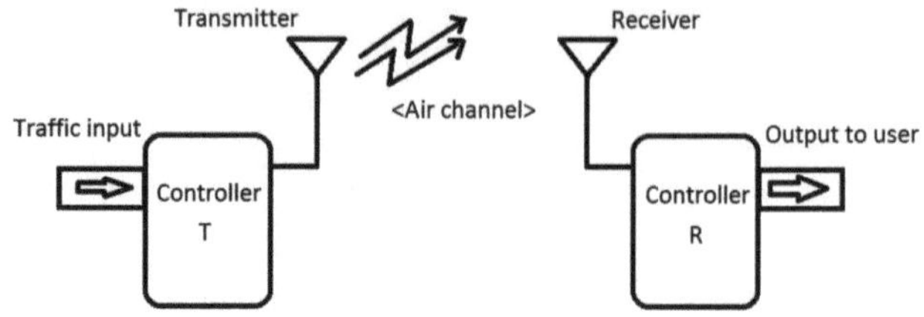

Fig. 1 Basic structure of the communication system

2.1 Transmitter Unit

The transmitter unit consist of an Arduino Uno, a photosensor module, the XBee Pro 802.15.4 and its adapter. The XBee would be mounted onto the adapter for safer and easier connection. The adapter is powered via the 5 V pin and the ground pin of Arduino Uno. The Tx (transmitting) pin of the XBee adapter is connected to the Rx (receiving) pin of the Arduino. The Rx pin of the XBee adapter is connected to the Tx pin of the Arduino. The Arduino Uno only have one set of serial ports, which means only a pair of pins on the board is supported by the universal asynchronous receiver/transmitter (UART). The photosensor module is a light-controlled variable resister (LDR). It is also powered by the 5 V pin and ground of the Arduino Uno pin. The analogue output (A0) of the module is connected to the analogue input pin A0 to sample and determine its instantaneous value. The value will then be processed by the Arduino and repackaged into a line of data, which consist of a header, the state of traffic (if the green light of the vehicle

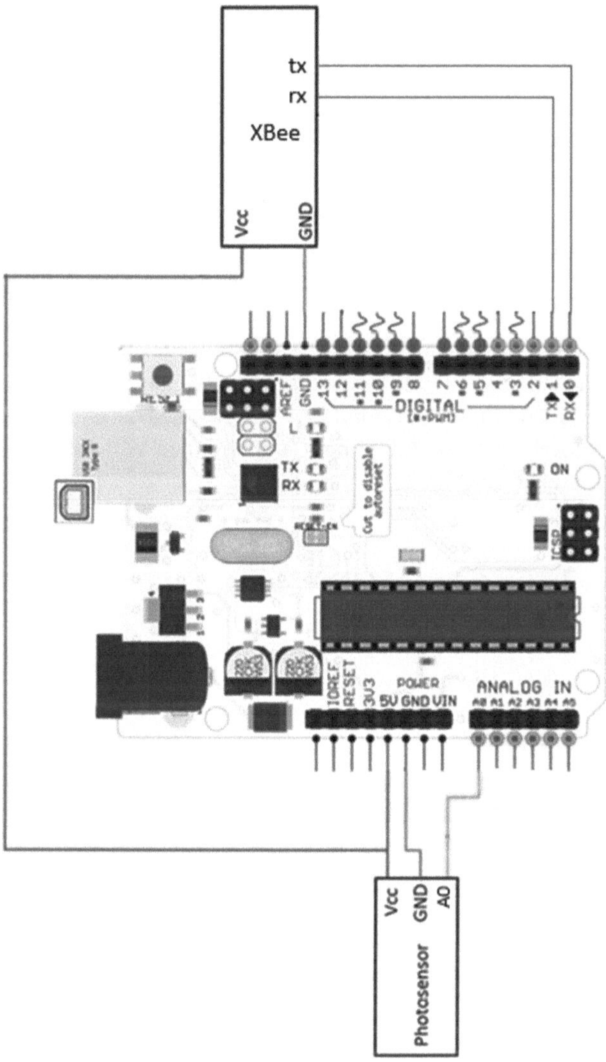

Fig. 2 Circuit diagram of the transmitter unit

is on), the amount of time for allocated for the user to cross and a checksum. The header and the checksum is use to ensure the data is read by the correct device and the data receive is accurate. This data is sent out via the mini antenna on the XBee. The circuit diagram of the transmitter unit is shown in Fig. 2.

This transmitter unit will be placed at the green light of the traffic light. Here, LDR detects changing state of the green light. There are two states involved here: (i) GREEN_ON_STATE which occurs when the LDR value is less than a threshold level. In this state, the green light of the traffic light is ON. The second, (ii) GREEN_OFF_STATE, occurs when the LDR value is greater than a threshold. In this state, the green light of the traffic light is OFF, and is safe to cross the road. This data is then transmitted to the receiver by using XBee Pro 802.15.4

2.2 Receiver Unit

The receiver part consists of the XBee and its adapter, Arduino Mega, the volume system and the audio buttons, Motor circuit and SD Card reader.

Figure 3 shows the connection of XBee and Aruduino Mega. The XBee is powered with the 5 V output and the ground pins of Arduino Mega via the XBee adapter. The serial output (Tx and Rx) is connected to pin 18 and 19 of Arduino Mega, which is the serial port 1. Pin 0 and pin 1 of Arduino Mega is used for debugging purposes. The RSSI is connected to the digital pin 6 of the Arduino Mega. The pin 6 of the Arduino board is set as an input. The output from the RSSI pin is a pulse width modulated (PWM) signal. As mentioned before this RSSI value is used to estimate the distance between the user and the traffic light.

Figure 4 shows the motor circuit connection. The transistors and diodes are safety features used to protect the Arduino Mega from damage. In this case, the load here is the motor. From the DC motor (rf-300fa-12350) datasheet, the maximum current draw is around 93 mA, which is more than two times a digital output pin could withstand. Therefore, the digital output pin is used with a transistor to control the motor. DC motor has the characteristic of an inductor when the motor is suddenly turn off. The motor current will not go to zero abruptly due to its inductive nature. By adding a diode, the reverse current is channelled back into the motor instead of letting it flow through the transistor, burning the transistor in the process. The motor is powered normally via the +5 V pin. However, when the digital pin input is low, the current in the motor still needs to

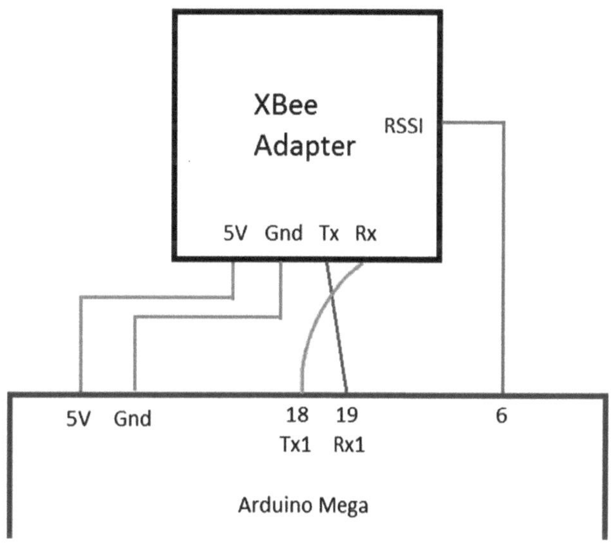

Fig. 3 Connection of the adapter with Xbee mounted to Arduino Mega

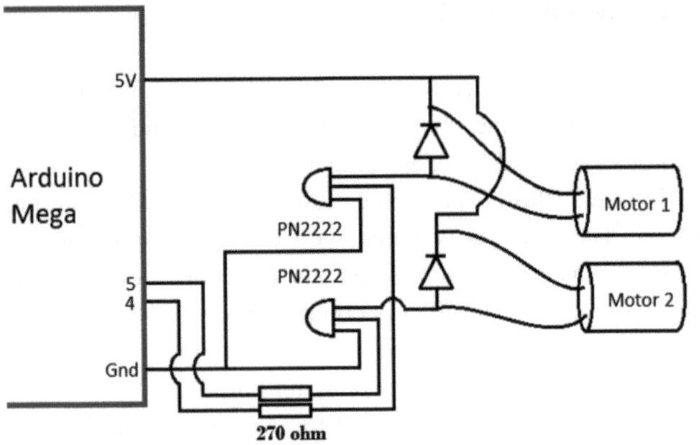

Fig. 4 Circuit diagram for motor control

flow due to its inductive nature. The diode acts as a short circuit across the motor, channeling the current away from getting into the transistor and back into the motor.

The audio system consists of two parts, the audio port and the SD card module connection. The SD card module is used to enable the Arduino to communicate with the SD card. The SD card is necessary for providing extra flash memory for the system to keep the voice files. Two buttons are used to control the volume of the voice that gives information to the user. One for volume increase and the other for volume decrease. This can be performed via software using the volume up and down functions provided by the TMRpcm library.

3 Testing and Calibration

To perform full test, an Arduino is connected to an LED and programmed to act as a traffic light. The LED is put on the surface of LDR at the transmitter. In order to ensure that the ambient light does not affect the performance of LDR at the transmitter, some calibration is needed. The LDR is put into a hollow tube so it is exposed only to the light form the traffic light.

There are two modes at the receiver: (a) DISTANCE_TRACKING mode and (b) TRAFFIC_LIGHT_STATUS mode.

4 Distance_Tracking Mode

In this mode, user is 'guided' to the traffic light and notified once arrived at the traffic light. This is done by using RSSI number obtained from RSSI pin (pin 6) of XBee that outputs a PWM signal representing this value. PWM is a method for generating analog signal by using digital source. The behaviour of PWM signals can be defined by duty cycle which is the percentage of one period/cycle in which signal is active.

RSSI is used to estimate distance of the user from traffic light. Larger RSSI numbers means that the user is closer to the traffic light, thus the larger the duty cycle of its square wave. A few ranges of RSSI number is set to indicate to the user the estimated distance from the traffic light. Different frequency of "beep" sound for each different ranges of RSSI number is used to indicate this distance to the user. Table 1 gives the range of RSSI values obtained in the experiment. RSSI number is obtained by using the Arduino PulseIn command function. PulseIn function waits for the RSSI_PIN (in this case digital pin 6 of Arduino Mega) to go HIGH, start timing, waits for the pin to go LOW then stop timing. The motor circuit will vibrate when the receiver is tracking the location of the traffic light and will stop vibrating once the user arrived at the traffic light.

5 Traffic_Light_Status Mode

In this mode, the user is assumed to have arrived at the traffic light. Here, the user will obtain information on the state of the traffic light, condition of the road and wait duration before crossing the road, via sound and vibration/rotation. Both sound and vibration/rotation will occur simultaneously. There are two different states in this mode:

a. GREEN_ON

At this state the user will hear the 'busy' audio every 3 s. Neither Motor1 nor Motor2 will vibrate or rotate during this state.

b. GREEN_OFF

At this state, information on the duration left before crossing the road will be sent out every 10 s, and at the last 5 s of waiting time, a countdown will be heard. Motor2 will rotate during this state while Motor1 will vibrate for the last 5 s before change of state. The state is then changed from GREEN_OFF to GREEN_ON.

DISTANCE_TRACKING mode and TRAFFIC_LIGHT_STATUS mode are differentiated by using the RSSI number. As mentioned before, one of the main problem is the instability of the RSSI number detected. There are times when the DISTANCE_TRACKING mode is suddenly turned on when already in TRAFFIC_LIGHT_STATUS mode. This is because the RSSI number suddenly decreases. The stability of the RSSI output can be improved by introducing another if-else function. The algorithm will only read RSSI number during DISTANCE_TRACKING mode and will set the RSSI number to be maximum when it is in TRAFIC_LIGHT_STATUS mode. This causes the

Table 1 Different range of RSSI number

Condition	Meaning
RSSI Value <30	Very far from traffic light
RSSI Value >30 && RSSI Value <40	Far from traffic light
RSSI Value >40 && RSSI Value <50	Near traffic light
RSSI Value >50	Arrived at traffic light

TRAFIC_LIGHT_STATUS mode to be independent of RSSI number.

6 Conclusion

In this project, a prototype of an electronic device for visually impaired and deaf-blind people has been designed and implemented. It is an electronic street crossing aid with an audio and motor vibration feedback attached to the walking stick. The prototype consist of two devices, the transmitter and receiver. The transmitter is placed at the green light of the traffic light and will detect status of the green light. The data from the transmitter is then transmitted to the receiver via transmitting XBee. Data is received at the receiver via receiving XBee. The strength of the received signal is used to estimate distance of the user from the traffic light and once the user arrived at the traffic light, he/she will received data on:

(i) Status of the traffic light
(ii) Duration that is safe for him/her to cross traffic light intersections
(iii) Duration five seconds before status change (from red to green).

At the present state, visually impaired or deaf-blind people who wish to cross traffic light intersections sometimes are unable to know the exact status of the traffic light and at the same time do not know how much time remains before the traffic change state because the data on pedestrian signals are usually in visual form. Thus this prototype is able to provide a low cost solution for the visually impaired and the deaf-blind to cross the traffic light intersection safely and independently.

However the receiver unit of the prototype are encased in a 15 cm x 8 cm x 8 cm box, while the transmitter unit encased in an 8 cm x 7 cm x 7 cm box. This indicates that the receiver unit is unreasonably large, compared to mobile phones. The reasons behind the large size of the receiver units are as follows:

- Arduino Mega is large, with its PCB measuring at 4 inch × 2.1 inch.
- The spring buttons are too long. The lower part of the button which is encased in the box is around 3 cm.
- 2 large DC motors that have a length of 2.5 cm and a diameter of 3 cm.
- Extra circuitries are done using thru-hole technology.

6.1 Future Work

There are a few numbers of aspects that can be improved for the future such as wearability of the receiver device by integrating its main components into a smaller devices. The accuracy on the location of the traffic light can be improved by incorporating image processing method to extract the location of a zebra crossing. In this project, the traffic light intersection is assumed to be on a one way traffic light, so a system for multiple traffic will be more practical, efficient and reliable.

References

1. Pradeep, V., Medioni, G., Weiland, J.: A wearable system for the visually impaired. In: 2010 Annual International Conference of the IEEE Engineering in Medicine and Biology Society (EMBC). IEEE, Buenos Aires (2010)
2. Aron, J.: Zebra crossing app finds safe places to cross. Retrieved September 9, 2014, from http://www.newscientist.com/blogs/onepercent/2011/09/zebra-crossing-app.html, last accessed 2014/09/01 (2011)
3. Shioyama, T., Uddin, M.: Detection of pedestrian crossings with projective invariants from image data. Meas. Sci. Technol. **15**(12), 2400–2405 (2004)
4. Huiying, S., Kee-Yip, C., Coughlan, J., Brabyn, J.: A mobile phone system to find crosswalks for visually impaired pedestrians. Technol. Disabil. **20**(3), 217–224 (2008)
5. Qing, L., Hernsoo, H., Youngjoon, H.: Top-view-based guidance for blind people using directional ellipse Model. Int. J. Adv. Robot. Syst. 101–13 (2013)
6. Pielot, M., Poppinga, B., Boll, S.: PocketNavigator: vibro-tactile waypoint navigation for everyday mobile devices. In: Proceedings of the 12th International Conference on Human Computer Interaction with Mobile Devices and Services (MobileHCI '10), ACM, New York, NY, USA, pp. 423–426 (2010)
7. Katz, B.F., Dramas, F., Parseihian, G., Gutierrez, O., Kammoun, S., Brilhault, A., Jouffrais, C.: NAVIG: guidance system for the visually impaired using virtual augmented reality. Technol. Disabil. **24**(2), 163–178 (2012)

8. Fernandes, H., Filipe, V., Costa, P., Barroso, J.: Location based services for the Blind Supported by RFID Technology. Procedia Comput. Sci. **27**, 2–8 (2014)
9. Bharath Kumar, M.R., Sibu, C.M.: Design for visually impaired to work at Industry using RFID technology. In: 2015 International Conference on Electronic Design, Computer Networks & Automated Verification (EDCAV), pp. 29–33 (2015) (29–30 Jan)
10. Keneko, Y., Hanada, T., Hirahara, Y., Kikuchi, Y.: Development of the navigation system for the Visually impaired. In: IEEE EMBS Asian-Pacific Conference on Biomedical Engineering, pp. 238–239 (2003) (20–22 Oct)
11. Takatori, N., Nojima, K., Matsumoto, M., Yanashima, K.: (2006,) Development of voice navigation system for the casually impaired by using IC tags. In: 28th Annual International Conference of the IEEE Engineering in Medicine and Biology Society, 2006. EMBS '06, pp. 5181–5184 (2006) (30 Aug–3 Sept)
12. Scott, A.C., Barlow, J.M., Guth, D.A., Bentzen, B.L., Cunningham, C.M., Long, R.: Nonvisual cues for aligning to cross streets. J. Vis. Impair. Blind. **105**(10), 648–661 (2011)
13. Scott, A.C., Barlow, J.M., Guth, D.A., Bentzen, B.L., Cunningham, C.M., Long, R.: Walking between the lines: nonvisual cues for maintaining headings during street crossings. J. Vis. Impair. Blind. **105**(10), 662–674 (2011)
14. Digi International: XBee/XBee-PRO RF Modules. IEEE 802.15.4 RF Modules (2009)

Fastest Gait Parameters Estimation Precision Comparison Utilizing High-Sensitivity and Low-Sensitivity Inertial Sensor

Amir Mukhriz Azman, Hirofumi Kuga, Koichi Sagawa, and Chikara Nagai

Abstract

The measurement system of stride length, toe's height and toe's angle at each step during fastest gait using inertial sensor is proposed. During fastest gait, acceleration and angular velocity of the toe exceed the measuring range of high-sensitivity inertial sensor. To estimate parameters of a gait, inertial sensor system that is composed of two triaxial accelerometer and two triaxial gyroscope with different measuring range each is developed. Sensor is attached to the right toe to observe the motion of fastest gait quantitatively. Subjects were asked to walk at their fastest speed while the inertial sensor measures the acceleration and angular velocity of the toe. To evaluate the effect of the proposed method, toe's trajectory are calculated using the high-sensitivity data only, low-sensitivity data only, and interpolation of both high-sensitivity and low-sensitivity data. Results are compared with reference obtained by optical motion capture system. As a result of the motion measurement of the toe during fastest gait, the precision of the estimation system is improved by applying the proposed method.

Keywords

Fastest gait • Inertial sensor • Gait analysis estimation • Stride length • Toe's angle • Toe's height

1 Introduction

As human gets older, memory loss may become a problem. Memory is affected by age, stress, tiredness, or certain illness and medications. Dementia is a syndrome associated with an ongoing decline of the brain and its abilities. There were studies which noticed the relationship between motor function and risk of dementia [1, 2]. Surveillance of gait condition is conducted as an indication of finding dementia symptoms and mild cognitive impairment [3]. The gait performance is evaluated using gait parameters such as gait speed and stride length.

The School of Medicine of Hirosaki University has conducted health examination called Iwaki Project for approximately 1000 citizens of Hirosaki city in Japan since 2005 as part of cohort study. The aim of this project is to increase the health level of residents in the city and extend the average healthy lifespan. 10 m fastest gait examination was originally conducted to investigate the sign of dementia by measuring maximum gait velocity. However, the previous method requires many staffs to measure the walking time with stopwatch by following the subjects and other gait parameters such as stride length, toe's angle, and toe clearance could not be measured. Meanwhile, researches on gait analysis utilizing inertial sensor have been conducted in recent years [4, 5].

Wearable Wireless Inertial Measurement Unit (WIMU) has been widely used for the quantitative evaluation of walking because it has been downsized and the measurement accuracy has been improved significantly. WIMU consists of a triaxial acceleration sensor and a triaxial gyroscope. Based on the data obtained from these two sensors, the time change

A. M. Azman (✉) · H. Kuga · K. Sagawa · C. Nagai
Hirosaki University, Bunkyo-Cho 3, Hirosaki-Shi, Aomori-Ken 036-8561, Japan
e-mail: h17ds201@hirosaki-u.ac.jp

F. Ibrahim et al. (eds.), *2nd International Conference for Innovation in Biomedical Engineering and Life Sciences*, IFMBE Proceedings 67, https://doi.org/10.1007/978-981-10-7554-4_13

of the position and orientation of the WIMU is reconstructed into three-dimensional space. The authors have developed WIMU and proposed 3D gait analysis algorithm for free walk of normal gait of speed about 1.4 m/s. However, when the walking speed goes as fast about 2.8 m/s, desirable analysis results could not be obtained. This is because it is relatively easy to distinguish the gait cycle into swing phase and stance phase in normal walking, but it is difficult to distinguish the gait cycle in high speed walking.

Conventional method uses inertial sensors with high sensitivity only [6]. We found out that some subjects walk fast enough to exceed the measuring ranges of the conventional sensors and it is impossible to analyze their data properly. In this study, to solve this problem, inertial sensor system composed of low-sensitivity sensor in addition to high-sensitivity sensor is introduced. Interpolation method of value of high-sensitivity sensor exceeding the measurement ranges with low-sensitivity sensor is also introduced. Furthermore, we propose integration interval determination method of the gait cycle that can be applied to fast gait. Using the proposed method, we perform fastest gait experiment and examine the measurement accuracy for each data collected using high-sensitivity sensor only, low-sensitivity sensor only, and from the interpolation method. Gait parameters such as stride length, toe's height, and toe's angle are estimated from the measured gait motion and then compared with the gait parameters obtained from motion capture system (MCS).

2 Theory

2.1 Calculation of the Foot Trajectory

Walking velocity is derived by integrating the acceleration data obtained from the sensor attached to the toe and further integration derives trajectory of the toe. Step-by-step foot trajectory during walking was calculated to minimize the effect of drift measured by the inertial sensor. This is done by integrating translational acceleration of the foot during each swing phase. The measured acceleration vector is represented by sensor-fixed coordinate system $\sum S$. Therefore for integration, acceleration vector should be transformed into the acceleration vector in the earth coordinate system $\sum E$ by defining orientation of the sensor coordinate system during foot-flat. Since the sensor comes to rest during each foot-flat, the direction of gravitational acceleration can be calculated and then subtracted from the measured acceleration vector of the sensor coordinate system.

Let the matrix representing the sensor orientation with respect to the earth coordinate system be $[\mathrm{i\ j\ k}] = {}^{E}E_{S}(t)$. The measured vector of acceleration and angular velocity represented in the $\sum E$ coordinate system ${}^{E}a(t)$ and ${}^{E}\omega(t)$ can be calculated from the vector of acceleration and angular velocity represented in the $\sum \mathrm{S}$ coordinate system ${}^{S}a(t)$ and ${}^{S}\omega(t)$ using the following equation.

$$ {}^{E}a(t) = {}^{E}E_{S}(t){}^{S}a(t) \tag{1} $$

$$ {}^{E}\omega(t) = {}^{E}E_{S}(t){}^{S}\omega(t) \tag{2} $$

In this study, the temporal change in the sensor orientation with respect to the base coordinate system during walking was estimated by time-integrating the angular velocity of the sensor from the time of foot-flat. ${}^{E}R_{S}(t)$ can be represented by the rotational matrix with respect to the earth coordinate system as follows.

$$ {}^{E}R_{S} = \begin{bmatrix} \cos\theta + \lambda^{2}(1-\cos\theta) & \lambda\mu(1-\cos\theta) - \nu\sin\theta & \nu\lambda(1-\cos\theta) + \mu\sin\theta \\ \lambda\mu(1-\cos\theta) + \nu\sin\theta & \cos\theta + \mu^{2}(1-\cos\theta) & \mu\nu(1-\cos\theta) - \lambda\sin\theta \\ \nu\lambda(1-\cos\theta) - \mu\sin\theta & \mu\nu(1-\cos\theta) + \lambda\sin\theta & \cos\theta + \nu^{2}(1-\cos\theta) \end{bmatrix} \tag{3} $$

$$ \begin{pmatrix} \lambda \\ \mu \\ \nu \end{pmatrix} = \frac{{}^{E}\omega}{{}^{E}\omega} \tag{4} $$

$$ \theta = \Delta\mathrm{t} \times^{E} \omega \tag{5} $$

Sensor orientation is updated every sampling period Δt by the following equation.

$$ E_{t+\Delta t} =^{E} R_{S}E_{t} \tag{6} $$

The matrix during a gait is updated by performing the above rotation conversion for every sampling period in the integration section.

Because the sensor mounted on the tip-toe comes to rest at each foot-flat, the translational acceleration must be zero. For integration, therefore gravitational acceleration is subtracted from the measured acceleration during gait. Acceleration during foot-flat ${}^{E}a(t_{0})$ can be assumed to be equal to the gravitational acceleration vector, and the translational acceleration vector is calculated as follows.

$$^{E}a(t) = ^{E}R_{S}(t)^{S}a(t) - ^{E}a(t_0) \quad (7)$$

In this study, acceleration vector is time-integrated from stance phase (t_0) to the next stance phase (t_{end}). At this time, since the integral error accumulates due to drifting, sensor noise, and misalignment of sensor system, estimated vertical velocity of the sensor at t_{end} hardly becomes zero. To resolve this problem so that the estimated velocity at t_{end} reaches zero, acceleration offset, a_e, subtracted from the vertical acceleration between t_0 and t_{end} can be estimated using the following equation.

$$0 = \int_{t_0}^{t_{end}} (a_z(t) - a_e)dt = \int_{t_0}^{t_{end}} a_z(t)dt - a_e(t_{end} - t_0) \quad (8)$$

where a_z is the vertical component of the gravity-compensated translational vector. We also assume that there is no slope on the walking surface and that the vertical displacement of the sensor at each foot-flat time is zero. So that the vertical displacement at t_{end} becomes zero, velocity offset, v_e, subtracted from the vertical velocity between t_0 and t_{end} can be estimated using the following equation.

$$0 = \int_{t_0}^{t_{end}} (v_z(t) - v_e)dt = \int_{t_0}^{t_{end}} v_z(t)dt - v_e(t_{end} - t_0) \quad (9)$$

where v_z is the vertical velocity of the sensor.

2.2 Determination of Integration Section

To accurately estimate the moving distance, it is necessary to distinguish the swing phase with the stance phase precisely to determine the integration section. In this study, to determine the integration section, we focus on the angular velocity in the dorsi-plantar-flexion direction of the foot. Figure 1 shows the relationship between the angular velocity in the dorsi-plantar-flexion direction of the foot and the movement of the foot. Point c_n is the minimum value of the angular velocity occurred when the foot is swing forward during the swing phase. Maximum point a_n of the angular velocity occurs before point c_n when the toe aparts from the support surface, and maximum point b_n of the angular velocity occurs after point c_n when the heel lands on the support surface. The integration section is determined by including the maximum values of point a_n and point b_n.

Integration section that minimizes the error of the estimated stride length during fastest gait was investigated. With the time of the swing phase as the reference, the time of the minimum value of the synthetic angular velocity at

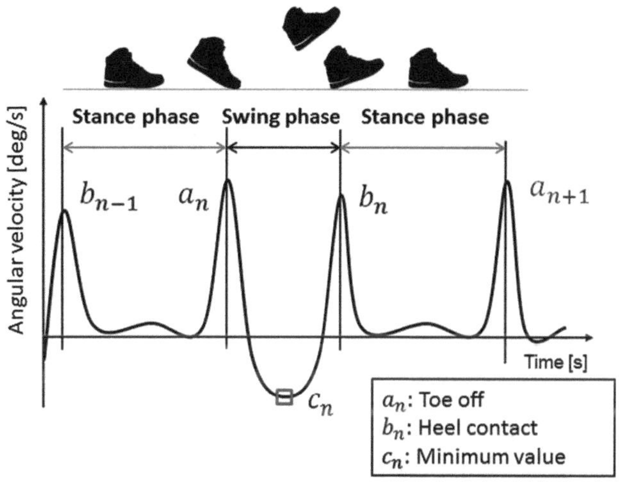

Fig. 1 Temporal changes of synthetic angular velocity of WIMU mounted on the tip-toe during walking. Swing phase starts with toe off a_n and ends with heel contact b_n. Minimum value c_n is taken in between the two peaks a_n and b_n

predetermined intervals before and after the toe's apart time is set as the start and end of the integration time. Minimum error is obtained when the start and end times of the integration section are set at -45 ± 1 and $140 \pm 10\%$ of the swing phase. The time when the minimal value of synthetic angular velocity found within the predetermined range immediately before and after the swing phase is selected as the start time and end time of the integration section. The minimal value immediately before the swing phase is assumed to be the moment when the tiptoe finishes the foot-flat and preparing to leaves the support surface. The minimal value immediately after the swing phase is assumed to be the moment when the tiptoe touches the support surface after the heel contact.

2.3 Gait Parameters

In this study, foot-trajectory was calculated off-line based on the measured sensor data. The elapsed time from a heel contact to the next heel contact of the ipsilateral leg is called the gait cycle and is divided to stance phase and swing phase. The stance phase refers to the period from the heel contact to the toe off and the foot is in contact with the support surface in this period. The swing phase refers to the time when the toe aparts from the ground until the next heel contact and the foot is away from the ground in this period.

The gait parameters derived in this study are stride length, toe's height, and toe's angle shown in Fig. 2. A stride is an operation from the contact of the heel of one side of the foot to the next contact of the heel on the same side to the support surface and the distance is defined as stride length. Applying double integration to vertical acceleration, toe's height is

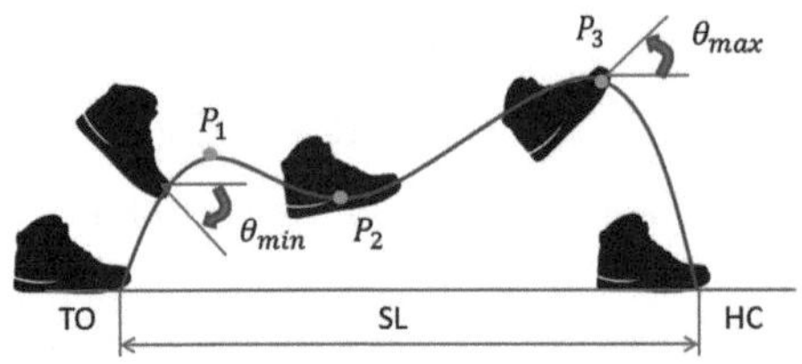

Fig. 2 Three gait parameters estimated using the sensor unit. Toe's height, toe's angle, and stride length are calculated from the measured data of WIMU and compared with corresponding MCS trajectory

obtained. Generally in normal gait, two maximum points P_1 and P_3 and a minimum point P_2 can be observed [7]. However in our experiment, several number of foot trajectory samples did not have the minimum point P_2. In this study, to evaluate the toe's height during fastest gait, we only calculated point P_3. The toe's angle is the amplitude of the angle of the toe direction from the floor and is defined by the difference between the maximum angle θ_{max} and the minimum angle θ_{min}.

3 Methods

3.1 Measurement System

During fast motion, large acceleration and angular velocity occur at the toe. In this case, high-sensitivity sensors could not measure these values because they are over the measurement ranges. To measure these large values accurately, we introduced sensor with wide measuring range (low-sensitivity sensor) in addition to conventional sensor which has small measuring range (high-sensitivity sensor). WIMU developed in this study consists of two sensor boards; one mounts high-sensitivity acceleration sensor and gyroscope (MPU-6050, InvenSense Inc, ±16 g, ±2000 dps), microprocessor unit (dsPIC33FJ128GP802, Microchip Technology), wireless communication module (Xbee, MaxStream, S1), and microSD card slot. The other board mounts low-sensitivity acceleration sensor (ADXL375, Analog Devices Inc, ±200 g) and low-sensitivity gyroscope (LPY4150AL, STMicroelectronics, ±6000 dps). Both boards are connected vertically to a switch board which consists of two controller switches, LEDs for information display, USB slot with charging function, and voltage regulator connected to lithium ion battery. Sampling frequency is taken at 1 kHz.

3.2 Interpolation Method

The sensor system used in the conventional method had only high-sensitivity sensor [6]. However, the conventional sensor system is only suitable to measure average walking with speed about 1.4 m/s. In case of fast gait, the acceleration and the angular velocity exceed the measurement range of the high-sensitivity sensor and therefore decrease the gait analysis accuracy. In this study, we attempt to cope with fast gait by introducing low-sensitivity sensor in addition to high-sensitivity sensor. For the interpolation method, value exceeding the measurement range of high-sensitivity sensor is interpolated with the low-sensitivity sensor. Acceleration threshold is set at 100 m/s^2 and angular velocity threshold is set at 1000 deg/s. When the measured value of high-sensitivity sensor exceeds the threshold, the value is interpolated using the measured value of low-sensitivity sensor.

4 Experiment

In this study, we introduced the utilization of low-sensitivity sensor in addition to high-sensitivity sensor. To quantitatively evaluate the proposal method of fast gait analysis, gait parameters are derived by applying high-sensitivity sensor, low sensitivity sensor, and the interpolation method. To evaluate the accuracy of the foot trajectory estimated using the stated method and the new integration section determination method, we asked 7 adult male participants aged 25.2 ± 7.9 years old to walk along a walkway at their fastest speed for 5 times. WIMU is fixed on the tiptoe with vinyl tape, and acceleration and angular velocity information during gait are recorded. We simultaneously measured the position of the markers attached to the WIMU using optical 3-dimensional MCS (Optitrack Prime 13, NaturalPoint Inc.) with six cameras placed on a side way capturing the walking course of length 7.0 m. Figure 3 shows schematic diagram of the experimental environment. Acceleration section and deceleration section are established so that fastest gait can be recorded on the limited measuring ranges of MCS. Data are sampled at frame speed 240 fps. Motion-captured trajectory is considered as a reference foot trajectory and used for

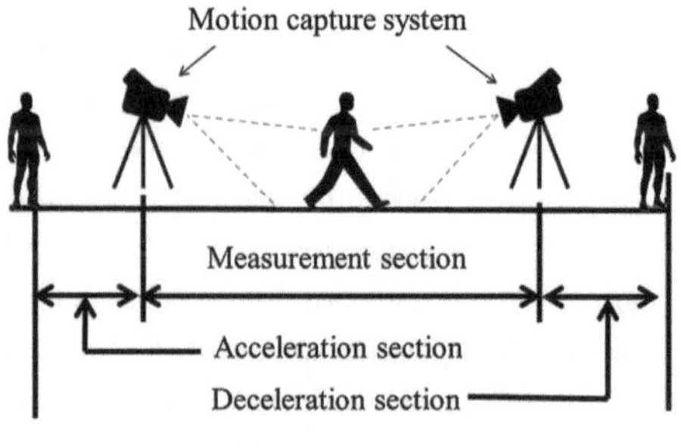

Fig. 3 Schematic diagram of gait experiment. Walking course is divided into 3 section; acceleration section, measurement section, and deceleration section

evaluation of estimation accuracy of each measuring method. From the reference and estimated foot trajectories, gait parameters such as toe's height, toe's angle, and stride length are calculated and compared.

5 Results

There are a total of 35 samples (7 subjects × 5 trials). The error estimation of toe's height, amplitude of the toe's angle, and stride length of 7 subjects (35 samples) using high-sensitivity data, low-sensitivity data, and interpolation method are summarized in Table 1. We did multiple comparisons between the estimation accuracy of high-sensitivity, low-sensitivity, and interpolation method for estimation error of angle amplitude, toe's height, and stride length. The results are as follows.

For amplitude of toe's angle, there are no significant differences of estimation error for any combination of measuring method. This indicates that interpolation method has the 2nd best amplitude of toe's angle estimation result.

For toe's height, there are significant differences when comparing the error of low-sensitivity with high-sensitivity and low-sensitivity with interpolation method. This shows that using low-sensitivity only for fastest gait analysis is not a good option when calculating toe's height.

For stride length, there is significant difference when comparing error of high-sensitivity with interpolation method. This proves that interpolation method has the best stride length estimation accuracy. Using high-sensitivity sensor only caused significant drops in estimation accuracy.

Overall, we can conclude that by applying interpolation method, compared to using high-sensitivity only or low-sensitivity only, the estimation error of gait parameters is improved.

6 Discussion

From the results of fastest gait experiment taken from the 7 subjects (35 samples), estimation of toe's angle amplitude shows that low-sensitivity measuring has the smallest error. However for the estimation of toe's height, high-sensitivity measuring shows the best result. In case of stride length estimation, interpolation method shows the best accuracy. The method of interpolating values exceeding the measurement range of high-sensitivity sensor by introducing low-sensitivity sensor is effective for the analysis of vertical foot trajectory during fastest gait. However, because of the effect of foot stamping in the direction of lateral acceleration, measured acceleration is not as smooth as the acceleration in the direction of gait and analysis accuracy drops. There might be useful information from the frequency of vibration recorded on acceleration sensor when the foot lands on the support surface during gait.

The accuracy of stride length during fastest gait using the present system applying interpolation method was approximately −15 mm, or −0.7%. A study conducted by Kuga et al. [8] estimates walking distance of 135 subjects during 10 m fastest gait using WIMU with interpolation method. The accuracy of the estimated walking distance was reported about −7.0%, indicating that the estimation method in the present method were more accurate than in previous study. The reasons behind this are the number of test subjects and gait cycles were different and might be the reasons of the difference in the result. However, we believe the new sensor system introduced in this study may have been comparatively better suited for foot trajectory estimation because of its small compact size and differences in drift behavior. Furthermore, our experiment was conducted with reference to MCS indicating there is minimal random error compared to the previous study's method and the results are more trustworthy.

Ojeda and Borenstein [5] studied a navigation system using an IMU (triaxial accelerometer and triaxial gyroscope) attached to the lateral side of the foot. They performed an experiment with a subject walked at normal pace of about 1.0 m/s and brisk pace of about 1.8 m/s. The average errors with the gait distance estimation system are 0.3 and 0.8% for normal and fast gait respectively. Our experiment was performed with average walking pace of 3.0 m/s but still maintain similar levels of accuracy during brisk pace gait distance estimation of the previous study.

There is certainly room for improving the precision in reconstructing the foot trajectory, especially of the toe's height in the future study. However, the proposed system definitely successful in estimating the foot trajectory during fastest gait with reasonable accuracy, proving its usefulness

Table 1 Error in estimation of amplitude of toe's angle, toe's height, and stride length for each measuring method

		High-sensitivity	Low-sensitivity	Interpolation method
Angle amplitude	Mean ± SD (°)	−7.4 ± 3.7	−6.4 ± 3.6	−6.9 ± 3.7
Toe's height	Mean ± SD (mm)	5.1 ± 21.3	−9.5 ± 24.3	7.4 ± 20.2
Stride length	Mean ± SD (mm)	−29.9 ± 27.9	−24.3 ± 28.3	−15.14 ± 26.1

to monitor the foot trajectory and applied in identifying diseases related to gait characteristics.

Acknowledgements A part of this study was supported by JSPS KAKENHI Grant Number 25350663. The authors also wish to express their sincere gratitude to all the volunteers who participated in the gait experiment.

References

1. Abbott, R.D., White, L.R., Ross, G.W., Masaki, K.H., Curb, J.D., Petrovitch, H.: Walking and dementia in physically capable elderly men. JAMA **292**(12), 1447–1453 (2004)
2. Klamer, F., et al.: Aging fitness training and neurocognetive. Nature **400**, 418–419 (1999)
3. Verghese, J., Ayers, E., Barzilai, N., Bennett, D.A., Buchman, A.S., Holtzer, R., Katz, M.J., Lipton, R.B., Wang, C.: Motoric cognitive risk syndrome: multicountry prevalence and dementia risk. Neurology **83**(8), 718–726, 19 Aug 2014
4. Yang, S., Li, Q.: Inertial sensor-based methods in walking speed estimation: a systematic review. Sensors (Basel) **12**(5), 6102–6116 (2012)
5. Ojeda, L., Borenstein, J.: Non-GPS navigation for security personnel and first responders. J. Navig. **60**(3), 391–407 (2007)
6. Sagawa, K., Ohkubo, K.: 2D trajectory estimation during free walking using a tiptoe-mounted inertial sensor. J. Biomech **48**(10), 2054–2059 (2015)
7. Kitagawa, N., Ogihara, N.: Estimation of foot trajectory during human walking by a wearable inertial measurement unit mounted to the foot. Gait Posture **45**, 110–114 (2016)
8. Kuga, H., Sagawa, K.: Estimation of walking distance 10 meter fastest walk utilizing tiptoe-mounted inertial sensor. In: Proceedings of International Conference on Mechatronics and Information Technology (ICMIT2015), pp. 201–205. Sokcho, Korea, 02–04 Dec 2015

Changes of the Head Control Ability in Patients with Cervical Spondylotic Myelopathy

Chih-Hsiu Cheng, Wei-Li Hsu, Andy Chien, Jaw-Lin Wang, Dar-Ming Lai, Shwu-Fen Wang, and You-Syuan Yu

Abstract

This study aims to compare the head control ability in terms of the cervicocephalic kinesthetic sensibility between the elderly with and without cervical spondylotic myelopathy (CSM). Thirty CSM patients and thirty healthy elderly were recruited to perform the maximal neck extension as well as the maximal cervical range of motion (ROM) and head reposition tasks in neck flexion, extension, right and left rotation. Three measurements of the reposition accuracy were analyzed, including the constant error (CE), variable error (VE) and root mean square error (RMSE) of the reposition movements. Independent t test was used to investigate the difference of all measurements between the groups. CSM group showed less maximal neck strength and smaller cervical ROM compared with the healthy group with significant difference in cervical extension ($p < 0.001$) and right rotation ($p < 0.05$). During the reposition task, the CE of all movements showed significant difference between groups. The CSM group trended to show greater VE in all directions. The RMSE were greater in CSM group with significant difference in cervical extension and right rotation ($p < 0.05$). This study demonstrated that the CSM patients suffer from poor head control compared with that in the age-matched elderly. Those findings could help to facilitate the rehabilitation program targeting on the treatment of cervical problem in CSM patients.

Keywords

Head control • Spondylotic myelopathy • Elderly

C.-H. Cheng (✉)
School of Physical Therapy and Graduate Institute of Rehabilitation Science, Chang Gung University, Taoyuan, Taiwan
e-mail: chcheng@mail.cgu.edu.tw

C.-H. Cheng
Bone and Joint Research Center, Chang Gung Memorial Hospital, Linkou, Taiwan

W.-L. Hsu · S.-F. Wang
School and Graduate Institute of Physical Therapy College of Medicine, National Taiwan University, Taipei, Taiwan

A. Chien
Department of Physical Therapy and Graduate Institute of Rehabilitation Science, China Medical University, Taichung, Taiwan

J.-L. Wang
Institute of Biomedical Engineering, National Taiwan University, Taipei, Taiwan

D.-M. Lai
Department of Surgery, National Taiwan University Hospital, Taipei, Taiwan

Y.-S. Yu
Biomedical Sciences, Chang Gung University, Taoyuan, Taiwan

F. Ibrahim et al. (eds.), *2nd International Conference for Innovation in Biomedical Engineering and Life Sciences*, IFMBE Proceedings 67, https://doi.org/10.1007/978-981-10-7554-4_14

1 Introduction

Cervical spondylotic myelopathy (CSM) is a pathology of the cervical spinal cord which is caused by degenerative cervical spine. The pathology would result in different clinical manifestations including abnormal sensation, muscle weakness, neck pain or stiffness, decreased range of motion, clumsy hands, deteriorated postural control, and gait disorder [1]. Previous study showed that sensory information ascends through the posterior columns in the spinal cord regarding the proprioception of knee joint is impaired in patients with CSM. The vibration stimulus to the neck would disturb the eye-head coordination thought the cervico-ocular reflex and thus leads to postural control disturbance [2]. However, little is known about the effects of compromised neuromotor integrality on the head control ability in patients with CSM.

Kinesthetic sensibility, defined as the ability to judge the joint position, is important in the coordinated movements of the head, trunk, and extremities [3–5]. The examination of the position sense of the head and neck, also known as the cervicocephalic kinesthetic sensibility test, commonly includes the "head-to-neutral" test during position-matching tasks which has been reported to be a sensitive method to differentiate subjects with and without neck pain [6]. Dysfunction of kinesthetic sensibility characterized by increased movement irregularities [7] and movement errors [8] during reposition tasks has been shown in middle-aged subjects with chronic neck pain. Those characteristics have not been explored in patients with CSM given the evidence that they usually come with serve neck pain.

The purpose of this study was therefore to compare the head control ability in terms of the cervicocephalic kinesthetic sensibility between the healthy elderly and patients with CSM.

2 Method

Thirty CSM patients and thirty healthy elderly aged from fifty-five to seventy-five years old were recruited in this cross-sectional study. All patients were assessed by the same neurosurgeon, and the diagnosis of CSM was confirmed clinically if the sagittal plane T2WI of cervical spine magnetic resonance image showed absence of cerebrospinal fluid signal with or without cord edema, and/or atrophy on the level of pathology with bilateral hand numbness, and/or leg weakness with increased deep tendon reflexes. Subjects were excluded if they have spinal surgery previously, upper neuron disease, tumor, other disease influencing the balance, or are unable to stand more than one minute independently. Asymptomatic subjects were volunteers to this study and were excluded if they had any history of neck pain or neck orthopedic disorders. This study was approved by the institutional medical research ethics committee.

Subjects were asked to sit on a chair with head in the neutral position and hands on the thighs with an upright trunk posture. They performed maximal voluntary isometric contraction (MVIC) for the measurement of the neck muscle strength. It was performed against a fixed surface for 3 s in the posterior directions, that is the neck extension, and the isometric strength were measured by an S-type loadcell (STC-20 kg SE, Vishay, USA). There was a rest period of 2 min between each repetition to minimize the effect of fatigue.

After sufficient rest, the head reposition task was then performed as follows: (1) memorize the neutral head position and actively reach the maximal cervical range of motion (ROM) (2) close eyes to complete one maximal cervical flexion and five consecutive half range of movements in a neutral-target-neutral order with the comfortable speed (3) execute the cervical extension, right and left rotation in a similar manner. The outcome measures were the cervical maximal ROM and three measurements of cervicocephalic kinesthetic sensibility, including the constant error (CE), variable error (VE) and root mean square error (RMSE) of the reposition movements. During the experiment, head motion was recorded by an ultrasound-based motion capture system (CMS 10, Zebris Medical GmbH, Germany) which has been evidenced to be with good reliability and validity for the clinical applications. Independent t test was used to investigate the difference of all measurements between the groups. Significant level was set at $p < 0.05$.

3 Result

There was no significant difference in terms of age effect between groups ($p > 0.05$). CSM group showed less maximal *neck extension* strength *as well as* smaller cervical ROM compared with the healthy group with significant difference in cervical extension ($p < 0.001$) and right rotation ($p < 0.05$) (Table 1).

During the reposition task, the CE of all movements showed significant difference between groups. *The CE were negative (undershoot from the target position) in the CSM group opposite to the positive trend (overshoot) in the healthy group especially in the neck flexion and left side rotation. The CE of the CSM group were also significantly greater than that of the healthy group in the neck extension and right side rotation with similar reposition movements.* The CSM group trended to show greater VE in all directions

Table 1 Basic information of the elderly

	CSM group	Healthy group
Male/Female	20/10	9/21
Age (years)	62.6 ± 6.7	63.2 ± 4.0
Height (cm)	162.0 ± 8.0	159.0 ± 5.6
Weight (kg)	66.7 ± 9.9†	59.9 ± 6.7
Max Neck Ext (kg)	5.0 ± 3.4*	8.1 ± 1.6
Flexion ROM (°)	44.8 ± 10.3	45.3 ± 7.2
Extension ROM (°)	44.5 ± 10.7*	58.8 ± 10.6
Right rotation ROM (°)	55.5 ± 7.7†	62.7 ± 9.3
Left rotation ROM (°)	55.6 ± 6.3	59.0 ± 8.4

$^{\dagger}p < 0.05$, $*p < 0.001$

without significant difference. The RMSE were greater in CSM group with significant difference in cervical extension and right rotation ($p < 0.05$) (Fig. 1).

4 Discussion

This study was to explore if the characteristics of the head control in CSM patients alter compared with those in age-matched elderly. The major results were that: (1) CSM group showed less neck muscle strength and smaller cervical ROM compared with the healthy group; (2) CSM group mostly undershoot when performing the reposition tasks in all movement directions; (3) CSM group trended to show greater movement variability during the repetitive tasks; (4) CSM group showed poor position accuracy compared with the healthy group.

Good head control is the basis to govern the whole body for functional movements. Perception of the orientation of the head in space demands not only the contribution of vestibular and visual cues but also proprioceptive information from the cervical spine [9]. Conscious proprioception is also essential for proper joint function in activities of daily living or work-related tasks. The abnormal physiologic loads on the neck are believed to affect the cervical structures and compromise neck function. In the pathologic course of CSM, atrophy and neuronal loss in the anterior horn and intermediate zone develop, and then focal demyelinating and remyelinating processes occur in the cervical cord [10]. Previous study has shown that information about proprioception ascends through the dorsal columns in the spinal cord [11]. Therefore, the perception of the orientation of the head could be compromised. Besides, given the fact that gait impairment is one of the primary symptoms of CSM, the impairment of head position sense might also contribute to the gait disorder.

Altered vibratory sense and proprioceptive changes are often present in cases with chronic or severe myelopathy. Since the narrowing or compression is degenerative in nature, progression is slow and generates adaptive deficits in both the cord and periphery. Previous studies have revealed the position accuracy of the sagittal head-to-neutral tests in young healthy subjects (18–30 years old) and in middle-aged subjects with neck pain (40–65 years old) [12, 13]. Compared with previous studies, the CSM patients in this study showed great range of CE, VE, and RMSE. The perception of egocentric space in adults includes kinesthetic information from visual input and vestibular afferents, as well as abundant muscle spindles and mechanoreceptors of the facet joint capsules in the neck region. With normal visual and vestibular afferents, the peripheral proprioceptive inputs are relayed to the central nervous system for regulating joint movements through the activation of muscles [14]. The changes of position accuracy in this study may due to the changes in the cervical mechanoreceptor function. Accordingly, the diminished mechanoreceptor function, and the central misinterpretation of the proprioceptive inputs could lead to the loss of cervicocephalic kinesthetic acuity.

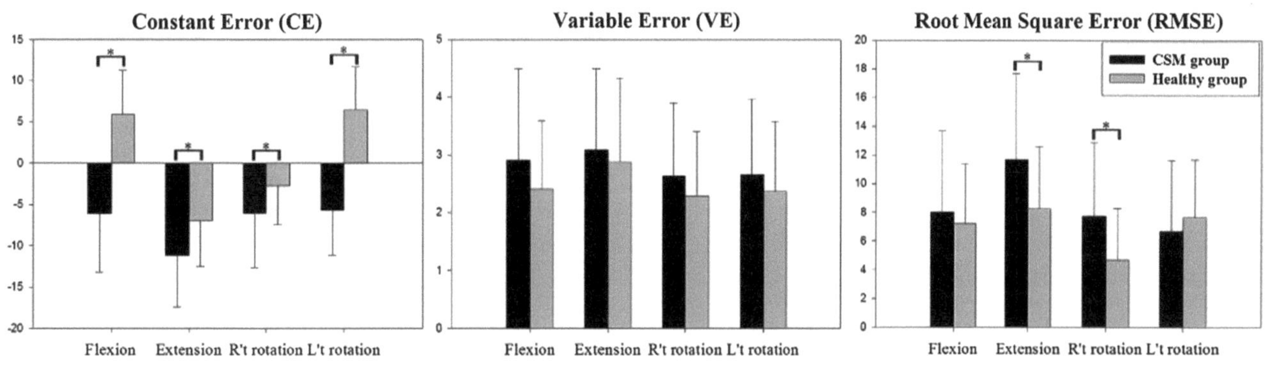

Fig. 1 The constant error, variable error, and root mean square error during the reposition task in flexion, extension, right and left rotation movements between the CSM group and healthy group. Asterisk indicates $p < 0.05$

5 Conclusion

The results showed that the CSM patients suffer from poor head control ability, in terms of neck strength, cervical range of motion, and position sense, compared with that in the age-matched elderly. With generally limited cervical ROM, the CSM patients exhibited problem of undershoot from the neutral position as well as poor repeatability and absolute position accuracy during the reposition tasks than the healthy elderly. Those findings could help to facilitate the rehabilitation program targeting on the treatment of cervical problem in CSM patients.

Acknowledgements This study was supported by research grants from the Ministry of Science and Technology, Taiwan (104-2221-E-182-078-MY3). None of the authors have any personal or institutional financial interest in drugs, materials, or devices described in this manuscript submission.

References

1. Salvi, F.J., Jones, J.C., Weigert, B.J.: The assessment of cervical myelopathy. Spine J **6**, 182S–189S (2006)
2. Andre-Deshays, C., Revel, M., Berthoz, A.: Eye-head coupling in humans. II. Phasic components. Exp. Brain Res. (Experimentelle Hirnforschung) **84**, 359–366 (1991)
3. Bouet, V., Gahery, Y.: Muscular exercise improves knee position sense in humans. Neurosci. Lett. **289**, 143–146 (2000)
4. Rix, G.D., Bagust, J.: Cervicocephalic kinesthetic sensibility in patients with chronic, nontraumatic cervical spine pain. Arch. Phys. Med. Rehabil. **82**, 911–919 (2001)
5. Rowe, M.J., Tracey, D.J., Mahns, D.A., et al.: Mechanosensory perception: are there contributions from bone-associated receptors? Clin. Exp. Pharmacol. Physiol. **32**, 100–108 (2005)
6. Kristjansson, E., Dall'Alba, P., Jull, G.: A study of five cervicocephalic relocation tests in three different subject groups. Clinical Rehabil. **17**, 768–774 (2003)
7. Sjolander, P., Michaelson, P., Jaric, S., et al.: Sensorimotor disturbances in chronic neck pain–range of motion, peak velocity, smoothness of movement, and repositioning acuity. Manual Ther. **13**, 122–131 (2008)
8. Lee, H.Y., Wang, J.D., Yao, G., et al.: Association between cervicocephalic kinesthetic sensibility and frequency of subclinical neck pain. Manual Therapy (2007)
9. Falla, D., Rainoldi, A., Merletti, R., Jull, G.: Myoelectric manifestations of sternocleidomastoid and anterior scalene muscle fatigue in chronic neck pain patients. Clin. Neurophysiol. **114**, 488–495 (2003)
10. Granata, K.P., Marras, W.S.: Cost-benefit of muscle cocontraction in protecting against spinal instability. Spine **25**, 1398–1404 (2000)
11. Revel, M., Minguet, M., Gregoy, P., et al.: Changes in cervicocephalic kinesthesia after a proprioceptive rehabilitation program in patients with neck pain: a randomized controlled study. Arch. Phys. Med. Rehabil. **75**, 895–899 (1994)
12. Lee, H.Y., Teng, C.C., Chai, H.M., et al.: Test-retest reliability of cervicocephalic kinesthetic sensibility in three cardinal planes. Manual Therapy **11**, 61–68 (2006)
13. Teng, C.C., Chai, H., Lai, D.M., et al.: Cervicocephalic kinesthetic sensibility in young and middle-aged adults with or without a history of mild neck pain. Manual Therapy **12**, 22–28 (2007)
14. Fasold, O., Heinau, J., Trenner, M.U., et al.: Proprioceptive head posture-related processing in human polysensory cortical areas. NeuroImage **40**, 1232–1242 (2008)

Investigation of the "Pendulum Effect" During Gait Locomotion Under the Novel Body Weight Support System and Counter Weight System

Tran Van Thuc and Shin-ichiroh Yamamoto

Abstract

In recent years, body weight support (BWS) systems have been applied and have played an important role in most of the gait training systems. Conventional BWS systems commonly use the rope–pulley mechanism such that the "pendulum effect" generated during gait training would cause discomfort to the subject. Our hypothesis was that the center of mass (COM) parameters in the lateral and vertical directions of the subject during walking under a weight support system would be modified. Moreover, the novel BWS system that was developed would reduce the "pendulum effect" during walking and exhibit an enhanced behavior in comparison with the counter weight system. Our findings revealed that the step width and the COM amplitude in the mediolateral and vertical directions in the cases using the new BWS and counter weight systems were significantly smaller than those for the normal walking case. The COM amplitudes obtained when using the new BWS were closer to the normal walking case and higher than the counter weight system.

Keywords

Body weight support system • COM amplitude • Step width

1 Introduction

In recent years, a body weight support (BWS) system has played an important role in gait training for patients recovering from a spinal cord injury or stroke. The use of a BWS system has been noticed in most of the gait training therapies, such as treadmill and ground walking. In the early stage of the therapy, the spinal cord injury or stroke patients are unable to carry their entire body weight because of their weak legs. Therefore, a BWS system is necessary to reduce most of the load on their lower limbs. During the gait training period, the sensory afferent nerves in the lower limbs are activated and conducted inward to the central nervous system, and a therapist could gradually reduce the unloading weight force support provided to the patient. In clinical trials, BWS systems showed enhanced effects in comparison to conventional therapies [1–3]. Currently, the development of a gait training system is usually combined with a BWS system in case of robotic leg orthosis; this combination would improve the effect of physiotherapy and reduce the labor cost [4, 5].

A BWS system commonly consists of three main parts: a main frame, an actuator, and a harness. The actuator is the core of the development of the BWS system because the actuator is the origin of the generated unloading force. Depending on the type of apparatus, one would consider three types of BWS systems, namely, the static system, passive system [6, 7], and active system [8–10]. The development of the actuator for the BWS systems has focused on maintaining the unloading force as a constant and easy adjustment of the generated force [7–10]. However, some actuators, such as a series of springs [7, 8], found it difficult to maintain a constant unloading force, while some other systems were bulky and complex such as a pneumatic cylinder with a cart [9], the mechatronic Lokolift BWS system [10], or a partial BWS system [11]. Moreover, most

T. Van Thuc (✉) · S. Yamamoto
Shibaura Institute of Technology, 3-5-7 Toyosu, Tokyo, Japan
e-mail: Nb13507@shibaura-it.ac.jp

F. Ibrahim et al. (eds.), *2nd International Conference for Innovation in Biomedical Engineering and Life Sciences*,
IFMBE Proceedings 67, https://doi.org/10.1007/978-981-10-7554-4_15

of the current BWS systems include a pulley structure that could generate a swing trunk effect similar to a “pendulum effect” when the subject walks [7–11].

In a previous study, a novel BWS system applying four pneumatic artificial muscles (PAMs) for body weight unloading, has been developed for gait training [12]. The new system had various advantages in comparison with conventional systems such as simplicity, low cost, and flexible adjustment of the unloading force. Furthermore, the new BWS system could implement a completely new strategy for gait training such that the generated unloading force could follow the movement of the center of pressure (COP). Recently, Dragunas et al. [13] have quantified the impacts of a BWS system on the stability during walking based on the step width. Aaslund et al. [14] considered the effect of a BWS system by assessing the acceleration of the center of mass (COM). The “pendulum effect” caused by a BWS system during walking, as mentioned above, however, has not been fully investigated. In this study, we investigated the effect of weight systems on the gait parameters when the subject walked on a treadmill using the new BWS system [12] and counter weight system. We hypothesized that the novel BWS system using PAMs could treat the pendulum effect by applying the new unloading force strategy to achieve an improved gait during locomotion.

2 Method

2.1 Experimental Protocol

Nine healthy subjects (all subjects were male of age = 24.2 ± 3.2 years (mean ± SD), height = 172.1 ± 6.1 cm, and weight = 61.5 ± 6.51 kg) with no prior disability participated in this experiment to validate the system. For each subject, we performed experiments with different unloading forces, mainly 30, 50, and 70%, using both the new BWS and counter weight systems, and compared them with normal walking. Each subject walked on the treadmill at a speed of 1 km/h for 60 s. The procedure was iterated for every trial as well as for all unloading force levels. The data collected from the four force sensors embedded under the treadmill were used to calculate the COP parameters. A motion capture system with six cameras was also used to collect data from 12 markers positioned at the ends of the bones of the lower limbs, pelvis, and shoulder of the subject. The data collected from the motion capture system were used to calculate the COM motion. The COM was considered extremely carefully for this research project because with the new BWS system or counter weight system, the COM pattern was significantly affected. The time series data collected from the sensors were difficult to use for analyzing the differences between the data for normal walking, the BWS system, and counter weight system. Therefore, to analyze the differences between the new BWS system and counter weight system for all the nine subjects, the gait parameters such as the COP amplitude in the mediolateral direction (step width) and COM amplitude (in both the mediolateral and vertical directions) were quantified by using two MATLAB programs.

2.2 Data Analysis

Two MATLAB routines have been developed to analyze the gait parameters such as step width (COP amplitude in the mediolateral direction) and COM amplitude in both the mediolateral and vertical directions. The first routine was to calculate the COP trajectory in one gait cycle such that we could calculate the step width. The key point to partition the COP data to single gait cycles was to determine the event that started the gait cycle (heel strike was considered). Figure 1 shows the analysis of the COP motion when the right heel strikes during gaiting. The moment when the COP began to switch from the left to the right side was defined as the right heel strike (also called as the heel-strike detection). The step width that was used as a parameter for quantifying the COP parameter, was defined as the distance between the two edges (left and right) of the COP trajectory. In the second routine of the calculation of the COM in one gait cycle, the event to detect the starting point of a new gait cycle was defined by the marker on the ankle joint going down. From that event (the coordination of the marker on the ankle joint moved to the lowest position), the time series

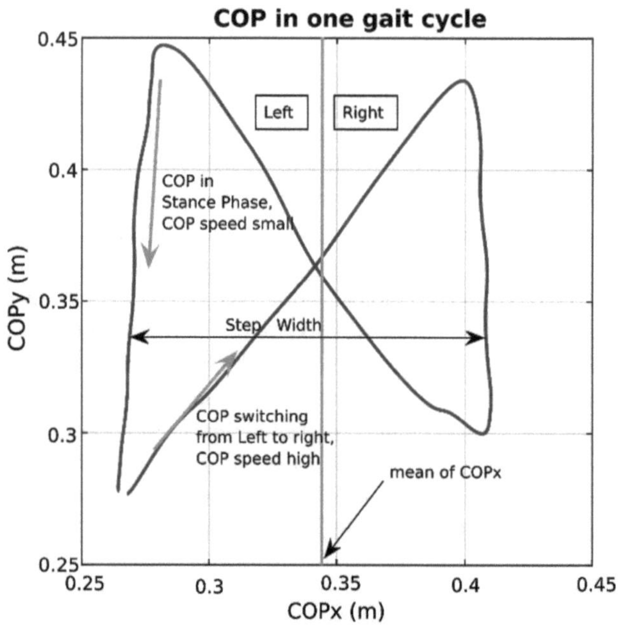

Fig. 1 COP in one gait cycle and definition of step width

data of the COM was partitioned to single gait cycles. The COM amplitude parameters in the mediolateral and vertical directions were quantified to represent the effect of the BWS system and counter weight system.

The data of the COP and COM amplitude parameters were standardized for all the nine subjects. The standardized procedure for the experimental data of each subject was conducted by using the formula provided below.

$$Standardized\,value = \frac{Parameter\,value\ (Normal, BWS, and\ CWS)}{Mean\,value\ (Normal)} \quad (1)$$

The standardized data were used to analyze the differences between the BWS sys-tem, counter weight system, and normal walking using one-way analysis of variance (ANOVA). The Tukey HSD test was used to investigate the difference between the BWS and counter weight systems. The procedure for the analysis was conducted based on the R statistical language using R Studio.

3 Result and Discussion

The step width (COP amplitude in the mediolateral direction) was considered to quantify the COP in the lateral direction. The data of the step width of all the nine subjects that were standardized to the normal walking case, were used to investigate the effect of the BWS and counter weight systems. Table 1 presents the data of the standardized step width in case of normal walking, the BWS system, and the counter weight system in which the unloading forces were equal to 30, 50, and 70% of the subject weight. Significant differences were found in the step widths for normal walking, the BWS system, and the counter weight system ($p < 0.001$). The mean values of the step widths of the BWS and counter weight systems were smaller than those for the normal walking case. A major difference in the step width for the BWS system and counter weight system was not found.

The differences in the amplitudes of the COMy (in the mediolateral direction) trajectories were considered. The standardized COMy data were used for analyzing the differences in the COMy amplitudes for both the BWS system and counter weight system using ANOVA. Figure 2 illustrates the comparison of the mean values of the BWS system and counter weight system at different weight support levels. We notice that increasing the weight support level reduces the amplitude of COMy during walking ($p < 0.001$). Significant differences are found between the mean values of the normal walking case and both the BWS and counter weight systems ($p < 0.001$). These results strongly confirm that the COM in the mediolateral direction is modified by using the unloading systems. However, we observe that the mean values obtained by using the BWS system were significantly higher than those obtained using the counter weight system ($p < 0.001$ at 30% weight support and $p < 0.05$ at 50% weight support). A noteworthy difference is not found between the mean values at 70% weight support ($p = 0.927$). The relative closeness of the mean values obtained using the BWS system with those of the normal walking case indicates that the new system exhibits an enhanced behavior in comparison with the counter weight system. The higher mean value of COMy when applying the BWS system compared with the counter weight system could be explained in terms of the unloading force that is applied to the trunk of the subject. Figure 3 illustrates the unloading force that is applied to the trunk of the subject in the counter weight system (subfigure A) and BWS system (subfigure B).

Table 1 Comparison of the step widths (of all the nine subjects) of the BWS system, the counter weight system, and normal walking by one-way ANOVA

Weight support	Method	Sample size	Mean	SD	95% CI
30%	NormalStd	139	1.0000[a]	0.0753	(0.9756, 1.0244)
	BwsStd	139	0.7971	0.1726	(0.7727, 0.8215)
	CwsStd	139	0.8012	0.1698	(0.7768, 0.8256)
50%	NormalStd	132	1.0000[a]	0.1131	(0.9749, 1.0251)
	BwsStd	132	0.7806	0.151	(0.7555, 0.8057)
	CwsStd	132	0.7567	0.1697	(0.7316, 0.7817)
70%	NormalStd	138	1.0000[a]	0.0695	(0.9644, 1.0356)
	BwsStd	138	0.9334	0.2223	(0.8979, 0.9690)
	CwsStd	138	0.9109	0.2852	(0.8753, 0.9464)

NormalStd, BwsStd, and CwsStd are the data of the step width standardized for normal walking, the BWS system, and the counter weight system, respectively. The significance level was set at 0.05. [a]represents the significant difference $p < 0.001$

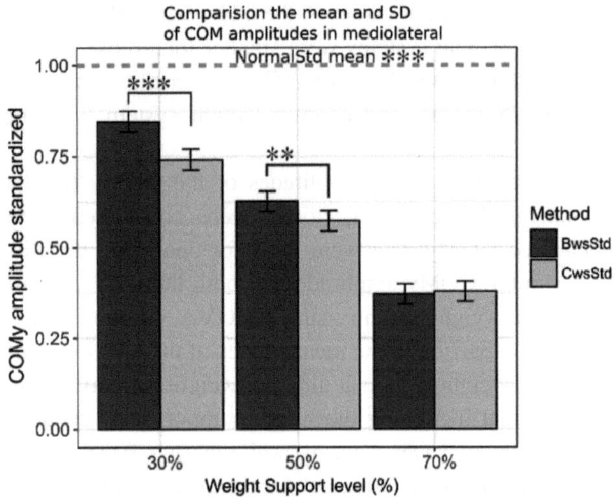

Fig. 2 Comparison of the standardized COMy (in the mediolateral direction) amplitudes by the BWS system and counter weight system at 30, 50, and 70% weight support. *** represents significant value $p < 0.001$, ** represents significant value $p < 0.05$

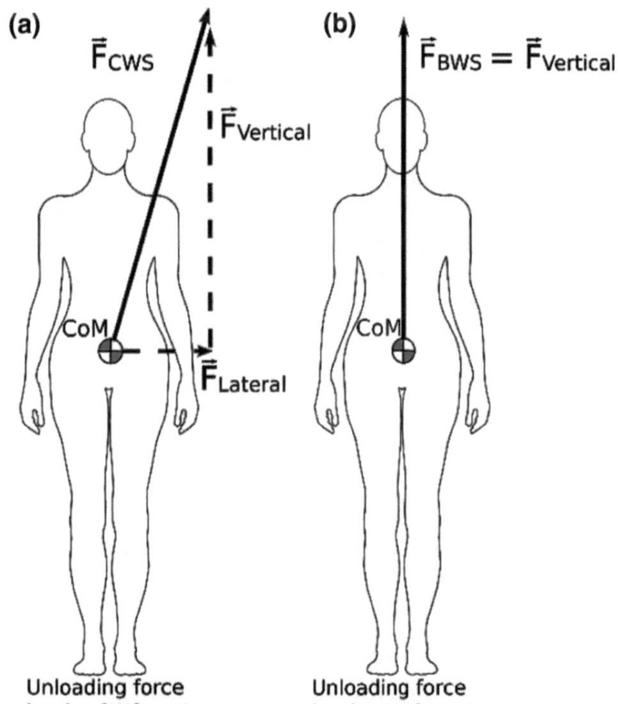

Fig. 3 Illustration of the unloading force applied to the trunk of the subject during walking under the weight support systems

In the counter weight system, the lateral part of the unloading force tends to prevent the motion of the COM during walking and pulls the COM in the mediolateral direction toward the center axis. This result is similar to the "pendulum effect" because in the pendulum mechanism, there always exists a lateral force. This "pendulum effect" could cause discomfort to a subject during walking and modify the gait parameters of the subject. However, in the BWS system, the lateral part of the unloading force would be small because the unloading force in this case always tends to follow the motion of the COP during walking, such that the effect of the lateral unloading force is reduced.

Figure 4 presents a comparison of the mean values the COM in the vertical direction (COMz) for both the BWS and counter weight systems. We can notice that the mean values of the COMz amplitudes of both the BWS and counter weight systems are smaller than those for the normal walking case. Increasing the weight support levels in the BWS system slightly decreases the COMz amplitude. A significant difference in the mean values is observed only for the 30 and 70% weight support cases ($p < 0.001$). There is no major difference in the pair of 30 and 50% weight supports ($p = 0.105$) and the pair of 50 and 70% weight supports ($p = 0.169$). With the counter weight system, a drastic difference in the means is not be found between the cases of 30, 50, and 70% weight supports. However, one can observe that for the counter weight system, at a high level of weight support the variance of the amplitude is extremely higher than the variances at a lower weight support level and at all weight support levels for the BWS system ($\sim$0.41 vs. 0.18). The higher variance of the COMz amplitude in the case of the counter weight system at a high level of weight support is representative of the additional oscillation of the COM in the vertical direction to the normal COMz. Furthermore, this extra oscillation would lead to a stronger influence of "the pendulum effect" on the COM gait parameters. Significant differences in the mean values are found between the normal walking case and both the BWS

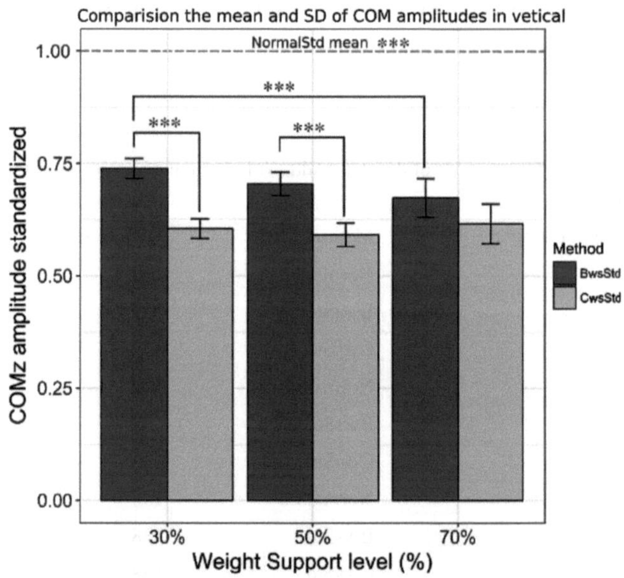

Fig. 4 Comparison of the standardized COMz (in the vertical direction) amplitudes by using the BWS and counter weight systems at 30, 50, and 70% weight support by using ANOVA. *** represents significant value $p < 0.001$

and counter weight systems ($p < 0.001$). This result confirms that the COMz amplitude is modified when subject uses the unloading system. Furthermore, note-worthy differences are noted in the cases using the BWS and counter weight systems at low- and mid-level unloading forces (30 and 50% weight support) ($p < 0.001$). This result also demonstrates the improved behavior of the new BWS system com-pared with the counter weight system because the COM amplitude of the BWS sys-tem is closer to the normal walking case. At a high level of the unloading force (70% weight support), the difference of COMz amplitudes in the cases using the BWS and counter weight systems is not found ($p = 0.164$). This result could be explained that in terms of the strong “pendulum effect” that influences the locomotion gait, so that the motion of the “virtual position” of the unloading force could not compensate the strong deformation of the COM motion.

4 Conclusion

In this study, the “pendulum effect” of a new BWS system and the counter weight system was investigated. The results agreed with a previous study, exhibiting that the gait parameters of the subject were modified by the weight support system [13, 14]. In the clinical trials, the gait parameters such as the step width and COM amplitude in the mediolateral and vertical directions when using the BWS and counter weight system were smaller in comparison with the normal walking case. The stronger reduction of the COM amplitude could be because of the stronger influence of the “pendulum effect” of conventional BWS systems that apply the rope–pulley mechanism. The improved behavior of the novel BWS system could be observed because the corresponding gait parameters were more similar to the normal walking case than to those of the counter weight system.

Acknowledgements This work was supported by JSPS KAKENHI Grant Number JP25282161.

References

1. Finch, L., Barbeau, H., Arsenault, B.: Influence of body weight support on normal human gait: development of a gait retraining strategy. Phys. Ther. **71**(11), 842–855, discussion 855–856 (1991)
2. Lovely, R.G., Gregor, R.J., Roy, R.R., Edgerton, V.R.: Effects of training on the recovery of full-weight-bearing stepping in the adult spinal cat. Exp. Neurol. **92**(2), 421–435 (1986)
3. Visintin, M., Barbeau, H., Korner-Bitensky, N., Mayo, N.E.: A new approach to retrain gait in stroke patients through body weight support and treadmill stimulation. Stroke **29**(6), 1122–1128 (1998)
4. Colombo, G., Joerg, M., Schreier, R., Dietz, V.: Treadmill training of paraplegic patients using a robotic orthosis. J. Rehabil. Res. Dev. **37**(6), 693–700 (2000)
5. Hornby, T.G., Zemon, D.H., Campbell, D.: Robotic-assisted, body-weight-supported treadmill training in individuals following motor incomplete spinal cord injury. Phys. Ther. **85**(1), 52–66 (2005)
6. He, J.P., Kram, R., McMahon, T.A.: Mechanics of running under simulated low gravity. J. Appl. Physiol. (1985). **71**(3), 863–870 (1991)
7. Kram, R., Domingo, A., Ferris, D.P.: Effect of reduced gravity on the preferred walk-run transition speed. J. Exp. Biol. **200**(Pt 4), 821–826 (1997)
8. Franz, J.R., Riley, P.O., Dicharry, J., Allaire, P.E., Kerrigan, D.C.: Gait synchronized force modulation during the stance period of one limb achieved by an active partial body weight support system. J. Biomech. **41**(15), 3116–3120 (2008)
9. Gazzani, F., Fadda, A., Torre, M., Macellari, V.: WARD: a pneumatic system for body weight relief in gait rehabilitation. IEEE Transactions on Rehabilitation Engineering **8**(4), 506–513 (2000)
10. Frey, M., Colombo, G., Vaglio, M., Bucher, R., Jorg, M., Riener, R.: A novel mechatronic body weight support system. IEEE Trans. Neural Syst. Rehabil. Eng. **14**(3), 311–321 (2006)
11. Lu, Q., Liang, J., Qiao, B., Ma, O.: A new active body weight support system capable of virtually offloading partial body mass. IEEE/ASME Trans. Mechatron. **18**(1), 11–20 (2013)
12. Thuc, T.V.: Yamamoto, S.-i.: development of a body weight support system using pneumatic muscle actuators: Controlling and validation. Adv. Mech. Eng. **8**(12), 1687814016683598 (2016)
13. Dragunas, A.C., Gordon, K.E.: Body weight support impacts lateral stability during treadmill walking. J. Biomech. **49**(13), 2662–2668 (2016)
14. Aaslund, M.K., Moe-Nilssen, R.: Treadmill walking with body weight support effect of treadmill, harness, and body weight support systems. Gait Posture **28**(2), 303–308 (2008)

Balance Ability Affects Dual-Task Strategy in Children with Attention Deficit Hyperactivity Disorder

Wen-Lan Wu, Jia-Hroung Wu, Ying-Yi Chen, Chia-Hsin Chen, Chih-Chung Wang, and Jing-Min Liang

Abstract

This study tested the dual-task strategy adopted by children with attention deficit hyperactivity disorder (ADHD) and determined how individual balancing ability affects dual-task performance. Nineteen children with ADHD (ADHD group) and 17 typically developing children (control group) were compared while they performed two concurrent (motor–cognitive) tasks. The motor task entailed walking on a balance beam, and the cognitive task entailed repeating 3 numbers, heard as they walked, in reverse order. Performance in the dual task was compared with that in each separate task. The children with ADHD demonstrated significantly poorer step accuracy and response accuracy compared with the controls. The control group had a negative modified attention allocation index (mAAI), whereas the ADHD group had a positive mAAI. In addition, significant relationships were determined between balance skill and the dual-task effect (DTE) on sway speed, step accuracy, and response accuracy in the ADHD group. The results indicated that, in contrast to the controls, the ADHD group assigned priority to maintaining their walking pattern while abandoning the memory task. The ADHD group's individual balance dysfunctions were associated with dual-task gait abnormalities. The higher the individual balancing ability was, the smaller the dual-task gait disturbances.

Keywords

Attention-deficit/hyperactivity disorder • Dual task • Balance • Working memory

1 Introduction

Attention deficit hyperactivity disorder (ADHD) is one of the most common childhood onset psychiatric disorders that affects 3–5% of school age children [1]. Various studies have shown that children with ADHD perform worse than healthy developed children in tasks of gross and fine motor control [2–8], balance skills [9–15], movement planning and execution, as well as in adapting their task performance to environmental requirements [16–19]. For these between-group comparisons, the average effect size for partial Eta-squared η^2 was 0.43 ± 0.17 (ranged from 0.02 to 0.66).

The control of balance and motor coordination involves the vestibular, visual, and somatosensory systems which jointly detect the orientation of both the head and body [20]. As sensory information decreases, people need to make more effort in allocating attention to balance [21, 22]. It has been reported that sensory integration and/or inhibition of excessive movement [23] are impaired in children with ADHD; hence, they have balance dysfunctions. Balance dysfunctions that exist in children with ADHD have the potential to further affect their gait [5, 6, 9, 12]. Although it has been verified that ADHD will affect stride time

W.-L. Wu (✉) · Y.-Y. Chen · J.-M. Liang
Kaohsiung Medical University, Kaohsiung, Taiwan
e-mail: wenlanwu@kmu.edu.tw

J.-H. Wu
Hsiuping University of Science and Technology, Taichung, Taiwan

C.-H. Chen · C.-C. Wang
Kaohsiung Medical University Hospital, Kaohsiung, Taiwan

F. Ibrahim et al. (eds.), *2nd International Conference for Innovation in Biomedical Engineering and Life Sciences*,
IFMBE Proceedings 67, https://doi.org/10.1007/978-981-10-7554-4_16

variability [5], perception of gait speed [6], body coordination [9], and balance maintain [12] during walking, the actual correlation between dysfunctions in balance control and gait is still unknown.

Performing dual tasks simultaneously is known to affect gait performance. For instance, balance skills are impaired when participants are engaged with concurrent mental tasks [10, 24, 25]. These concurrent mental tasks limit the attention allocated to the posture and gait control resulting in poorer performance [10]. Therefore, aside from congenital balance deficits, reduced attention to the walking task as a result of poor attention allocation may be a reason why children with ADHD are more prone to falls [26].

However, dual-tasks may have an especially detrimental effect in those with attention deficits. Problems with attention are core symptoms for children with ADHD, and a high proportion (approximately 47%) of children with ADHD have difficulties with gross motor skills [27]. This complicates the study of the role of the cognitive domain in gait. Previous studies of children with ADHD have shown that they become more cautious under dual task conditions, for example, by switching to a more secure strategy to control their body sway and walking more rhythmically under dual-task walking conditions [5]. They will also exhibit a rigid postural control strategy by decreasing their center of pressure sway amplitude and area under upright standing dual task conditions [28], and by decreasing the sway path with increasing time for concurrent mental tasks [29] during standing. In addition, Chen et al's study has found that children with ADHD demonstrated an increased step width in the obstacle crossing dual task condition [10]. Based on the above mentioned various response strategies observed in dual tasks, previous ADHD research has shown that the children's attention allocation depended greatly on demand setting and what type of secondary task they were performing [30]. We suspected that a balance beam walking task would interfere with a cognitive task much more than a level walking task would. If attention allocation by children with ADHD depended majorly on demand for safety, we suspected that children with ADHD would maintain their walking pattern by sacrificing the memory task during balance beam walking. Thus, the aims of this study were to examine which priority strategy is adopted by the ADHD group when processing this difficult dual-task. In addition, we tested how individual balance ability affects dual task performance. The research hypotheses were: (1) the dual-task strategy adopted by children with ADHD differs significantly from that of normal children, the first priority for children with ADHD walking on the balance beam is to maintain walking pattern by sacrificing the memory task and (2) there are interactions between dual task performances and individual balancing ability in children with ADHD, who have poorer balance ability would demonstrate poorer dual task performance.

2 Method

2.1 Participants

Fourteen boys and five girls (ADHD group), with a mean age of 9 years and 7 months (SD = 1.61 year), diagnosed with ADHD by doctors in Child and Adolescent Psychiatry Department at the local hospital or university hospital and without other combined syndromes, such as autism participated in this study. Ten ADHD subjects were treated with medicines for symptom control. However, all these subjects took drug holidays during the study. In addition, 17 typically developing children (Control group, 11 boys, 6 girls, with a mean age of 10 years and 6 months, SD = 1.13 year), without ADHD symptoms or any other neuromuscular symptoms, were recruited from a local school to participate in the study. The cognitive ability was evaluated by Raven's Coloured Progressive Matrices (CPM) test, the raw scores of which were converted to age-standardised percentiles and the percentiles transformed to IQs. There were no significant differences in anthropometric data and mean estimated IQ for two groups. Participation in this study was anonymous and participants knew that they could withdraw at any time and there was no compensation for participation. The study was approved by the Institutional Review Board of Kaohsiung Medical University Chung-Ho Memorial Hospital, and the guardians on behalf of the children gave informed consent in writing before children participated in the study.

2.2 Apparatus

To assess active balance, the balance beam (2 m × 10 cm) was placed at a vertical height of 45 cm from a stable surface. A six-camera motion capture system (Qualisys AB, Sweden) was set up along the balance beam to measure the three-dimensional position of reflective markers attached to the participant, using a sampling frequency of 100 Hz. Lightweight retro-reflective markers were attached to the skin of the participants over the following bony landmarks: bilaterally over the anterior superior iliac spine (ASIS), the sacrum, and each side of the lateral mid-thigh, the lateral femoral condyle of the thigh, the lateral malleolus, the calcaneus, the fifth metatarsal head, the acromion, the lateral epicondyle of the humerus, and the radial and the ulnar styloid processes.

2.3 Procedures

There were two different test conditions: single task (balanced beam walking or mental task), and dual task (balanced beam walking plus mental task). Each participant performed 3 trials under each condition in a random order. For the single task condition of balanced beam walking, participants were required to walk barefoot heel-to-toe on a 2-m balance beam with their hands on their waists. In each trial they had to walk from one end of the balance beam to the other at their fastest speed. For the single task condition of the mental task, participants had to sit on a chair and repeat in reverse order 3 numbers they heard as soon as possible (16 triads of numbers were used). For the dual task condition, participants had to perform the two tasks mentioned above simultaneously; that is, they had to repeat the numbers while walking back and forth on the balance beam at their fastest speed with no rest in between each trial. In this study, all participants were instructed to walk at their fastest speed in all conditions. The interval between two mental tasks was 3.5 s. The numbers were edited using STIM^2 software (Compumedics Neurosacn, USA) and broadcast continuously by an external speaker. Prior to the formal test, participants were familiarized with balance beam walking task by presenting them a simple association task to ensure that they would be able to maintain their equilibrium to avoid falling during balance beam walking.

2.4 Measures

Balance was measured with the Qualisys Track Manager Software (Qualisys AB, Sweden). All marker data were low-pass filtered using a Butterworth filter with a cut-off frequency of 6 Hz, and interpolated with a maximum gap fill of 10 frames using a 3rd degree polynomial. An 11-segment, full-body model (forearms, upper arms, head + trunk, thighs, shanks, feet) was implemented in Visual3D, and the three-dimensional position data of the center of mass (COM) were calculated within Visual3D. We defined the anterior-posterior direction as the X and the direction of medio-lateral as the Y of the global coordinate system. Gait speed (GS) was calculated by dividing anterior-posterior displacement of COM over time. Sway speed (SS) was calculated by dividing medio-lateral displacement of COM over time. Step accuracy (SA) was calculated by dividing the number of accurate steps by the total number of steps taken. An accurate step was defined as the step length within $\pm 15\%$ of that measured under static posture for each participant. Step length was the distance between the lateral malleolus of the preceding foot at the heel strike and the lateral malleolus of the other foot. Response accuracy (RA) regarding the mental tasks was recorded by the experimenter and calculated by dividing the number of accurate responses by the total number of responses made. Two retro-reflective markers were placed at the top of the balance beam 0.5 m from the edge of each side as our target distance. Five steps were picked out to calculate gait speed, sway speed, and step length when walking the target distance. If the participant clearly did not walk heel-to-toe or released his/her arms from the waist, then the trial would be identified as a failed trial. Only the successful trails were recorded. Data of any participant with more than three failed trials would be excluded from the analysis.

2.5 Data Analyses

The dual task effect (DTE) which represents a relative measure of change was calculated as follows:

$$\frac{\text{Dual task} - \text{Single task}}{\text{Single task}} \times 100\%$$

Decrease in step and response accuracy in the dual task condition indicated a decrement in performance, while increase represented improved performance. An increase in sway speed indicated a decrement in performance and vice versa. In addition, the modified attention allocation index (mAAI) was employed to measure the trade-offs within tasks [31]. The mAAI for "walking focus" was calculated as gait speed DTE minus response accuracy DTE. A positive value represented a shift toward the instructions given to the participants regarding their tasks, while a negative value indicated a shift away from the instruction [32].

2.6 Statistical Analysis

Independent two sample t-tests was used in this study to compare the following parameters: the anthropometric data, mean estimated IQ, CRS-R score, and the gait parameters (SS DTE, SA DTE, RA DTE, and mAAI) between the two groups. This study also used a 2×2 mixed-design ANOVA with a fixed factor of group (ADHD group and control group), and a repeated factor of task (single task and dual task) to determine the effect of two independent variables, group and task on three gait parameters (GS, SS, SA) and one cognitive (RA) dependent variable. All analyses were done with SPSS 19. Significance level was set at $p < 0.05$.

Table 1 The gait speed (GS), sway speed (SS), step accuracy (SA), response accuracy (RA), and walking focus mAAI in the ADHD and control groups

	ADHD		Control	
	Single task	Dual task	Single task	Dual task
GS (mm/s)	112.60 (39.22)	109.98 (33.18)	123.94 (23.28)	102.57 (18.23)
SS (mm/s)	16.49 (5.11)	17.23 (5.39)	17.34 (5.08)	14.97 (5.32)
SA (%)[b]	80.11 (15.13)	72.21 (24.27)	94.71 (8.27)	94.18 (6.49)
mAAI (%)[a]	–	16.21 (45.88)	–	−14.46 (26.36)
RA (%)[b]	65.16 (24.11)	58.89 (25.67)	81.41 (20.07)	78.59 (21.25)

[a]Denotes p value <0.05 for a statistical difference between the groups
[b]Denotes p value <0.01 for a statistical difference between the groups

Table 2 The dual task effect (DTE) for sway speed (SS), step accuracy (SA) and response accuracy (RA) in both groups

(%)	ADHD	Control
SS DTE[a]	6.17 (23.90)	−13.42 (17.85)
SA DTE	−8.44 (31.32)	0.47 (8.98)
RA DTE	−13.91 (27.41)	−1.84 (23.95)

[a]Denotes p value <0.05 for a statistical difference between the groups

3 Results

The results of the 2 × 2 mixed-design ANOVA (Table 1) showed a statistically significant main effect for the group variable in the SA (F1, 34 = 33.65, $p < 0.01$, $\eta_p^2 = 0.50$) and RA (F1, 34 = 8.19, $p < 0.01$, $\eta_p^2 = 0.19$). The differences indicated a lower SA and RA for the ADHD group compared with the control group. No statistically significant interaction effect was found for the group and task and no significant difference was found between the two task conditions in any parameters. In addition, the walking focus mAAI was −14.46% in the control group, indicating a shift away from instruction, while it was 16.21% in the ADHD group, representing a shift toward our instruction (walk as fast as possible). The difference between the mAAI means is statistically significant (t = 2.42, $p = 0.02$, Cohen's d = 0.82).

Table 2 presents the dual task effect (DTE) in both groups. A significant difference was found for SS DTE (t = 2.76, $p = 0.01$, Cohen's d = 0.93) between the two groups. There were no significant differences between the groups in SA DTE, and RA DTE (t = 1.13, $p = 0.25$, Cohen's d = 0.39; t = 1.40, $p = 0.17$, Cohen's d = 0.47, respectively). A negative value of SS DTE was found in the control group and a positive value of SS DTE was found in the ADHD group. This indicates that the control group had a decrement in SS when performing the dual task. The ADHD group, however, had an increment in SS when performing the dual task.

4 Discussion

In the present study, lower step accuracy for the ADHD group compared with the control group was found. This finding was consistent with previous reports that children with ADHD tended to walk without rhythmicity by increasing the stride time variability [5] and express a more impulsive behavior [6] compared to the controls. Further analysis revealed that children with ADHD exhibited poor response accuracy relative to the control group. The supporting evidence from previous literature showed that, under a working memory task, auditory stimuli alone may induce the largest magnitude attention loss in children with ADHD [33] and their poor adaptation of behavior to environmental demands [18] worsens their performance, resulting in poor response accuracy in the ADHD group compared to normal children. However, an unexpected result was that the present study did not find any difference between the 2 conditions (single and dual task). This may be because a simple storage short-term memory task, backward digit span task, used in this study cannot make a significant impact on postural stability when walking over a balance beam. At the same time, response accuracy under dual-task condition is only slightly poorer than that at the single task condition that do not reach the level of statistical significance.

The mAAI values in Table 1 suggest that most of the control group decreased their gait speed to shift attention to the cognitive task with a negative mAAI value, while the ADHD group still maintained their gait speed with a positive mAAI value. In other words, gait performance during the dual task was apparently influenced by attention demands among the control participants. However, due to the attention deficits, the ADHD group ignored the distraction of the cognitive stimuli, leading to continued focus on walking performance. Our observations in this study indicate that the control group would pause or slow down gait speed to respond to the cognitive tasks. This pattern was absent in the ADHD group which displayed similar gait speed in both single and dual task conditions. Therefore, it is plausible to infer that the ADHD group did not change their motor behavior when a cognitive task was presented; therefore, they could still maintain their "walking focus". The result supports our first hypothesis that children with ADHD will maintain their walking pattern by sacrificing the memory task. However, it is interesting that their gait pattern was still disturbed as evidenced by a significantly greater sway speed (Table 2).

As noted in the introduction, in this study of children with ADHD, we tested two hypotheses. The results support the two hypotheses. Key findings of the present study include: (1) the dual-task strategy adopted by children with ADHD differed significantly from that of normal children, the ADHD group gave prioprity to maintain their walking pattern, on the contrary abandon the memory task; (2) there were interactions between dual task performances and individual balancing ability in children with ADHD, a better individual balancing ability, the smaller dual-task gait disturbances. Traditionally, rehabilitation programs emphasize training balance under single-task conditions. According to study findings we suggest that interventions to improve dual-task balance performance are essential components of fall prevention programs for the ADHD children.

To sum up, we found that the ADHD group maintained their walking pattern, whereas they sacrificed the memory task under dual task condition. Nevertheless, their gait pattern was still disturbed as evidenced by a significantly greater sway speed in the working memory task condition and the level of dual-tasking postural disturbance was associated with individual balancing ability.

Acknowledgements This work was supported by the National Science Council, Taiwan (NSC 97-2320-B-037-004-MY3).

References

1. Parker, J., Wales, G., Chalhoub, N., Harpin, V.: The long-term outcomes of interventions for the management of attention-deficit hyperactivity disorder in children and adolescents: a systematic review of randomized controlled trials. Psychol. Res. Behav. Manag. **6**, 87–99 (2013)
2. Chen, Y.Y., et al.: Timing perception and motor coordination on rope jumping in children with attention deficit hyperactivity disorder. Phys. Ther. Sport **14**(2), 105–109 (2013)
3. Cho, H., Ji, S., Chung, S., Kim, M., Joung, Y.S.: Motor function in school-aged children with attention-deficit/hyperactivity disorder in Korea. Psychiatry Invest. **11**(3), 223–227 (2014)
4. Flapper, B.C., Houwen, S., Schoemaker, M.M.: Fine motor skills and effects of methylphenidate in children with attention-deficit-hyperactivity disorder and developmental coordination disorder. Dev. Med. Child Neurol. **48**(3), 165–169 (2006)
5. Leitner, Y., et al.: Gait in attention deficit hyperactivity disorder: effects of methylphenidate and dual tasking. J. Neurol. **254**(10), 1330–1338 (2007)
6. Papadopoulos, N., McGinley, J.L., Bradshaw, J.L., Rinehart, N.J.: An investigation of gait in children with attention deficit hyperactivity disorder: a case controlled study. Psychiatry Res. **218**(3), 319–323 (2014)
7. Piek, J.P., Pitcher, T.M., Hay, D.A.: Motor coordination and kinaesthesis in boys with attention deficit-hyperactivity disorder. Dev. Med. Child Neurol. **41**(3), 159–165 (1999)
8. Pitcher, T.M., Piek, J.P., Hay, D.A.: Fine and gross motor ability in males with ADHD. Dev. Med. Child Neurol. **45**(8), 525–535 (2003)
9. Buderath, P., et al.: Postural and gait performance in children with attention deficit/hyperactivity disorder. Gait Posture **29**(2), 249–254 (2009)
10. Chen, Y.Y., et al.: Comparison of dynamic balance under dual and simple task conditions in children with attention deficit hyperactivity disorder. J. Phys. Ther. Sci. **24**(8), 633–637 (2012)
11. Ghanizadeh, A.: Commentary to: postural and gait performance in children with attention deficit/hyperactivity disorder. Gait Posture **29**(4), 661 (2009)
12. Konicarova, J., Bob, P., Raboch, J.: Balance deficits and ADHD symptoms in medication-naive school-aged boys. Neuropsychiatric Dis. Treat. **10**, 85–88 (2014)
13. Mao, H.Y., Kuo, L.C., Yang, A.L., Su, C.T.: Balance in children with attention deficit hyperactivity disorder-combined type. Res. Dev. Disabil. **35**(6), 1252–1258 (2014)
14. Shum, S.B., Pang, M.Y.: Children with attention deficit hyperactivity disorder have impaired balance function: involvement of somatosensory, visual, and vestibular systems. J. Pediatr. **155**(2), 245–249 (2009)
15. Wang, J., Wang, Y., Ren, Y.: A case-control study on balance function of attention deficit hyperactivity disorder (ADHD) children. J. Peking Univ. Health sci. **35**(3), 280–283 (2003)
16. Schoemaker, M.M., Ketelaars, C.E., van Zonneveld, M., Minderaa, R.B., Mulder, T.: Deficits in motor control processes involved in production of graphic movements of children with attention-deficit-hyperactivity disorder. Dev. Med. Child Neurol. **47**(6), 390–395 (2005)
17. Tseng, M.H., Henderson, A., Chow, S.M., Yao, G.: Relationship between motor proficiency, attention, impulse, and activity in children with ADHD. Dev. Med. Child Neurol. **46**(6), 381–388 (2004)
18. Vallesi, A., D'Agati, E., Pasini, A., Pitzianti, M., Curatolo, P.: Impairment in flexible regulation of speed and accuracy in children with ADHD. J. Int. Neuropsychol. Soc. **19**(9), 1016–1020 (2013)
19. Wang, H.Y., Huang, T.H., Lo, S.K.: Motor ability and adaptive function in children with attention deficit hyperactivity disorder. Kaohsiung J. Med. Sci. **27**(10), 446–452 (2011)
20. Deliagina, T.G., Zelenin, P.V., Beloozerova, I.N., Orlovsky, G.N.: Nervous mechanisms controlling body posture. Physiol. Behav. **92** (1–2), 148–154 (2007)
21. Bucci, M.P., Seassau, M., Larger, S., Bui-Quoc, E., Gerard, C.L.: Effect of visual attention on postural control in children with attention-deficit/hyperactivity disorder. Res. Dev. Disabil. **35**(6), 1292–1300 (2014)
22. Shumway-Cook, A., Woollacott, M.: Attentional demands and postural control: the effect of sensory context. J. gerontol. Ser. A Biol. Sci. Med. Sci. **55**(1), M10–16 (2000)
23. Zang, Y., Gu, B., Qian, Q., Wang, Y.: Objective measurement of the balance dysfunction in attention deficit hyperactivity disorder children. Chin. J. Clin. Rehabil. **6**, 1372–1374 (2002)
24. Blanchard, Y., et al.: The influence of concurrent cognitive tasks on postural sway in children. Pediatr. Phys. Ther. **17**(3), 189–193 (2005)
25. Pellecchia, G.L.: Postural sway increases with attentional demands of concurrent cognitive task. Gait Posture **18**(1), 29–34 (2003)

26. DiScala, C., Lescohier, I., Barthel, M., Li, G.: Injuries to children with attention deficit hyperactivity disorder. Pediatrics **102**(6), 1415–1421 (1998)
27. Verret, C., Gardiner, P., Beliveau, L.: Fitness level and gross motor performance of children with attention-deficit hyperactivity disorder. Adap. Phys. Act. Quart. **27**(4), 337–351 (2010)
28. Shorer, Z., Becker, B., Jacobi-Polishook, T., Oddsson, L., Melzer, I.: Postural control among children with and without attention deficit hyperactivity disorder in single and dual conditions. Eur. J. Pediatr. **171**(7), 1087–1094 (2012)
29. Wu, W.L., et al.: Influence of working memory task and time on postural control of children with attention deficit hyperactivity disorder. J. Phys. Ther. Sci. **26**(3), 345–347 (2014)
30. Reimer, B., Mehler, B., D'Ambrosio, L.A., Fried, R.: The impact of distractions on young adult drivers with attention deficit hyperactivity disorder (ADHD). Accid. Anal. Prev. **42**(3), 842–851 (2010)
31. Siu, K.C., Woollacott, M.H.: Attentional demands of postural control: the ability to selectively allocate information-processing resources. Gait Posture **25**(1), 121–126 (2007)
32. Kelly, V.E., Janke, A.A., Shumway-Cook, A.: Effects of instructed focus and task difficulty on concurrent walking and cognitive task performance in healthy young adults. Exp. Brain Res. **207**(1–2), 65–73 (2010)
33. Alderson, R.M., et al.: Working memory deficits in boys with attention deficit/hyperactivity disorder (ADHD): an examination of orthographic coding and episodic buffer processes. Child Neuropsychol. **21**(4), 509–530 (2015)

Effects of Haptic Sensory Input of a Fluttering Cloth on Static and Dynamic Postural Control

Kazushige Oshita and Sumio Yano

Abstract

This study investigated the effect of haptic sensory input provided by a fluttering cloth (FC) on postural control. Twenty-one healthy men were randomly assigned to a static or dynamic balance task group. Participants in the static task group performed a single-leg stance for 30s, while those in the dynamic task group performed a functional reach test under three different conditions, with their eyes closed. In the first condition, participants wore only half- or short-tights (HT) during each task. In the second and third conditions, they wore HT and a cloth wrapped snugly around the waist, or HT and an FC wrapped around the waist extending to the lower leg, respectively. Both balance test results were significantly improved in the FC-condition than in the HT-condition. The effect size of improvement was large in the dynamic task, but small in the static task. Therefore, the effect of clothing on balance control was more pronounced in the dynamic than in the static task. These results suggest that postural control might be enhanced by a garment's shape or material.

Keywords

Light touch • Postural sway • Functional reach test • Garment

1 Introduction

Previous studies have reported that additional haptic sensory input by a light finger touch decreases postural sway in various postures [1–5]. Jeka and Lackner [1] observed that touching the index finger on a fixed pedestal (force level between finger and the pedestal are insufficient to provide mechanical body support; <1 N) resulted in decreased postural sway during an upright stance. This phenomenon suggests that haptic sensory input through the light touching of an object during postural control primarily provides information about the relative changes in the body segments. Therefore, this haptic sensory input helps an individual perceive their own body orientation [2, 4].

The maintenance of both static (i.e. postural sway during quiet stance) and dynamic (i.e. maintaining a stable position while a person performs a prescribed movement) control are important for daily living. Hageman et al. [6] reported that, although static balance does not decrease in the elderly until remarkable functional declines occur, dynamic balance decreases much earlier. Riemann et al. [7] and Sell [8] suggested a shift away from the use of static balance testing towards dynamic balance testing, since it may be more functional and applicable to healthy, physically active individuals. Therefore, from the view point of applying light touch effects to daily activities, lightly touching a real cane increases postural stability during the quiet stance [9] and the functional reach test (FRT) [10], a widely used dynamic balance test.

Although the effects of light touch is of particular interest to researchers, it is unusual for an individual to stand and

K. Oshita (✉)
Department of Sports Science, Kyushu Kyoritsu University, 1–8 Jiyugaoka, Yahatanishi, Kitakyushu, 807–8585, Japan
e-mail: monofinswim@gmail.com

S. Yano
Graduate School of Human Development and Environment, Kobe University, 3–11 Tsurukabuto, Nada, Kobe, 657–8501, Japan

F. Ibrahim et al. (eds.), *2nd International Conference for Innovation in Biomedical Engineering and Life Sciences*,
IFMBE Proceedings 67, https://doi.org/10.1007/978-981-10-7554-4_17

lightly touch an object while performing daily activities. In the case of lightly touching an object, such is likely to impede the performance of certain types of human movements. Therefore, studies should focus on the ability to develop light touch effects for more useful applications during daily activities. A previous study found that passive haptic input to the leg also reduced postural sway [11]. If haptic input through the passive contact of a cloth can also provide a light touch effect, it could be used to improve human movement. As people typically wear garments, the garment (cloth) might provide a light touch effect. Thus, the purpose of the current study was to investigate the effects of haptic sensory input produced by a cloth on static and dynamic balance. However, there are many kinds of garments, and the garment that is effective enough to provide a light touch effect should be considered. Therefore, to obtain fundamental data regarding the cloth that can provide the light touch effect, a soft cloth that will not interfere with human movement (fluttering cloth) wrapped around the waist and extending to the lower leg during standing was used.

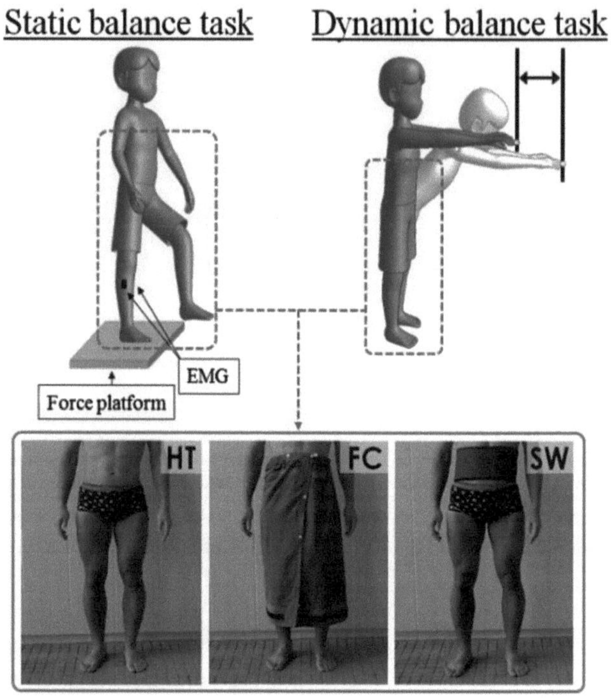

Fig. 1 Schematic of the each balance task

2 Method

2.1 Participants

Data were obtained from 21 healthy men (age: 21 ± 2 years) with no current or previous medical history of neural, muscular, or skeletal disorders. The participants were randomly assigned to static balance task (n = 10) and dynamic balance task (n = 11) groups. Before commencement of the study, all participants were informed of the study purpose, and provided informed consent. The study was approved by the Human Ethics Committee of the Graduate School of Human Development and Environment, Kobe University.

2.2 Experimental Setup (Fig. 1)

In the static balance task, postural sway and lower limb muscular activity were evaluated during the single-leg stance (SLS). SLS was performed on each participant's preferred leg with their eyes closed. The SLS tasks were performed under three different conditions: wearing only half- or short-tights (HT), wearing half- or short-tights and a fluttering cloth wrapped around the waist (L3–L5 point) extending to the middle point of the lower leg (FC), and wearing half-tights and a snuggly wrapped cloth around the waist only (SW) (Fig. 1).

Before the SLS, maximum voluntary muscle contraction (MVC) was performed in the plantar flexion and dorsiflexion directions, as described in a previous study [9]. The participants were asked to perform each MVC twice for 5-s each with a 10-s inter-test pause. They were allowed to reject an effort that they deemed not "maximal." Postural sway during SLS was evaluated by center of foot pressure (COP) using a force platform (T.K.K. 5810; Takei Scientific Instruments Co. Ltd., Japan). Data were recorded at 100-Hz using an analog-to-digital converter (AI-1608AY-USB; CONTEC, Japan). To assess the COP, a 10-s in the middle period of each SLS was chosen for the COP analysis. The average velocity of the COP trajectory (V-COP) was calculated as the total path length of COP trajectory divided by the calculated time.

To assess muscle activity, surface electromyography data were collected from the gastrocnemius-lateralis (GAS) and tibialis-anterior (TA) muscles. Bipolar electrodes connected with an electromyogram amplifier (ID2PAD; Oisaka Electronic Equipment Ltd., Japan) were set over the TA and GAs at a 2-cm inter-electrode distance. The data were acquired at a sampling frequency of 1 kHz using a data logger with an analog-digital converter (LP-MS1002; Logical Product Co., Japan), and were processed using SPCANA (ver. 4.92) wave-form analysis software. The root mean square of the electromyography data (RMS-EMG) in each task was calculated after band-pass filtering (1–500 Hz). For the MVC tasks, the RMS-EMG was calculated every 200 samples (0.2-s) and the maximum value was used as the each MVC value. In SLS tasks, a 10-s in the middle period was selected to be used in calculating the RMS-EMG. These RMS-EMG were normalized to the MVC value for the evaluation of muscle activity (% MVC). Further, the co-contraction between TA and GAS was calculated using co-contraction index (CCI), as the following Eq. [12]:

$$CCI = \left(\frac{\%MVC_{lower}}{\%MVC_{higher}}\right) \times \left(\%MVC_{lower} + \%MVC_{higher}\right)$$

where %MVC lower and higher means the average normalized RMS-EMG of GAS or TA activities.

For the dynamic balance task, the FRT was conducted as described in a previous study [10]. Each participant stood with the lateral aspect of their arms parallel to the wall. The participants were then asked to place their feet shoulder width apart, rise their arms until they were parallel to the floor, and keep their eyes closed (initial position). The participant was subsequently told to begin to extend their arms (including torso) as far forward as possible and then return to an upright position, without taking any steps while keeping the arms level with the acromion.

Kinematic data were acquired during the FRT using a digital camera (AW100; Nikon, Tokyo, Japan) at a frame rate of 120-fps, and the analysis was performed using the digitizing software program Kinovea (ver. 0.8.15; French). Kinematic data were recognized from a reflective marker fixed on the nail bed of the middle finger. Further, the FRT score was measured by the horizontal distance of the reflective marker from the initial position to the farthest reach attained.

2.3 Protocol

The participants in the static balance task group performed three SLS tasks with a three-minute rest period in between each. The first task was conducted in the HT condition, while the second and third tasks were performed in the FC or SW conditions, randomized across the participants. All participants were asked to remove all footwear, and stand on a force platform. Then, the experimenter instructed each participant to close his eyes, and raise his non-preferred leg straight forward about 10-cm from the platform. The participants maintained the posture for 30-s.

The participants in the dynamic balance task group performed three FRT trials with a three-minute rest period between trials. The first test was performed in the HT condition, while the second and third tests were in the FC or SW conditions, respectively, randomized across the participants. Each FRT was conducted two times, and the highest FRT score (meaning the one with the longest displacement) was selected for inclusion in the analysis.

Before each test, participants were allowed a five-minute practice session in the HT condition to familiarize themselves with the requirements.

2.4 Statistical Analysis

The statistical analysis was performed using a one-way repeated-measures analysis of variance (ANOVA), and significant differences among the HT, FC, and SW conditions in each balance task were evaluated with post hoc multiple comparisons using Tukey's test. Values of $P < 0.05$ were considered statistically significant. These analyses were performed using J-STAT (ver. 12.5) software. In addition to the significance testing, effect sizes (Cohen's *d*) of the three conditions were calculated.

3 Results

Regarding the static balance task, Table 1 shows the V-COP and muscle activity. The ANOVA result revealed a significant difference in V-COP among all three conditions. Although the post hoc test observed that the V-COP in the FC condition was significantly lower than that in the HT condition, no significant difference was ob-served between such in the HT and SW conditions. However, the effect sizes were small (HT vs. FC, $d = 0.28$; HT vs. SW, $d = 0.11$). Therefore, although postural sway decreased when the participant was worn with a fluttering cloth, the effect was small.

No significant differences were observed in the average TA or GAS activity, whereas a significant difference in CCI among the conditions was observed. Further, although the post hoc test revealed that CCI was significantly lower in the FC condition than in the HT condition, no significant difference was observed between the HT and SW conditions. However, the effect sizes with respect to all three conditions were small (HT vs. FC, $d = 0.47$; HT vs. SW, $d = 0.10$). Therefore, although activities using the ankle-joint muscles are not affected by the presence of SW, co-contraction of the ankle-joint muscles significantly decreases in the presence of FC.

Table 2 shows the FRT scores of the dynamic balance task. The ANOVA revealed a significant difference in FRT score among all three conditions. Although the post hoc test revealed that the FRT score was significantly higher in the FC condition than in the HT condition, no significant difference was observed between the HT and SW conditions. Furthermore, although the effect size of the FC condition

Table 1 Effect of cloth condition on static balance control

	HT	FC	SW	ANOVA	Versus HT (Tukey test/effect size)	
					FC	SW
V-COP (mm/s)	66.8 ± 38.7	56.9 ± 31.8	62.7 ± 36.9	$P < 0.05$	$P < 0.05$	N.S.
					$d = 0.28$	$d = 0.11$
EMG-TA (% MVC)	30.6 ± 11.3	26.4 ± 9.92	31.0 ± 9.76	$P = 0.07$	–	–
					$d = 0.40$	$d = 0.03$
EMG-GAS (% MVC)	30.5 ± 13.2	26.1 ± 12.7	29.4 ± 13.5	N.S.	–	–
					$d = 0.34$	$d = 0.08$
CCI (% MVC)	39.9 ± 13.0	33.9 ± 12.3	38.5 ± 14.1	$P < 0.01$	$P < 0.01$	N.S.
					$d = 0.47$	$d = 0.10$

Values are means ± standard deviation

Table 2 Effect of cloth condition on dynamic balance control

	HT	FC	SW	ANOVA	Versus HT (Tukey test/effect size)	
					FC	SW
FRT (cm)	43.8 ± 3.14	46.5 ± 3.03	43.5 ± 2.62	$P < 0.01$	$P < 0.01$	N.S.
					$d = 0.90$	$d = 0.11$
FRT(cm)/ Height (cm)	0.26 ± 0.02	0.28 ± 0.01	0.26 ± 0.01	$P < 0.01$	$P < 0.01$	N.S.
					$d = 1.00$	$d = 0.12$

was larger than that of the HT condition (d = 0.90), that of the SW condition was smaller than that of the HT condition (d = 0.11). These results were constant when the FRT score (cm) was normalized to the participant's height (cm). Therefore, the presence of an FC increased balance control in both the static and dynamic balance tasks. Further, although this effect was large in the dynamic balance task, it was small in the static balance task.

4 Discussion

The present study investigated the effects of light touching through a fluttering cloth on static and dynamic balance control. Although postural sway in the static dynamic balance task was significantly decreased by the presence of a fluttering cloth, the effect size was small. Additionally, the mean decrease in postural sway was about 15% (66.8–56.9 mm/s) in this study, while a previous study [9] that investigated the same single-leg stance task reported that postural sway was decreased by about 20% after light touching of the hand to a cane (53.1–42.8 mm/s). Therefore, the effect of cloth on static balance control might be smaller than that of active light touching of an object. Nagano et al. [13] reported that the effect of light touching one's own legs on postural sway was smaller than that of lightly touching an external fixed object. They suggested that touching an external fixed object provides "absolute" mechanical information between the fingers and the touched object, which is more helpful than touching one's own body (which provides "relative" information). In this study, because the cloth was not fixed, it could not provide absolute information. However, the novel finding of this study is that static balance control can be improved by the presence of a cloth when an external fixed object is not available, even if the touch effect is small.

Previous studies on the mechanisms of light touch effect on upright stance [9, 14] suggest that light touch effect is influenced by co-activation of the agonist (plantar flexor) and antagonist (dorsiflexor) muscles. Increased co-activation of agonist and antagonist muscles increases the rigidity of postural control [15, 16]. Such a stabilization strategy was observed during threatening conditions (i.e. when the platform height is low or high and the toes are positioned at or away from the edge) in healthy normal individuals while standing [17, 18]. However, haptic sensory input through touching can provide information about one's own body orientation [2, 4] and allows one to stand more easily, as compared with in the no touching condition. Therefore, co-activation of the ankle muscles might be decreased by haptic sensory information. Although the mean decrease in CCI rate in this study was 6% (39.9–

33.9% MVC), a previous study [9] that investigated the same method reported that CCI decreased by about 10% after light touching of the hand to a cane (35.4–26.5% MVC). Therefore, from the muscle activity viewpoint, the cloth had a smaller effect on the static balance task than active light touching of some object as observed in postural sway.

The effect of the fluttering cloth on balance control was observed in the static and dynamic balance tasks in this study. We conducted FRT as a dynamic balance task, which requires smooth body movements while participants maintain their posture. Postural control rigidity induced by greater muscle co-contraction reduces the degrees of freedom of movement [19]. Nagai et al. [20] also found that various balance training improves dynamic balance tests (including FRT) were associated with decreased muscle co-contraction in the ankle joint. Therefore, this might imply that the light touch phenomenon is also effective in the dynamic task. Although the fluttering cloth had a small effect in the static balance task, a larger effect was observed in the dynamic balance task. Postural control seems relatively more difficult during the dynamic balance task than during the static balance task, during which the sensory information from the cloth became relatively more important. It is also possible that the dynamic balance task is accompanied by larger movements than those in the static balance task. This means that increased contact between the skin and cloth provides increased haptic sensory information. Regarding these points, further studies are needed to clarify the details of cloth vs. haptic input; i.e. the movement of the cloth, contact force between the cloth and the skin, and awareness of the cloth.

A future study will also need to reveal whether the present results can be applied to actual activities requiring balance control, such as sports, as well as fall prevention. However, the present results, which demonstrated improvements in both static and dynamic balance tasks by the presence of a fluttering cloth, suggests a potential new use for clothing (or garments) to enhance balance control using light touch effects.

5 Conclusion

This study investigated the effects of haptic sensory input through the wearing of a fluttering cloth on static and dynamic balance control. Postural control was significantly improved by an FC during the static and dynamic balance tasks in the present study. During FC use, co-contraction of the TA and GAS was significantly decreased. The effect size of postural control was larger in the dynamic than the static task. Therefore, the effect of cloth on balance control may be more useful in the dynamic task than in the static task. These results suggest that FC use can provide a haptic sensory cue that assists postural control mechanisms by enhancing the perception of one's own body orientation. Therefore, postural control may be enhanced by garments of certain shapes or materials.

Acknowledgements This work was supported by Grant-in-Aid for Young Scientists (B) (15K21553).

References

1. Jeka, J.J., Lackner, J.R.: Fingertip contact influences human postural control. Exp. Brain Res. **100**, 495–502 (1994)
2. Jeka, J.J.: Chapter 4: Touching surfaces for control, not support. In: Rosenbaum, D.A., Collyer, C.E. (eds.) Timing of Behavior: Neural, Psychological, and Computational Perspectives, pp. 89–105. The MIT Press, Cambridge (1998)
3. Riley, M.A., Stoffregen, T.A., Grocki, M.J., Turvey, M.T.: Postural stabilization for the control of touching. Hum. Mov. Sci. **18**, 795–817 (1999)
4. Ishigaki, T., Imai, R., Morioka, S.: Cathodal transcranial direct current stimulation of the posterior parietal cortex reduces steady-state postural stability during the effect of light touch. Neuroreport **28**, 1050–1055 (2016)
5. Oshita, K., Yano, S.: Influence of light finger touch on postural stability during upright stance with cold-induced plantar hypoesthesia. Conf. Proc. IEEE Eng. Med. Biol. Soc. **2017**, 2526–2529 (2017)
6. Hageman, P.A., Leibowitz, J.M., Blanke, D.: Age and gender effects on postural control measures. Arch. Phys. Med. Rehabil. **76**, 961–965 (1995)
7. Riemann, B.L., Caggiano, N.A., Lephart, S.M.: Examination of a clinical method of assessing postural control during a functional performance task. J. Sport Rehabil. **8**, 171–183 (1999)
8. Sell, T.C.: An examination, correlation, and comparison of static and dynamic measures of postural stability in healthy, physically active adults. Phys. Ther. Sport **13**, 80–86 (2011)
9. Oshita, K., Yano, S.: Effect and immediate after-effect of lightly gripping the cane on postural sway. J. Physiol. Anthropol. **35**, 14 (2016)
10. Oshita, K., Yano, S.: The effect of lightly gripping a cane on the dynamic balance control. Open Biomed. Eng. J. **9**, 146–150 (2105)
11. Menz, H.B., Lord, S.R., Fitzpatrick, R.C.: A tactile stimulus applied to the leg improves postural stability in young, old and neuropathic subjects. Neurosci. Let. **406**, 23–26 (2006)
12. Greenwood, N.L., Duffell, L.D., Alexander, C.M., McGregor, A. H.: Electromyographic activity of pelvic and lower limb muscles during postural tasks in people with benign joint hypermobility syndrome and non hypermobile people. A pilot study. Man. Ther. **16**, 623–628 (2011)
13. Nagano, A., Yoshioka, S., Hay, D.C., Fukashiro, S.: Light finger touch on the upper legs reduces postural sway during quasi-static standing. Mot. Control **10**, 348–358 (2006)
14. Oshita, K., Yano, S.: Effects of light finger touch to the upper legs on postural sway and muscle activity during quiet standing. Conf. Proc. IEEE Eng. Med. Biol. Soc. **2013**, 7459–7562 (2013)
15. Tang, P.F., Woollacott, M.H.: Inefficient postural responses to unexpected slips during walking in older adults. J. Gerontol. **53**, M471–M480 (1998)
16. Wu, G.: Age-related differences in Tai Chi gait kinematics and leg muscle electromyography: a pilot study. Arch. Phys. Med. Rehabil. **89**, 351–357 (2008)

17. Carpenter, M.G., Frank, J.S., Silcher, C.P.: Surface height effects on postural control: a hypothesis for a stiffness strategy for stance. J. Vestib. Res. **9**, 277–286 (1999)
18. Carpenter, M.G., Frank, J.S., Silcher, C.P., Peysar, G.W.: The influence of postural threat on the control of upright stance. Exp. Brain Res. **138**, 210–218 (2001)
19. Tucker, M.G., Kavanagh, J.J., Barrett, R.S., Morrison, S.: Age-related differences in postural reaction time and coordination during voluntary sway movements. Hum. Mov. Sci. **27**, 728–737 (2008)
20. Nagai, K., Yamada, M., Tanaka, B., Uemura, K., Mori, S., Aoyama, T., Ichihashi, N., Tsuboyama, T.: Effects of balance training on muscle coactivation during postural control in older adults: a randomized controlled trial. J. Gerontol. **67**, 882–889 (2012)

Design and Development of Low Cost Hand Exoskeleton for Rehabilitation

Vickneswari Durairajah, Suresh Gobee, Waleed Rauf, Kok Sin Ngie, and John Hong Aun Lim

Abstract

The proposed project is a low cost hand exoskeleton that is design and developed for rehabilitation while the safety rules and regulation kept in mind. The aim of the project is to rehabilitate the stroke patients who is suffers from stroke on the one side of the body. This system to rehabilitate the patient to gain mobility in the fingers where a sound hand is used to control the unsound hand. The sound hand is consists of flex sensors attached to the leather glove on MCP, PIP and IP joints while the unsound hand consists of servo motors. The microcontroller used is Arduino Nano and the software used is Arduino interfaced with MATLAB for the GUI. The communication is wireless with the help of Bluetooth. The testing results were taken to be 92.75% efficient.

Keywords

Exoskeleton • Control hand • Stroke hand • Rehabilitation • Flex sensor • DoF • Microcontroller and servo motors

1 Introduction

Stroke has been always among the most common leading disabilities which is caused for long term worldwide, it includes hemiplegia which is associated with abnormal coordination and activation, and it also has side effects like dexterity loss, weakness of muscle and precision. These are being one of the most common and huge contribution towards the disabilities. Recovery regarding the upper section of the body like arm, hand, and finger is considered very challenging. Among 100 patients only 38 of them partially recovered and regain dexterity in the arm after the stroke which results 38% of the project as success. The further the hand and fingers are concerned the task become more challenging and demanding and has very limited chance of improvement so far as proved [1].

The need of rehabilitation after the stroke for these patients is compulsory in order to regain their lost abilities of hand fingers with the passage of time. Rehabilitation process is compulsory in order to regain the muscle movement and it should be done within first 6 months in order to regain up to 70%. Robotic hand exoskeleton for rehabilitation is one of the solution which is independent of physiotherapist [2, 3].

Recently many of the projects have been done on rehabilitation of hand fingers. As per each project it has faced many drawbacks and there is a gap of improvement. The main limitation during the projects were the precise bending movements of the fingers and their flexion and extension. The chances of recovery are more if the system is more efficient and precise [4].

The development of this project is to aid the rehabilitation process to revive the fingers and thumb by way of flexion and extension by utilizing the Arduino Nano microcontroller to receive feed-back as well as to control servo motors by using Flex sensors which detects the finger bending. A graphical user interface (GUI) is put together based on the information collected from the Flex sensors to give a real-time biofeedback simulation of a performed flexion and

V. Durairajah (✉) · S. Gobee · W. Rauf · K. S. Ngie · J. H. A. Lim
School of Engineering, Asia Pacific University, Kuala Lumpur, Malaysia
e-mail: vicky_nesa@apu.edu.my

F. Ibrahim et al. (eds.), *2nd International Conference for Innovation in Biomedical Engineering and Life Sciences*, IFMBE Proceedings 67, https://doi.org/10.1007/978-981-10-7554-4_18

extension exercise of each control hand and stroke hand fingers. The project is limited to 2 DoF and is more focused on fingers. Furthermore, the proposed exoskeleton is to be designs in such a way that it can fit anyone without any assistance and the size is made flexible for most of the hands and the comfort level is maximum as its consists of leather glove. Grabbing and grasping would be possible as well.

2 Methodology

The methodology of developed hand exoskeleton is shown in Fig. 1. The Sects. 2.1–2.4 explains the methodology Fig. 1 in detailed manner.

2.1 Design Architecture (Exoskeleton)

The design consists of two exoskeletons. First one is the control (left) hand which consists of flex sensors attached to leather glove is assumed to be sound hand. Second one is the stroke (right) hand which consists of servo motors attached to leather glove is assumed to be unsound hand. In designing the exoskeleton safety measures also has been taken as per the environment factors.

2.2 Circuit Design (Electronics)

Flex sensors used for flex reading (voltage difference) from flex sensors. The voltage divider is form between resistors and flex sensors in which the Vin is divided by the ration of two resistors. The output voltage is 75% Vin due to the 22 k resistor and the flex sensor is straight. The voltage increased to 83% Vin when the sensor is bent. The flex sensor works in a single direction.

Bluetooth HC-05 is used for wireless communication control between flex sensors and actuators. Bluetooth HC-05 has been installed on Control and Stroke Hand.

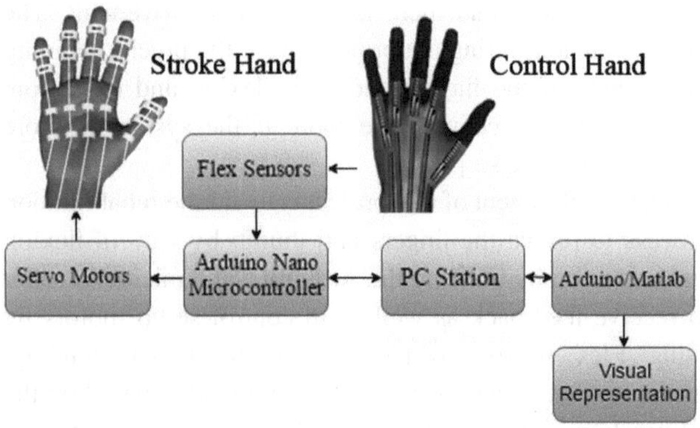

Fig. 1 System block diagram

The actuators proposed in the project are the servo motors which are attached to each finger of the stroke hand with the help of nylon string. The main objective of the servo motors is the flexion and extension of the stroke hand. The actuation is controlled using flex sensors interfaced with Arduino microcontroller. The actuation is programmed in such a way that when the control hand fingers moves up to certain angle the servo motors makes the stroke hand fingers move accordingly. The limitations are defined for servo motors in the program according to each finger and the thumb of different sizes.

2.3 Rehabilitation Exercise

The proposed hand exoskeleton has the potential to rehabilitate the stroke patient's fingers and the thumb using certain exercise, the project proposed the different angle movement of each finger and thumb of the stroke hand for proper rehabilitation. The patients can do the movement exercise in steps, from first stretched finger position of stoke hand to the second stretch position simultaneously. The angle movement for all the fingers and thumb is from 0°–90° and 90°–120°. The nylon string will pull the fingers up to 120° and the elastic band is to bring back the fingers to its initial position. The second exercise is to grab and grasp various items and check the precision of the fingers.

2.4 Computer Interface (GUI)

The Graphical User Interface (GUI) of the system is done using Arduino software interfaced with MATLAB. The GUI shows the real-time simulation of the control hand's flex sensors reading and stroke hand servo motors movement graphically for each finger and thumb with time.

3 Design and System Implementation

3.1 Design

3D model of the exoskeleton is shown in Fig. 2, the exoskeleton is anthropometrics designed. The design of control hand consists of flex sensors attached to the stitched cloth. The reason for stitching the cloth for flex sensors is to have proper movement to achieve the stability of the flex sensors as it is known that the flex sensors are very sensitive in nature, also to stabilize the flex readings. The design of stroke hand consists of PVC, Nuts, and Nylon String and servo motors.

The mechanical stroke hand mostly consists of PVC. The reason for choosing PVC material is to have stable and fixed

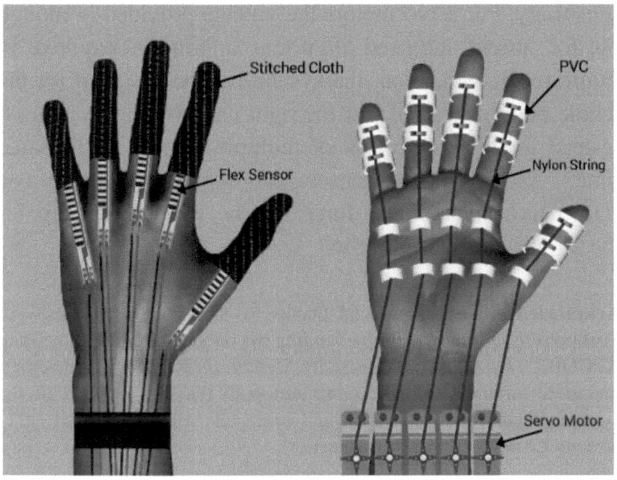

Fig. 2 3D model of exoskeleton

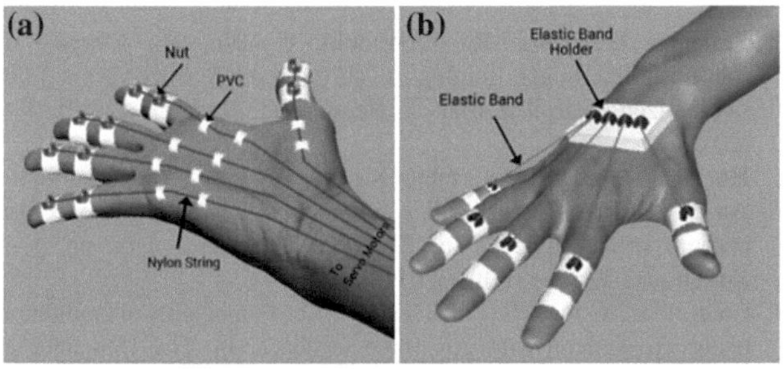

Fig. 3 **a** Nylon string and **b** elastic band stretching and position

movements for the joints like MCP, PIP and DIP. The mechanical stretching of the fingers in stroke hand is done using Nylon String. The Nylon String is fixed on top of the nut and passes through the second nut and 2 PVC bridges which are fixed just to have smooth movement for the finger and to avoid the collision between the strings as shown in Fig. 3a.

The elastic bands are used to stretch the stroke hand from the opposite side of Nylon String. They work as a spring to bring all the fingers and thumb back to their original position if the nylon string is released from the other side and the servo is not stretching the fingers towards it. The elastic bands are attached with the help of wire holders on PVC material rings which were cut earlier for stroke hand. The elastic band is attached to the nearest ring for the sake of more pull and initializing the position as shown in Fig. 3b.

3.2 Prototype

The prototype of hand exoskeleton for rehabilitation is built with proper equipment as shown in Fig. 4. For safety purposes the soft materials like leather gloves are used. All the limitation were defined as to avoid any kind of injury or

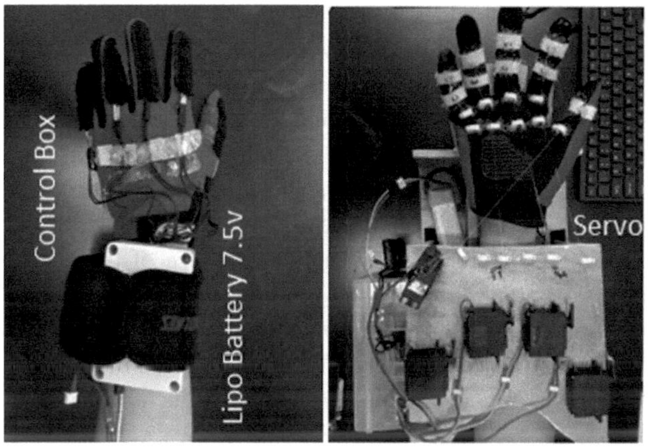

Fig. 4 Control and stroke hand exoskeleton prototype

errors. The components used are inexpensive and efficient. Flex sensors and microcontroller with Bluetooth are attached to both Control and Stroke hand in a compact size box which can be attached and unattached easily. Servo motors are attached to wood strip which is further attached to the stroke hand. The prototype is portable.

4 Results and Discussion

There are total of 2 tests done. Five volunteers (3 males and 2 females) participated in the tests.

As stated in Fig. 5 the efficiency was tested as following; Test 1 which is the flex readings for all the fingers. The test is done for all 5 volunteers. The initial value and the final value for each finger of all 5 volunteers are the same. For Pinky finger the range (mm) is 45–48 for angle 0°–90° and 82–89 for angle 90°–120°. Similarly for Ring finger the ranges are 35–41 and 62–70. For Middle its 31–35 and 55–60, whereas for index and thumb, the values are 32–35, 52–57 and 22–42, 53–59. The maximum flex sensor values from pinky to thumb are 89, 70, 60, 57 and 59 respectively. The maximum difference is 8 between all the fingers of all the volunteers except for the thumb of the volunteer 1, the reason is the injury on the thumb which could not move properly during the testing. The overall efficiency is 92% if 8% of the error is considered is obtain from the average of the practical efficiency. For the volunteer 2 and 4 which are

	Theoretical efficiency	Practical efficiency
Test 1	100%	92%
Test 2	100%	93%
Test 3	100%	89%
Test 4	100%	97%
Final	100%	92.75%

Fig. 5 Efficiency of the system

female the flex values are quite similar due to the same hand size. The accuracy for these volunteers is 98%.

Test 2 which is servo motor readings for moving the stroke hand. The test is done for all 5 volunteers. The maximum error difference is 15° which is almost 1.8 times of flex sensor readings. The angle difference is almost the same which ranges from 0°–156°, 2°–113°, 3°–104°, 0°–93° and 0°–104° for Pinky to Thumb. The size of the hand is inversely proportional to the bending. For servo motor the maximum values from pinky to thumb 160°, 126°, 108°, 102° and 106° respectively. The error for servo motor is 7% (1.8 times of flex sensor readings) which gives 93% accuracy.

5 Conclusion

Conclusion, the main objective of this system was achieved to be low-cost. With this system being low-cost it serves a very important role in serving the least fortunate people who are left unserved and under-serve. This system is able to give these people a fighting chance in getting the rehabilitation they need to recover from their disabilities. The design is simple, comfortable, portable and fits to most of the human hand. The developed hand exoskeleton successfully able to rehabilitate the fingers and the thumb of the patients suffering from stroke on the right side of the body. Furthermore the Flex sensors are attached to the control (left) hand and provides data successfully. The stroke (right) hand actuate the servos and makes the stroke fingers move. In limitations, the initial value of the flex sensors fluctuates a lot due to its sensitivity. For servo motors the voltage provided is enough but for current, the need of current amplifier is required. In future recommendation, the system can be designed for the whole body instead of just the right part. Wi-Fi can be used instead for the connection of multiple devices at the same time. Accelerometer sensor can be used for 3-axis (DoF) movement. Flexi force can be used on the tip for the strong gripping purpose.

Acknowledgements A special thanks to Asia Pacific University of Technology and Innovation for funding the project. Similarly, thanks to APCORE (Asia Pacific University Center of Robotic Engineering) members for their valuable contribution to the development of the system. Finally, thanks to each and every one who contributed either directly or indirectly to the project.

References

1. Sustanto, E., Tong, R., Ockenfeld, C., Ho, N.: Efficacy of robot-assisted finger training in chronic stroke survivor: a pilot randomized-controlled trial. J. Neuro Eng. Rehabil. **12**(42), 1–9 (2015)
2. Ho, N., Tong, K., Hu, X., Fung, K., Wei, X., Rong, W., Susanto, E.: An EMG-driven exoskeleton hand robotic training device on chronic stroke subjects. In: ETH Zurich Science City, pp. 1–5. Zurich and 29 June–1 July 2011
3. Chan, C.L., Gobee, S., Vickneswari, D.: Finger grip rehabilitation using exoskeleton with grip force feedback. In: The 15th International Conference on Biomedical Engineering Springer International Publishing, pp. 520–523. (2014)
4. Gujrati, A., Sing, K., Khushboo, Soral, L., Ambikapathy.: Hand-talk gloves with flex sensors: a review. Int. J. Eng. Sci. Invention **2**(4), 43–46 (2013)

Design and Development of Upper Limb Soft Exoskeleton for Rehabilitation

Suresh Gobee, Vickneswari Durairajah, and G. Mugilan

Abstract

Current physiotherapy sessions are expensive or not available readily for patients and elderly. The main aim of this project is to design and develop a wearable exoskeleton which could be made practical to rehearse wrist rehabilitation exercises. In the ever advancing robotics field, pneumatic muscles are a cost reliable alternative actuator for locomotion and manipulation for a master-slave biophysical system. Implementing the flex sensors as primary sensory tool on the sound limb to detect angular deflection of the wrist to assist radial deviation, ulnar deviation, extension and flexion motions, while the pneumatic muscles operate the unsound limb. The overall system efficiency was measured based on testing done with several human subjects, all using specifically fabricated pneumatic muscles of variable total length but a fixed active length of 14 cm. the active length of the pneumatic muscle is the key feature of any pneumatic muscle as this feature determines the contraction length that translates to the angular wrist movement. Furthermore, the system was able to be monitored aptly using a GUI programmed using LabVIEW and interfaced to Arduino. The testing results were taken to be 55.72% overall efficient.

Keywords

Soft exoskeleton • Assistive motion • Low cost • Pneumatic muscle • Rehabilitation and flex sensor

1 Introduction

Generally, exoskeletons are made to provide structural support or rigidity to the human's body externally. Majority of exoskeleton are focused in the military field, mainly to give better strength and higher durability to the soldier. There are also research on exo-skeleton used for rehabilitation purpose [1, 2]. The more intricate approach to the common rigid exoskeleton is the soft exoskeleton. Unlike the predecessor versions, this soft exoskeleton behaves more like a clothing rather than an armour. These soft exoskeletons are called in short as exosuits. In addition to that, the exosuit are being proposed vastly in the medical field as well, with various aims [3]. There are multiple ways of designing and programming the exosuit. The challenge would be to design the exosuit in a way that it doesn't restrict natural human movement and of course ultimately be beneficial to the wearer for their specified rehabilitation programmes.

Rehabilitation is basically the act of restoring the ability to work or perform accordingly, where the objective is to slowly recover the affected limbs and gain as motion control over the limbs. Moreover, different rehabilitation procedures will have to be constructed for patients with varying stroke constraints [4]. This context is mostly related to those with muscular problems. Rehabilitation may be given to the human depending on the area of severity. Such patients are just one of the 15 million people affected from strokes and cardiovascular problems worldwide, according to the World Health Organization (WHO) [5]. Forms of rehabilitation are

S. Gobee (✉) · V. Durairajah · G. Mugilan
School of Engineering, Asia Pacific University of Innovation and Technology, Technology Park, 57000 Bukit Jalil, Kuala Lumpur, Malaysia
e-mail: suresh.gobee@apu.edu.my

F. Ibrahim et al. (eds.), *2nd International Conference for Innovation in Biomedical Engineering and Life Sciences*,
IFMBE Proceedings 67, https://doi.org/10.1007/978-981-10-7554-4_19

physiotherapy, occupational therapy and balance therapy among many more. However, this methods are often time consuming and the therapies given needs to be manual altered to suit the individual's preference.

The main idea behind my proposed project is discussed based on an alternative low cost rehabilitation method. The concept design is done based on virtual work theory. The exosuit designed will be made to fit from the wrist to the forearm. The actuation produced should be sufficient to perform several preliminary test in order to pass the requirements.

2 Methodology

The methodology for the soft exoskeleton is displayed in Fig. 1, shows the rehabilitation exercises for the wrist. The sound hand motion will be detected via flex sensors and the data will be interpreted to replicate the motion on the unsound hand using adequate pneumatic muscle contraction. The precision of the replicated motion will be monitored using an IMU sensor on the unsound hand, while the GUI provides the visual representation of the sensor performances.

2.1 Rehabilitation Exercise

The wrist exercise involves flexion, extension, abduction (radial deviation) and adduction (ulnar deviation). The subject can choose to carry out any one of the exercise or both according to their intent. These exercises will be done using the sound hand. Figure 2 illustrates the average deviation angle for each motion from rest state.

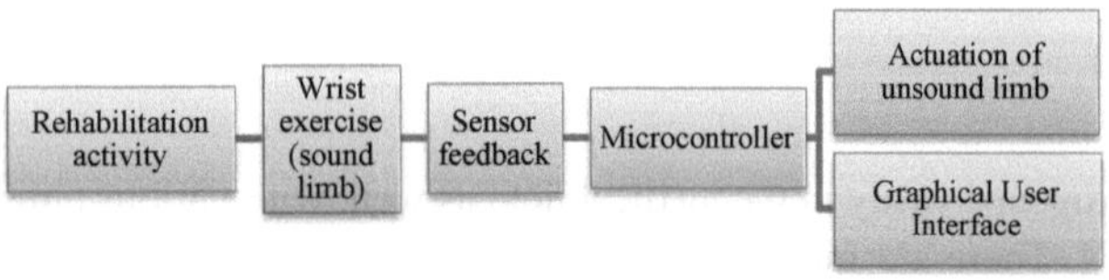

Fig. 1 System block diagram

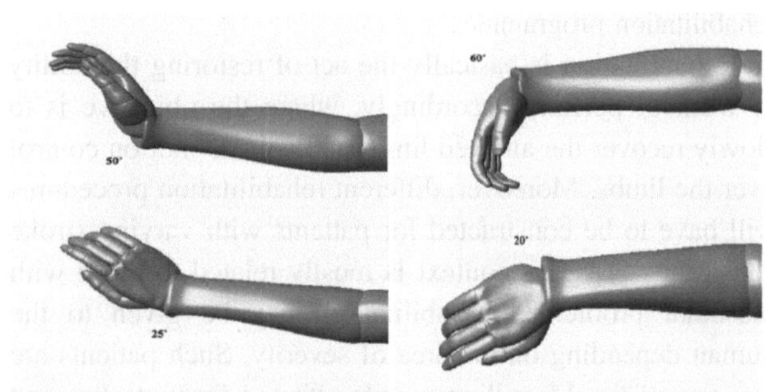

Fig. 2 (Clockwise from top-left) Extension, flexion, adduction and abduction [2]

2.2 Design Architecture (Soft Exoskeleton)

The design consists of a dual arm exoskeleton. The sound (left) hand consists of 4 2.2″ flex sensors attached to a compressive arm sleeve that stretches from palm to elbow. The unsound (right) hand consists of 4 pneumatic muscles acting as the primary actuators and 1 IMU sensor attached to the remaining of the pair of arm sleeve. The exoskeleton was designed keeping in mind the wearable robot concept as well as safety measures necessary for stroke patients especially when self-administration is enabled.

2.3 Flex Sensors 2.2″ (SEN-1024)

Basically the flex sensor is variable resistor that operates through bending. Each flex sensor requires a voltage supply of 5 V and requires a single analog output pin. It is also coupled with a resistor, to form a voltage divider for detection of the sensor. A standard 2.2″ flex sensor has 25 kΩ resistance when straight and 45–120 kΩ resistance depending on bend radius. The flex sensor works in a single direction, hence a single flex sensor can be used for detection of a single motion at a time.

2.4 IMU Sensor (MPU 6050 GY-521)

The IMU is basically the integration of a gyroscope and an accelerometer, which is used to measure the distance and orientation of the object it is attached to. There are several movements of the sound limb that the soft exoskeleton has to replicate. The motions involve a range of distances and angles, which needs to be identified correctly. The IMU sensor is a comprehensive motion sensor as it is capable of detecting displacement and angular deviation. The sensor operates with based on the angular velocity and linear acceleration data acquired. Most importantly it has an inbuilt magnetometer, which plays an important role for yawing angles (parallel to gravitational plane) are to be detected.

2.5 Pneumatic Muscle

The major hardware involved, the pneumatic muscles which primarily actuates via contraction. The self-fabrication of the pneumatic muscle was made practical using trigonometric calculations as the linear actuation of the pneumatic muscle needed to be converted to angular displacement to achieve the wrist motions. The size of the pneumatic muscles were determined by using a flex sensor to determine initial and final wrist angles after respective exercise routines. Using the angular data and sine rule function, the theoretical active lengths of pneumatic muscles were deduced. There were bound to be a minor range of uncertainty as the angles were rounded off to the nearest degree.

2.6 Computer Interface (GUI)

The Graphical User Interface (GUI) of the system was constructed using the NI LabVIEW 2013 software. The LINX source code was used to bridge the LabVIEW and Arduino Uno microcontrollers using serial protocol. The GUI shows the real-time performance of all the flex sensors and IMU sensor. The data can be used to graphically monitor and ease the calibration of the sensors used for each motion separately.

3 Design and System Implementation

3.1 Design

The fabrication of the pneumatic muscle is significant for the outcome of this system. Firstly, a length pneumatic tube was inserted 2 cm into the latex tube at one end. Then the latex tube was inserted into the braided sleeve. A hose clamp was inserted about 1 cm on the end with the pneumatic tube insertion. The hose clamp was tighten as much as possible. Then a keychain ring was attached to the clamp using cable ties. The latex tube and braided sleeve were aligned on the other end and folded for 1 inch. Another keychain ring was inserted at the fold before another hose clamp was inserted to lock down the fold. A single strap of Velcro belt was into the desired keychain ring and stitched. The steps above were repeated for all four pneumatic muscles as shown in Fig. 3. The Velcro belt strap could be stitched on either end of the semi-circle keychain ring. The pneumatic muscles can be adjusted between 20 and 27 cm in total length to fit the particular user.

3.2 Prototype

Figure 4 shows the fully assembled system as both extent of the hardware and software developed were merged in order to complete the proposed soft exoskeletal system. The exercise done on the left wrist has been successfully detected via flex sensor and mimicked on the right wrist using pneumatic muscle contraction which is then validated with the IMU sensor. Furthermore, two separate Arduino circuitry had to be built in order to cope with the programming extent of the flex sensors and IMU sensor. The positioning of the IMU sensor was also strategically relocated so that a single IMU sensor was sufficient to detect both the yawing and pitching motions of the unsound limb. Moreover, both circuits are attached with individual LCD displays to provide direct angular data to the user so that the user doesn't need to rely solely on the PC's serial monitor to obtain the data. In a whole, four flex sensors and a 20 × 4 LCD will be connected to the left arm which is the sound limb and on the right arm which is the unsound limb, a single IMU sensor and a 16 × 2 LCD will be mounted. Colour coded LEDs were used to signify certain operations. The extensive hardware also include the pressure control valves, solenoid valves and compressed air supply. The prototype is designed as table-top setup.

3.3 Working Algorithm

Figure 5 displays the schematic of the pneumatic muscle with respect to the wrist joint. The work done by the pneumatic muscle results in ΔL in terms of force, while the corresponding work done on the wrist joint results in $\Delta\theta_L$ in terms of torque. This mechanism is known as the virtual work [6]. The force applied by the pneumatic muscle will be negative as it contracts in length.

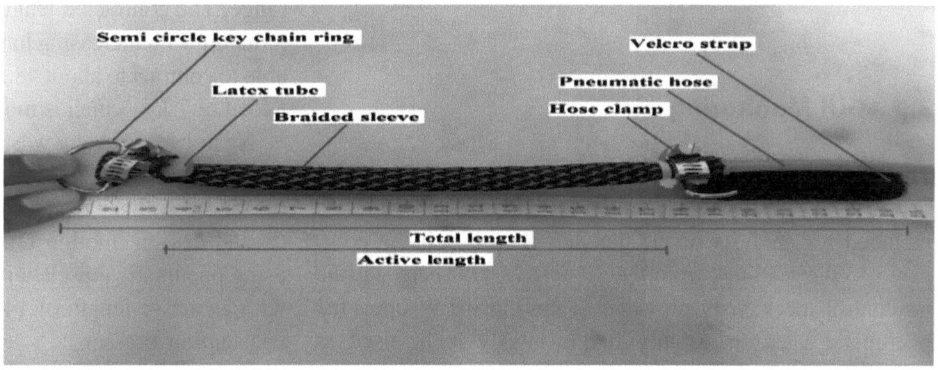

Fig. 3 Self-fabricated pneumatic muscle

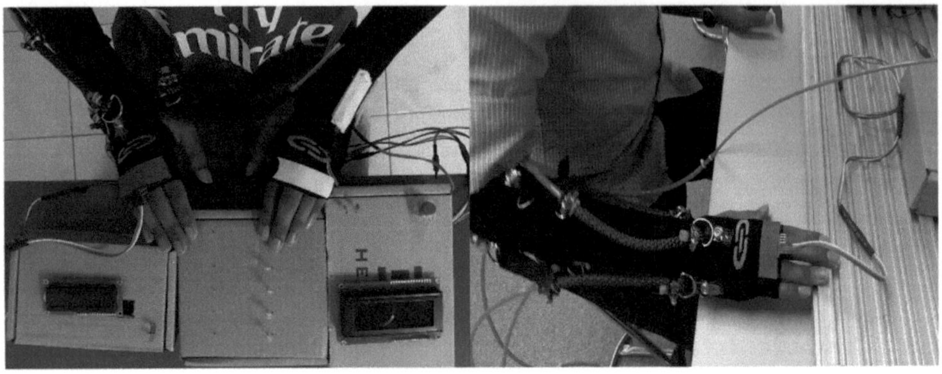

Fig. 4 Final prototype developed

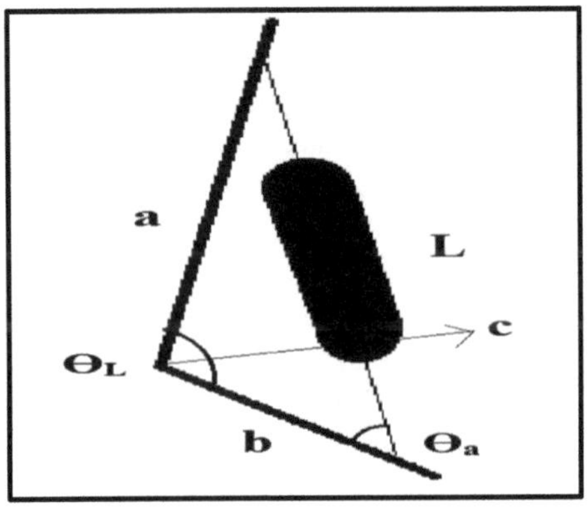

Fig. 5 Schematic of wrist joint with pneumatic muscle

$$Work\,done(pneumatic\,muscle) = -F * \Delta L$$
$$Work\,done(wrist\,joint) = T * \Delta\theta_L$$

Once that was done, the force created by the pneumatic muscle had to be addressed. Generally force is the product of area and pressure. However the area of the pneumatic muscle was tough to validate. Hence, the approximate volume of the pneumatic muscle was instead calculated. Moreover, the average value of braided thread angle was resolved to be 25.31° as well as the theoretical number of turns of the thread is sorted to be 3.776 [7]. This value can be used to calculate thread length and number of turns of the thread, thus validating the diameter of the pneumatic muscle for a known length L.

3.4 GUI Monitoring

The GUI design on the LabVIEW software is altered for better data analysis and comparison. Four graphs will be included for analysis of the flex sensors and will be compared to the readings from the 20 × 4 LCD. There will be no control mechanism provided to the LabVIEW upon the pneumatic actuators as the GUI is merely to be used to monitor and control the sensor data obtained. For example, if the flex sensor readings of the Arduino differs greatly from the GUI, appropriate recalibration of the sensors can be carried out before any component failure occurs. The data could even be logged into an Excel spreadsheet if required. Adding to that, data from the IMU sensor will also be recorded and displayed in graphs to monitor signal fluctuations. The two main data to be extracted are from the accelerometer and gyroscope of the IMU sensor that will be displayed in two separate graphs with 3 linearized signal each. The readings obtained are instantaneous as the calculations are done within an infinite loop on the software.

4 Results and Discussion

The proposed soft exoskeleton system was put through a total of five experimental tests were carried out using 5 human subjects. The tests were related to identifying the contraction rate of the fabricated pneumatic muscles, the pneumatic muscle loading range, the dimensions of the human wrist, the deflection angles during a wrist motion and the calibration of the flex sensors as shown in Table 1.

The average mass of human arm is 4.7–6.5% of the body mass. The test was carried out using increasing loads from 0.5 to 7.0 kg with constant air pressure of 3.0 MPa. Figure 6 displays the end result of contracted length with increasing load and without load. The pneumatic muscle with 14 cm active length was chosen as it had the largest contracted length range. The contracted length of the pneumatic muscle began reducing with increasing load. The pneumatic muscle was still able to contract to about 48.57% of the initial contraction with a load of 7 kg, which is much heavier than most human arm.

Next, every subject was required to carry out 3 repetitions of the four wrist motions, starting with radial deviation (motion A), ulnar deviation (motion B), extension (motion C) and flexion (motion D). The experiment was carried out using pneumatic muscle with adjustable length of 21–27 cm, with an active length of 14 cm and supplied with 3 MPa of constant air pressure.

Table 1 Efficiency rate for each motion based on each subject

Δ Angle (%)		Subject 1	Subject 2	Subject 3	Subject 4	Subject 5	Efficiency (%)		Subject 1	Subject 2	Subject 3	Subject 4	Subject 5
Motion A	*1*	39.0	87.8	45.7	52.3	97.7	Motion A	A	67.9	79.4	39.7	51.5	89.8
	2	80.0	68.8	30.2	50.5	80.6		V					
	3	84.7	81.8	43.4	51.7	91.0		G					
Motion B	*1*	7.3	54.0	46.0	89.3	79.6	Motion B	A	48.6	49.0	44.8	49.5	86.0
	2	58.6	45.2	39.0	21.4	93.0		V					
	3	79.8	47.7	49.4	37.7	85.4		G					
Motion C	*1*	51.5	43.7	35.3	22.3	51.0	Motion C	A	50.9	35.2	37.0	21.1	69.1
	2	51.6	29.8	35.0	19.6	77.1		V					
	3	49.7	32.1	40.9	21.5	79.2		G					
Motion D	*1*	88.2	64.5	43.0	45.8	49.6	Motion D	A	85.5	53.1	52.9	40.9	62.5
	2	72.9	38.3	60.3	36.3	72.9		V					
	3	95.3	56.6	55.4	40.7	65.1		G					

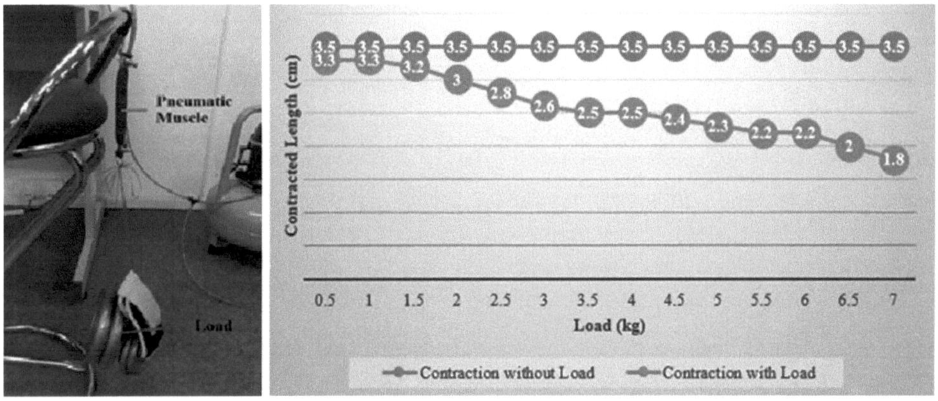

Fig. 6 Pneumatic muscles extension with load and without load

The measured angular readings were averaged to obtain efficiency after readings collected from 3 repetitions for each wrist motions. The highest efficiency for motion A, motion B and motion C was attained for subject 5 at 89.8, 86.0 and 69.1% respectively, while motion D registered 85.5% highest accuracy for subject 1. The effectiveness for each motion varies since each motion have varying kinematics. In a whole, the prototype has an overall efficiency of 55.72% taking into account the difference among the human subject's arm-wrist proportion, pneumatic muscle lengths and type of motion.

5 Conclusion

In a nutshell, the soft exoskeleton provides comfort to the wearer with its stretchable design and subjects minimal load on the wearer's arm. Using simple sensor integration and low cost pneumatic actuator, the proposed system is a great step going forward in the wearable robot field of study. Moving on, the real-time monitoring and controlling of the soft exoskeleton was developed to assist the user to obtain performance data and statistics for better manageability of the soft exoskeleton system.

Based on this research, limitations were present in the compressive arm sleeve used. The flexibility of the material allowed too much stretching to occur when the pneumatic muscles are pressurized at higher pressure. This results in the compressive force generated by the pneumatic muscle to being reduced by the overly stretching sleeve. This affects the performance of the system.

The enhancement that could lift the versatility of this proposed system is the inclusion of a compact and mobile pneumatic supply. With this approach, the soft exoskeleton may be worn indoors and even outdoors. The entire system would be more sufficient, rather than being just a table top setup.

Acknowledgements A special thanks to Asia Pacific University of Technology and Innovation for funding the project. Similarly, thanks to APCORE (Asia Pacific University Center of Robotic Engineering) members for their valuable contribution to the development of the system. Finally, thanks to each and every one who contributed either directly or indirectly to the project.

References

1. Rahman, M.H., Rahman, M.J., Cristobal, O.L., Saad, M., Kenné, J. P., Archambault, P.S.: Development of a whole arm wearable robotic exoskeleton for rehabilitation and to assist upper limb movements. Robotica **33**(1), 19–39 (2015)
2. Ganesan, Y., Gobee, S., Durairajah, V.: Development of an upper limb exoskeleton for rehabilitation with feedback from emg and imu sensor. Procedia Comput. Sci. **76**(1), 53–59 (2015)
3. Yap, H.K., Lim, J.H., Nasrallah, F., Goh, J.C., Yeow, R.C.: A soft exoskeleton for hand assistive and rehabilitation application using pneumatic actuators with variable stiffness. In: IEEE international conference on robotics and automation (ICRA), pp. 4967–4972. Washington, 26–30 May 2015
4. National Stroke Association of Malaysia: http://www.nasam.org. Last Accessed 03 Sept 2016
5. Mackay, J., Mensah, G.A., Mendis, S., Greenlund, K.: The atlas of heart disease and stroke. World Health Organization (2004)
6. Varga, Z., Moučka, M.: Mechanics of pneumatic artificial muscle. J. Appl. Sci. Thermodyn. Fluid Mech. **3**(2), 6–11 (2009)
7. Force and Torque. http://www.shadmehrlab.org/book/force_and_torque/forcetotorque.htm. Last Accessed 18 Jan 2017

Part III

Biosensing and Life Sciences

Association of Rainfall and the Occurrence of Pathogenic *Leptospira* spp. in Recreational Stream Water, Hulu Langat, Selangor

K. Dzulaikha, M. Y. Nurul Yuziana, J. J. Maizatulriah, and A. W. Marfiah

Abstract

Leptospirosis is a human disease caused mainly by *Leptospira interrogans*. Global Leptospirosis burden is significant especially in a tropical climate country like Malaysia. However, inadequate detection in environments and recreational areas has affected the awareness of the disease among the public community. In this study, the presence of pathogenic *Leptospira* spp. in recreational stream water samples from Sg. Tekala, Hulu Langat was described. A total of 126 water samples were collected randomly in bi-weekly time frame from February to July 2016 and tested for pathogenic *Leptospira* spp. using a PCR assay targeting the *lipL32* gene. To assess the link between the presence of *Leptospira* spp. and rainfall, data on daily rainfall during six months of sampling have also been recorded from Malaysian Meteorological Department. The pathogenic *Leptospira* spp. were detected in 27/126 (21.4%) of samples. Statistical evaluation using Spearman's Correlation Test on the association of *Leptospira* detected samples and rainfall data has shown a significant negative relationship, where a probability of pathogenic *Leptospira* spp. to be detected in water was higher at the time of low rainfall. This was the first report on the association of rainfall and the occurrence of pathogenic *Leptospira* spp. in recreational water. The connotation between the rainfall and presence of pathogenic *Leptospira* spp. in recreational water suggested the needs to develop a systematic approach for *Leptospira* spp. monitoring especially during dry season.

Keywords

Pathogenic *leptospira* and rainfall • PCR • Recreational water

1 Introduction

Leptospirosis is an emerging infectious disease, caused by pathogenic *Leptospira* spp. which is endemic with a major health impact in Malaysia. Generally, the genus *Leptospira* is divided into two major *Leptospira interrogans* species consist of all pathogenic strains and *Leptospira biflexa* comprising the saprophytic strains [1, 2].

A broad range of animals could serve as reservoir hosts, from rodents to domestic animals like dogs, pigs and cattle to wild animals include chiropterans [3], marsupials [3, 4], squirrels [3, 5], mongoose [3, 4, 6], monkeys [5] and bats [7]. Tropical climate characteristic of Malaysia with high seasonal rainfall and warm temperature has lengthen the viability of *Leptospira* spp. in the environment [5]. Chances of the pathogenic *Leptospira* spp. in the environment to infect humans through skin abrasions, cuts or mucous membrane are high during recreational activities such as swimming, rafting, barbequing and picnicking at the recreational area. Poor hygienic condition at the recreational area

K. Dzulaikha · J. J. Maizatulriah · A. W. Marfiah (✉)
Division of Water Resources and Environmental System, Faculty of Civil Engineering, Universiti Teknologi MARA, Shah Alam, Malaysia
e-mail: marfi851@salam.uitm.edu.my

M. Y. Nurul Yuziana
School of Biosciences, Faculty of Science and Technology, Universiti Kebangsaan Malaysia, Selangor, Malaysia

F. Ibrahim et al. (eds.), *2nd International Conference for Innovation in Biomedical Engineering and Life Sciences*,
IFMBE Proceedings 67, https://doi.org/10.1007/978-981-10-7554-4_20

(usually from barbequing and picnicking activities) has been identified as a major attraction for rodents and wild animals to come and pollute the area with their urines which in turn pose risk to *Leptospira* spp. infection [8].

Several PCR-based methods for detection of pathogenic *Leptospira* spp. in sera, urine and blood samples after DNA extraction have been reported in the last decade [1, 4, 9, 10]. However, study on direct detection by PCR of pathogenic *Leptospira* spp. in water and environment are very few. This study assessed the relative performance of LipL32-PCR for the detection of pathogenic *Leptospira* spp. using simple pre-treatment method before proceeding with PCR techniques.

Sungai Tekala is amongst the popular recreational areas in Hulu Langat, Selangor. Situated 30 km from main city of Kuala Lumpur, this recreational site is visited by almost 100–200 visitors during weekends. There is no data on the occurrence of pathogenic *Leptospira* spp. at Sungai Tekala recreational area has been reported. However, it was reported that 32% of Leptospirosis cases in Malaysia were contributed from recreational areas [11]. Inspection by Public Health Laboratory of Ministry of Health Malaysia in June 2016 suggested that 11% of 71 recreational areas were detected positive for Leptospira contamination. However, this result did not guarantee that negatively tested sites were completely free from the infection [11]. The objective of this study was to provide a data on the presence of pathogenic *Leptospira* spp. and its relationship with rainfall to provide a systematic approach for monitoring of pathogenic *Leptospira* spp. in recreational areas. This kind of information is required to effectively implement a program to prevent the transmission of Leptospirosis.

2 Materials and Methods

2.1 Primer Design and Specificity Test for *LipL32 Gene*

The Forward and Reverse Primers in this study were generated based on 819 bp *LipL32* full gene sequence using PrimerExplorer V5 software available at http://primerexplorer.jp/e/. The primers specificity towards *LipL32* gene was demonstrated in silico using primer-BLAST program available at http://blast.ncbi.nlm.nih.gov/Blast.cgi. Results from primer-BLAST specificity tool linked to NCBI databases showed the Forward primer 5′-AAGCATACTATCTC TATGTTTGG-3′ and Reverse primer 5′-TTGGTCAGG CATAATCGC-3′ have generated alignments exclusively against *LipL32* gene and no matches were found for other environmental microbes or genes. For further confirmation, the primers specificity was also tested with other environmental species including *Burkholderia cepacia, B. thailandensis, B. pseudomallei, Bacillus subtilis, B. macerans, B. circulans, B. megaterium, Staphylooccus sp., S.epiderminis, S. heamolyticus, E. coli,* and *Shinella granuli.*

2.2 Optimization of LipL32-PCR Using Synthetic DNA of LipL32 *Gene*

The control template for LipL32-PCR optimization was constructed synthetically using *LipL32* gene incorporated into pIDTSMART_lipL32 plasmid (IDT, USA). The annealing temperature of LipL32-PCR was optimized using Gradient PCR (Bio Rad, California, USA) ranging from 52 to 58.3 °C.

2.3 Samples Collection

The stream water samples were obtained from Sungai Tekala recreational area located at Semenyih district in Langat River Basin, Selangor. A total of nine water samples were collected during each sampling day. Samples were collected at the areas with characteristics such as shaded areas, low current or stagnant water and near animal footprints. The sampling was done by dipping a 500 ml sterile water container directly into the water. The samples were then transported to the laboratory in ice and processed within 24 h of samples collection. The samples were collected randomly in bi-weekly time frame from February to July 2016 and tested for pathogenic *Leptospira* spp. using LipL32-PCR.

2.4 Water Samples Preparation

Five hundred milliliters of sample water were filtered through 1.5 µm (PFTE type, 47 mm diameter, Supelco, MO, USA) with a sterile Sterifil® Aseptic System and Filter Holder (Sigma-Aldrich, USA) and a vacuum unit to remove all fine particles in water samples included mud and debris. Leptospires (motile spiracle shape and 0.2–0.3 µm in size) has the ability of burrowing penetrated this filter and were collected in the filtrate. The filtrate was then re-filtered them using 0.20 µm membrane filter (Nitrocellulose, 25 mm diameter, GVS, USA) with sterile Pall® reusable syringe filter holder (Pall, USA). After filtration, the filter membrane was removed from the filter holder and placed in a 1.5 ml sterilized microfuge tube containing 500 µl of cell lysis buffer [025% Triton X-100, 10 mM M Tris (pH 8.0) and 1 mM EDTA]. The tube was then heated at 95 °C for 10 min in order to lyse the cells. The concentrated lysate was then subjected directly to PCR amplification.

2.5 Spiked Sample

The spiked sample for internal control of the test and a whole sample processing was simulated using *E.coli*-LipL32 (host carrying the pIDTSMART_LipL32 gene) to mimic alive bacteria in recreational water. The *E.coli*-LipL32 surrogates (0.5 µm in size) were expected to pass the first filtration process by pressure and captured on the 0.20 µm filter during the second filtration process, same as Leptospires. The spiked sample of the experiment was prepared by adding 10 µl of 10^4 cfu/ml *E.coli*-LipL32 into 500 ml of stream water sample to get a total of 20 cfu/ml *E.coli*-LipL32 and followed by water sample preparation method.

2.6 LipL32-PCR of Recreational Water Samples

The samples were firstly prepared according to water samples preparation. The LipL32-PCR amplifications were performed in 25 µl reaction containing 10X GoTaq PCR buffer (Promega, USA), 2 mM dNTPs (New England Biolabs, USA), 40 mM $MgCl_2$ (Promega, USA), 0.1 U Go*Taq* DNA polymerase (Promega, USA), 10 pmol/µl Forward and Reverse primers (Integrated DNA Technology, USA) and 2.5 µl DNA source (samples or spiked). The program for LipL32-PCR started with initial denaturation at 95 °C for 5 min, followed by 35 cycles of denaturation at 95 °C for 15 s, annealing at (55.9 °C) for 20 s and elongation at 72 °C for 30s and then final elongation for 2 min at 72 °C. The PCR products were run in 2% agarose gel and stained using the SYBR Safe (Bio Rad, California, USA).

2.7 Rainfall Data Collection

The data on daily rainfall for Semenyih district were requested from Meteorological Department of Malaysia. The rainfall data were then aligned with the PCR results according to the day of sampling. The graphs on the association of rainfall (in a mean of 7 days prior sampling date) and occurrence of positive samples were plotted using Microsoft Excel 2016.

2.8 Statistical Analysis

The statistical analysis of this study was performed using SPSS Statistic Data Editor version 23. Spearman's Rank Correlation Analysis was used to evaluate the strength of the relationship between rainfall and the occurrence of pathogenic *Leptospira* in Sg. Tekala recreational water. The equation for Spearman's correlation coefficient, rho (ρ) is given by:

$$\rho = 1 - \frac{6\sum d_i^2}{n(n^2 - 1)} \quad (1)$$

where di = difference in paired ranks and n = number of detected samples.

The rainfall was calculated in a mean of seven days before the sampling day to give a collective view of rainfall pattern. Both variables were ranked separately where the scales 1–5 indicated the dense of rain (mean of rainfall) and the number of detected samples (occurrence). The result of Spearman's Correlation Analysis was indicated in a range between −1 and +1.

3 Results

3.1 Optimum *LipL32* PCR Amplifications

Although, the annealing temperature for optimum amplification of *LipL32* showed that the annealing temperatures between 52.7 and 58.3 °C were suitable for LipL32 PCR amplification because all of them showed positive band. However, the annealing temperature of 55.9 °C showed slightly thick band and good consistency in further PCRs. Therefore, this temperature has been chosen for the next whole experiment. The expected size of PCR product appeared in the agarose gel was 238 bp as shown in Fig. 1.

3.2 Samples Collection and PCR Amplification

Results showed 27 from 126 samples were positive for pathogenic *Leptospira* spp. The LipL32-PCR amplification was tested in triplicates in the different PCR runs. Example of results for 21st February 2016 has shown 5 samples were tested positive out of 9 samples collected as shown in Fig. 2.

3.3 Spearman's Correlation Analysis

The results of Spearman's Correlation Analysis can be divided into three categories; positive association, negative association or no association between variables. The negative (−) results show a negative relationship (the higher score on X associated with the lower score on Y). Positive (+) results indicate a positive relationship (the higher score on X associated with the higher score on Y). No predictable results indicate no association exists. Results on the Spearman's Correlation Test were given in Table 1. To give a better view of association trend of *Leptospira* occurrence in recreational water and rainfall pattern, a graph in Fig. 3 was plotted.

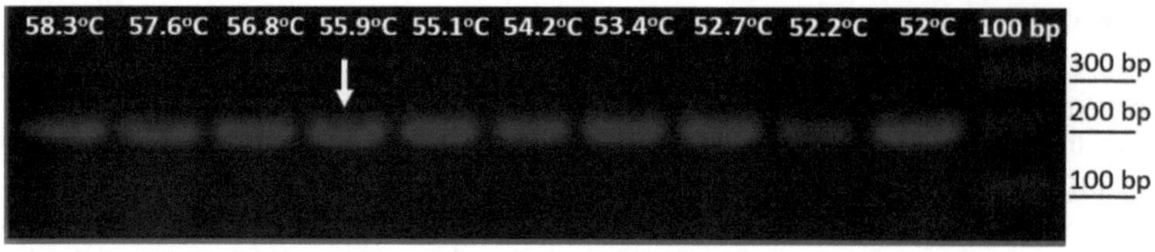

Fig. 1 The optimization of annealing temperature for LipL32-PCR amplification. The annealing temperature for the PCR reaction was optimized using gradient PCR ranging from 52 to 58.3 °C. The optimum annealing temperature for LipL32-PCR amplification chosen for this study was 55.9 °C when this temperature showed slightly thick band and good consistency

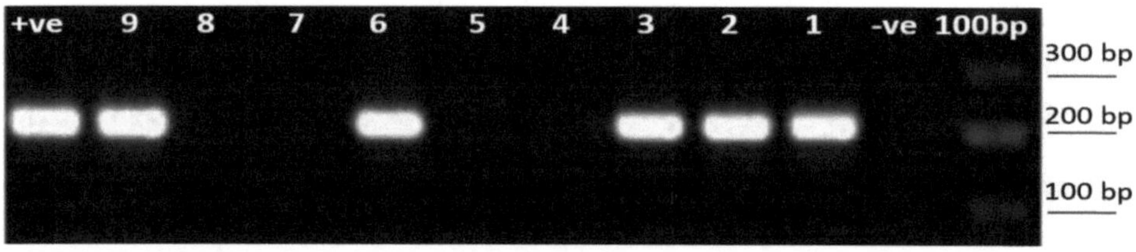

Fig. 2 Results of LipL32-PCR amplification of samples collected on 21st February 2016. The PCR products of samples and controls showed bands at around 238 bp. Lane 1–9 represents the band produced from LipL32-PCR amplification of nine samples collected randomly along the stream. Lane −ve is negative control (non-template control) and lane +ve is the spiked samples using *E.coli*-LipL32 surrogates

Table 1 The Spearman's correlation test showed a negative association between the dense of rainfall in a week and the occurrence of pathogenic *Leptospira* spp.

			Rainfall (mm)	Occurrence (no.)
Spearman's rho	Rainfall (mm)	Correlation coefficient	1.000	−0.718**
		Sig. (2-tailed)	–	0.004
		N	14	14
	Occurrence (no.)	Correlation coefficient	−0.718**	1.000
		Sig. (2-tailed)	0.004	–
		N	14	14

**Correlation is significant at the 0.01 level (2-tailed)

4 Discussion

The PCR method is rapid, sensitive, specific and does not necessarily require the initial isolation and culture of *Leptospira* [10, 12]. Only a few reported studies can detect *Leptospira* directly from environmental samples using PCR. A serious problem in the use of PCR to directly detect microorganisms in environmental samples is the potential presence of natural compounds such as humic acids which can inhibit a necessary step in the PCR method [13, 14]. This limitation on direct detection of pathogenic *Leptospira* spp. in environment usually re-directing the researcher to culture the concentrated water sample by centrifugation in Ellinghausen-McCullough-Johnson-Harris (EMJH) plate until it grows and follows by PCR for confirmation [8, 15]. This method indeed is time-consuming. In environmental waters, saprophytic *Leptospira* bacteria are always naturally presented in high concentrations in streams while pathogenic *Leptospira* are believed to be presented sporadically and at much lower concentrations [5]. The growth of saprophytic *Leptospira* consistently predominates and interferes with the recovery of pathogenic *Leptospira*. Moreover, the long incubation period (3–4 weeks and sometimes extended incubation period until 16 weeks) [4, 16] required for the culture of *Leptospira* has prevented the useful application of the data on pathogenic *Leptospira* spp. contamination. This fact has denied the ability of culturing method to be useful to recreational areas especially when it comes to life-threatening disease like Leptospirosis.

Removing process of all fine particles and mud's were done in the first filtration step using 1.5 µm pore size. The recovery of *Leptospira* later was done in the second filtration step where the 0.2 µm membrane filter would retain the *Leptospira* that sizes of around 0.2–0.3 µm in diameter. A nearly 100% recovery of *Leptospira* using 0.22 µm of membrane filter have been reported by several studies on

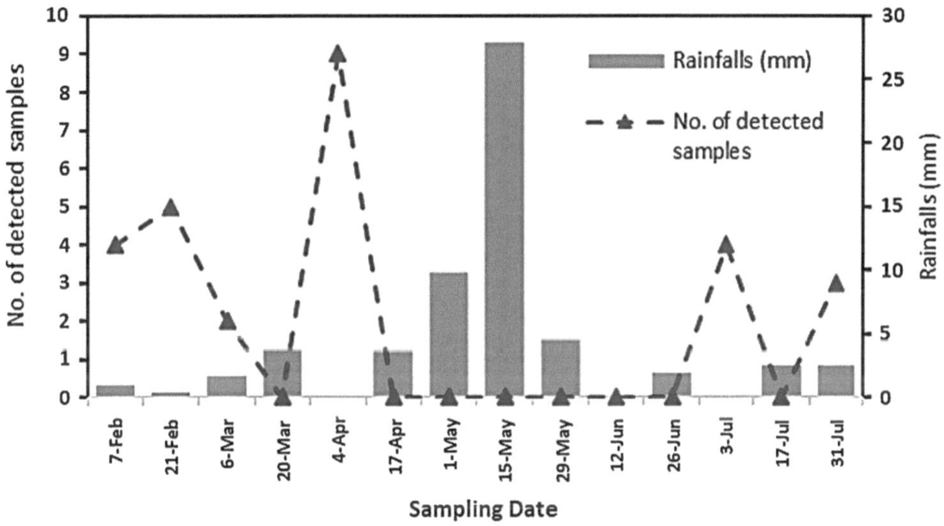

Fig. 3 The graph showed the association between the mean of rainfall in a week before sampling and the presence of pathogenic *Leptospira* spp. in recreational water. The highest number of *Leptospira* were detected during a mean of rainfall is 0. While during heavy raining season between 17th April and 26th June 2016, zero (0) number of *Leptospira* was detected. It is predictable that the presence of *Leptospira* would be higher during low raining season, and lower during heavy raining season. A total of 27 samples were detected positive *Leptospira* giving 21.4% from 126 total samples

filtration method of *Leptospira* from water samples [17, 18]. This step also passed through the un-wanted humic acid and other natural compound that could possibly retard the PCR amplification proses [17]. The subsequently lysis step during samples processing would release the DNA of trapped *Leptospira* and other similar sizes of bacteria and viruses. However, high specificity primers used in LipL32-PCR would selectively discriminate the other non-LipL32 *Leptospira* targets.

Rainfall data for Semenyih district, Hulu Langat were collected from Malaysian Meteorological Department and compared to the presence of *Leptospira* in stream water. Statistical analysis using Spearman's Correlation Test showed there was a significant negative relationship between rainfall and the occurrence of pathogenic *Leptospira* spp. in Sg. Tekala recreational water. Result $\rho\ (12) = -0.718$, $p = 0.001$ as shown in Table 1 gave conclusion that relationship between both variables were strong.

The trend on the occurrence of pathogenic *Leptospira* versus rainfall means for seven days prior sampling was demonstrated on a graph in Fig. 3. The transportation of microbial and other contaminants in stream water are controlled by the flow of water, therefore, rainfall is the key factor of this circumstance [19]. Our findings suggested that peak timing of occurrence of pathogenic *Leptospira* spp. was the time where the rainfall occasion was very low. The low rainfall or dry season in tropical climate country like Malaysia is also associated with the rising in daily temperature exceeding 32–35 °C [20]. This condition, in turn, raises the average temperature of the water body. The warm temperature of the water during this period favors the incubation and stimulates the replication of pathogenic *Leptospira*. Moreover, the stream water is also the source of water for other wild animals, especially during hot weather. The wild animals, which most of them are reservoir hosts, have made the situation worse when they might also contaminate the stream with urines containing the pathogen. In this situation, a higher concentration of pathogenic *Leptospira* in stream water are expected. A low streamflow at this period of time induces the accumulation of this bacteria thus increases their concentration level in the water body. Our findings showed no appearance of pathogenic *Leptospira* during the peak time of rainfall. This result endorsed high rainfall event would directly increase the streamflow thus, dilute the level of pathogenic *Leptospira* in water. High rainfall or storm water was also thought to flush all the microbes including pathogenic *Leptospira* subsequently run off into surface waters [19].

Statistical analysis showed that probability of pathogenic *Leptospira* spp. to be detected in water was higher at the time of low rainfall and no incidence of heavy rainfall within a week prior sampling time. This evaluation could give an overall picture that the availability of *Leptospira* in stream water was depending on the rainfall occasion. Thus, this has given a suggestion for a regular monitoring of pathogenic *Leptospira* spp. to be done during the dry seasons.

5 Conclusion

The connotation between the rainfall and presence of pathogenic *Leptospira* spp. in recreational water suggest the needs to develop a systematic approach for *Leptospira* spp. monitoring especially during low rainfall season. The predictable occurrence of pathogenic *Leptospira* together with a good detection method and regular monitoring would serve as a preventive measure at the recreational area, where more lives could be saved.

Acknowledgements The authors would like to thank the Ministry of Higher Education, Malaysia for the financial support of this study through **FRGS/1/2016/STG05/UITM/02/1**.

References

1. Vital-Brazil, J.M., Balassiano, I.T., de Oliveira, F.S., Pereira, M.M., Costa, A.D.D.S., Hillen, L.: Multiplex PCR-based detection of leptospira in environmental water samples obtained from a slum settlement. Mem. Inst. Oswaldo Cruz Rio Janeiro. **105**(3), 353–355 (2010)
2. Lin, X., Chen, Y., Lu, Y., Yan, J., Yan, J.: Application of a loop-mediated isothermal amplification method for the detection of pathogenic Leptospira. Diagn. Microbiol. Infect. Dis. **63**(3), 237–242 (2009)
3. Thayaparan, S., Robertson, I.D., Fairuz, A., Suut, L., Abdullah, M. T.: Leptospirosis, an emerging zoonotic disease in Malaysia. Malays. J. Pathol. **35**(2), 123–132 (2013)
4. Bharti, A.R., et al.: Leptospirosis: a zoonotic disease of global importance. Lancet Infect. Dis. **3**(12), 757–771 (2003)
5. Azali, M.A., Yean Yean, C., Harun, A., Aminuddin Baki, N.N., Ismail, N.: Molecular characterization of Leptospira spp. in environmental samples from North-Eastern Malaysia revealed a pathogenic strain, leptospira alstonii. J. Trop. Med. **2016** (2016)
6. Fujioka, R.S., Asahina, A.Y.: Development of a feasible method to detect pathogenic Leptospira bacteria in environmental waters. Honolulu (HI): Water Resources Research Center, University of Hawaii at Manoa. WRRC-98-09 (1998)
7. Bunnell, J.E., Hice, C.L., Watts, D.M., Montrueil, V., Tesh, R.B., Vinetz, J.M.: Detection of pathogenic Leptospira spp. infections among mammals captured in the Peruvian Amazon basin region. Am. J. Trop. Med. Hyg. **63**(6), 255–258 (2000)
8. Ismail, S., et al.: A study on the presence of pathogenic Leptospira spp. in environmental water samples obtained from selected recreational areas in Terengganu, Malaysia. Res. J. Pharm. Technol. **7** (2014)
9. Sachse, K., Nat, R.E.R., Frey, J.: PCR detection of microbial pathogens. Methods Mol. Biol. **216**(5), 235–242 (2003)
10. González, S., Geymonat, J.P., Hernández, E., Marqués, J.M., Schelotto, F., Varela, G.: Usefulness of real-time PCR assay targeting lipL32 gene for diagnosis of human leptospirosis in Uruguay. J. Infect. Dev. Ctries. **7**(12), 941–945 (2013)
11. Cheng, N.: Waspada Risiko Virus Kencing Tikus Di Kawasan Rekreasi. Mstar Online
12. Lucchesi, P.M.A., Arroyo, G.H., Etcheverría, A.I., Parma, A.E., Seijo, A.C.: Recommendations for the detection of Leptospira in urine by PCR. Rev. Soc. Bras. Med. Trop. **37**(2), 131–134 (2004). https://dx.doi.org/10.1590/S0037-86822004000200003
13. Rutjes, S.A., Van Den Berg, H.H.J.L., Lodder, W.J., De Roda Husman, A.M.: Real-time detection of noroviruses in surface water by use of a broadly reactive nucleic acid sequence-based amplification assay. Appl. Environ. Microbiol. **72**(8), 5349–5358 (2006)
14. Tsai, Y.L., Olson, B.H.: Rapid method for separation of bacterial DNA from humic substances in sediments for polymerase chain reaction. Appl. Environ. Microbiol. **58**(7), 2292–2295 (1992)
15. Rawlins, J., et al.: Molecular detection of Leptospiral DNA in environmental water on St. Kitts. Int. J. Environ. Res. Public Health **11**(8), 7953–7960 (2014)
16. Fujioka, R., Wilson, R.: Development of a method to selectively isolate pathogenic Leptospira from environmental samples. Water Sci. Technol. **31,** 275–282 (1995)
17. Walker, M.: Method development and preliminary applications for *Leptospira Spirochetes* in water samples. Off. Res. Dev. (2008)
18. Kaboosi, H., Razavi, M.R., Al, A., Noohi, S.: Efficiency of filtration technique for isolation of *Leptospires* from surface waters: role of different membranes with different pore size and materials. Afr. J. Microbiol. Res. **4**(9), 671–676 (2010)
19. Jagai, J.S., Griffiths, J.K., Kirshen, P.K., Webb, P., Naumova, E. N.: Seasonal patterns of gastrointestinal illness and streamflow along the Ohio River. Int. J. Environ. Res. Public Health **9**(5), 1771–1790 (2012)
20. Amirabadizadeh, M., Huang, Y.F., Lee, T.S.: Recent trends in temperature and precipitation in the Langat River basin, Malaysia. Adv. Meteorol. **2015**, 1–16 (2015)

Optimal EEG Channel Selection for Vascular Dementia Identification Using Improved Binary Gravitation Search Algorithm

Noor Kamal Al-Qazzaz, Sawal Hamid Bin Mohd Ali, Siti Anom Ahmad, and Javier Escudero

Abstract

The aim of the present study was to select optimal channels that may help in detecting the abnormalities in the electroencephalogram (EEG) of vascular dementia (VaD) patients. Spectral entropy $(SpecEn)$, approximation entropy $(ApEn)$ and permutation entropy $(PerEn)$ have been extracted from the EEG background activity of 5 VaD, 15 patients with stroke-related mild cognitive impairment (MCI) and 15 healthy control subjects during a working memory (WM) task. EEG artifacts were removed using automatic independent component analysis and wavelet denoising technique (AICA-WT). In order to reduce the computational time, improved binary gravitation search algorithm (IBGSA) channel selection was used to find the most effective EEG channels for VaD patients' detection. Eight channels were found suitable to extract EEG markers that help to detect dementia in the early stages. Moreover, *k*-nearest neighbors (*k*NN) was used after the IBGSA technique. The IBGSA technique increased the *k*NN classification accuracy from 86.67 to 90.52%. These results suggest that IBGSA consistently improves the discrimination of VaD, MCI patients and control normal subjects and it could be a useful feature selection to help the identification of patients with VaD and MCI.

Keywords

Improved binary gravitation search algorithm • Electroencephalography • Dementia • Channels selection

1 Introduction

Dementia is considered as the fourth most common brain disorder in the elderly population. It is identified by progressive impairment in memory and thinking whose impairment is severe enough to cause a decline in the patient activity of daily living. Mild cognitive impairment (MCI) is common after stroke [1, 2]. MCI is the loss of cognitive function, particularly attention, executive function and memory [3]. Clinically, mild cognitive impairment (MCI) is considered as the traditional stage between early normal brain cognition and late severe dementia [4]. Efforts are being made to detect dementia early before substantial mental decline occurs. Therefore, it is important to develop neurophysiological markers sensitive to dementia.

N. K. Al-Qazzaz · S. H. B. M. Ali
Department of Electrical Electronic & Systems Engineering, Faculty of Engineering & Built Environment, Universiti Kebangsaan Malaysia, Selangor, Malaysia

N. K. Al-Qazzaz
Department of Biomedical Engineering, Al-Khwarizmi College of Engineering, Baghdad University, Baghdad, Iraq

S. A. Ahmad (✉)
Department of Electrical and Electronic Engineering, Faculty of Engineering, Universiti Putra Malaysia, Selangor, Malaysia
e-mail: noorbmemsc@gmail.com

S. A. Ahmad
Malaysian Research Institute of Ageing (MyAgeing), Universiti Putra Malaysia, Selangor, Malaysia

J. Escudero
Institute for Digital Communications School of Engineering, The University of Edinburgh, Edinburgh, UK

F. Ibrahim et al. (eds.), *2nd International Conference for Innovation in Biomedical Engineering and Life Sciences*,
IFMBE Proceedings 67, https://doi.org/10.1007/978-981-10-7554-4_21

Electroencephalogram (EEG) is a neurophysiological signal that reflect the electrical activity of the brain [5]. An urgent need for a specific, accurate and cost-effective method to diagnose dementia in the early stages exists. In physiology, the extracted features from EEG signals provide a concise representation that shows the distribution of an EEG signal in using entropy based features. Therefore, EEG features are key to detect interesting information related to cognitive and memory performance [6]. As attention and executive function that associated with working memory (WM) are the most affected domains, WM was considered in this study. WM is the ability to maintain and manipulate information for brief periods (10–15 s up to 60 s). WM is considered as a temporary memory that can store approximately 7 ± 2 items [7, 8]. Studies on EEG signal processing have been conducted to identify the brain activity patterns involved in cognitive process and memory [9].

Feature selection is a process of choosing a subset of features from the original set of features forming patterns in a given dataset. The importance of feature selection is to reduce the problem size and search space for learning algorithms and to improve the quality and speed of classification.

Channel selection is a kind of feature selection, which can be characterized as a special filter [10] used for selecting related-feature channels and removing irrelevant or noisy channels [11]. Many approaches have been proposed for EEG channel selection, for instance the mutual information technique [12], sparse common spatial pattern (SCSP) algorithm to remove noisy or irrelevant channels without affecting the classification accuracy [13], and recursive channel elimination (RCE) method, which could disregard task-irrelevant channels [10]. Although the approach of channel selection can provide the advantages of removing irrelevant channels or selecting few meaningful EEG features to improve the classification performance.

2 Methods and Materials

In order to extract meaningful information from EEG signal of dementia patients' that help to develop valuable markers for early detection and identification of vascular dementia (VaD) and stroke-related mild cognitive impairment (MCI) patients' EEG, the recorded EEG needs successive signal processing and analysis stages. Figure 1 illustrates the block diagram of the proposed method.

2.1 Subjects and EEG Recording Procedure

EEGs were obtained from 35 participants (5 VaD: 2 women and 3 men; age = 64.6 ± 4.8 years, mean ± standard deviation, SD), 15 stroke-related MCI patients (10 women and 5 men; age = 60.26 ± 7.77 years, mean ± standard deviation, SD), and 15 normal records (8 women and 7 men; age = 60.06 ± 5.21 years, mean ± standard deviation, SD) fulfilling the criteria of probable AD. The stroke patients were recruited from the stroke ward of the Pusat Perubatan Universiti Kebangsaan Malaysia (PPUKM), the Medical center of the National University of Malaysia. The VaD patients were recruited from the PPUKM Neurology clinic. The control subjects had no previous history of mental and neurological abnormalities. All control subjects, MCI, and VaD patients underwent cognitive evaluation, including Mini-Mental State Examination [14] and Montreal Cognitive Assessment [15]. The MMSE scores for these VaD and stroke-related MCI patients were 14.8 ± 1.92 and 20.2 ± 5.63 (mean ± SD), respectively. Add to that the MoCA scores for these VaD and stroke-related MCI patients were 13.2 ± 2.38 and 16.13 ± 5.97 (mean ± SD), respectively. The MMSE and MoCA scores for the control subjects were 29.6 ± 0.73 (mean ± SD) and 29.06 ± 0.88 (mean ± SD), respectively.

The EEG datasets were recorded using the NicoletOne system (V32), which was designed and manufactured by VIASYS Healthcare Inc., USA. Nineteen electrodes, as well as ground and system reference electrodes, were positioned on the scalp using cap electrodes according to the 10–20 international system (i.e., Fp2, F8, T4, T6, O2, Fp1, F7, T3, T5, O1, F4, C4, P3, F3, C3, P3, Fz, Cz, and Pz). All experiment protocols were approved by the Human Ethics Committee of the National University of Malaysia. Signed informed consent forms were also obtained from the participants. The session started with a 0.5 s fixation cue when the subjects were asked to be relaxed and avoid the movement activity as much as possible. A simple WM task was then performed, during which the subjects were asked to memorize five words for 10 s. Afterward, they were asked to remember these words with their eyes closed, and the EEG data were recorded. After 60 s, the patients were asked to open their eyes and enumerate all words that they could remember [4, 16].

2.2 AICA-WT Denoising Technique

In the present study, the AICA-WT denoising technique is used to decompose a set of recorded EEG signal into its underlying components. The FastICA algorithm [17] was used because of its simplicity, fast convergence and efficiency to decompose the recorded EEG as in Eq. 1:

$$\mathbf{x}(t) = \mathbf{A}\mathbf{s}(t) \tag{1}$$

where $\mathbf{x}(t)$ is the output vector, A is the mixing matrix, $\mathbf{s}(t)$ is the input vector [18]. The artifactual independent components (ICs) were identified using several statistical metrics

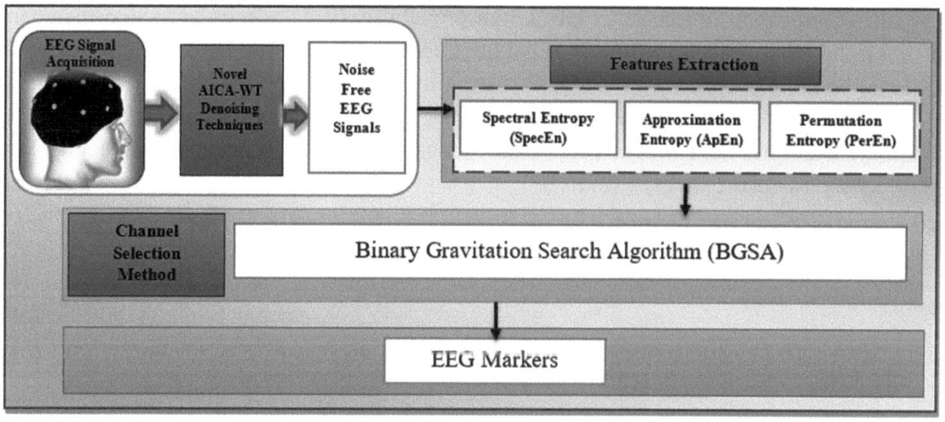

Fig. 1 The block diagram of the proposed method

and were arranged into a new dataset to be ready to pass through the wavelet (WT) Denoising technique. Afterward, the corrected ICs were returned back to the EEG dataset to be reconstructed and extract features in the following stage [19].

2.3 Features Extraction

To characterize the VaD, MCI patients' and control subjects' EEG, Spectral entropy (*SpecEn*), approximation entropy (*ApEn*) and permutation entropy (*PerEn*) features were estimated for the EEG datasets of 15 healthy normal subjects and 15 stroke-related MCI patients. Features were computed for 19 channels with a total length of 15,360 samples divided into six epochs; the length of each epoch was 2560 data points (one segment represents 10 s) of EEG recordings during WM tasks.

SpecEn has been used to distinguish dementia patients EEGs from the normal age-match subjects [18, 20, 21]. In order to estimate the *SpecEn*, the PSD was normalized to a scale from 0 to 1 to get normalized PSD (PSD_n) so that $\sum PSD_n(f) = 1$, afterwards, *SpecEn* is computed as in [18].

ApEn is computed using the algorithm presented in [22]. For our analysis, *ApEn* is computed with a run length of epochs, m = 2 and tolerance, r = 0.2 × SD, where SD is the standard deviation [23].

PerEn is calculated as in [24]. In this study, Given that the brain can investigate the complex dynamic information, entropies including *PerEn* with embedded dimension d = 3 and time delay l = 1 [24].

Therefore in this paper, the total number of features can be computed by multiplying the number of channels by the number of the used feature. The feature matrix was consisted of (90 × 57), where (15 subjects × 6 epochs) = 90 for the healthy control subjects and MCI patients was observations and (3 features × 19 channels) = 57 attributes, while that of the feature matrix for the VaD patients was (30 × 57), where (5 VaD × 6 epochs) = 30 observations and (3 features × 19 channels) = 57 attributes. In this study, the minority class was represented by the VaD patients. A synthetic oversampling technique (SMOTE) was applied to overcome the data imbalance [25].

2.4 Channel Selection Using IBGSA

The improved binary gravitation search algorithm (IBGSA) algorithm has been used to detect the most optimal channels and reduce the amount of information. GSA is an effective optimization algorithm which was designed based on Newtonian laws of gravity and motion and it was introduced for solving binary-valued problems in [26]. In order to find the optimal EEG channels with the aid of IBGSA, N objects (agents) are defined for the algorithm. This group of objects creates the initial population. In this research, each object is considered as a binary vector with the dimensionality of 19. The mentioned dimensionality is equal to the total number of EEG channels. The *i* th object can be considered as the following vector. The main goal is finding the object which has produced the best fitness value. In this study, the classification accuracies for each set of EEG channels are considered as the fitness values for the objects and it can be computed using Eq. 2 [27]:

$$fit_i = \omega_1 \times accu_i + \omega_2 \times \left[1 - \frac{\sum_{j=1}^{p} f_j}{p}\right] \qquad (2)$$

where ω_1, ω_2 are two predefined weight factors, ω_1 is the weight factor for the classification accuracy of the *k*-nearest neighbors (*k*NN) classifiers respectively determined by the 10-fold cross-validation (CV) method; $accu_i$ is the 1-NN classification accuracy; ω_2 is the weight factor for the number of selected features and f_j is the value of feature mask. The weight factor of accuracy can be adjusted to a high value (such as 100%) if accuracy is the most important.

The object with high fitness value has a high probability of affecting the other objects' positions of the next iteration, so it should be set appropriately [14, 15]. The $accu_i$ is obtained by Eq. 3, in which *corr* represents the number of correctly classified examples and *incorr* represents the number of incorrectly classified examples [27]:

$$accu_i = \frac{corr}{corr + incorr} \times 100\% \quad (3)$$

3 Results and Discussion

3.1 Channel Selection

IBGSA was used with 8 iterations and the initial population of 35 objects to find the optimal channels. Then the classification process was completed by using 10-fold CV.

In order to find the channels that are selected in different runs of channel selection algorithm, the IBGSA was performed for 35 independent runs in each subject. Then the numbers of each selected channel was counted for all of the 19 channels independently. Table 1 shows the optimal channels that were detected with the aid of IBGSA. Moreover, Figure 2 shows the channel selection frequency in 35 independent runs.

From Fig. 3, it can be recognized that the most frequent channels are the following: F7, F8, T3, T4, C3, C4 and Pz which were selected in most of the subjects. These channels cover the frontal lobe and temporal lobes in left and right hemispheres and the midline. One of the functions of the frontal and the temporal lobes are concerned with working memory and its consolidation to convert into long term memory. Figure 3 illustrates the selected frontal (F7 and F8) and temporal channels (T3 and T4) in both of the hemispheres were covered with the mentioned

Table 1 Channels selection using IBGSA

Control	Selected channels using IBGSA	No. of ch	MCI	Selected channels using IBGSA	No. of ch
1	Fp2, F8, T4, T6, O2, Fp1, F7, T3, O1, C4, F3, P3, Pz	13	1	Fp2, F8, O2, F7, T3, T5, O1, C4, C3, P3, Cz	11
2	Fp2, F8, T4, T6, O2, Fp1, F7, T3, O1, C4, F3, P3, Pz	13	2	F8, T4, T6, F7, T5, O1, F4, P4, P3, Fz, Cz, Pz	12
3	Fp2, T4, F7, T3, O1, F3, C3, P3, Fz, Cz, Pz	11	3	F8, T4, O2, Fp1, T3, T5, O1, F4, C4, C3, Cz, Pz	12
4	Fp2, F8, T4, T6, O2, Fp1, F7, T5, O1, F4, C4, P4, P3, Fz, Cz, Pz	16	4	Fp2, F8, T4, T6, Fp1, F7, T3, F4, C4, C3, P3, Fz, Cz, Pz	14
5	Fp2, F8, T4, Fp1, F7, T3, T5, F4, P3, Cz	10	5	F8, T4, T6, O2, Fp1, F7, T5, F4, C4, P4, P3, Fz, Cz	13
6	Fp2, F8, T4, Fp1, F7, T3, O1, F4, C3, P3, Cz	13	6	F8, T4, F7, F4, C4, P4, F3, C3, Cz, Pz	10
7	Fp2, F8, T4, Fp1, F7, T3, O1, F4, C3, P3, Cz	11	7	Fp2, F8, T4, T6, F7, T3, O1, C4, P4, P3, Pz	11
8	Fp2, F8, T4, F7, T3, O1, F4, C4, P4, F3, P3, Fz, Cz, Pz	14	8	Fp2, F8, T6, O2, Fp1, T3, T5, O1, F4, C3, Cz, Pz	12
9	Fp2, F8, T4, T6, O2, Fp1, F7, T3, O1, F4, P3, Pz	12	9	F8, T3, T5, O1, F4, C4, P4, F3, C4, P3, Cz	12
10	T6, F7, F4, F3, C3, Cz, Pz	7	10	F8, T3, T5, O1, F4, C4, P4, F3, C5, P3, Cz	11
11	Fp2, F8, T4, T6, F7, O1, F4, C4, P4, F3, C3, Cz, Pz	13	11	Fp2, F8, Fp1, F7, O1, P4, F3, Fz, Cz, Pz	10
12	Fp2, F8, T4, O2, F7, T3, O1, F4, P4, F3, C3, Fz, Cz, Pz	14	12	T4, T6, F7, T3, T5, O1, F4, P4, C3, P3, Cz, Pz	12
13	Fp2, F8, T6, O2, Fp1, F7, T5, F4, C4, P4, Fz, Cz	12	13	Fp2, F8, T4, Fp1, F7, T3, O1, F3, P3, Fz	10
14	F8, T4, O2, Fp1, F7, O1, C3, Cz, Pz	9	14	F8, T4, T6, O2, O1, C4, P4, C3, P3, Pz	10
15	Fp2, T4, T6, O2, Fp1, F7, T3, T5, O1, F4, C4, C3, P3, Fz, Cz, Pz	16	15	Fp2, T4, O2, T3, T5, O1, F4, P4, C3	9

VaD	Selected channels using IBGSA	No. of ch
1	F8, F7, T5, O1, F4, C3, P3, Fz, Cz, Pz	10
2	F8, T4, T6, Fp1, T3, O1, F4, C4, F3, C3, P3, Cz	12
3	Fp2, F8, T4, O2, F7, T5, O1, F4, C4, C3, Pz	11
4	Fp2, F8, T4, T6, F7, T3, O1, F4, C3, P3, Fz	11
5	F8, T4, Fp1, F7, T3, O1, C4, P3, Fz	9

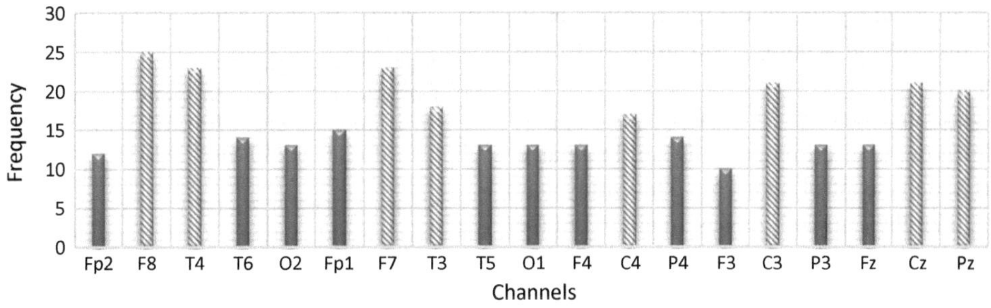

Fig. 2 Channel selection frequency in 35 independent runs

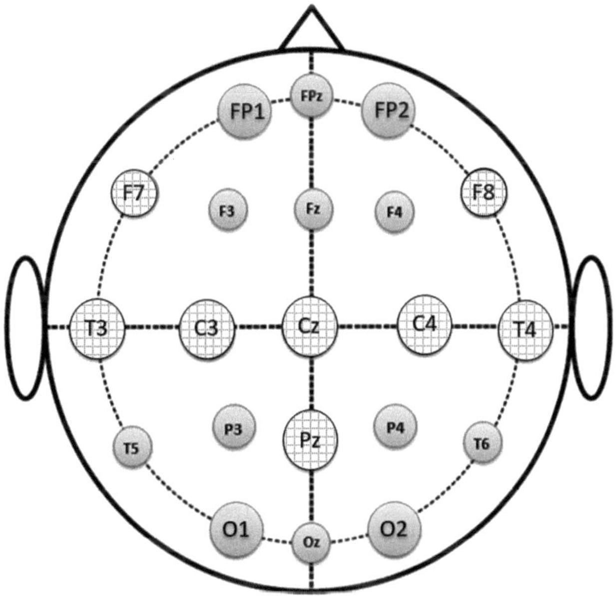

Fig. 3 Location of the 19 EEG channels

Table 2 The table of confusion matrix calculations for multi-class classification using *k*NN without IBGSA technique

Confusion matrix	VaD (%)	MCI (%)	Control (%)
VaD	93.33	6.67	0
MCI	5.56	93.33	1.11
Control	16.67	10	73.33

Table 3 The table of confusion matrix calculations for multi-class classification using *k*NN with IBGSA technique

Confusion matrix	VaD (%)	MCI (%)	Control (%)
VaD	67.78	31.11	1.11
MCI	7.78	88.89	3.33
Control	3.33	6.67	90

electrodes while other related parts of the brain are covered with channels C3, C4 and Pz. The selected channels obtained results confirm that it is almost achievable to detect the specific brain areas which are related to working memory.

4 Conclusion

The confusion matrix of the *k*NN classifier without and with using the IBGSA technique are presented in Tables 2 and 3 respectively. From Tables 2 to 3, IBGSA improves the classification over all accuracy for all the three groups as in the *k*NN classification overall accuracy was improved from 86.67 to 92.52%.

Thus, *k*NN was included in the study to support multi-class classification, to discriminate VaD, stroke-related MCI patients and healthy control subjects'. This study had several limitations, like the sample size was small and in the future, an additional analysis with a large database should be performed.

EEG plays an important role in brain activity analysis. This study is focused on EEG signal analysis and processing to provide optimal channels and valuable markers which help in improving detection dementia early. AICA-WT technique has been used as a denoising technique. *SpecEn*, *ApEn* and *PerEn* have been used as a features to test the irregularity in VaD and MCI patients. The IBGSA algorithm was used to clarify the optimal EEG channels that may help in detecting dementia early. Therefore, the most frequent channels can be recognized in the frontal, central and temporal areas. Finally, EEG could be a valuable marker for inspecting the background activity in the identification of patients with stroke-related MCI.

References

1. McVeigh, C., Passmore, P.: Vascular dementia: prevention and treatment. Clin. Interv. Aging **1**(3), 229 (2006)
2. Ruitenberga, A., et al.: Incidence of dementia: does gender make a difference? Neurobiol. Aging **22**, 575–580 (2001)
3. Cumming, T.B., Marshall, R.S., Lazar, R.M.: Stroke, cognitive deficits, and rehabilitation: still an incomplete picture. Int. J. Stroke **8**(1), 38–45 (2013)
4. Al-Qazzaz, N.K., et al.: Cognitive impairment and memory dysfunction after a stroke diagnosis: a post-stroke memory assessment. Neuropsychiatric Dis. Treat. **10**, 1677 (2014)
5. Davidson, P.R., Jones, R.D., Peiris, M.T.: EEG-based lapse detection with high temporal resolution. IEEE Trans. Biomed. Eng. **54**(5), 832–839 (2007)
6. Klimesch, W.: EEG alpha and theta oscillations reflect cognitive and memory performance: a review and analysis. Brain Res. Rev. **29**(2), 169–195 (1999)
7. Chellappan, K., et al.: Post-stroke brain memory assessment framework. In: 2012 IEEE EMBS Conference on Biomedical Engineering and Sciences (IECBES), IEEE (2012)
8. Baddeley, A.: Working memory. Science **255**(5044), 556–559 (1992)
9. Jeong, J.: EEG dynamics in patients with Alzheimer's disease. Clin. Neurophysiol. **115**(7), 1490–1505 (2004)
10. Schröder, M., et al.: Robust EEG channel selection across subjects for brain-computer interfaces. EURASIP J. Appl. Sig. Process. **2005**, 3103–3112 (2005)
11. Lal, T.N., et al.: Support vector channel selection in BCI. IEEE Trans. Biomed. Eng. **51**(6), 1003–1010 (2004)
12. Lan, T., et al.: Channel selection and feature projection for cognitive load estimation using ambulatory EEG. Comput. Intell. Neurosci. **2007**, 8–8 (2007)
13. Arvaneh, M., et al.: Optimizing the channel selection and classification accuracy in EEG-based BCI. IEEE Trans. Biomed. Eng. **58**(6), 1865–1873 (2011)
14. Folstein, M.F., Folstein, S.E., McHugh, P.R.: Mini-mental state. A prac-32 (1998)
15. Smith, T., Gildeh, N., Holmes, C.: The montreal cognitive assessment: validity and utility in a memory clinic setting. Can. J. Psychiat. **52**(5), 329 (2007)
16. Al-Qazzaz, N., et al.: Selection of mother wavelet functions for multi-channel eeg signal analysis during a working memory task. Sensors **15**(11), 29015 (2015)
17. Hyvarinen, A.: Fast and robust fixed-point algorithms for independent component analysis. IEEE Trans. Netw. **10**(3), 626–634 (1999)
18. Escudero, J., et al.: Blind source separation to enhance spectral and non-linear features of magnetoencephalogram recordings. Application to Alzheimer's disease. Med. Eng. Phys. **31**(7), 872–879 (2009)
19. Al-Qazzaz, N.K., et al.: Automatic artifact removal in EEG of normal and demented individuals using ICA–WT during working memory tasks. Sensors **17**(6), 1326 (2017)
20. Escudero Rodríguez, J.: Applications of blind source separation to the magnetoencephalogram background activity in alzheimer's disease. Universidad de Valladolid (2010)
21. Hornero, R., et al.: Spectral and nonlinear analyses of MEG background activity in patients with Alzheimer's disease. IEEE Trans. Biomed. Eng. **55**(6), 1658–1665 (2008)
22. Richman, J.S., Moorman, J.R.: Physiological time-series analysis using approximate entropy and sample entropy. Am. J. Physiol. Heart Circ. Physiol. **278**(6), H2039–H2049 (2000)
23. Lake, D.E., et al.: Sample entropy analysis of neonatal heart rate variability. Am. J. Physiol-Regul. Integr. Comp. Physiol. **283**(3), R789–R797 (2002)
24. Morabito, F.C., et al.: Multivariate multi-scale permutation entropy for complexity analysis of Alzheimer's disease EEG. Entropy **14**(7), 1186–1202 (2012)
25. Chawla, N.V., et al.: SMOTE: synthetic minority over-sampling technique. J. Artif. Intell. Res, 321–357 (2002)
26. Ghaemi, A., et al.: Automatic channel selection in EEG signals for classification of left or right hand movement in brain computer interfaces using improved binary gravitation search algorithm. Biomed. Signal Process. Control **33**, 109–118 (2017)
27. Xiang, J., et al.: A novel hybrid system for feature selection based on an improved gravitational search algorithm and k-NN method. Appl. Soft. Comput. **31**, 293–307 (2015)

Fabrication of Low-Resistance Carbon MEMS/NEMS as Material for Biosensing

M. F. Zulkeflee, F. Ibrahim, and M. Madou

Abstract

This paper presents the fabrication of low-resistance carbon MEMS/NEMS as material for biosensing. The carbon source is the negative photoresist SU8 epoxy. At room temperature the SU8 is in liquid form, making it easier to mix with nanoparticles. The SU8 epoxy is mixed with single-walled-Carbon-Nanotubes (swCNTs) and stirred for 48 h. The experimental variable was the weight percentage (wt%) of swCNTs in the SU8 epoxy. The controlled variables were the layer thickness and experimental environment such as the room's humidity and furnace model. The silicon wafer was used as substrate because it can withstand high temperatures during the pyrolysis step. The spin coater was used for coating the mixture onto the substrates uniformly. The spinning process was repeated for sample design 2 and 3 to obtain the double layer characteristic. Samples were exposed to ultra-violet (UV) radiation for 40 s. This process is called photolithography. The samples were heated on the hotplate before (pre-bake) and after (post-bake) the photolithography step for solvent evaporation. SU8 developer solution and acetone were used to remove the un-exposed SU8 for pattern developer. Finally, the samples were carbonized at 900 °C maximum in an oxygen free environment (purified Nitrogen gas (N_2) gas 99.9995%). The experimental result showed that the lowest possible resistance value was 25 Ω for the sample design 2 and 3 at 0.01 wt% swCNTs.

Keywords

Carbon-MEMS/NEMS • Photolithography • SU8

M. F. Zulkeflee · F. Ibrahim (✉) · M. Madou
Department of Biomedical Engineering, Faculty of Engineering, University of Malaya, Kuala Lumpur, Malaysia
e-mail: fatimah@um.edu.my

M. F. Zulkeflee · F. Ibrahim · M. Madou
Department of Biomedical Engineering, Faculty of Engineering, Centre for Innovations in Medical Engineering, University of Malaya, Kuala Lumpur, Malaysia

M. Madou
Department of Biomedical Engineering, University of California, Irvine, USA

M. Madou
Department of Mechanical and Aerospace Engineering, University of California, Irvine, USA

1 Introduction

Carbon-micro/nano-electro-mechanical-systems (or C-MEMS/NEMS) is not a new term in micro/nano technology especially in the making of mechanical and electrical applications. C-MEMS permit a wide variety of interesting new MEMS and NEMS application that employ structures having a wide variety of shapes, resistivity and mechanical properties [6]. The advantages of using carbon as the precursor are that the carbon resource are widely-available and environmental-friendly.

The SU8 epoxy is the common precursor of the glassy carbon and glass-like carbon. The SU8 epoxy was selected

F. Ibrahim et al. (eds.), *2nd International Conference for Innovation in Biomedical Engineering and Life Sciences*, IFMBE Proceedings 67, https://doi.org/10.1007/978-981-10-7554-4_22

due to its high aspect ratio, versatility and ease of patterning via lithography technique. The lithography or photolithography process is the patterning of the SU8 by using ultra-violet (UV) radiation and with or without templates. In the laboratory, our group successfully produced thin films of SU8 with layer thickness of 10 nm and array gap of 10 nm. Smaller gaps in the array structure result in larger surface area which is good for surface-to-DNA contact for bio-sensor applications [9]. The large surface area also promotes the uniform shrinkage of the SU8 during pyrolysis.

Other sources of carbon include carbon-nanotubes (CNTs), graphene, hydro-carbon gas, nano-cellulose and polymers. CNTs and graphene can be produced by using chemical vapor deposition (CVD). The CVD technique requires the presence of natural gas precursors (hydro-carbon gas) such as methane and propane gas. The type of gases controls the yield of the depositions. The drawbacks of this technique is the inability to directly pattern the carbon layer according to the desired design. CNTs have been previously introduced into SU8 solutions to lower the resistance across the flat surface [9, 11]. It also has exceptional mechanical, electrical and thermal properties and their incorporation into polymer matrices can potentially lead to multifunctional composites with greatly enhanced properties [4]. Some of the SU8-based application which have been published include the microfluidic system [10], cantilever sensor system [5] and optic biosensor [3].

The pyrolysis or carbonization of SU8 epoxy needs to be in an inert environment. The SU8 epoxy will turn into ash in the presence of oxygen in the environment, as the carbon elements in the SU8 epoxy tend to react with oxygen to form carbon dioxide and water vapor. Nitrogen, argon and hydrogen are examples of oxygen-free environments. In the previous study, an increase in the ramping rate improved the quality of the glassy carbon obtained [2]. In this experiment the ramping rate were set to 10 and 20 °C/min to save resources e.g. gas and electricity during the pyrolysis step.

In order to create portable and handy medical devices, the components need to be lightweight and smaller in scale. The C-MEMS/NEMS technique is able to produce three-dimensional carbon structures as environmental friendly sensor. The drawbacks of this technique were the cracking and peel-off of samples during the pyrolysis step. To overcome the problem, swCNT was added to SU8 to form a nano composite solution and a double layer structure was fabricated. Hence, in this study, a novel double layer and swCNTs was prepared to increase the conductivity and reduce the rate of sample peel-off and cracks.

2 Methodology

2.1 Substrate and Sample Preparation

The samples were prepared based on three designs; design 1, design 2 and design 3. The base design was the 'original design' which is only coated by pure SU8. The first design consisted of a single layer of SU8 and swCNTs mixture while the second and third design are double layer. The layers consist of plain SU8 layer and the SU8 and swCNTs mixture layer. For second design the pure SU8 was coated onto of the mixture layer while the third design the mixture was coated onto of the pure SU8 layer (Fig. 1).

The silicon wafer was used as the substrate. The pre-cleaned silicon wafer was heated overnight at 120 °C in the vacuum oven. The swCNTs was stirred with the SU8 for 48 h at 100RPM in the room temperature. According to the previous research publication the 48 h was the time needed make the mixture homogeneous at the uniform speed. Different ratios of the swCNTs and SU8 mixture were prepared according to Table 1.

The mixture was uniformly spread onto the substrate by using the spin coating machine. The acceleration of the spin coat was set to be 500RPM and spin for 30 s at 4000 RPM. According to the Microchem SU8 datasheet, an estimated layer thickness of 13 µm was formed according to the experimental setup and parameters. The coated substrate was pre-baked and post-baked at on the hot plate. The UV light exposure duration for the lithography process was set at 40 s for all designs. If the UV is under-exposed, the bottom sections of the SU8 will not harden and this may cause the samples to dissolve and wash away during the cleaning. For these cases, over-exposure will not significantly affect the samples as the samples is a flat surface [12].

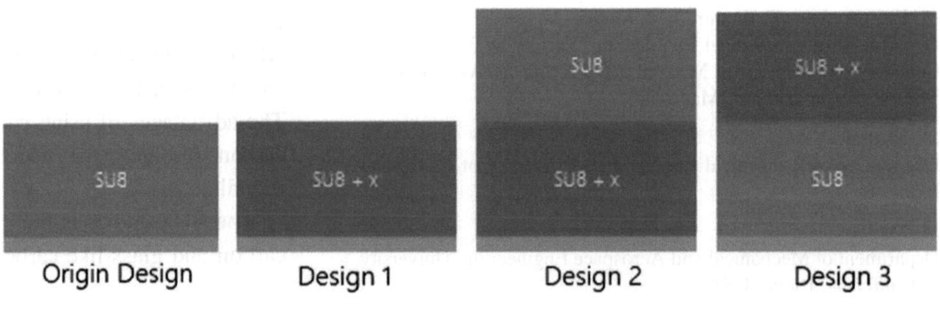

Fig. 1 The sample designs of the flat surface carbon

Table 1 The weight percentage (wt%) of the swCNTs in the mixture

Sample	wt% swCNTs
1	10.0
2	5.0
3	1.0
4	0.5
5	0.1
6	0.01

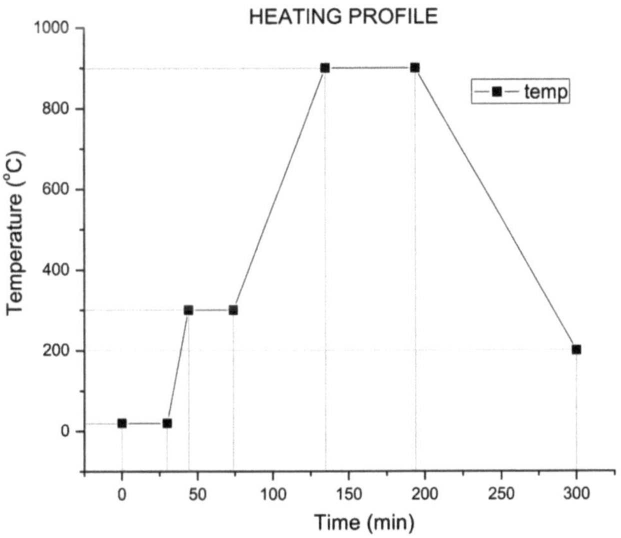

Fig. 2 The heating profile used for this fabrication of carbon biosensor

To create the second layer for sample design 2 and 3, the spin coating process was used. The pure SU8 was used to coat the mixture layer of sample design 2 while mixture epoxy was used to coat the SU8 layer of sample design 3. The same acceleration and RPM of the coating technique was used to ensure the samples have the same thickness.

2.2 Pyrolysis of the Sample

The samples were pyrolyzed in an inert atmosphere (Purified N_2 gas 99.9995%). The flowrate was set to be 2LPM from initial stage to the cooling stage at 200 °C. The heating profile used for pyrolysis as shown in Fig. 2.

Referring to above Fig. 2 the heating profile is divided into a few stages; heating (ramping), climax and cooling stages. The first ramping rate was set to be 20 °C/min for room temperature to 300 °C and 10 °C/min for 300–900 °C. The second ramping rate was set lower to avoid the sample cracks and peel-off. [7].

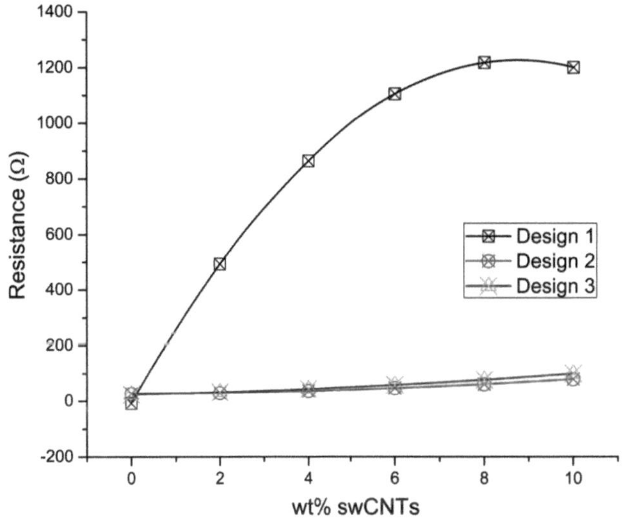

Fig. 3 The graph relationship of wt% swCNTs and resistance

3 Result and Discussion

Figure 3 shows the graph of resistance versus the weight percentage of swCNTs used for the experiment. The baseline resistance of pure SU8 was 250 Ω as measured on the 'original design' sample (refer to Fig. 1). The addition of 0.5 wt% swCNTs caused the resistance to reduce to 50 Ω. At 0.5 wt% swCNTs the resistance of samples design 2 and 3 were lower than the samples of design 1 which were 30 Ω and 25 Ω respectively. Samples prepared using design 1 were prone to sample peel-off during pyrolysis compared to sample design 2 and 3. At 0.01 wt% and 0.1 wt% swCNTs the resistance of sample design 2 and 3 were same which was 25 Ω. This indicated that 0.01 wt% swCNTs was sufficient to produce the lowest resistance value in this study. The value changed to 30 Ω for sample design 2 and 25 Ω for sample design 3 at 0.5 wt%. Based on the observation of the study it was found that the addition of the swCNTs and the double layer helped to reduce the resistance from 250 Ω to 25 Ω. The single layer carbon flat surface was prone to suffer from peel-off and cracks compared to the double layer carbon flat surface. The 13 μm was too thin and doubling the layer thickness to 26 μm helped to strengthen the surface binding to the substrate.

Sample peel-off and cracks during pyrolysis tends to happen if the surface of the silicon wafer is defective. The other effects causing the peel-off and cracks are the composition of the mixture and the thickness of the layer. Homogenous composition of mixture is important to produce spin-coated layers with even thickness during spin-coating. A homogenous composition of SU8 and swCNTs mixture was achieved by stirring the mixture for 48 h or more. The uniform thickness caused the uniform thermal expansion and shrinkage across the surface and finally reduced the cracks which also reduced the sample peel-off.

4 Conclusion

The experiment was carried out to determine the effects of the swCNTs and double layers for the flat surface carbon-based sensor. This is because of the swCNTs act as catalyst and also improve the conductivity of the carbon surface. The double layer also increased the conductivity and reduced the peel-off effects during the pyrolysis. Sample design 3 showed the best result with the lowest resistance value of 25 Ω.

Acknowledgements The authors would like to thank for the research funds and support from Malaysian Ministry of Higher Education Transdisciplinary Research Grant Scheme (TRGS) TR002-2014B grant, Ministry of Science Technology and Innovation (MOSTI) Science Fund SF-020-2014 and University of Malaya Flagship grant FL001A-14AET. We thank Dr. Teh Swe Jyan for her assistance in proofreading this manuscript. The authors would like to dedicate this work to the late Prof Dr. Sharifah Bee Abdul Hamid, former Director of Nanotechnology & Catalysis Research Centre (NANOCAT), University of Malaya.

References

1. Crivello, J.V.: The discovery and development of onium salt cationic photoinitiators. Polym. Sci. **37** (1999)
2. Hassan, Y.M., Caviglia, C., Hemanth, S., Mackenzie, D.M.A., Alstrom, T.S., Petersen, D.H., Keller S.S.: High temperature SU-8 pyrolysis for fabrication of carbon electrodes. J. Anal. Appl. Pyrol. (2017) http://dx.doi.org/10.1016/j.jaap.2017.04.015
3. Indrajit, B., Roshan, M., Bikash, D.C., Arun, M., Chalapathi, K.: SU-8 Polymer Based waveguide biochemical sensor for medical diagnostic application. Paper presented at the international conference on optic and photonics, Chandigarh, India (2009)
4. Jiang, L., Spearing, S.M., Monclus, M.A., Jennett, N.M.: Formation and mechanical characterisation of SU8 composite films reinforced with horizontally aligned and high volume fraction CNTs. Compos. Sci. Technol. **71**(2011), 1301–1308 (2011)
5. Johansson, A., Calleja, M., Rasmussen, P.A., Boisen, A.: SU-8 Cantilever sensor system with integrated readout. Sens. Actuator **A123–124**(2005), 111–115 (2005)
6. Madou, M.: C-MEMS and C-NEMS. Paper presented at the international conference MEMS, NANO and Smart System (ICMENS 2004). Alberta, Canada (2004)
7. Pramanick, B., Ibrahim, F., Thiha, A.: Carbon nanowire fabrication: C-MEMS. In: Madou, M.J., Perez-Gonzalez, V.H., Pramanick, B. (eds.) Carbon: the next silicon?. Momentum Press, New York (2015)
8. Rodrigo, M.D.: SU-8 Photolithography as a toolbox for carbon MEMS. Micromachines **5**, 766–782 (2014). https://doi.org/10.3390/mi5030766
9. Swati Sharma, M.M.: Micro and nano patterning of carbon electrodes for bioMEMS. Bioinspired, Biomimetric, Nanobiomater. **1**(BBN4), 252–256 (2012). https://doi.org/10.1680/bbn.12.00010
10. Yang, R., Wang, R.: A numerical and experimental study on gap compensation and wavelength selection in UV-lithography of ultra-high aspect ratio SU-8 microstructures. Sens. Actuators **110**
11. Chen, W., Beidaghi, M., Penmatsa, V., Bechtold, K., Kumari, L., Li, W.Z., (Peggy) Wang, C.: Integration of carbon nanotubes to C-MEMS for on-chip supercapacitors. IEEE Trans. Nanotechnol. **9** (2010)
12. Zhang, J., Tan, K.L., Hong, G.D., Yang L.J., Gong H.Q.: Polymerization optimization of SU-8 photoresist and its applications in microfluidic systems and MEMS. J. Micromech. Microeng. **11** (2001)

Evaluation of Depth Perturbation Method for Localizing Fluorophore Inside a Pork Chuck Meat Sample

Ichiro Sakuma, Tuo Zhou, Keiichi Nakagawa, and Etsuko Kobayashi

Abstract

Depth perturbation method for improvement of a mesoscopic epifluorescence tomography (MEFT) we previously developed was evaluated in an experimental system using biological samples. We conducted ex vivo experiments using pork chuck meat to verify the depth perturbation concept in a relatively realistic environment. Fluorescence images were acquired with/without a thin optical phantom (perturbator with known optical scattering property) onto the phantom. Fluorescence intensity variation resulting from the depth perturbation was used for estimating three dimensional location of fluorophores. Estimated fluorophore central depths (placed at depth from 3–9 mm) are consistent with the expected values. Absolute estimation errors of the center position of fluorophores placed inside the phantom were 0.3–0.5 mm (on average). Slight optical heterogeneity did not bring great estimating errors.

Keywords

Mesoscopic epifluorescence tomography • Depth perturbation • Fluorophore localization

1 Introduction

Mesoscopic epifluorescence tomography (MEFT) can detect fluorophore distribution in reflectance geometry at depths of several millimeters [1]. For improvement of MEFT, we previously proposed depth perturbation method [2]. Fluorescence images are acquired with/without a thin optical phantom onto the medium surface. Fluorescence intensity variation resulting from the depth perturbation is used for estimating three dimensional location of fluorophores since relative decreases in the excitation light fluence rate, emission fluence rate, and, fluorescence intensity at an observing point due to insertion of the perturbator are smaller for a larger depth of the fluorophore. We conducted ex vivo experiments using pork chuck meat to verify the depth perturbation concept in a relatively realistic environment to evaluate its feasibility in biological tissue.

I. Sakuma (✉) · T. Zhou · K. Nakagawa · E. Kobayashi
School of Engineering, The University of Tokyo, 7-3-1, Hongo, Bunkyo-Ku, Tokyo, 113-8656, Japan
e-mail: sakuma@bmpe.t.u-tokyo.ac.jp

2 Materials and Methods

2.1 Epi-Fluorescence Continuous Wave Tomography (ECWFMT) System

A custom made epi-fluorescence continuous wave tomography (ECWFMT) system as shown in Fig. 1 was used [2]. A fiber pigtailed laser diode (785 nm, LP785-SF20, Thorlabs, USA) was used as the light source. The X-Y motorized stages. A notch filter, band pass filter, two polarizing beam splitters (PBS), and a dichroic mirror (DM) rejected ambient light and reflection of excitation light. Fluorescence was collected by an electron multiplying charge coupled device (EMCCD) camera (resolution 1000 × 1000, ADT-100, Flovel, Japan).

2.2 Resolving Depth by Depth Perturbation [2–4]

A thin optical phantom with known optical properties is used as a depth perturbator. By superposing the perturbator onto a

F. Ibrahim et al. (eds.), *2nd International Conference for Innovation in Biomedical Engineering and Life Sciences*,
IFMBE Proceedings 67, https://doi.org/10.1007/978-981-10-7554-4_23

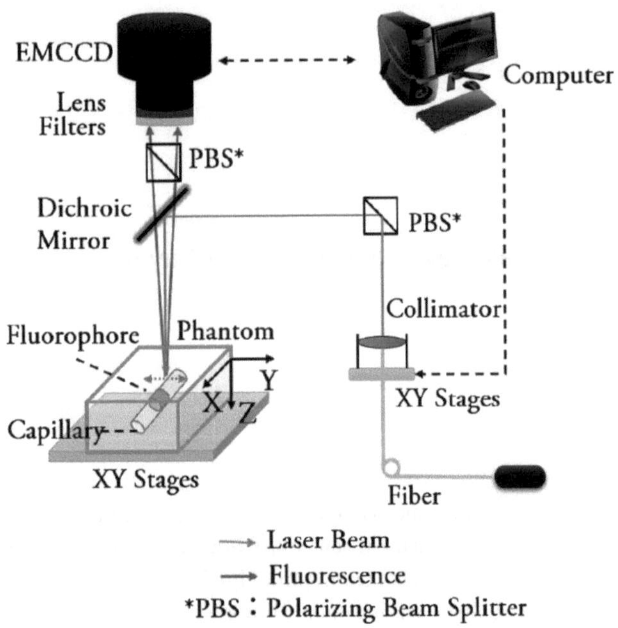

Fig. 1 Schematic of the epi-fluorescence continuous wave tomography system [2]

Table 1 Optical Coefficients of the perturbator and the phantom: each value was the statistics of measurements on five randomly selected locations

	Perturbator		Pork chuck	
Wavelength (nm)	785	830	785	830
μ_s' (mm^{-1})	0.44	0.37	0.25	0.22
μ_a (mm^{-1})	0	0	0.02	0.02

sample, the depth of a fluorescent object inside the sample is deliberately changed. Fluorescent signals are measured before and after the perturbation. It is considered that the fluorophore depth can be determined according to the variations of the measurements. The estimated fluorophore central location was then utilized as a new constraint in the reconstruction of fluorophore 3D distribution to ensure unbiased results. Then a piecewise spatially varying regularization method was applied. The detailed algorithm is available in references [2–4].

2.3 Biological Phantom

Pork meat was also used as a biological phantom. A small-size (around 1.5 mm × 2.0 mm × 1.5 mm = 4.5 mm^3) PDMS based ICG probe was fixed inside the tissue as fluorophore inclusion. For depth perturbation. A silicon perturbator made of PDMS and TiO_2 with diameter of 45 mm and thickness of 1.0 mm was used. In these ex vivo experiments, the number of covered meat pieces is 1, 2, 3, and 4 sequentially. The resultant fluorophore central depth is 2.6, 4.7, 6.4, and 8.8 mm respectively. For each depth, the measurements were repeated for five times. Location and depth of the PDMS based ICG probe was estimated using the depth perturbation method for each experiment. Optical properties of the perturbator and the phantom are shown in Table 1 (Fig. 2).

3 Results and Discussion

Estimated fluorophore's central depths are consistent with the expected values. Absolute estimation errors are 0.3–0.5 mm (on average), which were slightly larger than those obtained in experiments using phantoms were made of PDMS. TiO_2

Fig. 2 Experimental setups for depth perturbation method

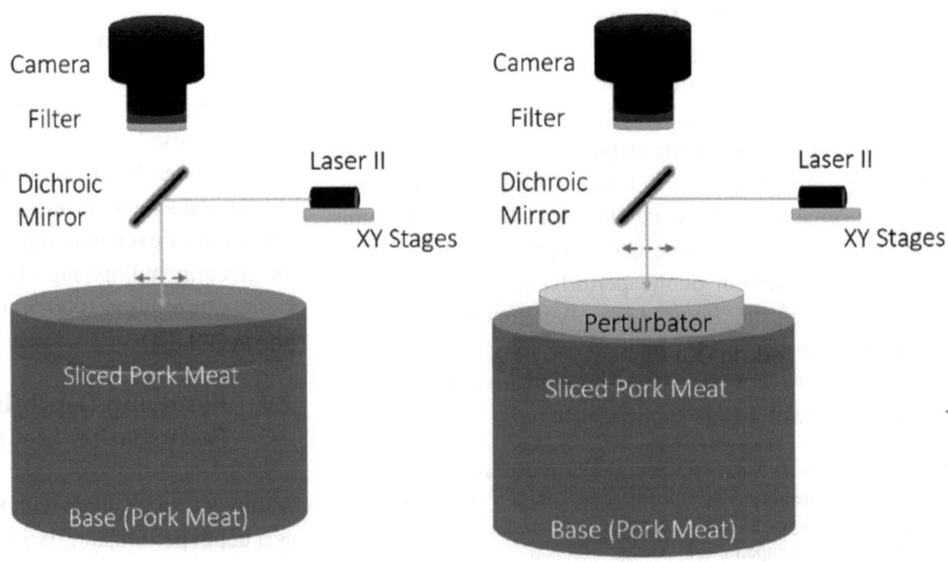

Table 2 Estimated depths of the center of fluorophores in the ex vivo experiments

	Expected depth (mm)			
	2.6	4.7	6.4	8.8
Estimation (mm)	3.0 ± 0.1	4.2 ± 0.2	6.3 ± 0.3	8.5 ± 0.4
Abs. error (mm)	0.4 ± 0.1	0.5 ± 0.2	0.3 ± 0.1	0.3 ± 0.4

(optical scattering material) mixed with ink (optical absorber) to provide tissue-like scattering and absorption coefficients (Data are not shown. They are available in [4].) These increases in errors are attributable to optical heterogeneity of the tested sample. A little of fat and tiny slits can be observed on the tissue surface. These heterogeneities apparently lead to unexpected light propagation. Similarly, the distribution of intensity ratios resulting from depth perturbation also deviates from the theoretical ratio model. However, owing to the averaging effect using thousands of data points for data fitting, this heterogeneity did not lead to unacceptable results. Uneven surface of the biological phantom also deteriorated estimation accuracy (Table 2).

In conclusion, judging from the on the basis of the results of the ex vivo experiments, the proposed methods proved its potential feasibility in biological tissue. Slight optical heterogeneity did not bring great estimating errors.

Acknowledgements This research is partly supported by a grant for Translational Systems Biology and Medicine Initiative (TSBMI) from the Ministry of Education, Culture, Sports, Science and Technology of Japan.

References

1. Björn, S., Ntziachristos, V., Schulz, R.: Mesoscopic epifluorescence tomography: reconstruction of superficial and deep fluorescence in highly-scattering media. Opt. Express **18**, 8422–8429 (2010)
2. Zhou, T., Ando, T., Nakagawa, K., Liao, H., Kobayashi, E., Sakuma, I.: Localizing fluorophore (centroid) inside a scattering medium by depth perturbation. J. Biomed. Optics **20**, 017003 (2015)
3. Zhou, T., Ando, T., Liao, H., Kobayashi, E., Sakuma, I.: Piecewise weighted Tikhonov regularization for reconstructing fluorophore distribution in tissue using mesoscopic epifluorescence tomography. Adv. Biomed. Eng. **2**, 84–94 (2013)
4. Zhou, T.: Researches on epi-fluorescence molecular tomography based on depth perturbation and spatially varying regularization, Ph. D. Dissertation, School of Engineering, The University of Tokyo, Tokyo, Japan (2015)

An Effect of Magnetic Beads to *Boesenbergia rotunda* Antioxidant Activity Using Photoprotective Microfluidic CD

Nurhaslina Abd Rahman, Fatimah Ibrahim, M. Mahdi Ainehvand, Rohana Yusof, and Marc Madou

Abstract

Microfluidics compact disc (CD) system has offer many advantages to biological and chemical analysis. It simplifies the procedures and miniaturize the sample and reagent volume. Beads has been used in the microfluidics CD technology in order to enhance the mixing of the reagents or sample. However, the effect of the beads need to be monitor to ensure the effectiveness, biocompatibility and corrosion protection of the beads. This paper has presented the advantage of photoprotective microfluidic CD for *Boesenbergia rotunda* antioxidant activity and the effect of magnetic bead in enhancing the reaction time of the mixing. The results have shown that, with the magnetic beads usage, the uniformity time of the liquid mixing in the reaction chamber are faster than the non-beads reaction chamber. However, in terms of the stability, the beads have been found to give an interference to the *Boesenbergia rotunda* 2,2-diphenyl-1-picrylhydrazyl (DPPH) antioxidant activity. To solve this issues, we would recommend an inert coating of the beads to prevent a chemical reaction between the beads and the free radical's reagents. The proposed method has minimized human handling in the DPPH plant antioxidant activity by using microfluidics CD with imitation of the real test environment in the conventional method (photoprotective). Consequently, the laborious repetitive routine in the laboratories, which is one of the factor in pre-analytical error has been reduced. This system would be a great advantage to the future laboratories techniques, as it can be applied as a point of care testing and be used in the small laboratories.

Keywords

Boesenbergia rotunda • Photoprotective microfluidics CD • DPPH antioxidant activity
Free radicals

N. Abd Rahman · F. Ibrahim
Department of Biomedical Engineering, Faculty of Engineering, University of Malaya, 50603 Kuala Lumpur, Malaysia
e-mail: nurhaslina89@yahoo.com

N. Abd Rahman · F. Ibrahim (✉) · M. M. Ainehvand · M. Madou
Centre for Innovation in Medical Engineering, Faculty of Engineering, University of Malaya, 50603 Kuala Lumpur, Malaysia
e-mail: fatimah@um.edu.my

M. M. Ainehvand
School of Engineering and Sciences, Tecnologico de Monterrey, Av. Eugenio Garza Sada 2501 Sur, 64849 Monterrey, NL, Mexico
e-mail: m.aeinehvand@itesm.mx

R. Yusof
Department of Molecular Medicine, Faculty of Medicine, University of Malaya, 50603 Kuala Lumpur, Malaysia
e-mail: rohana@um.edu.my

M. Madou
Department of Biomedical Engineering, University of California, Irvine, Irvine, 92697, USA
e-mail: mmadou@uci.edu

M. Madou
Department of Mechanical and Aerospace Engineering, University of California, Irvine, Irvine, 92697, USA

F. Ibrahim et al. (eds.), *2nd International Conference for Innovation in Biomedical Engineering and Life Sciences*,
IFMBE Proceedings 67, https://doi.org/10.1007/978-981-10-7554-4_24

1 Introduction

Micro total analysis system (μTAS) is a system that offer the reduction of sample volume and provide simplification of biological or chemical assay. Microfluidics is a system that manipulate the fluids mechanics, control the steps and process sequence of the biological and chemical analysis in a small scale of sample and reagents [1]. Its offer many important advantages such as sample process parallelization, sequential loading and batch-mode mixing. Microfluidics is a part of μTAS that offer simplicity in a complex procedures such as polymerase chain reaction (PCR) and enzyme link immune assay (ELISA) making it suitable for point of care applications or small laboratory to operate the system [1]. Microfluidics technology has been applied to many areas and integrated many disciplines for examples biology, engineering, chemistry and physics. From the design fabrication to procedures exploitations, microfluidics has been a promising system for future laboratory practice.

Pre-analytical errors have cause many false positive results, which leads to unreliable outcome and wrong test interpretations. One of the major cause of the pre analytical error is, the human prone error when handling the sample [2]. Human tends to do mistakes, because of numerous test procedures, large sample sizes, left handed person, many pipetting steps and fatigues may cause confusions and thus delivered inaccuracy and ambiguous results [3]. Furthermore, the musculoskeletal and upper limb disorder have been reported as an occupational hazard and major ergonomics problems among the health care professional [4, 5]. This disorder is happening due to manual repetitive pipetting that involves routine motion of the thumb during extracting and dispensing the liquids which leads the muscles, tendons and articular joints injuries to the hand, wrist, and the shoulders.

Centrifugal microfluid also have been referred as microfluidic compact disc (CD) is a microfluidics device that taken in a form of the compact disc. This technology has been applied in many complex procedures such as fluorescence immune assay (FIA), Loop Mediated Isothermal Amplification (LAMP), ELISA and PCR [6, 7]. The integration of microfluidic principle into the centrifugal platform have automate, simplified and miniatures the experimental procedures in a size of a compact disc. The all in one CD have offer great advantages for the health care professionals in terms of time, parallelization of assay procedures and sample processing [8]. Meanwhile, passive valve is the simplest valve techniques used in the microfluidics CD. It does not require any energy to operate and simply depends on the speed of centrifugal force of the microfluidics CD to control the liquids flows [9]. However, there is a way to enhance the mixing in the passive valve which is by using the external forces such as magnetic beads in the chamber. The magnetic force would supply an actuation movement of magnetic beads or particles. The centrifugal and the kinetic motion can be applied in the microfluidics system for separation, transportation and mixing of the solution [10].

Currently, there are many studies of centrifugal microfluidic CD that applied the magnetic beads. For example, in raw blood analysis, the cancer cells have been separated by using the integration of the magnetic beads into the system [11]. In [12], beads have been used to break the cell membrane layers, by collisions, frictions and shearing between the beads, cell membrane and the chamber wall. While in [13], the beads has been used in the nucleic acid extraction by lysing the cells.

In this paper, we present an automation of the plant antioxidant activity assay by using a photoprotective microfluidics CD and magnetic beads. The photoprotective features have been applied to protect the antioxidant properties and prevent the deterioration of the free radical's reagents used in the procedures. The proposed method has demonstrated sample sequential loading, mixing and parallelization of the procedures. This study will also monitor the effectiveness of the external force on passive valve in the designed microfluidic CD on the plant antioxidant activity. *Boesenbergia rotunda* has been chosen as a plant material as it has been reported to have high antioxidant activity [14]. With this work, pipetting steps has almost been reduced which will be beneficial to many laboratorians and health care professional.

2 Methodology

This section will be discussing all the procedures involved in conducting the experiments. The experiments are divided into three sub-section. The first part is the plants extraction preparations, the second part is the microfluidic CD construction, and the third part is the experimental setup of *Boesenbergia rotunda* DPPH antioxidant activity by using microfluidics CD.

2.1 Plant Extraction Preparation

The reagents and instruments used for the plant extraction are, 2,2-diphenyl-1-picrylhydrazyl (DPPH) (Sigma Aldrich), Ethanol and Methanol (Merck), Mili-Q ultra-pure water, and rotary evaporator (R-114, Buchi). The *Boesenbergia rotunda* was collected in Tapah, Perak, Malaysia and is chosen in the experiments due to its high antioxidant activity among the local herbs [14]. The plant is cut into to small pieces and dried in the oven at 40 °C. After that, it is pulverized by using laboratory grinder and the powder was soaked in the 95% of ethanol for 48 h. The solutions are then being filtered by using the filter paper and the solvents are removed by using a rotary evaporator (Buchi rotavapor R-114). The compound is then freeze dried for further use.

2.2 Microfluidics CD Construction

The materials used for the microfluidics CD constructions are black polymethyl methacrylate (PMMA) and pressure sensitive adhesives (PSA) (FLEXcon, USA), The black microfluidic CDs for the antioxidant assay was designed by using computer aided design software (AutoCAD). It consists of 5 layers; 3 layers of black PMMA and 2 layers of PSA in between the black PMMA layer (see Fig. 1). The uppermost layers, middle layer and the bottom layer of the microfluidic CD was made from the black PMMA. However, the middle layer was engraved with the main of microfluidics channel features. All the microfluidics channel features were designed in duplication and the carving was done by using Computer Numerical Control (CNC) machine (VISION 2525, by Vision Engraving and Routing Systems, USA). Meanwhile, connecting holes and micro features on the PSA layers were cut by using a cutting plotter machine (GCC P2-60/PUMA II, by GCC, Taiwan). Each component of the photoprotective microfluidic CD was pressed bound by using custom made press roller system.

Meanwhile, in Fig. 2, the schematic diagram of the engraved channel and features of the photoprotective microfluidic CD are portrays. The photoprotective microfluidic CD was design in duplicate for the concentration of 25, 50, 75 and 100 mg/ml. In the first set, the beads were not used in the reaction chamber, and vice versa in the second chamber. The beads with a diameter of 1.75 mm are used to creates and enhances the liquid mixing with a random movement.

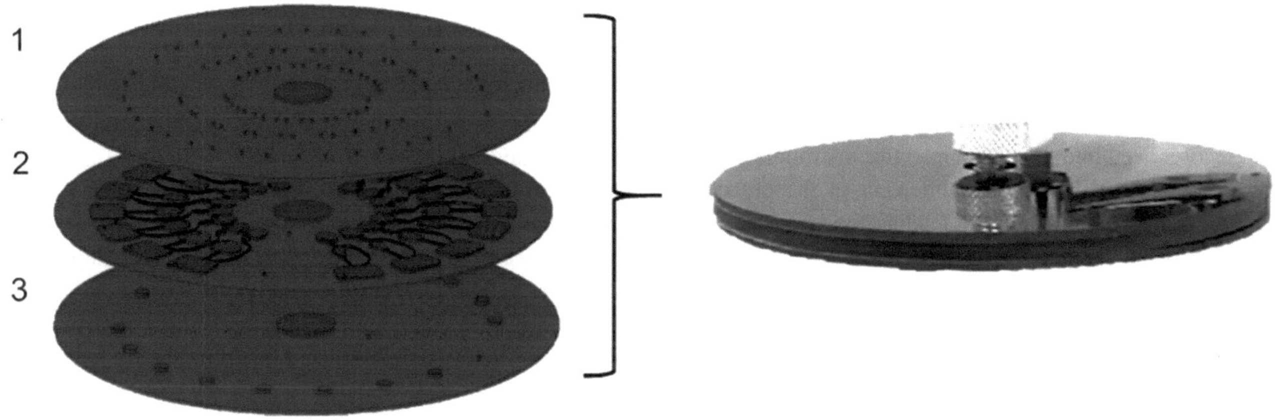

Fig. 1 Construction of the photoprotective microfluidic CD. Figure shows the three layers of black PMMA used in the microfluidic CD fabrication

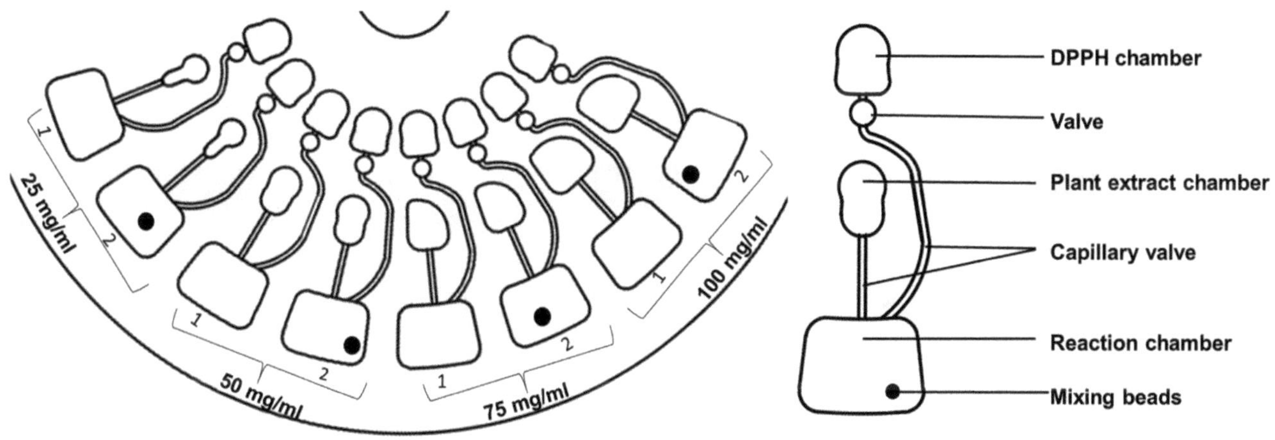

Fig. 2 The channel and features of photoprotective microfluidic CD

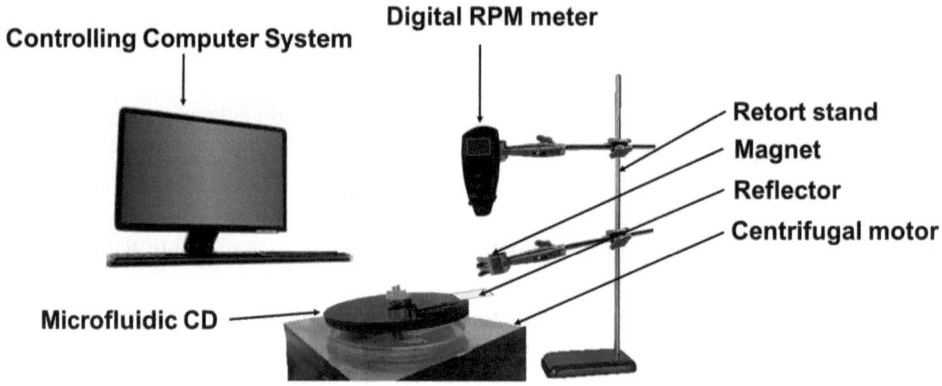

Fig. 3 Complete experimental setup of centrifugal microfluidic CD

2.3 *Boesenbergia rotunda* DPPH Antioxidant Activity by Using Microfluidics CD

Figure 3 is depicting the drawing of an actual whole experimental setup. It is consisting of a microfluidics centrifugal platform and the microfluidic CD, controlling computer system, digital rpm meter and magnet attached retort stand. The magnet is used as an external factor and to create magnetic force and actuates the movement of the beads inserted in the reaction chamber of photoprotective microfluidics CD. The conventional method of DPPH assay [15] have been modified and incorporated into the photoprotective microfluidics CD to allow miniaturization and simplification of the procedures. The experiments are started by loading the respective liquid into designated chamber on the photoprotective microfluidics CD. Consecutively, the CD are loaded into the centrifugal platform and the speed is increases slowly up to 1100 rpm until the end of experiments.

3 Results and Discussion

In this section, the results of the microfluidic CD sequential loading and mixing and the effect of the beads are shown.

3.1 *Boesenbergia rotunda* Antioxidant Activity in Photoprotective Microfluidics CD

Figure 4 shows the results of *Boesenbergia rotunda* antioxidant activity in the photoprotective microfluidic CD. The CD have minimized the human handling and automate the plant antioxidant activity process. The steps in Fig. 4a shows the liquid has been loaded to the designated chamber and the beads have been inserted in the second duplicated chamber. In Fig. 4b–f all the process of sequential loading and mixing is in automation by the centrifugal microfluidics platform. After all the designated liquid in Fig. 4a have been

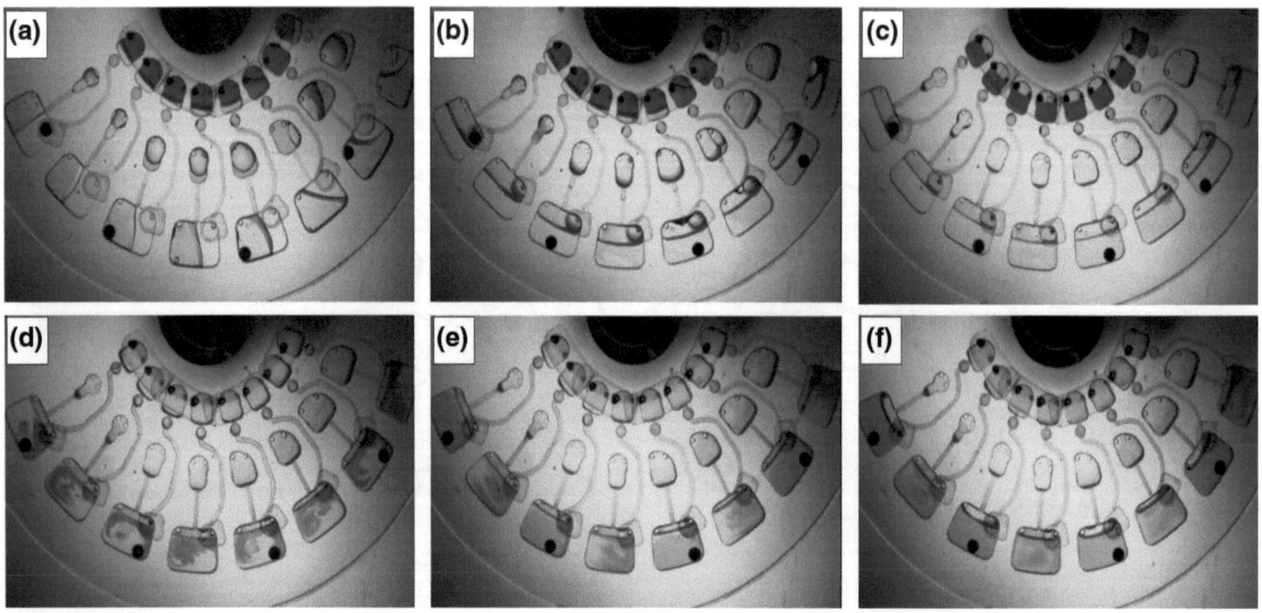

Fig. 4 Pictures shows the results of the sequential loading and mixing of the photoprotective microfluidic CD. **a** initial loading of the respected liquid. **b** plant extracts are slowly moving out from the its chamber to the reaction chamber. **c** the plant extract chambers have been emptied. **d** DPPH liquid moving out from its chamber to the reaction chamber. **e** DPPH chambers have been emptied. **f** mixing of all liquid in the reaction chamber

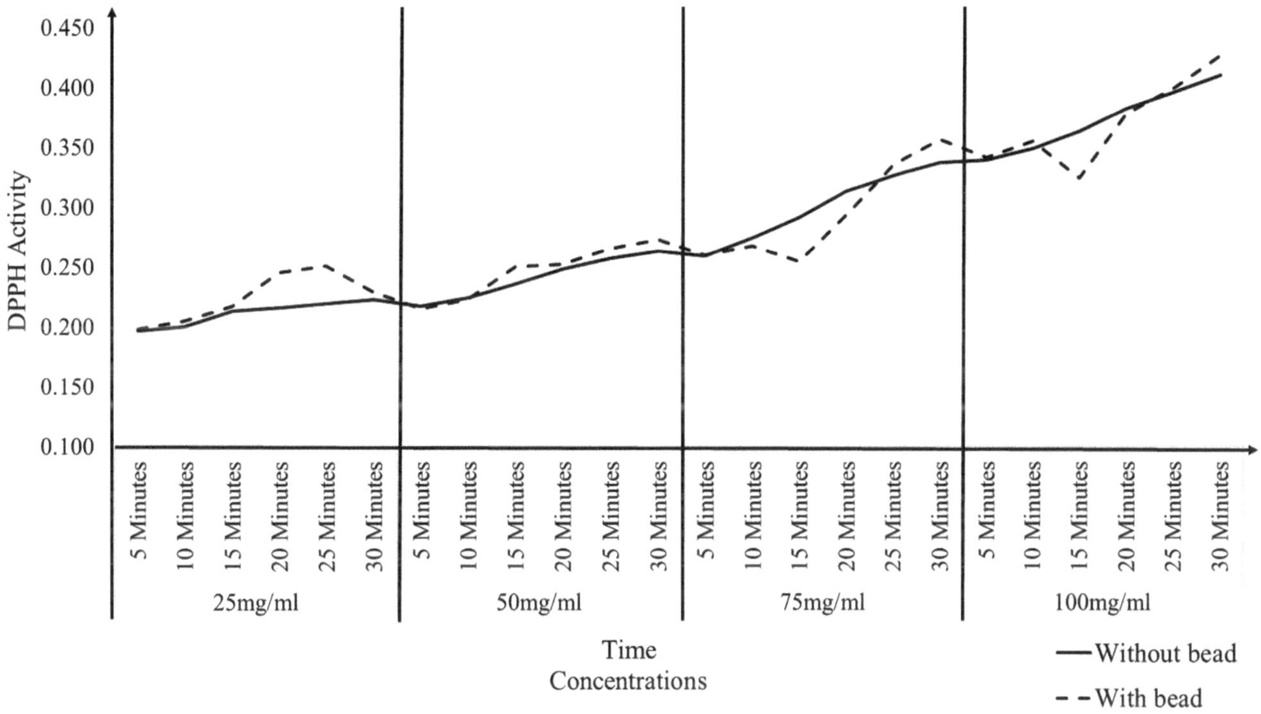

Fig. 5 Comparisons of beads effect of *Boesenbergia rotunda* antioxidant activity in photoprotective microfluidic CD

loaded, the speed is increased slowly, this steps is done to apply centrifugal force that move the liquid from its chamber. As can be seen in Fig. 4b, plant extracts are slowly flowing out from the its chamber to the reaction chamber as the speed is increased. In Fig. 4c the plant extract chambers have shown to be emptied and the DPPH liquid moving out and emptied from its chamber to the reaction chamber (see Fig. 4d, e). The final process is taken place in the reaction chamber where all the liquids are allowing to react until the end of experiments (30 min). It has been shown that the beads used in the second duplicate chamber shows faster uniformity of liquid with only 15 s, while without beads it takes 1 min and 20 s for the liquid to mix properly. As the CD spinning, the beads will moved towards the external magnet, this active mixing has fasten the liquid uniformity time by Strokes drag force that creates turbulence in the reaction chamber [16].

Conforming to the Fig. 5, the beads reading is showing inconsistence reading of DPPH activity compared to the absence of beads. At first, the initial reading of the DPPH activities is almost the same as in 5 and 10 min for both with beads and without beads, however, as can be clearly seen in the Fig. 5, the reading have deviate and being inconsistence in 15 min. Moreover, with the beads application, the colour of the control solution and the liquids in the reaction chamber have change compare to the absence of the beads (see Fig. 6). This situation is may cause by the chemical reaction between the beads and the DPPH solution. In this experiment, the DPPH is a free radical agent that have been use to evaluate the effectiveness of *Boesenbergia rotunda* to reduce the free radicals. With the presence of the metal ions from the beads, it may cause the instability of the reading [17]. The magnetic beads must be designed with suitable measures according to the assay compatibility to ensure the biocompatibility, stability and protection against corrosion [18].

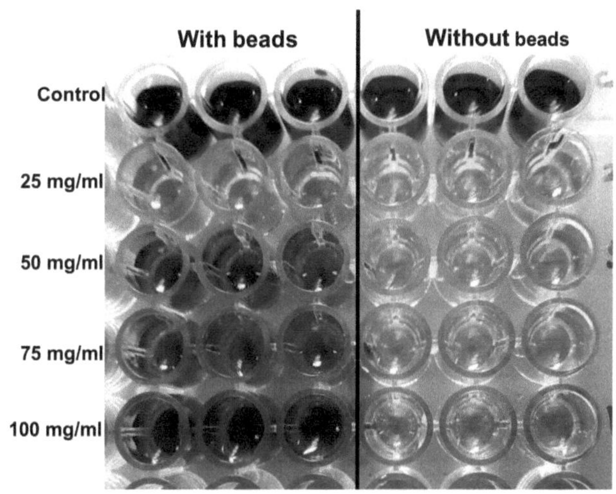

Fig. 6 Comparisons of colour changes in 96 well plate by the presence of beads

4 Conclusion

In this study, we have accomplished to performed an experiment on *Boesenbergia rotunda* plant antioxidant activity on centrifugal microfluidic platform. The conventional methods have been modified and incorporated into the microfluidics CD. Furthermore, in this experiments, the sequential loading and mixing of the procedures have been automated by using the centrifugal microfluidic platform, which makes less human interference in the analytical procedures. The used of the photoprotective microfluidics CD have also offer miniaturization, imitation of the real test environment such in the conventional method and reduce the use of pipetting techniques which can lead to evaporations of the reagents. In this experiments, we have also introduced the beads as an additional mixing enhancer in the CD. The result shows that, the beads have been a good benefit as mixing enhancer compared to the non-beads liquid mixing. The time taken of the beads usage in the reaction to reach liquids uniformity is faster compared to the non-beads chamber. Nevertheless, in terms of the chemical stability, the beads have been interfering with the DPPH activity reading. The DPPH activity results are unstable and color changes have occurred to the final mixed solutions. The instability results of the beads application may have caused by the beads metal ions intervention with the DPPH free radicals. With this, we would suggest an inert polymer coating of the beads to prevent a chemical interaction of the free radical agent and beads. On the other hands, the speed of the centrifugal platforms can also be increase as an alternative of non-beads usage. This photoprotective microfluidics CD can be a good platform for any other photo-sensitive assay and antioxidant assay.

Acknowledgements This research is supported by University of Malaya Research Grant (UMRG: RP022B-14AFR) and Postgraduate Research Grant (PPP)-PG333-2016A. Nurhaslina Abd Rahman would also thank Shah Mukim Udin, Abkar Ahmad Sayed, Karunan Joseph and Faizan Qamar for their technical, graphic help and moral support.

References

1. Whitesides, G.M.: The origins and the future of microfluidics. Nature **442**(7101), 368 (2006)
2. Hammerling, J.A.: A review of medical errors in laboratory diagnostics and where we are today. Lab. Med. **43**(2), 41–44 (2015)
3. Hawkins, R.: Managing the pre-and post-analytical phases of the total testing process. Ann. Lab. Med. **32**(1), 5–16 (2012)
4. Holm, J.W., Mortensen, O.S., Gyntelberg, F.: Upper limb disorders among biomedical laboratory workers using pipettes. Cogent Med. **3**(1), 1256849 (2016)
5. Agrawal, P.R., et al.: Work related musculoskeletal disorders among medical laboratory professionals: a narrative review. Int. J. Res. Med. Sci. **2**(4), 1262–1266 (2017)
6. Sayad, A.A., et al.: A microfluidic lab-on-a-disc integrated loop mediated isothermal amplification for foodborne pathogen detection. Sens. Actuators B Chem. **227**, 600–609 (2016)
7. Ducrée, J., et al.: The centrifugal microfluidic bio-disk platform. J. Micromech. Microeng. **17**(7), S103 (2007)
8. Focke, M., et al.: Centrifugal microfluidic system for primary amplification and secondary real-time PCR. Lab Chip **10**(23), 3210–3212 (2010)
9. Kong, L.X., et al.: Lab-on-a-CD: a fully integrated molecular diagnostic system. J. Lab. Autom. **21**(3), 323–355 (2016)
10. Kakaç, S., et al.: Microfluidics Based Microsystems: Fundamentals and Applications. Springer, Berlin (2010)
11. Mark, D., et al.: Microfluidic Lab-on-a-Chip Platforms: Requirements, Characteristics and Applications, in Microfluidics Based Microsystems, pp. 305–376. Springer, Berlin (2010)
12. Kim, J., et al.: Cell lysis on a microfluidic CD (compact disc). Lab Chip **4**(5), 516–522 (2004)
13. Siegrist, J., et al.: Validation of a centrifugal microfluidic sample lysis and homogenization platform for nucleic acid extraction with clinical samples. Lab Chip **10**(3), 363–371 (2010)
14. Abdelwahab, S.I., et al.: The methanolic extract of *Boesenbergia rotunda* (L.) Mansf. and its major compound pinostrobin induces anti-ulcerogenic property in vivo: possible involvement of indirect antioxidant action. J. Ethnopharmacol. **137**(2), 963–970 (2011)
15. Saha, M.R., et al.: Effect of *Acacia catechu* (Lf) willd. On oxidative stress with possible implications in alleviating selected cognitive disorders. PLoS ONE **11**(3), e0150574 (2016)
16. Grumann, M., et al.: Batch-mode mixing on centrifugal microfluidic platforms. Lab Chip **5**(5), 560–565 (2005)
17. Gallego, S.M., Benavides, M.P., Tomaro, M.L.: Effect of heavy metal ion excess on sunflower leaves: evidence for involvement of oxidative stress. Plant Sci. **121**(2), 151–159 (1996)
18. Gijs, M.A.: Magnetic Particle Handling in Microfluidic Systems, in Microfluidics Based Microsystems, pp. 467–480. Springer, Berlin (2010)

Assessment of Local Heating Effect on the Cutaneous Microcirculatory Vasodilatation

You-Heng Su, Jia-Jung Wang, Yue-Ting Tsai, and Wei-Kung Tseng

Abstract

The skin provides a conveniently accessible site to assess the peripheral microvascular function which may be used as an early index of allover cardiovascular system. In the study, the laser Doppler flowmetry (LDF) was applied to investigate the vasodilatation of microcirculation on the foot bottom during a local heating protocol in 21 healthy participants. The temperature was heated up to 44 °C in one minute and maintained such temperature through the 30-min heating period. Both a 10-min record of LDF signals during the baseline and a 30-min record during the heating period were registered and then analyzed using the wavelet transform. For the 21 participants, the amplitudes of the initial peak [208 ± 64 perfusion unit (PU)] and the plateau (279 ± 80 PU) of the LDF signals were significantly higher during the local heating period than the baseline (85 ± 25 PU). Meanwhile, the composite vasodilatation indexes for the initial peak and the plateau were 154 ± 65 and 239 ± 93%, respectively. Also found was a significant decline in the spectral power density percentage associated with the endothelial nitric oxide-dependent metabolic activity band during the heating period, as compared with the baseline. Those findings may help to assess the skin microcirculation dysfunction in patients with diabetes or hypertension in future clinical applications.

Keywords

Local heating • Microvascular function • Laser doppler flowmetry

1 Introduction

The phenomenon of peripheral microcirculatory irregularity may probably come into view before serious cardiovascular disease is happened. Therefore, it is necessary to search for some reliable indexes to help physicians to early diagnose the dysfunction of microcirculatory function under certain experimental protocols [1, 2]. Thus, the goal of the study was to investigate microvascular vasodilatation under local heating protocol.

2 Methods

This study included twenty-one healthy and non-smoking participants (age: 25 ± 9 years, body mass index: 22 ± 3.2 kg/m^2). Due to its convenience and applicability [3], the laser Doppler flowmetry (LDF) was utilized in the study to measure the microvascular perfusion on the foot bottom close to the foot thumb in all participants who were in the supine position. The room temperature was controlled at 27–28 °C. The measuring site was heated up to 44 °C in one minute and maintained at such temperature. A 10-min record and a 30-min record of the LDF signals during the

Y.-H. Su (✉) · J.-J. Wang · Y.-T. Tsai
Department of Biomedical Engineering, I-Shou University, Kaohsiung, Taiwan
e-mail: fabien1028@gmail.com

W.-K. Tseng
Department of Cardiology, E-Da Hospital, Kaohsiung, Taiwan

F. Ibrahim et al. (eds.), *2nd International Conference for Innovation in Biomedical Engineering and Life Sciences*, IFMBE Proceedings 67, https://doi.org/10.1007/978-981-10-7554-4_25

Table 1 The spectral power density percentage (SUM%)

	CB	RB	MB	SB	EB_1	EB_2
Baseline	5 ± 4	6 ± 4	18 ± 10	26 ± 9	15 ± 13	30 ± 9
Heat period	13 ± 8	9 ± 4*	19 ± 9	21 ± 6	13 ± 12	25 ± 11*

* $p < 0.05$

baseline and during the local heating period were registered, respectively, and later analyzed using the wavelet transform.

3 Results

For the 21 subjects, the LDF amplitudes of the initial peak [Q_I: 208 ± 64 perfusion unit (PU)] and the plateau (Q_P: 279 ± 80 PU) were significantly higher during the local heating period than the baseline (Q_B: 85 ± 25 PU) ($p < 0.001$). Also, the composite vasodilatation indexes for the initial peak [$(Q_P - Q_B) / Q_B$] and the plateau [$(Q_P - Q_B) / Q_B$] were found to be 154 ± 65 and 239 ± 93%, respectively.

Table 1 shows the comparison of the spectral power density percentage (SUM%) in six characteristic bands between the baseline and heating period in the 21 participants. We found that SUM% corresponding to the cardiac activity band (CB) and respiratory activity band (RB) were significantly greater during the heating period than the baseline ($p < 0.05$). In addition, SUM% corresponding to the endothelial nitric oxide-dependent metabolic activity band (EB_2) was significantly less during the heating period, as compared with the baseline ($p < 0.05$). Between the heating period and the baseline, however, there were no differences in SUM% corresponding to the myogenic activity band (MB), and the sympathetic activity band (SB) and the endothelial nitric oxide-independent metabolic activity band (EB_1).

4 Discussion and Conclusion

The present study shows that there are significant difference in SUM% of some characteristic bands, such as the cardiac activity band, the respiratory activity band and the endothelial nitric oxide-dependent metabolic activity band, during the 30-min local heating period, as compared with the 10-min baseline in the healthy subjects. In the beginning of the heating period, the endothelial nitric oxide-dependent metabolic activity is believed to be increased [4]. The nitric oxide concentration may be higher in the initial heating period than the baseline, and its concentration may be decreased in the latter portion of the heating period due to metabolic consumption. This may partially explain why a lower average SUM% of EB2 in the whole 30-min course of heating is present. It is still not clear, unfortunately, why both the SUM% averages of CB and RB are greater during the heating period than the baseline. Those findings may be useful for the assessment of the skin microcirculation dysfunction in patients with diabetes, hypertension, and overweight.

Acknowledgements The study was supported by MOST 105-2221-E-214 -012 -MY3, Taiwan.

References

1. Iredahl, F., Löfberg, A., Sjöberg, F., Farnebo, S., Tesselaar, E.: Non-Invasive measurement of skin microvascular response during pharmacological and physiological provocations. PLoS One **10**(8), e0133760 (2015). https://doi.org/10.1371/journal.pone.0133760
2. Roberts, K.A., van Gent, T., Hopkins, N.D., Jones, H., Dawson, E. A., Draijer, R., Carter, H.H., Atkinson, C.L., Green, D.J., Thijssen, D.H.J., Low, D.A.: Reproducibility of four frequently used local heating protocols to assess cutaneous microvascular function. Microvasc. Res. **112**, 65–71 (2017)
3. Rousit, M., Cracowski, J.L.: Non-invasive assessment of skin microvascular function in humans: an insight into methods. Microcirculation **19**(1), 47–64 (2012)
4. Metzler-Wilson, K., Kellie, L.A., Tomc, C., Simpson, C., Sammons, D., Wilson, T.E.: Differential vasodilatory responses to local heating in facial, glabrous and hairy skin. Clin. Physiol. Funct. Imaging **32**(5), 361–366 (2012)

Effects of Oral Glutathione Precursors' Supplementation on Human Glutathione Level

Asdani Saifullah Dolbashid, Mas Sahidayana Mohktar, Wan Safwani Wan Kamarul Zaman, Nur Rasyidah Hasan Basri, Mohd Faiz Azmi, Sakunie Sawai, and Mohd Yusof Hasif Ilyasa

Abstract

Glutathione (GSH) is a tripeptide of cysteine, glutamate, and glycine. These amino acids made the synthesis of GSH possible and act as a major cellular antioxidant in the body. The production of GSH in the body depends on several factors; such as dietary intake and cysteine availability. With the turnover rate of endogenous GSH can be as high as 65% per day, more consumers has turned to GSH supplement to maintain or boost up their GSH level. GSH supplement available in the market usually contains GSH precursor/s that are said to be able to increase GSH level in the body. In this study, the oral GSH supplement used contains all three precursors of GSH; cystine replacing cysteine, glutamate, and glycine. Subjects were asked to consume the supplement according to the instructed dosage in order to measure the effects of oral GSH precursors' supplement on human GSH level after the three-month study period ends. The results from this study suggest that the oral GSH precursors' supplement increases GSH level across the test groups after the three-month study period.

Keywords

Glutathione • Antioxidant • Supplement

1 Introduction

The most abundant non-protein thiol, Glutathione (GSH) is present in all mammalian tissues. The highest concentration of GSH is in liver, with around 1–10 mM concentrations. GSH is synthesized from glutamate, cysteine, and glycine and the process is catalyzed by sequentially by two cytosolic enzymes, γ-glutamylcysteine synthetase and GSH synthetase [1]. Apart from its function as an antioxidant, GSH is a key determinant of redox signaling, detoxification of xenobiotics, and modulates cell proliferation, apoptosis, immune function, and fibrogenesis [2]. GSH exists in two forms; thiol-reduced and disulfide-oxidized (GSSG) [3]. GSH is the predominant form as it comprised of >98% of total GSH in the body [4].

Studies have been done to find correlation between GSH level and general health. In one such study, it was reported that high GSH level accompanies excellent physical and mental health in women aged 60–103 years [5]. Another study reported preserved hepatic GSH and improved liver function after glutamine; one of the GSH precursors was introduced in the diet [6]. Children with severe edematous malnutrition have been shown to have their GSH synthesis rate and concentration restored after being supplemented with cysteine during early treatment of the disease [7].

A. S. Dolbashid · M. S. Mohktar · W. S. W. K. Zaman
N. R. H. Basri · M. F. Azmi · S. Sawai
Department of Biomedical Engineering, Faculty of Engineering, University of Malaya, Kuala Lumpur, Malaysia

M. S. Mohktar (✉) · W. S. W. K. Zaman
Faculty of Engineering, Center for Innovation in Medical Engineering, University of Malaya, Kuala Lumpur, Malaysia
e-mail: mas_dayana@um.edu.my

M. Y. H. Ilyasa
Chakra We Care Resources Sdn Bhd, Petaling Jaya, Selangor, Malaysia

F. Ibrahim et al. (eds.), *2nd International Conference for Innovation in Biomedical Engineering and Life Sciences*, IFMBE Proceedings 67, https://doi.org/10.1007/978-981-10-7554-4_26

Enzymes involved in GSH metabolism are regulated at transcriptional, translational, and posttranslational level [8]. GSH is involved in many metabolic processes, apart from its key role in antioxidant defense system [9]. The study of GSH is very important at the moment as it can be used as possible target for medicine and pharmacology purposes [9], thus improving and saving the lives of many people.

In this study, the effectiveness of a GSH precursors' supplement was tested on 20 test subjects. The GSH supplement tested contains three amino acids that are needed for GSH synthesis to happen; cystine replacing cysteine, glutamate, and glycine. L-cysteine is reduced from L-cystine in the cells, and it is one of rate-limiting precursor amino acids for GSH synthesis [10–12]. The study was done for 12 weeks, and measurements were recorded in 4-week intervals. The GSH level was obtained by measuring GSH in the blood plasma of the subjects. The results were then compared with the control group and analyzed.

2 Materials and Methods

2.1 Materials

The oral glutathione precursors' supplement in this study was obtained from Chakra We Care Resources Sdn Bhd (Petaling Jaya, Selangor, Malaysia). The supplement comes in powder form; five gram per sachet. The GSH assay kit was purchased from Sigma-Alrich (St. Louis, MO, USA). All reagents used were of analytical grade.

2.2 Study Design

This experimental study was comprised of three groups. A total of 20 subjects were involved in this study, aged ranging from 18 to 65 years old. The first group was used as a control. Subjects in control group were the ones with normal Body Mass Index (BMI) while those categorized by BMI as overweight and obese were randomly assigned into the two other groups (Group 1 and Group 2). For control group, measurements were taken at Week 0 (baseline) and after three-month study period (Week 12). Subjects in Group 1 and Group 2 were asked to consume the oral glutathione precursors' supplement each day during the 12-week study period. Group 1 consumed a single dose while Group 2 consumed two doses of the supplement each day. For Group 1 and Group 2, measurements of GSH level were taken during Week 0 (baseline), Week 4, Week 8, and Week 12.

2.3 Blood Collection and Preparation of Samples

Blood was collected from subject and centrifuged at $2500\times g$ for 10 min at room temperature. The plasma obtained was then separated into another tube for GSH assay.

2.4 GSH Assay

UV-visible spectrophotometer (Tecan Group Ltd, Männedorf, Switzerland) was used to obtain the value of glutathione concentration inside each sample. The readings were measured in 1–5 min interval (total of 6 measurements) at a wavelength of 412 nm. Results were calculated as follows:

$$\text{nmoles GSH per ml of sample} = \frac{\Delta A_{412}/\text{min(sample)} \times \text{dil}}{\Delta A_{412}/\text{min(sample)} \times \text{dil}}$$

ΔA_{412}/min(sample) = slope generated by sample (after subtracting the values generated by the blank reaction).
ΔA_{412}/min(1 nmole) = slope calculated from standard curve for 1 nmole of GSH.
dil = dilution factor of original sample.
vol = volume of sample in the reaction in ml.

3 Results

After the 12-week study period was finished, the results were plotted in Fig. 1.

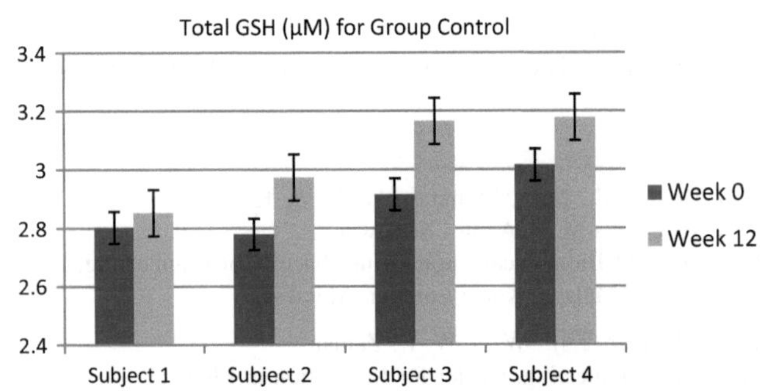

Fig. 1 Total GSH value for Control group. Each bar represents means ± SE

For Control group, the GSH level were measured only during Week 0 and Week 12 as the subjects did not consumed the GSH supplement. It was shown that the GSH level has increased at Week 12. Although the GSH level had increased, it was still considered in the normal GSH range [5, 13].

Figure 2 showed total GSH values for Group 1. It was observed that no significant increase of GSH level occurred at Week 4. All subjects showed increased GSH level at Week 8, with one subject recorded a higher than average increase in his/her GSH level. In Week 12, three subjects recorded increased GSH level while five subjects recorded decreased GSH level.

Figure 3 shows total GSH values for Group 2. On Week 4, three subjects recorded increased GSH level while the three others recorded decreased GSH level. It also has to be noted that one subject in this group has higher than average baseline GSH value. Week 8 recorded increased GSH level in two subjects and decreased GSH level in another four subjects. At Week 12, increased GSH level was observed on all subjects in the group.

Figure 4 shows the comparison between GSH values between all groups in this study. From the chart it was observed that control group has higher baseline GSH level compared to Group 1 and Group 2. The increased GSH level in control group can be attributed to several factors; and has been explained in the discussion section. For Group 1, the subjects showed increased GSH level up until Week 8, where it decreased slightly at Week 12. Group 1 shows an increase of ±0.5 μM in GSH values after the 12-week study was completed.

Group 2, while having a higher baseline GSH level compared to Group 1, recorded decreased GSH level in Week 4 and Week 8. On Week 12 the GSH level increased, however the value was still lower when compared to Group 1 of the same week. A five percent increase of GSH level was observed in control group at Week 12. Group 1 GSH level recorded a nineteen percent increase at Week 12, while Group 2 recorded a sixteen percent increase at Week 12.

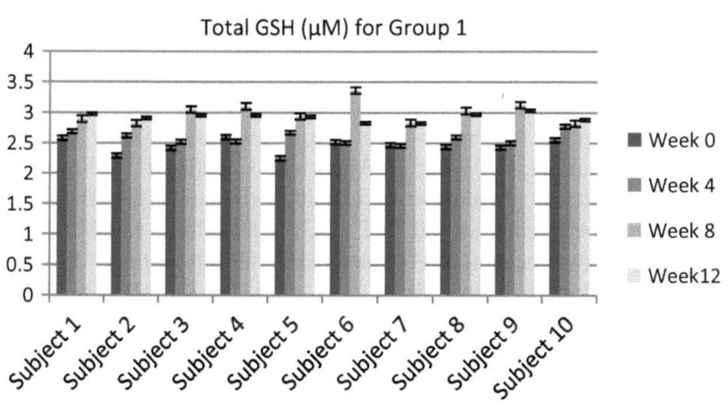

Fig. 2 Total GSH values for Group 1. Each bar represents means ± SE

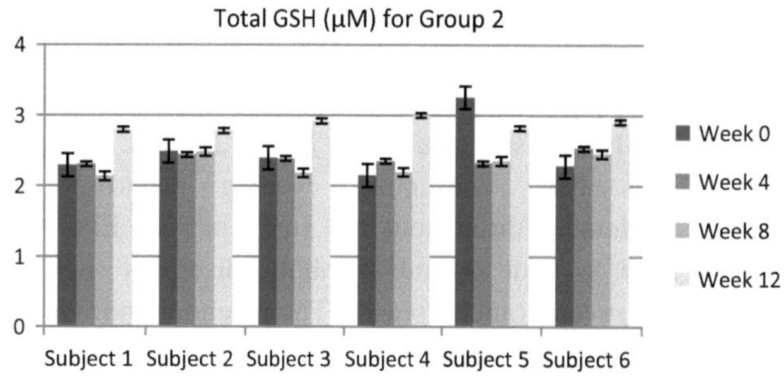

Fig. 3 Total GSH values for Group 2. Each bar represents means ± SE

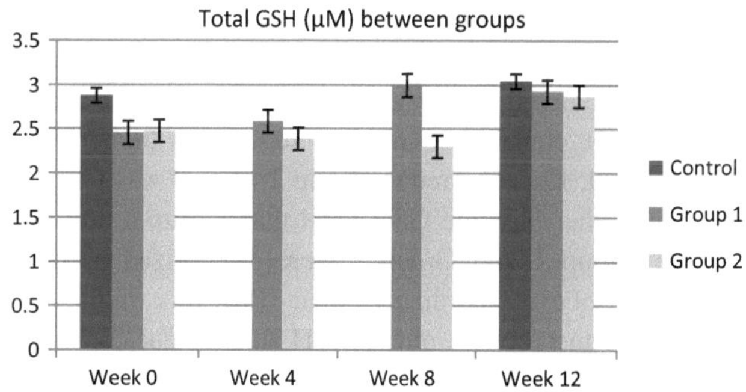

Fig. 4 Total GSH values between groups. Each bar represents means ± SE

4 Discussion

The glutathione precursors' supplement used in this study contains three building blocks of glutathione; glycine, cystine replacing cysteine, and glutamine. On the biochemical activity, several investigators have shown that the GSH synthesis is mostly regulated by the availability of glutamate-cysteine ligase (GSL) and cysteine, which the latter has been shown to have a parallel pathway with GSH synthesis [2, 14]. Thus, cysteine has been recognized as a rate-limiting factor in GSH synthesis [9]. Cystine enhanced the Cys/CySS (cysteine/cystine) redox state and in turn reduces oxidative stress and improves delivery of substrate [15]. It makes sense to consume GSH precursors' supplement as the turnover rate of GSH endogenously is 65% per day, which means all GSH is completely utilized in only 36 h [1, 16].

Based on GSH level recorded by the control group, it shows all subjects in control group has increased glutathione level after three months interval in which the group did not consume any glutathione supplement within the time period. The fluctuation in GSH level of patient is normal; however the significant increase in three out of the four control patients can be attributed to several factors. One subject in the control group has been taking Vitamin C intermittently which causes the level of GSH to increase by ±0.2 μM. Vitamin C has been known to help the body produces

Glutathione [17–19]. A subject has consumed whey protein throughout the study period. Whey protein contains glutamine and cysteine which acted as precursors to glutathione molecules, thus raising the GSH level [20–22]. Another subject has been consuming evening primrose oil (EPO) supplement throughout the study period. EPO are usually used to treat several metabolic disorders such as eczema and several other diseases. The potential antioxidant capacity of EPO has been shown to increase GSH level in previous studies [23, 24]. This may explain the significant increase of GSH level by ±0.2 μM for that particular subject.

On the other hand, results obtained in the two test groups (Group 1 and Group 2), shows that GSH level increased significantly after the three-month study period (Week 12). It also has to be noted that Group 2 consumed twice the dosage that Group 1 consumed. Group 1 reported a higher GSH level at Week 12 compared to Group 2 when it was Group 2 who has higher baseline GSH level than Group 1. As subjects in Group 1 and Group 2 were categorized as overweight and obese according to their BMI, this might have affected the metabolism of the GSH supplement consumed too. The subjects in Group 1 and Group 2 recorded between 16 and 19% increase in GSH level, compared to only five per cent increase in control group. This suggests the effectiveness of the oral glutathione precursors' supplement used in this study. However, clear correlation between the amount of dosage taken and its effects in raising GSH level cannot be made yet as the sample size is still small at the moment.

5 Conclusion

The complex relation of metabolic activities and how it affects the GSH level in human is still unclear at the moment. For now, the results suggest the effectiveness of the oral GSH precursors' supplement in increasing GSH level of the tested group. As time progress and a higher number of subjects will be tested and analysed from this study, it is hoped that a clearer correlation between the dosage of the GSH supplement taken and GSH level obtained can be formed in the future.

Acknowledgements This research was supported and funded by Chakra We Care Resources Sdn Bhd, Project No. PV004-2017.

References

1. Wu, G., Fang, Y.-Z., Yang, S., Lupton, J.R., Turner, N.D.: Glutathione metabolism and its implications for health. J. Nutr. [Internet] [cited 14 Jul 2017] **134**(3):489–492. Available from: http://www.ncbi.nlm.nih.gov/pubmed/14988435 (2004)
2. Shelly, C., Lu, M.D.: Glutathione synthesis. Biochim. Biophys. Acta [Internet]. Elsevier B.V. **1830**(5):3143–3153. Available from: http://dx.doi.org/10.1016/j.bbagen.2012.09.008 (2014)
3. Kaplowitz, N., Aw, T.Y., Ookhtens, M.: The regulation of hepatic glutathione. Annu. Rev. Pharmacol. Toxicol. [Internet]. Annual Reviews 4139 El Camino Way, P.O. Box 10139, Palo Alto, CA 94303-0139, USA; 28 Apr 1985 [cited 2017 Jul 28] **25**(1), 715–744. Available from: http://www.annualreviews.org/doi/10.1146/annurev.pa.25.040185.003435 (1985)
4. Forman HJ, Zhang H, Rinna A.: Glutathione: overview of its protective roles, measurement, and biosynthesis. Mol. Aspects Med. [Internet] [cited 28 Jul 2017] **30**(1–2), 1–12. Available from: http://www.ncbi.nlm.nih.gov/pubmed/18796312 (2009)
5. Lang, C.A., Mills, B.J., Lang, H.L., Liu, M.C., Usui, W.M., Richie, J., et al.: High blood glutathione levels accompany excellent physical and mental health in women ages 60 to 103 years. J. Lab. Clin. Med. **140**(6), 413–417 (2002)
6. Hong, R.W., Rounds, J.D., Helton, W.S., Robinson, M.K., Wilmore, D.W.: Glutamine preserves liver glutathione after lethal hepatic injury. Ann. Surg. [Internet]. Lippincott, Williams, and Wilkins [cited 28 Jul 2017] **215**(2), 114–119. Available from: http://www.ncbi.nlm.nih.gov/pubmed/1546897 (1992)
7. Badaloo, A., Reid, M., Forrester, T., Heird, W.C., Jahoor, F.: Cysteine supplementation improves the erythrocyte glutathione synthesis rate in children with severe edematous malnutrition. Am. J. Clin. Nutr. **1–3**, 646–652 (2002)
8. Pócsi, I., Prade, R.A., Penninckx, M.J.: Glutathione, altruistic metabolite in fungi. In: Advances in Microbial Physiology [Internet] [cited 28 Jul 2017], pp. 1–76. Available from: http://www.ncbi.nlm.nih.gov/pubmed/15518828 (2004)
9. Lushchak, V.I.: Glutathione homeostasis and functions: potential targets for medical interventions. J. Amino Acids [Internet] **2012**, 736837. Available from: http://www.hindawi.com/journals/jaa/2012/736837/ (2012)
10. Hosoya, K.-I., Tomi, M., Ohtsuki, S., Takanaga, H., Saeki, S., Kanai, Y., et al.: Enhancement of L-cystine transport activity and its relation to xCT gene induction at the blood-brain barrier by diethyl maleate treatment. J. Pharmacol. Exp. Ther. **302**(1), 225–231 (2002)
11. Sinha-Hikim, I., Shen, R., Paul Lee, W.-N.N., Crum, A., Vaziri, N.D., Norris, K.C.: Effects of a novel cystine-based glutathione precursor on oxidative stress in vascular smooth muscle cells. AJP Cell. Physiol. [Internet] [cited 9 Aug 2017] **299**(3), C638–642. Available from: http://www.ncbi.nlm.nih.gov/pubmed/20592243 (2010)
12. Sinha-Hikim, I., Shen, R., Kovacheva, E., Crum, A., Vaziri, N.D., Norris, K.C.: Inhibition of apoptotic signalling in spermine-treated vascular smooth muscle cells by a novel glutathione precursor. Cell. Biol. Int. [Internet] [cited 9 Aug 2017] **34**(5), 503–511. Available from: http://www.ncbi.nlm.nih.gov/pubmed/20121705 (2010)
13. Fan, Y., Yu, J., Kang, W., Zhang, Q.: Effects of glutamine supplementation on patients undergoing abdominal surgery. Chin. Med. Sci. J. [Internet] Chin. Acad. Med. Sci. **24**(1), 55–59. Available from: http://dx.doi.org/10.1016/S1001-9294(09)60060-2 (2009)
14. Diaz Vivancos, P., Wolff, T., Markovic, J., Pallar, F.V., Foyer, C.H.: A nuclear glutathione cycle within the cell cycle. Biochem. J. [Internet] [cited 14 Jul 2017] **431**, 169–178. Available from: http://www.biochemj.org/content/431/2/169.full-text.pdf (2010)
15. Sinha-Hikim, I., Sinha-Hikim, A.P., Shen, R., Kim, H., French, S.W., Vaziri, N.D., et al.: A novel cystine based antioxidant attenuates oxidative stress and hepatic steatosis in diet-induced obese mice. Exp. Mol. Pathol. [Internet] [cited 9 Aug 2017] **91**(1), 419–428. Available from: http://www.ncbi.nlm.nih.gov/pubmed/21570964 (2010)

16. Lyons, J., Rauh-Pfeiffer, A., Yu, Y.M., Lu, X.M., Zurakowski, D., Tompkins, R.G., et al.: Blood glutathione synthesis rates in healthy adults receiving a sulfur amino acid-free diet. Proc. Natl. Acad. Sci. U S A [Internet] **97**(10), 5071–5076. Available from: http://www.pubmedcentral.nih.gov/articlerender.fcgi?artid=25783&tool=pmcentrez&rendertype=abstract (2000)
17. Özkaya, D., Naziroğlu, M., Armağan, A., Demirel, A., Köroglu, B. K., Çolakoğlu, N., et al.: Dietary vitamin C and E modulates oxidative stress induced-kidney and lens injury in diabetic aged male rats through modulating glucose homeostasis and antioxidant systems. Cell. Biochem. Funct. **29**(4), 287–293 (2011)
18. Park, S.: The effects of high concentrations of vitamin C on cancer cells. Nutrients. **5**(9), 3496–3505 (2013)
19. Wang, Z., Xiao, Y., Chen, W., Tang, K., Zhang, L.: Increased vitamin C content accompanied by an enhanced recycling pathway confers oxidative stress tolerance in Arabidopsis. J. Integr. Plant Biol. **52**(4), 400–409 (2010)
20. Bounous, G.: Whey protein concentrate (WPC) and glutathione modulation in cancer treatment. Anticancer Res. **20**, 4785–4792 (2000)
21. Kent, K.D., Harper, W.J., Bomser, J.A.: Effect of whey protein isolate on intracellular glutathione and oxidant-induced cell death in human prostate epithelial cells. Toxicol. Vitr. [Internet] **17**(1), 27–33. Available from: http://www.ncbi.nlm.nih.gov/pubmed/12537959%5Cn http://ac.els-cdn.com/S0887233302001194/1-s2.0-S0887233302001194-main.pdf?_tid=4f550018-02e3-11e3-b43f-00000aab0f26&acdnat=1376266286_5e138993e488bee748586e3ddfbb33c4 (2003)
22. O'Keeffe, M.B., FitzGerald, R.J.: Antioxidant effects of enzymatic hydrolysates of whey protein concentrate on cultured human endothelial cells. Int. Dairy J. [Internet]. Elsevier Ltd. **36**(2), 128–135 (2014) Available from: http://dx.doi.org/10.1016/j.idairyj.2014.01.013
23. De La Cruz, J.P., Quintero, L., Galvez, J., Villalobos, M.A., Sánchez de la Cuesta, F.: Antioxidant potential of evening primrose oil administration in hyperlipemic rabbits. Life. Sci. [Internet] [cited 14 Jul 2017] **65**(5), 543–555. Available from: http://www.ncbi.nlm.nih.gov/pubmed/10462081 (1999)
24. Vasiljevic, D., Veselinovic, M., Jovanovic, M., Jeremic, N., Arsic, A., Vucic, V., et al.: Evaluation of the effects of different supplementation on oxidative status in patients with rheumatoid arthritis. Clin. Rheumatol. **35**(8), 1909–1915 (2016)

Review on the Usage of Swarm Intelligence in Gene Expression Data

Nurhawani Ahmad Zamri, Bhuvaneswari Thangavel, Nor Azlina Ab Aziz, and Nor Hidayati Abdul Aziz

Abstract

This paper presents a review of the recent usage of swarm intelligence for optimizing feature selection in microarray data focusing on its application for cancer detection and classification. The feature selection technique is used in the analysis of microarray so that only useful data is trained for further analysis and prediction. The process of feature selection would affect the effectiveness of the classification. This is due to the enormous quantity of genes being expressed at the same time. An optimized feature selection would ensure a high accuracy of classification. Swarm intelligence has been effective in solving feature selection and classification problems. This paper also gives overview on the sources of microarray data which are used in the literature.

Keywords

Swarm intelligence • Microarray • Feature selection • Classification

1 Introduction

Swarm intelligence (SI) is a computational model which is commonly inspired by the social behavior seen in nature. It is based on the collective behaviors resulting from the local interactions of individual agents with each other and their environment. The agents act in a coordinated way even in the absence of external controller or coordinator.

SI can be applied in a variety of fields including resource management such as nurse scheduling problem [1], information technology like data clustering [2] and engineering such as path tracking of autonomous mobile robot [3]. Application of SI in the field of bioinformatics as optimizers for feature selection algorithms is the focus of this paper. The usage of SI in this field had shown good results [4].

Feature selection in microarray data is a dimensionality reduction problem. Feature selection plays a crucial role in ensuring efficient analysis of the high dimensionality microarray data. Numerous studies have shown that most genes measured in DNA microarray experiment are not contributing to the accuracy of classification [5]. Therefore selection of genes that highly expressed the disease is essential.

Generally, microarray data is represented in a matrix where rows corresponds to genes and columns corresponds to different samples such as experiment conditions or tissues. Therefore, a cell represents an expression of a gene for a sample. Under various conditions, the transcription levels of genes in an organism is being measured and gene expression is built up. The expression level of a particular gene is expressed as number. The general procedure of DNA microarray analysis can be divided into two main steps as illustrated in Fig. 1. The first step is the selection of the most relevant features. There are numerous feature selection methods that can be applied to the microarray data which can be categorized into filter, wrapper, embedded and hybrid. Alternatively, SI can be applied as optimizers to the feature selection techniques to further improve the selection process.

N. Ahmad Zamri · B. Thangavel (✉) · N. A. Ab Aziz · N. H. Abdul Aziz
Faculty of Engineering & Technology, Multimedia University, Jalan Ayer Keroh Lama, 75450 Melaka, Malaysia
e-mail: t.bhuvaneswari@mmu.edu.my

F. Ibrahim et al. (eds.), *2nd International Conference for Innovation in Biomedical Engineering and Life Sciences*, IFMBE Proceedings 67, https://doi.org/10.1007/978-981-10-7554-4_27

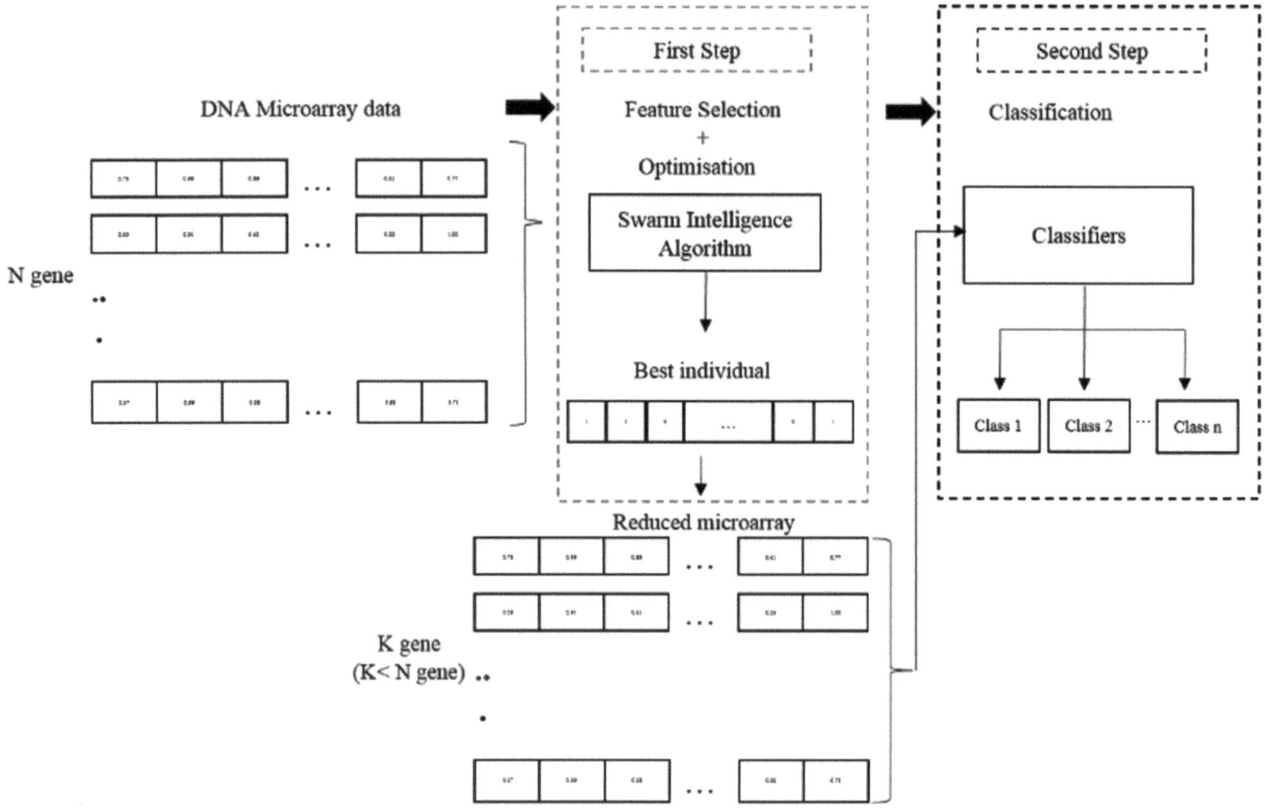

Fig. 1 General procedure of microarray analysis

The best individuals are referred to the genes that best describe the disease. In filter techniques, the interaction with classifier is ignored. A feature relevance score is calculated and low scoring features are eliminated. Then, the feature subset with high scores will be the input to the classifier afterwards. On the other hand, in wrapper techniques, once the best set of genes is obtained, it will be used as training samples for the classifier at the second step to achieve the best classification results. Whereas for embedded techniques, the searching for an optimal subset of features is included with the classification model. A hybrid technique usually combined the feature selection techniques for example a filter and wrapper approach. The evaluation is normally measured in terms of accuracy, specificity and sensitivity.

The structure of the paper is explained as follows: General algorithm of SI and SI optimizers that are frequently used in microarray's feature selection process, such as Particle Swarm Optimization and Artificial Bee Colony Optimization is described in Sect. 2. Section 3 describes the related works done by research community on feature selection using swarm intelligence approach. Section 4 discusses the data source used by the works reviewed in the paper. Finally, Sect. 5 concludes the paper.

2 SI Optimizers

The review presented in this paper is based on eleven works published between 2010 and 2016. It is observed that PSO, ACO and ABC are the popular choice of swarm intelligence algorithms applied to microarray data. Figure 2 shows summary of works reviewed according to the swarm intelligence algorithm used. Based on the literature, PSO is the most popular option with six works and this is followed by

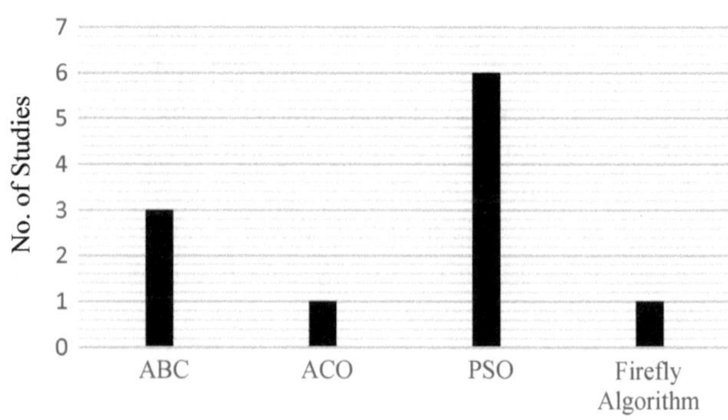

Fig. 2 Recent trends of optimization algorithm used in DNA microarray data

three ABC based research, and one each for ACO and Firefly Algorithm.

The SI algorithms follow similar framework (Fig. 3), where the search for optimal solution is conducted by swarm of agents. Each of the agents holds a candidate solution. The search starts with random initialization of the agents according to the problem. This is followed by evaluation of the quality of the candidate solutions proposed by the agents. The third step is the generation of new candidate solutions, which differs from one algorithm to another. The generation of new candidate solution follows the algorithms' sources of inspiration. For example, in PSO this step mimics the fish schooling and bird flocking behaviour, while the bees swarming behavior is mimicked in ABC, ants foraging is mimicked in ACO and fireflies' flash intensity attraction is mimicked in firefly algorithm. The candidate solutions' quality evaluation and generation of new candidate solutions are repeated until stopping condition is met.

2.1 Particle Swarm Optimization (PSO)

Particle Swarm Optimization (PSO) was originally proposed by James Kennedy and Russell Eberhart in 1995 [6]. PSO is a population-based stochastic optimization technique inspired by the social practices watched over animals for example bird flocking and fish schooling. The term particle was used to notate the PSO's search agents [7]. It also refers to population members which are mass-less and volume-less or with an arbitrarily small mass or volume.

A particle, *i*, in the swarm searches for the solution in the high-dimensional problem space using four vectors, its current position, x^i, which represents a solution, the particle's best found position, p^i, the best position found by its neighbourhood from the start of the search, p^g, and lastly its velocity, v^i. The velocity and position at *k*th iteration is updated using the equations below:

$$v_k^i = v_{k-1}^i + {}^{c_1}r_{1,k}(p_k^i - x_{k-1}^i) + {}^{c_2}r_{2,k}(p_k^g - x_{k-1}^i) \quad (1)$$

$$x_k^i = x_{k-1}^i + v_k^i \quad (2)$$

In Eq. (1), c_1 and c_2 represent cognitive and social parameters, while r_1 and r_2 represent random numbers between 0 and 1. PSO has been successfully applied in many areas other than bioinformatics including industrial applications [8], electromagnetic [9] and power systems [10].

```
Initialization
Evaluate candidate solutions' quality
Generate new candidate solutions
Check for stopping condition
```

Fig. 3 Pseudo code of SI algorithm

2.2 Artificial Bee Colony (ABC)

Karaboga and Basturk proposed Artificial Bee Colony (ABC) in 2007 [11]. ABC is influenced by the collective behaviour of honey bees to find food sources around the hive. The colony of artificial bees in ABC algorithm contains three groups of bees which are the employed bees, onlookers and scouts.

Employed bees bring loads of nectar to the hive. The employed bees share the information with onlooker bees which stay in the hive. The scout bees search for new food source in the surrounding area of hive. Once an onlooker bee and scout bee select a food source they become employed. The food source is chosen based on a probability value of the following equation:

$$p_i = \frac{fit_i}{\sum_{n=1}^{N} fit_n} \quad (3)$$

where N represents number of food sources and fit_i represents the fitness value of the solution. On the other hand, an employed bee is turned into a scout or onlooker bee when their food sources become empty. A new candidate of food source is calculated by using:

$$x_i^j = x_{\min}^j + rand(0,1)^* (x_{\max}^j - x_{\min}^j) \quad (4)$$

where x_i^j represents the position of the food source and $x_{\max}^j$ and $x_{\min}^j$ represents the lower bound and upper bound of the *j* dimension respectively. The possible solution to a problem under consideration is represented by the food source in ABC algorithm in an iterative search process. In ABC algorithm, exploration is done by scout bees whereas the employed and onlooker bees are responsible for exploitation.

3 Feature Selection Using SI

In this section, the related works in feature selection is reviewed. The related works reviewed here are grouped according to the optimization algorithms used.

3.1 PSO Based

PSO has been used extensively within these five years as referred to Fig. 2. This might due to the easy implementation of PSO as it requires few parameters to adjust. The search is decided by the speed of the particle. During the development of several generations, only the most optimist particle can transmit information onto the other particles [12]. Thus, with this optimization ability, problem can be completed easily.

Particle Swarm Optimization (PSO) was used in [13] for producing an optimized feature subset. The proposed method is tested on four cancer datasets; leukemia, colon, Diffuse Large B-Cell Lymphoma (DLBCL) and breast cancer. The use of PSO during feature selection improves the classification accuracy compared to without using PSO with classification accuracy more than 80%. In [13], Support Vector Machine (SVM) and K-Nearest Neighbourhood (KNN) were used as the classifier.

PSO was also employed in [14]. In [14], PSO and adaptive KNN technique are combined and used for classification of cancer subgroups. Even with least amount of genes in the subset, the classification accuracy still can be improved due to the fitness function of PSO that has been formulated in this work. This is achieved by selecting the optimized value of neighbours, k for KNN classifier. A proper value of k would help the formation of suitable numbers of neighbourhood to be explored. Thus, dataset can be more accurately classified. Less number of genes are appropriately formed by the proposed method from a dimensionally large microarray gene expression data. Results shows that it achieved high classification accuracy on blind test samples.

Hamming distance is applied in [15] compared to normal Euclidian distance measurement. Two different objects may look-alike in an enormous feature space, therefore, the work claimed that Euclidian distance might not be suitable for high dimensional data. The important features subset was selected by the velocity update of particle in binary PSO framework (HDBPSO). In this work, the velocity is updated by the adapted Hamming distance as a proximity measure. The results are evaluated based on validity indices and classification accuracies. Binary data were used in the proposed method for the comparison with GA and multi-objective GA. The results shows HDBPSO performs excellently as compared to others.

In [16], a multi-objective Binary Particle Swarm Optimization (BPSO) algorithms is introduced for feature selection from high dimensional gene expression data. Dimensionality reduction took place at the first stage by which normalization, discretization and conversion of data to binary distinction table occurred. Then, the selection of feature subset occurred at the second stage done by BPSO. Standard classifiers SVM and KNN were used to validate the selected feature subset in terms of classification accuracy.

An integration of PSO searching algorithm and C4.5 decision tree classifiers called PSODT was introduced in [17]. The important genes were identified using PSO algorithm and the fitness of the selection is evaluated using C4.5. The results shows high classification accuracy with PSODT compared to stand alone C4.5 classifier.

A Hybrid Particle Genetic Swarm Optimization (PGSO) is proposed in [18] to optimize the selected features to efficiently classify either normal or early or different stages of ovarian cancer. The basic framework of this proposed work is PSO. However, to improve the PSO performance, Genetic Algorithm (GA) is used as a local optimizer at each iteration. Operation such as selection, crossover and mutation are applied to the initial created particles. The results show an accuracy of 98% with two datasets when using multiclass SVM classifier compared to ANN, 95% and Naïve Bayes, 93%.

3.2 ABC Based

Artificial Bee Colony algorithm was chosen for gene selection problems in several works. Its popularity is also contributed by its few parameters compared to other optimization algorithms. Both global and local search can be conducted by ABC. Hence, the probability of finding the optimal is significantly increased. In [19], ABC algorithm was used to find a subset of genes which is then used to train different Artificial Neural Network (ANN); Multilayer Perceptron (MLP), Radial Basis Function Neural Network (RBF) and SVM.

A hybrid algorithm called Genetic Bee Colony (GBC) algorithm is introduced in [20]. GBC integrates the ABC algorithm and Genetic Algorithm (GA). In this proposed method, the replacement rate is increased to improve the movement speed. This is done by increasing the number of scout bee from one to two. A mutation operator from GA is also adopted to upgrade the exploitation process of the algorithm. This GBC selection approach is combined with SVM as the classifier. The proposed method shows efficient results in terms of classification accuracy.

Minimum Redundancy Maximum Relevance (mRMR) approach combined with ABC algorithm (mRMR-ABC) is introduced in [21]. mRMR which can be used for continuous and discrete datasets is implemented to measure the relevancy and redundancy of features. This technique can determine the promising features of the gene and facilitate the classifier to be trained accurately. The classification accuracy of SVM classifier will determine the fitness value of the ABC algorithm implemented in this work. The algorithm changes its solution to the new solution when the new fitness value is better than the old fitness values, otherwise it stays in its solution.

Table 1 Data source and accuracy using SI algorithms

Technique	Classifier	Dataset/accuracy (%)	Type of cancer class (binary/multiclass)	Source of data
PSO [13]	SVM	DLBCL (100)	Multiclass	[26]
		Leukemia (100)	Multiclass	
		Breast cancer (100)	Binary	
		Colon (97.50)	Binary	
PSO [17]	Decision tree	11_Tumors (97.82)	Multiclass	[27]
		14_Tumors (74.60)	Multiclass	
		9_Tumors (74.00)	Multiclass	
		Brain_Tumor1 (57.03)	Multiclass	
		Brain_Tumor2 (86.06)	Multiclass	
		Leukemia2 (100)	Multiclass	
		SRBCT (92.94)	Multiclass	
		DLBCL (92.55)	Multiclass	
		5Prostate_Tumor (94.31)	Binary	
		Lung cancer (100)	Binary	
PSO [14]	KNN	SRBCT (100)	Multiclass	[28]
		MLL (100)	Multiclass	
		ALL_AML (97.0588)	Binary	
PSO and GA [18]	SVM	Ovarian cancer (98)	Multiclass	[29]
Multi-objective PSO [16]	KNN, Bayes family function based classifiers and tree based classifiers	Leukemia (>90)	Multiclass	[30]
		Lymphoma (>90)	Multiclass	[31]
		Colon (>90)	Multiclass	[32]
Binary PSO with haming distance [15]	KNN	Leukemia (100)	Multiclass	[30]
		Lymphoma (100)	Multiclass	[31]
		Colon (100)	Multiclass	[32]
ABC and GA [20]	SVM	Lymphoma (98.48)	Multiclass	[27]
		Leukemia2 (95.83)	Multiclass	
		SRBCT (96.38)	Multiclass	
		Leukemia1 (96.43)	Binary	
		Colon (95.64)	Binary	
		Lung (99.50)	Binary	
ABC and mRMR [21]	SVM	Lymphoma (96.96)	Multiclass	[27]
		Leukemia2 (96.12)	Multiclass	
		SRBCT (96.30)	Multiclass	
		Leukemia1 (95.83)	Binary	
		Colon (94.17)	Binary	
		Lung (98.95)	Binary	

(continued)

3.3 ACO Based

ACO is originally developed to solve discrete combinatorial optimization problems such as the travelling salesman problem (TSP). Many of bioinformatics problems like the sequence alignment, gene mapping and feature selection of the gene in microarray data are similar to TSP. This makes ACO suitable for optimization of bioinformatics problems.

Table 1 (continued)

Technique	Classifier	Dataset/accuracy (%)	Type of cancer class (binary/multiclass)	Source of data
ABC [19]	ANN	ALL_AML (91.9)	Binary	[33]
		DLBCL-NIH (59)	Binary	[34]
		Breast cancer (62.1)	Binary	[35, 36]
		Prostate tumor (86.9)	Binary	[37]
ACO and ABC [22]	Fuzzy	ALL_AML (98.7)	Binary	[33]
		Colon (99.5)	Binary	[38]
		Lymphoma (98.5)	Binary	[39]
Firefly Algorithm [24]	NA	National Cancer Institute (NCI) dataset—leukemia, melanoma, lung, colon, central nervous system, ovarian, renal, breast and prostate cancers (NA)	Multiclass	[40]
		Lung dataset (NA)	Binary	

A hybrid Ant Bee Algorithm (ABA) which combines Ant Colony Optimization (ACO) with Artificial Bee Colony (ABC) algorithm in fuzzy expert system for encoding the solution variables using a modified form of representation was introduced in [22]. The optimal rule set of the combinatorial optimization is formed by the implementation of ACO in proposed work. The representation of the membership function as continuous number is done by ABC. The capability of the method is determined using several gene expression data sets which include rheumatoid arthritis versus controls (RAC) and rheumatoid arthritis versus osteoarthritis (RAOA), leukemia, type 2 diabetes, lymphoma and colon cancer. Receiver Operating Characteristic (ROC) analysis have been done to every datasets in the proposed work. The value of area under ROC curve is used to compare the performance of proposed ABA with other algorithm. ABA is reported to have the best value when compared to BCGA, RCGA, PSO and GA.

3.4 Firefly Algorithm Based

Firefly algorithm has the advantage of high convergence rate and robustness [23]. In [24], a multi objective firefly algorithm technique for multiclass gene selection is introduced. The method optimizes the multiple fire flies in the multiclass type of microarray datasets to select the genes. This technique is compared with existing gene selection methods and the outcomes shows that this technique achieves high classification accuracy with less complexity than the existing methods.

4 Microarray Data

This section is devoted to the details of microarray data that are present in the literatures. Gene expression data can either be unlabeled, fully labeled or partially labeled. Data that are marked with some meaningful labels or classes is the labeled data. Unlabeled data on the other hand contain the features without the presence of labels with any explanation or information.

The selection of feature subset that make used of labeled data is called supervised feature selection. Unlabeled data are used in unsupervised feature selection and semi supervised used the semi labeled data. In this review paper, supervised feature selection is mainly involved. A labeled microarray data can be classified into two types of the dataset which are binary and multiclass. Binary data normally consists of normal and malignant tissues which have been used to separate healthy patients form cancer patients. On the other hand, multiclass data is used to distinguish different type of tumors [25] in which the classification task becomes more complicated. The works reviewed in the previous section are employed for both types of data. The type and source of data that have been used in the literatures are illustrated in Table 1. Majority of the microarray data are publicly available for the research community.

Based on Table 1, there are few works done with the same benchmark dataset. The works in [15] and [16] utilized different techniques on feature selection and classification of the same dataset. Two different objects may look-alike in an enormous feature space, therefore, [15] claimed that Euclidian distance might not be suitable for high dimensional data. Therefore, [15] proves that implementation of the Hamming distance in feature selection managed to increase the accuracy when compared to normal Euclidean distance done in [16].

There are three works [17, 20] and [21] that used the same dataset [27] in which [20] and [21] are from the same author but with different techniques. Based on Table 1, the average accuracy for all the dataset used in [20, 21] achieved more than 90%. There is only slight difference of accuracy between the two techniques. PSO is applied in [17] with more dataset compared to [20, 21]. The work done in [17] also managed to classify the dataset to more than 90% accuracy except for Brain_Tumor1 which only managed to get 57.03% accuracy.

5 Conclusion

Classifying cancer using microarray's gene expression data is challenging because microarray has a high dimensional data and low sample dataset. Hence, many techniques have been suggested to address this problem. The ability of swarm intelligence algorithms to eliminate irrelevant genes and identify the informative genes has become a major interest among researchers. In this paper, usage of PSO, ABC, ACO and Firefly swarm intelligence algorithms in DNA microarray data analysis is reviewed. Based on the related works on feature selection, the optimization algorithms managed to improve the accuracy of classification algorithms with minimum number of selected genes for binary and multiclass type of microarray data.

Acknowledgements We would like to thank Multimedia University for their assistance and this work is supported by FRGS grant (FRGS/1/2015/TK04/MMU/03/2).

References

1. Mutingi, M., Mbohwa, C.: A Fuzzy-based particle swarm optimization algorithm for nurse scheduling. In: Proceedings of the World Congress on Engineering and Computer Science, pp. 22–24. (2014)
2. Chuang, L., Lin, Y., Yang, C., Swarm, A.P., Pso, O.: An improved particle swarm optimization for data clustering. In: Proceedings of the International MultiConference of Engineers and Computer Scientist, pp. 1–6. (2012)
3. Ali, R.S., Almousawi, A.K.: Design an optimal PID controller using artificial bee colony and genetic algorithm for autonomous mobile robot. Int. J. Comput. Appl. **100**(16), 8–16 (2014)
4. Larran, P., Saeys, Y.: Gene expression—a review of feature selection techniques in bioinformatics. Bioinformatics **23**(9), 2507–2517 (2007)
5. Golub, T.R., Slonim, D.K., Tamayo, P., Huard, C., Gaasenbeek, M., Mesirov, J.P., Coller, H., Loh, M.L., Downing, J.R., Caligiuri, M.A., Bloomfield, C.D., Lander, E.S.: Molecular classification of cancer: Class discovery and class prediction by gene expression monitoring. Science AAAS **286**(5439), 531–537 (1999)
6. Kennedy, J., Eberhart, R.: Particle swarm optimization. In: Proceedings of IEEE International Conference on Neural Networks, pp. 1942–1948 (1995)
7. Christian Blum, D.M.: Swarm Intelligence—Introduction and Applications. Springer, Berlin (2008)
8. Jérôme, L.J.D., Onwunalu, E.: Application of a particle swarm optimization algorithm for determining optimum well location and type. Comput. Geosci. **14**(1), 183–198 (2010)
9. Matekovits, L., Mussetta, M., Pirinoli, P., Selleri, S., Zich, R.E.: Improved PSO algorithms for electromagnetic optimization. In: IEEE antennas and propagation society international symposium, pp. 33–36. (2005)
10. Cristian, D., Barbulescu, C., Kilyeni, S., Popescu, V.: Particle swarm optimization techniques. Power systems applications. In: 2013 6th international conference on human system interactions (HSI), Sopot, pp. 312–319. (2013)
11. Karaboga, D., Basturk, B.: A powerful and efficient algorithm for numerical function optimization: artificial bee colony (ABC) algorithm. J. Global Optim. **39**, 459–471 (2007)
12. Bai, Q.: Analysis of particle swarm optimization algorithm. Comput. Inf. Sci. **3**(1), 180–184 (1998)
13. Sahu, B., Mishra, D.: A novel feature selection algorithm using particle swarm optimization for cancer microarray data. Procedia Eng. **38**, 27–31 (2012)
14. Kar, S., Das Sharma, K., Maitra, M.: Gene selection from microarray gene expression data for classification of cancer subgroups employing PSO and adaptive K-nearest neighborhood technique. Expert Syst. Appl. **42**(1), 612–627 (2015)
15. Banka, H., Dara, S.: A Hamming distance based binary particle swarm optimization (HDBPSO) algorithm for high dimensional feature selection, classification and validation. Pattern Recogn. Lett. **52**, 94–100 (2015)
16. Sekhara, C., Annavarapu, R., Dara, S., Banka, H.: Original article: cancer microarray data feature selection using multi-objective binary particle swarm optimization algorithm. EXCLI J. **15**, 460–473 (2016)
17. Chen, K., Wang, K.-J., Tsai, M.-L., Wang, K.-M., Adrian, A.M., Cheng, W.-C., Yang, T.-S., Teng, N.-C., Tan, K.-P., Chang, K.-S.: Gene selection for cancer identification: A decision tree model empowered by particle swarm optimization algorithm. BMC Bioinform. **15**, 0–9 (2014)
18. Yasodha, P., Anathanarayanan, N.R.: Analysing big data to build knowledge based system for early detection of ovarian cancer. Indian J. Sci. Technol. **8** (2015)
19. Garro, B.A., Rodríguez, K., Vázquez, R.A.: Classification of DNA microarrays using artificial neural networks and ABC algorithm. Appl. Soft Comput. **38**, 548–560 (2016)
20. Alshamlan, H.M., Badr, G.H., Alohali, Y.A.: Genetic Bee Colony (GBC) algorithm: A new gene selection method for microarray cancer classification. Comput. Biol. Chem. **56**, 49–60 (2015)
21. Alshamlan, H., Al-Ohali, Y., Badr, G.: mRMR-ABC: A hybrid gene selection algorithm for cancer classification using microarray gene expression profiling. Hindawi **2015**, (2015)
22. Ganesh kumar, P., Rani, C., Devaraj, D., Victoire, T.A.A.: Hybrid ant bee algorithm for fuzzy expert system based sample classification. IEE Trans. Comput. Biol. Bioinform. **11**(2), 347–360 (2014)

23. Pal, N.S.: Robot path planning using swarm intelligence: A survey. Int. J. Comput. Appl. **83**(12), 5–12 (2013)
24. Manoharan, G.V., Shanmugalakshmi, R.: Multi-objective firefly algorithm for multi-class gene selection. Indian J. Sci. Technol. **8**, 27–34 (2015)
25. Bolon-Canedo, F.H.V., Sanchez-Marono, N., Alonso-Betanzos, A., Benitez, J.M.: A review of microarray datasets and applied feature selection methods. Inf. Sci. **282**, 111–135 (2014)
26. GEDatasets. http://sdmc.lit.org.sg/GEDatasets/. Last accessed 11 Jan 2017
27. Microarray cancer datasets. http://www.gems-system.org. Last accessed 11 Jan 2017
28. Gene Expression Datasets. http://research.nhgri.nih.gov/microarray/Supplement/. Last accessed 11 Jan 2017
29. Normalized gene expression data. http://tcga-data.nci.nih.gov/. Last accessed 11 Jan 2017
30. Leukaemia. http://www.genome.wi.mit.edu/MPR. Last accessed 11 Jan 2017
31. Lymphoma. http://llmpp.nih.gov/lymphoma/data/figure1/figure1.cdt. Last accessed 11 Jan 2017
32. Colon Cancer. http://microarray.princeton.edu/oncology. Last accessed 11 Jan 2017
33. Bloomfield, C.D., Lander, E.S., Golub, T.R., Slonim, D.K., Tamayo, P., Huard, C., Gaasenbeek, M., Mesirov, J.P., Coller, H., Loh, M.L., Downing, J.R., Caligiuri, M.A.: Molecular classification of cancer: class discovery and class prediction by gene expression monitoring. Science **286**, 531–537 (1999)
34. Rosenwald, L.M.S.A., Wright, G., Chan, W.C., Connors, J.M., Campo, E., Fisher, R.I., Gas-coyne, R.D., Muller-Hermelink, H. K., Smeland, E.B., Giltnane, J.M., Hurt, E.M., Zhao, H., Averett, L., Yang, L., Wilson, W.H., Jaffe, E.S., Simon, R., Klausner, R.D., Powell, J., Duffey, P.L.: The use of molecular profiling to predict survival after chemotherapy for diffuse large-b-cell lymphoma. N. Engl. J. Med. **346**, 1937–1947 (2002)
35. van de Vijver, M.J., He, Y.D., van 't Veer, L.J., Dai, H., Hart, A. A., Voskuil, D.W., Schreiber, G.J., Peterse, J.L., Roberts, C., Marton, M.J., Parrish, M., Atsma, D., Witteveen, A., Rutgers, E. T., Glash, A., Delahaye, L., van der Velde, T., Bartelink, H., Rodenhuis, S., Bernards, R., Rutgers, E.T., Friend, S.H.: A gene-expression signature as a predictor of survival in breast cancer. N. Engl. J. Med. **347**, 1999–2009 (2002)
36. van 't Veer, L.J., Dai, H., van de Vijver, M.J., van de Vijver, M.J., He, Y.D., Hart, A.A., Mao, M., Peterse, H.L., van der Kooy, K., Marton, M.J., Witteveen, A.T., Schreiber, G.J., Kerkhoven, R.M., Roberts, C., Linsley, P.S., Bernards, R., Friend, S.H.: Gene expression profiling predicts clinical outcome of breast cancer. Nature **415**, 530–536 (2002)
37. Singh, D., Febbo, P.G., Ross, K., Jackson, D.G., Manola, J., Ladd, C., Tamayo, P., Renshaw, A.A., D'Amico, A.V., Richie, J.P., Lander, E.S., Loda, M., Kantoff, P.W., Golub, T.R., Sellers, W.R.: Gene expression correlates of clinical prostate cancer behavior. Cancer Cell **1**, 203–209 (2002)
38. Alon, U., Barkai, N., Notterman, D.A., Gish, K., Ybarra, S., Mack, D., Levine, A.J.: Broad patterns of gene expression revealed by clustering of tumor and normal colon tissues probed by oligoneu-clotide arrays. Proc. Nat'l Acad. Sci. USA **96**, 6745–6750 (1999)
39. Alizadeh, A.A., Eisen, M.B., Davis, R.E., Ma, C., Lossos, I.S., Rosenwald, A., Boldrick, J.C., Sabet, H., Tran, T., Yu, X., Powell, J.I., Yang, L., Marti, G.E., Moore, T., Hudson Jr., J., Lu, L., Lewis, D.B., Tibshirani, R., Sherlock, G., Chan, W.C., Greiner, T. C., Weisenburger, D.D., Armitage, J.O., Warnke, R., Levy, R., Wilson, W., Grever, M.R., Byrd, J.C., Botstein, D., Brown, P.O., Staudt, L.M.: Distinct types of diffuse large b-cell lymphoma identified by gene expression profiling. Nature **403**, 503–511 (2000)
40. National Cancer Institute Homepage. https://www.cancer.gov/. Last accessed 11 Jan 2017

Support Vector Machine Classification of EEG Nonlinear Features for Primary Insomnia

Haslaile Abdullah, Chanakya Reddy Patti, Chamila Dissanyaka, Thomas Penzel, and Dean Cvetkovic

Abstract

Primary insomnia is a term used to describe a subtype of insomnia that constitutes the disorder itself and is not a consequent to any other psychiatric or sleep disorder. Hitherto, there is no clear objective markers from Polysomnography (PSG) signal to characterize insomnia. Although linear methods like spectral analysis of EEG frequency bands have been used to detect physiological arousal in patients with insomnia, these methods may not be sufficient enough to extract valuable information and detect abnormalities in the signals. The EEG signal itself originate from a complex neuronal activity in the brain, therefore the use of nonlinear measures may show some hidden information that could better explain the activation of this hyperarousal. The aim of the present study is to classify the primary insomnia patient from the healthy based on the supervised learning machine technique of SVM and the usage of nonlinear features of EEG signal. The classification result by using SVM achieved an overall of 83% of accuracy, 85 and 80% of sensitivity and specificity respectively.

Keywords

Support vector machine • EEG • Primary insomnia

1 Introduction

Primary insomnia is a term used to describe a subtype of insomnia that constitutes the disorder itself and is not a consequent to any other psychiatric or sleep disorder. In contrast, a secondary insomnia is known as comorbid insomnia, which is usually associated with new episodes of depression. Hitherto, there is no clear objective markers from Polysomnography (PSG) signal to characterize insomnia. PSG studies on insomnia patients generally show abnormalities, such as prolonged latency to sleep onset, frequent arousals, and reduced amounts of total sleep [1].

Functional neuroimaging studies have reported a similar findings that insomnia patients have shown higher whole brain metabolism which failed to decline in activity from wake to sleep and resulted in greater Non-Rapid Eye Movement (NREM) sleep [2]. EEG studies of linear spectral analysis have focused in the high frequency range of 16–50 Hz during NREM sleep. Increased beta activity and reduced delta activity have been observed in primary insomnia [3, 4]. The finding from all these studies suggests that symptoms of heightened physiological and cognitive arousal (hyperarousal) are common characteristics that could differentiate insomnia from healthy. In recent years, various methods have been applied to extract different features in the sleep EEG to characterize different pathological conditions. Although linear methods like spectral analysis of EEG frequency bands have been used to detect physiological arousal in patients with insomnia, these methods may not be

H. Abdullah (✉)
Razak School of Engineering and Advanced Technology, Universiti Teknologi Malaysia, Kuala Lumpur, Malaysia
e-mail: haslaile.kl@utm.my

C. R. Patti · C. Dissanyaka · D. Cvetkovic
RMIT University, Melbourne, Australia

T. Penzel
Charité - Universitätsmedizin Berlin, Berlin, Germany

F. Ibrahim et al. (eds.), *2nd International Conference for Innovation in Biomedical Engineering and Life Sciences*, IFMBE Proceedings 67, https://doi.org/10.1007/978-981-10-7554-4_28

sufficient enough to extract valuable information and detect abnormalities in the signals. The EEG signal itself originate from a complex neuronal activity in the brain, therefore the use of nonlinear measures may show some hidden information that could better explain the activation of this hyperarousal.

A nonlinear technique like Largest Lyapunov Exponent has shown a better prediction of epileptic seizure [5]. In another study of epilepsy EEG signal, four entropy features namely Approximate Entropy, Sample Entropy, Phase Entropy 1 and 2 were fed to several classifier for an automatic detection of epilepsy and has reached a high overall accuracy [6]. Furthermore, the use of EEG nonlinear features such as Detrended Fluctuation Analysis, Lyapunov Exponent, Higuchi Fractal and Correlation Dimension have increased overall classifier accuracy when compared to a linear power bands features to discriminate depressed patients from the healthy [7]. The application of nonlinear features for an automatic sleep stages scoring have also been reported in several studies [8, 9]. From the reported results, the nonlinear features can characterize and distinguish different sleep stages and was very suitable to be used as an input for an automatic sleep staging. Despite that, there are still very limited studies of nonlinear features for an automatic detection of insomnia. In [10], we have combined linear and nonlinear features from EEG and ECG signal, but the overall classification accuracy from neural network still low. Therefore, the aims of the present study is to classify the primary insomnia patient from the healthy based on the supervised learning machine technique of SVM and the usage of nonlinear features of EEG signal.

2 Method

2.1 Sleep Data

The PSG data were collected from ten healthy subjects and ten patients diagnosed for primary insomnia at Charité University (Berlin, Germany). The sleep stages N1, N2, N3 and REM were scored from a 30 s epoch. In the present study, EEG channel of C3–A2 with a sampling frequency of 200 Hz was used to extract the nonlinear measures. The mean and standard deviation for age and sleep parameters are summarized in Table 1.

Table 1 Subject age and sleep parameters

	Healthy	Primary insomnia
Age (years)	36.1 ± 13.46	35 ± 11.67
Total sleep time (min)	393.95 ± 31.15	372.30 ± 58.56
Sleep efficiency (%)	84.73 ± 7.86	77.24 ± 12.98
Stage wake (min)	67.75 ± 46.89	112.60 ± 69.34
Stage N1 (min)	72.36 ± 40.52	57.65 ± 19.65
Stage N2 (min)	46.65 ± 25.21	185.90 ± 36.77
Stage N3 (min)	210.75 ± 37.21	88.95 ± 44.84
Stage REM	68.80 ± 21.52	39.80 ± 22.20
Sleep latency (min)	15.53 ± 18.44	26.50 ± 25.39
REM latency (min)	115.28 ± 66.64	102.35 ± 45.92

2.2 Nonlinear Feature Extraction

2.2.1 Largest Lyapunov Exponent (LLE)

LLE is a quantitative measure of chaotic data that highly depends and sensitive to its initial condition [11]. The EEG signal is reconstructed in the phase space where two nearest neighboring points at time t and time 0 has distance of $\delta x_i(t)$ and $\delta x_i(0)$ respectively. The LLE is an average rate of divergence or convergence of these two points as described by Eq. (1). A positive value of LLE indicates an existence of chaos in the signal.

$$\lambda_i = \lim_{t\to\infty} \frac{1}{t} \log_2 \frac{\|\delta x_i(t)\|}{\|\delta x_i(0)\|} \tag{1}$$

2.2.2 Detrended Fluctuation Analysis (DFA)

DFA is a method to determine an existence of long range correlation in a non-stationary time series [12, 13]. For uncorrelated signal or white noise, the value of $\alpha = 0.5$. The values of $0.5 < \alpha \leq 1$ shows long-range correlation. The computational of DFA involves several steps as following:

1. The time series is integrated as

$$y(k) = \sum_{i=1}^{k} [x(i) - x_{ave}] \tag{2}$$

2. The integrated signal is divided equally to boxes of length, N
3. Apply a least-square line to fit each segment and denote the straight line as

$$y_n(k) \tag{3}$$

4. Detrend the time series with subtracting the local trend as

$$z(k) = \sum_{i=1}^{k} [y(k) - y_n(k)] \tag{4}$$

5. Calculate the root mean square of z(k) as

$$F(n) = \sqrt{\frac{1}{N}\sum_{k=1}^{N} z(k)} \quad (5)$$

6. Repeat all step for all time scales
7. Plot log F(n) versus log (n) to get the scaling exponent α, the slope of the log graph.

2.2.3 Sample Entropy (SampEnt)

SampEn is a regularity measure of signal where a lower value of SampEn indicates the time series is more self-similarity. This method is introduced in [14, 15] to overcome the shortcomings of calculation Approximate Entropy. The main advantage of SampEn is it does not count self- matches. Secondly, it is independence of record length which means it does not require a long data records. The SampEn is a calculation of negative natural logarithm conditional probability of two time series sequence with length, *N* for *m* dimension is remain similar within tolerance *r*. The tolerance *r* is set to 0.2 in our experiment as represented by Eq. (6).

$$SampEn\,(m,r,N) = \ln\left(\frac{A(m)}{B(m-1)}\right) \quad (6)$$

2.3 Supervised Learning Technique

2.3.1 Support Vector Machine (SVM)

SVM is a simple data classification technique which produces a model based on training data and further predicts the class values of the tested data. Firstly, the nonlinear features were scaled to [0, 1] and labeled as $(\mathbf{x_i}, \mathbf{y_i})$, $i = 1,\ldots, k$ where $x_i \in R^n$ and $y_i \in \{1, -1\}^k$. The training vectors $\mathbf{x_i}$ are mapped into a higher dimensional space by function ϕ and SVM will search a linear separating hyperplane with the maximal margin [16]. The Radial Basis Kernel (RBF) was chosen as our problem is non-linear by nature. Several values of RBF parameters were tested heuristically and found that C = 10 and $\gamma = 0.01$ was the optimized values for our study. The SVMTorch II was used for the implementation of supervised learning technique. For each nonlinear feature, there are 7500 samples which were divided into half for training (3750 samples) and another half for testing. In this two-class problem, the healthy and the patient group were coded as digit 0 and 1 respectively.

2.3.2 Classification Performance Evaluation

The performance of SVM was evaluated by measuring its sensitivity, specificity and accuracy which are defined as following:

$$\text{Sensitivity} = \frac{TP}{TP + FN} \quad (7)$$

$$\text{Specificity} = \frac{TN}{TN + FP} \quad (8)$$

$$\text{Accuracy} = \frac{TN + TP}{TP + TN + FP + FN} \quad (9)$$

where TP and TN represent the total number of True Positive and True Negative respectively. In addition, FP and FN are the total number of False Positive and False Negative counted during the classification [17].

3 Result and Discussion

We randomly used 50% of the nonlinear features data as training set and the remaining as testing set. The result from testing set we obtained from SVM classifier was summarized in Table 2.

A comparison of our result with other classification systems given in the literature is very difficult due to limited reported study on automatic detection of insomnia. But it can be seen from our study, a simple automatic detection by using three nonlinear features extracted from sleep EEG signal with SVM could achieved an overall of 83% of accuracy and 85 and 80% of sensitivity and specificity respectively. In [18], a linear feature based on a singular spectrum analysis (SSA) was extracted from sleep EEG of secondary type of insomnia patients. The feature was further classified into patient and control groups by using an artificial neural network with classification accuracy of 98%. In another study, the sleep onset period of normal and insomniac was characterized by the EEG frequency bands and linear discriminant analysis with 75% accuracy. Although both studies we referred have shown good performance on the automatic classification of insomnia, there are two key differences need to be highlighted. Firstly, both studies have

Table 2 SVM performance measures

Measures	Percentage
Sensitivity	85
Specificity	80
Accuracy	83

extracted a linear feature from spectral analysis of sleep EEG. The sleep EEG which exhibits a complex physiological activity is better described as non-stationary, noisy and chaotic signal [11]. Therefore, a nonlinear approach is far more superior to linear method. In our work, non-linear analysis was implemented to overcome the aforementioned limitation of linear methods. Secondly, in this study we analyzed EEG of primary insomnia patients which a subtype of insomnia that constitutes the disorder itself. In contrast, [18, 19] used EEG of secondary insomnia patients that is a sleeping disorder associated with psychological, psychiatric or medical problems.

4 Conclusion

In this study, we extracted non-linear features from a single channel sleep EEG signal of healthy and primary insomnia patients. We further implemented SVM for classification of primary insomnia and healthy by utilizing LLE, DFA and SampEn as an input to our classifier. Although the overall classification result we obtained still low, this study has successfully shown that a combination of three nonlinear features and SVM potentially can be used to develop an automatic detection system of primary insomnia.

Acknowledgements This research was funded under Research University Grant (Vot No. 15H54) from Universiti Teknologi Malaysia. The authors would like to thank the Research Management Centre of UTM and the Ministry of Education Malaysia for their financial support.

References

1. Vgontzas, A.N., Fernandez-Mendoza, J., Liao, D., Bixler, E.O.: Insomnia with objective short sleep duration: The most biologically severe phenotype of the disorder. Sleep Med. Rev. **17**(4), 241–254 (2013)
2. Nofzinger, E.A., Buysse, D.J., Germain, A., Price, J.C., Miewald, J.M., Kupfer, D.J.: Functional neuroimaging evidence for hyperarousal in insomnia. Am. J. Psychiatry **161**(11), 2126–2128 (2004)
3. Spiegelhalder, K., Regen, W., Feige, B., Holz, J., Piosczyk, H., Baglioni, C., Nissen, C.: Increased EEG sigma and beta power during NREM sleep in primary insomnia. Biol. Psychol. **91**(3), 329–333 (2012)
4. Svetnik, V., Snyder, E., Ivgy-May, N., Ma, J., Tao, P., Herring, W.: Comparison of EEG power spectra between patients with primary insomnia and good sleeper controls, accounting for the effects of age, gender, and sleep period, using a large compilation of PSG recordings from three clinical trials. Sleep Med. **14**, e278 (2013)
5. Shayegh, F., Sadri, S., Amirfattahi, R., Ansari-Asl, K.: A model-based method for computation of correlation dimension, Lyapunov exponents and synchronization from depth-EEG signals. Comput. Methods Programs Biomed. **113**(1), 323–337 (2014)
6. Hosseinifard, B., Moradi, M.H., Rostami, R.: Classifying depression patients and normal subjects using machine learning techniques and nonlinear features from EEG signal. Comput. Methods Programs Biomed. **109**(3), 339–345 (2013)
7. Acharya, U.R., Molinari, F., Sree, S.V., Chattopadhyay, S., Ng, K. H., Suri, J.S.: Automated diagnosis of epileptic EEG using entropies. Biomed. Signal Process. Control **7**(4), 401–408 (2012)
8. Koley, B., Dey, D.: An ensemble system for automatic sleep stage classification using single channel EEG signal. Comput. Biol. Med. **42**(12), 1186–1195 (2012)
9. Latchoumane, C.F.V., Jeong, J.: Quantification of brain macrostates using dynamical nonstationarity of physiological time series. IEEE Trans. Biomed. Eng. **58**(4), 1084–1093 (2011)
10. Abdullah, H., Penzel, T., Cvetkovic, D.: Detection of insomnia from EEG and ECG. In The 15th International Conference on Biomedical Engineering, pp. 687–690. Springer, Cham, 2014
11. Acharya, U.R., Faust, O., Kannathal, N., Chua, T., Laxminarayan, S.: Non-linear analysis of EEG signals at various sleep stages. Comput. Methods Programs Biomed. **80**(1), 37–45 (2005)
12. Lee, J.M., Kim, D.J., Kim, I.Y., Suk Park, K., Kim, S.I.: Nonlinear-analysis of human sleep EEG using detrended fluctuation analysis. Med. Eng. Phys. **26**(9), 773–776 (2004)
13. Peng, C.K., Havlin, S., Stanley, H.E., Goldberger, A.L.: "Quantification of scaling exponents and crossover phenomena in nonstationary heartbeat time series", Chaos: an interdisciplinary. J. Nonlinear Sci. **5**(1), 82–87 (1995)
14. Richman, J.S., Moorman, J.R.: Physiological time-series analysis using approximate entropy and sample entropy. Am. J. Physiol. Heart Circ. Physiol. **278**(6), 2039–2049 (2000)
15. Acharya, U.R., Molinari, F., Sree, S.V., Chattopadhyay, S., Ng, K. H., Suri, J.S.: Automated diagnosis of epileptic EEG using entropies. Biomed. Signal Process. Control **7**(4), 401–408 (2012)
16. Hsu, C.W., Chang, C.C., Lin, C.J.: A practical guide to support vector classification (2003)
17. Baratloo, A., Hosseini, M., Negida, A., El Ashal, G.: Part 1: simple definition and calculation of accuracy, sensitivity and specificity. Emergency **3**(2), 48–49 (2015)
18. Aydın, S., Saraoğlu, H.M., Kara, S.: Singular spectrum analysis of sleep EEG in insomnia. J. Med. Syst. **35**(4), 457–461 (2011)
19. Merica, H., Gaillard, J.M.: The EEG of the sleep onset period in insomnia: a discriminant analysis. Physiol. Behav. **52**(2), 199–204 (1992)

Prospects of Low-Cost Photometers for Colorimetric Serum Iron Concentration Determination

Arthur Foo Yen Tan, Noor Azrina Talik, Boon Kar Yap, and Prajindra Sankar Krishnan

Abstract

Iron is an essential element in the body that could cause harm to the body if there is deficiency or overload. Iron deficiency is a widespread nutritional disorder that affects the lives of a huge amount of people in the world and could lead to anemia if left untreated. Meanwhile, iron overload could lead to liver, heart problems or even death. Iron status is commonly diagnosed using sophisticated instruments, which are bulky and expensive. The need for a low-cost and portable iron body reader arises, as there are no such commercially available devices for this specific purpose. The serum iron concentration can be measured via colorimetric method using photometers. The colorimeter is commonly used to correlate the light absorbance and the concentration. This paper reviews the published works on developments of low-cost photometers. The possibility of developing a low-cost and portable colorimeter, specifically for measurement of serum iron was discussed.

Keywords

Serum • Iron • Colorimeter • Low-cost • Portable

1 Introduction

Iron is a transition metal, which is very important to the human body. While only in trace amount, it is vital for the fundamental biological functions including oxygen transport and cellular proliferation [1, 2]. It is also an important component of the hemoglobin which in turn forms erythrocytes, cytochromes and many enzymes [2, 3]. An adult body usually has around 3–4 g of iron, at which two thirds are for the formation of erythrocytes and recycled iron from the destruction of mature erythrocytes [2]. From that amount, only 1–2 g of iron can be absorbed each day by the intestines and circulated in the blood plasma by binding with the transferrin [3]. Excess iron is stored as ferritin, primarily in the liver, bone marrow and spleen [1].

According to the World Health Organization (WHO), iron deficiency is a global nutritional disorder affecting around 2 billion people, more commonly on infants, toddlers, women of childbearing age as well as pregnant and lactating women [1, 4, 5]. Iron deficiency refers to the reduction of iron stores in the body and is often associated with anemia because if left untreated, could lead to iron deficiency anemia (IDA) [4, 6]. The most common cause of iron deficiency is insufficient dietary iron and blood loss [1]. Iron deficiency can be treated and prevented by increasing intake of food rich with dietary iron. Iron supplements and iron fortification are among other common methods of treating iron deficiency. However, iron medication comes with side effects like epigastric discomfort, nausea, diarrhea as well as constipation if taken irregularly or at higher dosage [7].

Iron overload refers to the accumulation of iron that exceeds the body iron capacity such that it becomes toxic to the body [2, 8]. Since the body does not have an active iron excretion mechanism, absorption of more than 1–2 g of iron

A. F. Y. Tan · N. A. Talik · B. K. Yap (✉)
Electronic and Communications Department, College of Engineering, Universiti Tenaga Nasional, Selangor, Malaysia
e-mail: kbyap@uniten.edu.my

P. S. Krishnan
Electronic and Communications Department, College of Engineering, Universiti Tenaga Nasional, KM-7, Jalan Uniten-Ikram, 43000, Kajang, Selangor, Malaysia

F. Ibrahim et al. (eds.), *2nd International Conference for Innovation in Biomedical Engineering and Life Sciences*,
IFMBE Proceedings 67, https://doi.org/10.1007/978-981-10-7554-4_29

will cause iron overload leading to liver and heart failures. The accumulation of iron could be a result of bone marrow failures and of repeated transfusions in patients with anemia from genetic disorders such as thalassemia, sickle cell disease (SCD), and Diamond Blackfan syndrome [2].

Although iron is an important element for the body biological functions, iron deficiency or overload could potentially pose a serious health threat. Therefore, the need for iron body detection becomes ever important to control the issue of iron deficiency and iron overload [8]. Currently, there are no specific instruments to detect iron status in the body. Sophisticated instruments such as red cell analyzer and biochemistry auto-analyzer used to diagnose iron deficiency anemia are bulky and expensive. The development of a low-cost and portable iron reader has the potential to reduce and eliminate iron deficiency anemia in rural areas and even in developing countries [9]. Thus, this paper is intended to review previous selected works on low-cost photometers together with a prospect in developing low-cost colorimeter to measure body iron-level.

2 Light Absorption Spectroscopy Designs

Talik et al. [8] reported that colorimetric analysis using photometer could potentially be used as a low-cost iron body reader. The colorimeter is an example of a device used for colorimetry. The colorimeter is also known as the filter photometer where the filter is a monochromator [10]. It is commonly used to determine the absorption of a monochromatic light by a specific colored solution. Recently, many researchers have adopted colored LEDs as the light source as its mono-colored nature eliminated the need for a monochromator or a wavelength selective filter [11] (Fig. 1).

Light absorbance of a solution is described by Beer-Lambert law. Beer-Lambert law states that the absorbance of light is proportional to the concentration of the solution [11]. It forms the working basis of many light absorbance measurement devices.

$$A = -\log\frac{I}{I_o} = \varepsilon c l \quad (1)$$

The equation above is based on Beer-Lambert law. It states that the absorbance, A, is directly proportional to the molar absorptivity constant, ε, path length of the sample, l, and the concentration of the compound. I is denoted as the transmitted light intensity and I_o is the incident light intensity [11].

As an alternative for the commercialized expensive spectrophotometer, various works in developing low-cost colorimeters have been demonstrated [11, 13–15]. Colorimeters and spectrophotometers are normally used for colorimetric analysis. Table 1 shows a comparison of these two instruments, where the main difference is the spectral coverage.

2.1 Microcontroller Based Colorimetric Devices

Modern colorimetric devices normally use microcontrollers to simplify the measurement and control processes [11, 18–20].

2.1.1 Microcontroller-Based Colorimeter Using RGB LED

Suzuki et al. [19] developed a simple and portable colorimeter for on-site determination of nitrite and iron in river water. This dry-cell powered colorimeter utilizes an RGB LED as its light source and was driven by current-regulated diodes at each emitter to ensure optimum emission strength. Two photodiodes are used as the photodetector. One photodiode was used to detect the light transmitted through the sample and another was used for the correction of fluctuating emission strength. From the photodiodes, output signals were conditioned by operational amplifiers before passing through an analog to digital converter. The digital signal is then fed into the embedded controller for absorbance calculation and displayed onto an LCD.

The colorimeter was first tested on color dye before being used in the field. As a comparative study, the results of this colorimeter are compared with a conventional double beam spectrophotometer and a Flame Atomic Absorption Spectrometer (FAAS). When compared with the conventional spectrophotometer, the results of have smaller slopes and narrower linear range. This was because the wide spectral bandwidth of the RGB LED along with a large effect of stray light. However, it is deemed acceptable for field applications. The results for the determination of nitrogen as nitrite and determination of iron in river water samples also correlates well with the spectrophotometer and the FAAS.

Fig. 1 Structure of a colorimeter. Sourced from [12]

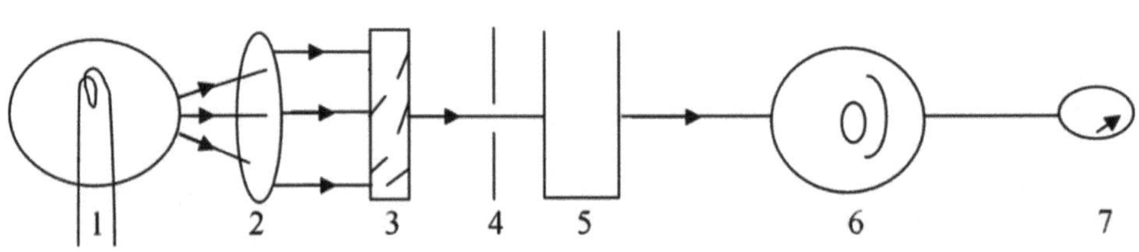

1-light source 2-collimating lens 3-monochromator 4-slit 5-sample cell 6-photoelectric cell 7-galvanometer

Table 1 Comparison between colorimeter and spectrophotometer [16, 17]

	Colorimeter	Spectrophotometer
Light source	Use a monochromatic light source which is obtained either by using a full spectrum light source and color filter, or a narrow spectrum light emitting diode (LED)	Use a full spectrum light source and diffraction grating or prism system to deliver monochromatic light over a range of wavelengths
Light detection	Generally use photodiodes	Use spectrometers
Light measurement method	Light transmittance and absorbance of colors of specific wavelength of a sample	Light transmittance and absorbance as a function of wavelength
Spectrum coverage	Visible spectrum	Ultraviolet, visible and infrared spectrum
Wavelength	Fixed	Variable

Kittipannyangam et al. [20] developed a battery-powered low-cost light absorbance measurement device that utilized an RGB LED as a replacement for the monochromator to reduce space. A phototransistor was chosen over the photodiode and the LDR due to the higher coefficient of determination. The phototransistor is connected to a PIC16F877A microcontroller, which acts as a calculator that displays the light absorbance reading on an LCD display. Data could also be transmitted to a computer via the serial port. The device costs 2300 JPY, which is approximately 21 USD or 88.33 MYR. There is also an Xbee communication interface.

The device works by measuring the light absorbance of identical colored solutions. The results of the measurement device were compared with that of a commercial spectrophotometer to determine its accuracy. The results showed that light absorbance of the red and blue colored solution correlated well with the results of the commercial spectrophotometer. However, the light absorbance of the green solution is slightly inaccurate. This issue is because the colored light and the photodetector did not match. Yet, the device managed to produce a coefficient of determination of less than 1% for all tests, which makes it suitable to replace the commercial spectrophotometer for education purposes. This device is not suitable for applications that require more details as the RGB LED is not able to identify its wavelength when the colors are mixed.

2.1.2 The SLIM Spectrometer

Cantrell et al. [18] developed the SLIM spectrometer that was evaluated using thionine as an analyte in a flow cell configuration. The measurement results were compared with that of a commercial spectrometer. SLIM stands for Simple, Low-power, Inexpensive and Microcontroller-based. The function of the SLIM spectrometer is to measure the light intensity across a sample. The SLIM spectrometer was not tailored for a specific analytical determination.

This battery-powered spectrometer design involves three different colored LEDs, namely red, blue and yellow-orange that covers a large portion of the visible spectrum. A PIC16F84 microcontroller acts as the brain of this spectrometer. The photodetector is a light to frequency converter, which measures light intensity as frequency output. It is fed as an input to the microcontroller, hence eliminating the need for a separate analog to digital converter circuit, which was absent in the PIC16 microcontroller. A serial time clock and a 128 kb EEPROM were used as a timekeeping device and a memory storage device for data logging, respectively.

This spectrometer was not configured to calculate absorbance. The data transmitted by the photodetector are simply pulses that are linearly related to the light intensity. Associated noise by room light fluctuations was removed by wrapping the spectrometer in black felt.

The SLIM spectrometer has some weaknesses. It is prone to noises because it is not light-tight. The noises can be seen in the dark signal level and low-level fluctuations in the ambient light signals. It also suffered nonlinearity in the results due to the polychromatic nature of the LEDs, which leads to inaccurate absorbance measurements.

However, the SLIM spectrometer does have its strengths. Its versatility allows it to be used in a variety of applications, such as replacing a commercial spectrometer in monitoring the redox state of an immobilized redox indicator, controlling a miniature pumping system and for teaching in schools. Being battery-powered and compact in size allows it to be portable. Most importantly, the total cost is only $25, rendering the choice very affordable.

2.1.3 Open Source Colorimeter and Photometer

To date, an open-source microcontroller for system control and data processing gained many attentions considering the cost and circuit simplicity. Arduino, is one of the examples for open source microcontroller used in developing low-cost instrument, costing under $50.

Yang et al. [21] used Arduino to develop two-channel photometer to measure pH of seawater. Their project demonstrated accuracy up to 0.01 unit pH. Grinias et al. [22]

used Arduino as a data acquisition device together with Python open-source software for chemical instrumentation education. Arduino was chosen, as it does not require any commercial license and is compatible with any PC or Mac computers.

Anzalone et al. [14] demonstrated the use of open sourced hardware and software to design and develop a colorimeter. The circuit components consist of an Arduino board, an LED with a peak emission of 606 nm and a light to frequency converter. The colorimeter case was 3D printed and the total cost was approximately 50 USD [14]. This colorimeter was designed to measure the chemical oxygen demand (COD). The method used requires a light source with a peak emission around 606 nm. Results yielded were compared with that from a commercial colorimeter. The open source colorimeter produced higher standard deviation but still within the acceptable range. The absorbance is also skewed lower than the commercial colorimeter because the LED has a peak emission of 620 nm, which is further than the required wavelength of 606 nm.

Further improvements could be made to this design to make it portable by using batteries or solar power. Other improvements include adding compartments for simultaneous testing and wireless capabilities.

2.2 Dual-Channel Colorimetric Devices

Most of the low-cost colorimeter use single-channel scheme for simplicity. There are also few works demonstrated using two-channels schemes in order to improve the sensitivity of the device and to eliminate errors [11, 15].

Jones et al. [15] used two different light sources to pair with one photovoltaic cell as photodetector and disposable polystyrene as sample containers. This colorimeter was used to test cobalt chloride solution with different concentration. A bridge circuit was constructed for the detection circuit and potentiometer was used to balance the bridge circuit. The absorbance of the samples was then compared with the blank sample.

Other than that, Bhadane et al. [11] demonstrated two-channel colorimeter based on a microcontroller developed using red, green or blue LEDs as light source. Colored LEDs produce monochromatic wavelengths and are able to replace the monochromator or the wavelength selective filter used by traditional colorimeters. The light was measured using a photodiode, which produced electric current proportional to the light intensity. The current then passes through a transimpedence amplifier to be converted into voltage before entering the input of the PIC18F4550 microcontroller. The output of the microcontroller was displayed on the 16 × 2 LCD display. It is powered by a DC 18 V 5A supply.

The dual channel design provides error correction. The error was due to the reagent absorbing light, which resulted in device inaccuracy. The design eliminated error by relating the sample solution with the non-absorbing reference solution. The concentration measurement of an unknown solution was carried out by using a calibration curve. The calibration curve was generated from the light absorption of solutions with known concentrations.

The design successfully measured cobalt concentration in a solution and is suitable to be used as a general-purpose colorimeter for routine laboratory work. LED based colorimeters provide less informative light transmittance data because it does not produce an absorbance output for all wavelengths. Instead, the absorbance data is broken down per wavelength. The cost of the colorimeter is approximately $25.

2.3 Low-Cost Colorimeter for Serum Iron Detection

Iron in serum can be measured using few techniques such as Atomic Absorption Spectrophotometry, colorimetric as well as potentiostatic coulometry [22–25]. In order to give a definitive diagnosis for human body iron level, red cell analyzer or biochemistry autoanalyzer are normally used. However, these commercialized instruments are bulky and expensive, hence limiting early stage diagnosis especially in poor countries that have limited access to sophisticated laboratory. Thus, portable and low-cost iron body reader is urgently needed [8]. To the author's best knowledge, not many reports demonstrate significant development for this specific diagnosis.

Sorouraddin et al. [13] developed a €2 colorimeter for serum iron determination. The circuit comprised of an LED, an LDR as a photodetector and two 1.5 V dry cells. The components were enclosed in a cell, coated with black masking tape. A digital multimeter was used to record the resistance of the LDR, which varies according to the transmitted light intensity.

Serum iron determination was carried out using two methods, Ferrozine kit and Ferene kit. A green LED with a spectral emission range of 520–605 nm was used to cover the proposed wavelengths for both Ferrozine and Ferene methods.

Results obtained by the colorimeter for both methods were compared with an automated clinical chemistry analyzer. The concentration of serum iron obtained by the colorimeter correlates well with the automated analyzer.

This colorimeter design is not only very affordable at €2, but it also requires smaller amount of samples for every test. Hence, it is suitable to be used for serum iron determination.

3 Conclusion and Prospects

Based on the review of published works, the design and development of a low-cost, portable and accurate colorimeter for measuring serum iron is feasible. A microcontroller could be used to simplify the calculation process. By connecting an LCD to the microcontroller, the design will be suitable for on-site applications. This design could be adapted to specifically be used for detection of serum iron by matching the spectral emission of the light source with the peak absorbance wavelength of the sample at 562 nm [26]. Applying a dual channel design provides more stable results [11, 15]. The device should cost no more than $100 (~430 MYR). The absence of a specialized iron reader in the market provides commercial opportunities. Further development is required to close the technological gap.

Acknowledgements This work is partially funded by UNIIG grant (J 510050621), 20161111FRGS and 20170101PRGS.

References

1. Wu, A.M., Lespernace, L.M.P., Bernstein, H.D.: Screening for iron deficiency anemia. Pediactrics Rev. **23**, 171–178 (2002). https://doi.org/10.1542/pir.23-5-171
2. Kohgo, Y., Ikuta, K., Ohtake, T., Torimoto, Y., Kato, J.: Body iron metabolism and pathophysiology of iron overload. Int. J. Hematol. **88**, 7–15 (2008). https://doi.org/10.1007/s12185-008-0120-5
3. Andrews, N.C.: Disorders of iron metabolism. N. Engl. J. Med. **341**, 1986–1995 (1999). https://doi.org/10.1056/NEJM199912233412607
4. Burke, R.M., Leon, J.S., Suchdev, P.S.: Identification, prevention and treatment of iron deficiency during the first 1000 days. Nutrients **6**, 4093–4114 (2014). https://doi.org/10.3390/nu6104093
5. WHO.: Nutrition: micronutrient deficiencies. Iron Defic. Anemia 1–2 (2013)
6. Camaschella, C.: Iron-deficiency anemia. N. Engl. J. Med. **372**, 1832–1843 (2015). https://doi.org/10.1056/NEJMra1401038
7. World Health Organization.: The global prevalence of anaemia in 2011. WHO Rep. **48** (2011). https://doi.org/10.1017/S1368980008002401
8. Talik, N., Yap, B.K., Lim, W.F.: A Review on Methods and Devices to Determine Body Iron Status, pp. 1–10 (2017) (unpublished work)
9. Lim, W.F., Yap, B.K., Lai, M.I., Talik, N., Nasser, A.A.: Iron Deficiency Anaemia : With the Conclusion of a Need for Iron Reader, pp. 1–7 (unpublished work)
10. Zaveri, M., Mukesh, J., Shinde, A.A.: Absorbance measurement by colorimeter. Int. J. Adv. Res. Comput. Sci. Softw. Eng. **3**, 874–878 (2013)
11. Bhadane, P.K.: Development of microcontroller based analyzer for the detection of chlorine in water. Int. J. Innov. Res. Sci. Eng. Technol. **3**, 10890–10897 (2014)
12. Udoji Itodo, A., Usman, A., Bashir Sulaiman, S., Ugbede Itodo, H.: Color matching estimation of iron concentrations in branded iron supplements marketed in Nigeria. Adv. Anal. Chem. Sci. Acad. Publ. **2**, 16–23 (2012). https://doi.org/10.5923/j.aac.20120201.04
13. Sorouraddin, M.H., Saadati, M., Samadi, A.: A simple and cheap device for colorimetric determination of serum iron. J. Chinese Chem. Soc. **57**, 56–61 (2010). https://doi.org/10.1002/jccs.201000009
14. Anzalone, G.C., Glover, A.G., Pearce, J.M.: Open-source colorimeter. Sensors (Switzerland) **13**, 5338–5346 (2013). https://doi.org/10.3390/s130405338
15. Jones, N.B., Riley, C., Sheya, M.S., Hosseinmardi, M.M.: A low-cost colorimeter. J. Med. Eng. Technol. **8**, 9–14 (1984). https://doi.org/10.3109/03091908409032066
16. Kim, J.-S., Kim, A.-H., Oh, H.-B., Goh, B.-J., Lee, E.-S., Kim, J.-S., Jung, G.-I., Baek, J.-Y., Jun, J.-H.: Simple LED spectrophotometer for analysis of color information. Biomed. Mater. Eng. **26**, S1773–S1780 (2015). https://doi.org/10.3233/BME-151478
17. Snizhko, D.V, Sushko, O.A., Reshetnyak, E.A., Shtofel, D.H., Zyska, T., Smolarz, A., Mussabekov, N., Kalizhanova, A.: Colorimeter based on color sensor. N. Karazin Natl. Univ. Vinnytsia Natl. Tech. Univ. Lublin Univ. Technol. **93** (2017). https://doi.org/10.15199/48.2017.05.19
18. Cantrell, K.M., Ingle, J.D.: The SLIM spectrometer. Anal. Chem. **75**, 27–35 (2003). https://doi.org/10.1021/ac026015s
19. Suzuki, Y., Aruga, T., Kuwahara, H., Kitamura, M., Kuwabara T., Kawakubo, S., Iwatsuki, M.: A simple and portable colorimeter using a red-green-blue light-emitting diode and its application to the on-site determination of nitrite and iron in river-water. Anal. Sci. Int. J. Jap. Soc. Anal. Chem. **20**, 975–977 (2004). https://doi.org/10.2116/analsci.20.975
20. Kittipanyangam, S., Abe, K., Eguchi, K.: Design of a measurement device explaining the relationship between the concentration of solution and the light absorbance for chemical education. In: 13th International Conference on Electrical Engineering/Electronics, Computer, Telecommunications and Information Technology (ECTI-CON 2016). https://doi.org/10.1109/ECTICon.2016.7561271
21. Yang, B., Patsavas, M.C., Byrne, R.H., Ma, J.: Seawater pH measurements in the field: a DIY photometer with 0.01 unit pH accuracy. Mar. Chem. **160**, 75–81 (2014). https://doi.org/10.1016/j.marchem.2014.01.005
22. Grinias, J.P., Whitfield, J.T., Guetschow, E.D., Kennedy, R.T.: An inexpensive, open-source USB Arduino data acquisition device for chemical instrumentation. J. Chem. Educ. **93**, 1316–1319 (2016). https://doi.org/10.1021/acs.jchemed.6b00262
23. Lötterle, J.: Measurement of the hemoglobin content of fresh blood by atomic-absorption-spectrophotometric iron determination (author's transl). Z. Rechtsmed. **85**, 283–288 (1980)
24. Zettner, A., Mensch, A.H.: The use of atomic absorption spectroscopy in hemoglobinometry: I. the determination of iron in hemoglobin. Am. J. Clin. Pathol. **48**, 225–228 (1967). https://doi.org/10.1093/ajcp/48.2.225
25. Dörner, K., Gustmann, H., Sippell, W.: A new method for the determination of serum iron: potentiostatic coulometry with the ferrochem 3050. Clin. Chem. Lab. Med. **19** (1981). https://doi.org/10.1515/cclm.1981.19.9.967
26. Lim, W.F., Abdullah, M., Yap, B.K., George, E., Lai, M.I.: Novel assay for efficient iron release from ferritin. In: 2014 IEEE Conference on Biomedical Engineering and Sciences (IECBES), pp. 132–136 (2014)

Decision Support for Parenteral Nutrition Supplementation in ICU Using Model-Based Glycemic Control Protocol

A. Abu-Samah, N. N. Razak, F. M. Suhaimi, U. K. Jamaluddin, A. M. Ralib, M. B. Mat-Nor, and G. Chase

Abstract

Nutrition therapy is part of the standard care given to all critically ill patients. In general, nutrition is administered as enteral nutrition (EN) and/or parenteral nutrition (PN). PN is given if the patients have contraindications to EN or as supplement if daily energy requirement cannot be achieved by EN alone. PN can be given as partial (dextrose solution only) or complete (include all macro- and micro-nutrients). The mode of nutrition therapy is influenced by several factors which include the need to maintain normoglycemia. A simulation is done to find the appropriate time to introduce PN while the patients are already on EN. In this context, a virtual study was conducted on 66 retrospectives critically ill patients' data using clinically validated insulin-nutrition model and Stochastic TARgeted (STAR) protocol. The results suggested that this protocol benefited critically ill patients in two-fold. This approach is not only useful in controlling per-patient normoglycemic level, but also able to recognize the time for PN supplement when patients become hypoglycemic. This serves as a potential decision support in the intensive care environment when healthcare providers faced with the complexity of dynamics between good glycemic control and optimized nutrition therapy.

Keywords

Nutrition decision support • Insulin • Parenteral nutrition • Enteral nutrition • Glycemic control

A. Abu-Samah (✉) · N. N. Razak
Department of Electronics and Communication Engineering, College of Engineering, Universiti Tenaga Nasional, Kajang, Malaysia
e-mail: Asma@uniten.edu.my

F. M. Suhaimi
Advanced Medical and Dental Institute, Universiti Sains, George Town, Malaysia

U. K. Jamaluddin
Faculty of Mechanical, Universiti Malaysia Pahang, Gambang, Malaysia

A. M. Ralib · M. B. Mat-Nor
Kulliyah of Medicine, International Islamic University, Islamabad, Malaysia

G. Chase
Department of Mechanical Engineering, University of Canterbury, Christchurch, New Zealand

1 Introduction

Adequate nutritional therapy in critically ill patients is highly recommended to ensure optimum clinical outcomes. The accumulative energy deficit in critically ill patients is commonly correlated with the duration of stay in the Intensive Care Unit (ICU). This is associated with an increased incidence of infectious complications and risk of death [1]. Critically ill patients are also exposed to stress induced hyperglycemia which contributes further to higher rate of mortality and morbidity. Insulin production which is crucial in maintaining normoglycemia depends on the complex interaction with nutrition therapy. Glucose coming from

F. Ibrahim et al. (eds.), *2nd International Conference for Innovation in Biomedical Engineering and Life Sciences*, IFMBE Proceedings 67, https://doi.org/10.1007/978-981-10-7554-4_30

nutrition is a major stimulator of insulin release, however, elevation of blood glucose occurs as the result of increased hepatic glucose production and reduced glucose utilization by peripheral tissues during stress of hospitalization [2, 3]. During acute illness, stress induced hyperglycemia can be exacerbated by excessive nutrition in the presence of high insulin resistance.

Current guideline recommends Enteral Nutrition (EN) as the default way to provide the nutrients within ICUs [4]. EN is the main solution to nutrition delivery for critically ill due to several advantages including decreased mortality, infectious morbidity and length of hospital care [5, 6]. It is commonly administered via nasogastric tube or under certain conditions, through percutaneous gastric cube. Standard enteral formulas consist of carbohydrates, protein and lipids. However, intolerance to EN may results in underfeeding and the consequences of feeding inadequacy [1]. In patients with high risk of malnutrition, Parenteral Nutrition (PN) is often started in those who are contraindicated to EN or supplemented in cases where EN is unable to provide total energy requirement. Improvements in nutrition formulation bring the use of intravenous administration of PN into highlight [7]. Ever since its introduction, medical healthcare providers are subjected to more questions regarding its use over EN. Controversial areas of recommendations include when is the appropriate time to introduce PN [8, 9] and how to improve glycemic control during PN [10]. To date, few prospective randomized studies have addressed support decision strategies of when to introduce PN after initial administration of EN.

Decisions regarding nutritional therapy in the critically ill is a complex subject, and even more when it is paired with the glycemic control. The timing of parenteral nutrition introduction in critically ill patients should also be considered as a potential strategy in reducing the risk of complications associated with stress induced hyperglycemia. The aim of this paper is to provide analysis on the decision support for introduction of PN in critically ill patients using retrospective data from University Malaya Medical Centre (UMMC).

2 Method

2.1 Clinical Data

Retrospective data of 66 critically ill patients (a cohort) from an intensive care unit in UMMC are analyzed. Data from this unit are collected to include patients who received nutrition therapy and require insulin infusion for their glycemic control during their stay in the ICU. Seven types of EN formulas (Diben, Glucerna, Jevity, Nepro, Novasource, Osmolite, and Peptamen) were used with with g/mL carbohydrate values range between 0.086 and 0.185. All patients who were included in this study received only Enteral feeding, stayed in the ICU for more than 24 h and had sufficient blood glucose measurement in an effort to capture the patients dynamic in a virtual trial [11, 12]. Minimum hours of stay in this data is 64 h and minimum BG measurement is 29. Our study is limited to patients with functional gastrointestinal tract to look at the timing for introduction to PN on per-patient basis.

2.2 Analysis Tools

Hyperglycemia is common in critical care, and tight control provides the reduction in mortality [13]. A physiological insulin-nutrition ICING model [14] which captures both insulin and exogenous nutritional inputs regulation is regarded as solution to model the dynamic between blood glucose control, insulin and nutrition for the critically ill patients in ICU. Using this model, clinical data from patients are used in two stages, (1) Data fitting to generate per-patient insulin sensitivity (SI) profiles and (2) Simulation using generated SI profiles. The simulation is realized with an established model-derived STAR protocol [15, 16], which determines insulin and nutrition interventions in real time based on identified patient specific SI, rather than strictly on blood glucose levels and/or its changes.

Hence, insulin and nutrition are given in balance, based on estimated response to the prior insulin and nutrition intervention. This is enabled by the protocols explicit knowledge of carbohydrate intake. The overall system matches the nutrition and exogenous insulin given to the body's patient-specific ability to utilize them, thus avoiding hyperglycemia. As a result, stability, and thus measurement frequency are a function of a patient's assessed insulin sensitivity. Figure 1 illustrates the operation of this protocol. The protocol is applied on the retrospective data from 66 patients for simulation. The BG target range of STAR and for the paper analysis is benchmarked at 4.4–8.0 mmol/L [16].

2.3 Dataset Preparation

3 datasets from the patients are created, distinguished with all patients receiving: (i) true clinical carbohydrate (g/mL) values respectively per received EN feed, and all patient receiving the same type of EN feed; (ii) with minimum value, (Nepro with 0.0863 g/mL) and (iii) with maximum value (Novasource with 0.185 g/mL). The datasets are first used to identify patient-specific SI profile via ICING model fitting function by applying iterative integral-based method [17]. Based on the SI profile and baseline measurement of

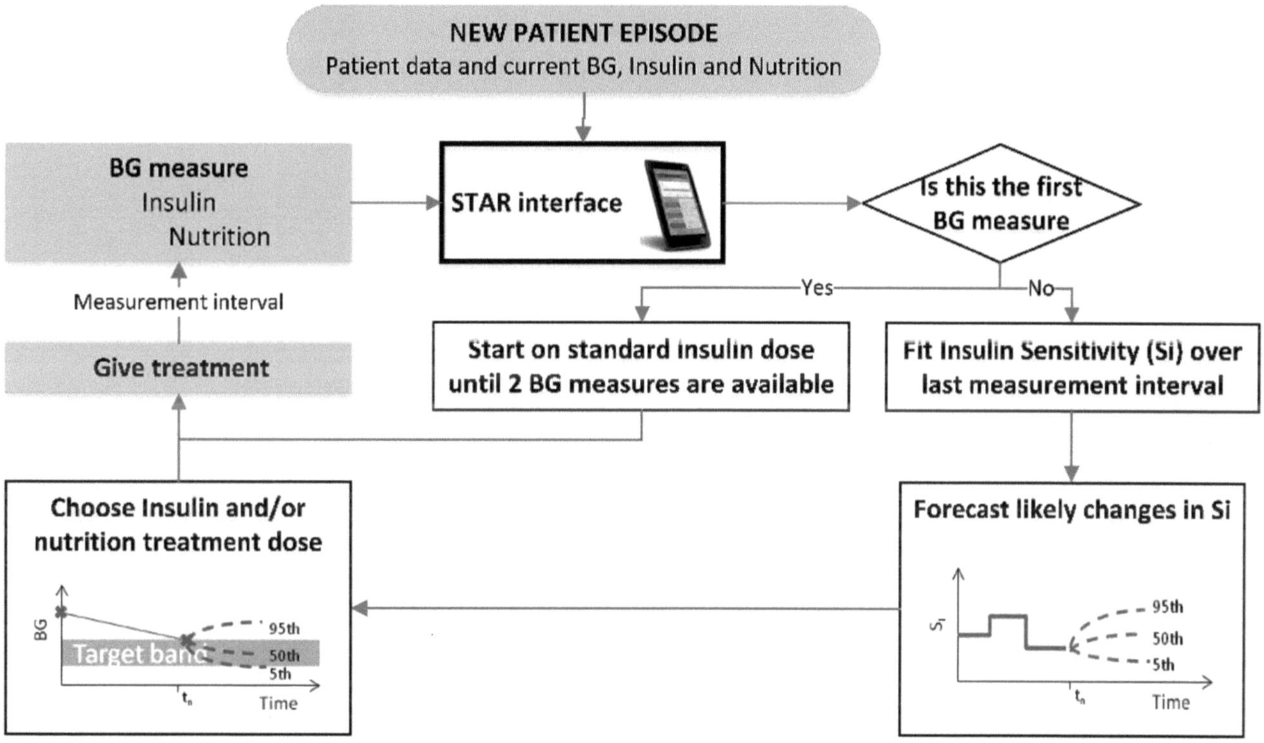

Fig. 1 Summary of STAR protocol operations. White boxes operations are integrated in the protocol's algorithm while gray boxes operations require actions by clinical actors

BG levels, nutrition and insulin doses, STAR protocol virtual trial is simulated to generate hourly BG response of the virtual patients and proposed feeds. The whole purpose of the analysis is to see the sufficiency of chosen EN in controlling the glycemic level in predefined target range while proposing PN only when it is necessary.

3 Results and Discussions

Efficacy of the STAR protocol incorporating patient specific SI profiles resulted in 48 patients across all dataset consistently receiving the same proposition of EN and PN despite the variation of EN values. Table 1 highlights the subset of 18 patients that have different proposed PN based on the g/mL carbohydrate value of EN. A maximum of 2 PN propositions per patient is detected between the 3 different cases.

A proof of concept for these differences from patient BUM022 simulated using true and max feed is presented in Fig. 2a, b. True clinical nutrition that this patient had include Osmolite (0.143 g/mL) and Peptamen (0.128 g/mL). The figures display the different proposition of insulin and nutrition feed, with highlight to PN bolus, 2 for the clinical case and 1 using the carbohydrate g/mL value. It can be seen that in both cases, PN is proposed upon sensing a drop to the

Table 1 Number of proposed PN bolus for patients according to different trials

Patient code	Number of proposed PN bolus(es)		
	True value (clinical data)	Maximum value	Minimum value
BUM002	–	2	–
BUM008	–	1	–
BUM009	1	–	–
BUM012	5	3	3
BUM016	1	–	–
BUM019	1	1	–
BUM021	1	1	–
BUM022	2	1	3
BUM024	1	4	–
BUM028	–	1	–
BUM036	–	1	–
BUM038	1	–	–
BUM040	5	2	5
BUM042	1	1	–
BUM053	9	7	9
BUM055	2	2	1
BUM058	–	2	–
BUM063	1	1	–

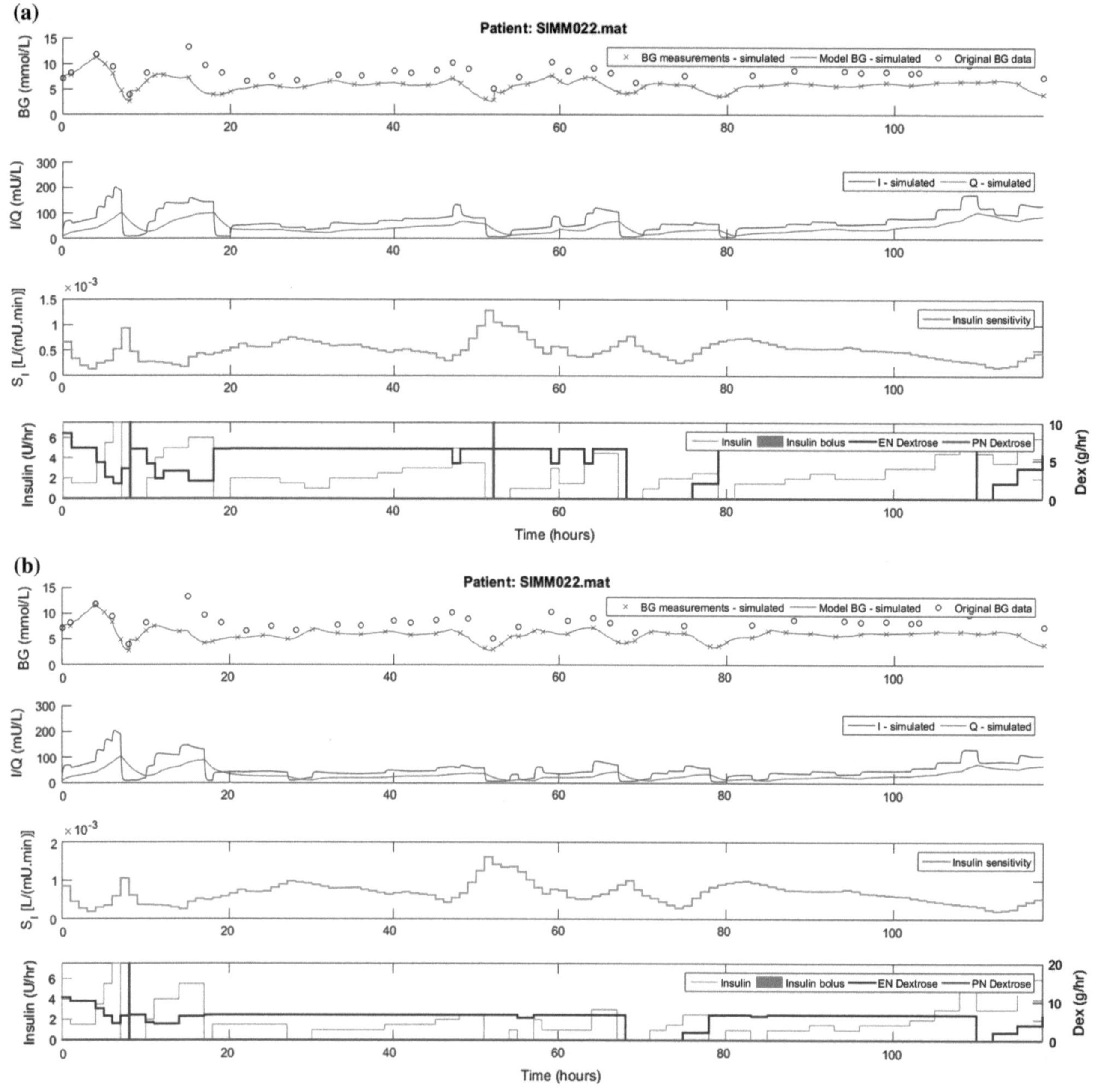

Fig. 2 Simulation's proof of concept for patient BUM002. The red PN dextrose spikes are the PN boluses

severe hypoglycemia limit ((BG < 2.2 mmol/L). Overall analysis on 66 patients results confirms the same observation.

The whole cohort simulation statistics using STAR protocol are summarized in Table 2. Overall, in the 'true clinical value' case, the simulation shows the highest percentage of BG level for the targeted band of 4.4–8.8 mmol/L, with 51.5%. This trial successfully limits hyperglycemia incidences (BG > 10 mmol/L) below 17% and a maximum of 5% of the risk of light hypoglycemia (BG < 4.0 mmol/L) [15, 18]. However it surprisingly records severe hypoglycemia incidences to 7, corresponding to 0.14% of the overall BG measurements.

In fact, American College of Chest Physician (ACCP) guidelines are used in STAR protocol to determine the patients' daily caloric goal intake and EN is proposed between 30 and 100% of this target [18, 19]. This higher than standard dose enables a lower workload for medical staffs in terms of length of treatment and number of intervention [16]. Further analysis from our trial is done by comparing the results with 'min' and 'max' cases. If only limited to glycemic control performance, maximum feed

Table 2 Statistical summary of simulation according to different trials

Whole cohort statistics	Nutrition g/mL values		
	True value in clinical data	All maximum	All minimum
Number of patients	66		
Total hours	8031	8033	8026
Number of BG measurements	5158	4994	5548
BG mean (geometric) in mmol/L	7.34	7.13	7.82
% BG StDev (geometric) in mmol/L	1.38	1.38	1.37
% BG within 4.4–6.5 mmol/L	26.46	30.28	20.96
% BG within 4.4–7.0 mmol/L	35.44	38.75	28.08
% BG within 4.4–8.0 mmol/L	51.51	54.30	43.58
% BG within 5.1–8.0 mmol/L	44.98	47.48	39.20
% BG within 8.0–10.0 mmol/L	26.11	24.49	30.01
% BG > 10.0 mmol/L	16.34	14.12	21.99
% BG < 4.4 mmol/L	6.05	7.11	4.43
% BG < 4.0 mmol/L	4.01	4.67	2.88
% BG < 2.2 mmol/L	0.21	0.18	0.22
Number of incidences with BG < 2.2 mmol/L	7	6	7
Median insulin rate [IQR] in U/h	5.0 [2.0–8.0]	4.0 [2.0–7.5]	6.0 [3.0–8.0]
Median glucose rate [IQR] in g/h	3.9 [1.4–6.5]	3.9 [1.9–6.6]	2.6 [1.7–5.0]

case seems to produce the overall better control results, suggesting a revision on the chosen type of feed for certain patients. As for the minimum case, the insulin median, 6.0 [3.0–8.0] U/h and BG mean, 7.82 mmol/L is highest. This is in contrary to the belief that the lower the carbohydrate feed, the lower the BG would be.

4 Conclusion

This paper demonstrates a potential use of insulin-nutrition model-based STAR protocol as a decision support to decide on PN supplementation over EN for critically ill patients towards maintaining normoglycemia. It is complemented with the proposition to consider a higher or lower nutrition dose to have better results in glycemic control. Its advantage is the personalized proposition of PN over EN as well as insulin dose per-patient.

The results of this simulation of well-defined cohort of critically ill patients are promising and could be validated in a well-designed clinical trial. However, there are limitations in this study. First, the protocol depends on the generation of well fitted insulin sensitivity in real time which requires a certain amount of initial BG, insulin and nutrition measurements. Validation in the first and later part of the protocol's performance needs to be done differently. Second, the STAR protocol provides recommendation for a balance between insulin and nutrition; however it does not ensure the optimality. Third, the proposition is based on results of glycemic control and insulin sensitivity. Furthermore, various factors that are important in choosing PN over EN in the same patient are not considered in the proposed trial. In clinical practice, the approach is potentially useful in assisting clinicians to adjust the doses of PN and EN when both modes are utilized in malnourished patients. We are looking for these potentials as our future perspectives.

Acknowledgements The Medical Research Ethics Committee, UMMC granted ethics approval for the audit, analysis and publication of this data. The authors gratefully acknowledge UMMC and IIUM for their provision of data for the case study. The authors also acknowledge Ministry of Higher Education and UNITEN under BOLD SCHEME for their support on our ongoing Research.

References

1. Alberda, C., Gramlich, L., Jones, N., Jeejeebhoy, K., Day, A.G., Dhaliwal, R., Heyland, D.K.: The relationship between nutritional intake and clinical outcomes in critically ill patients: results of an international multicenter observational study. Intensive Care Med. **35**(10), 1728–1737 (2009)
2. Dungan, K.M., Braithwaite, S.S., Preiser, J.C.: Stress hyperglycaemia. Lancet **373**(9677), 1798–1807 (2009)
3. Gosmanov, A.R., Umpierrez, G.E: Medical nutrition therapy in hospitalized patients with diabetes. Curr. Diabetes Rep. **12**(1), 93–100 (2012)
4. Kreymann, K.G., Berger, M.M., Deutz, N.E., Hiesmayr, M., Jolliet, P., Kazandjiev, G., Nitenberg, G., Van den Berghe, G.,

Wernerman, J., Ebner, C., Hartl, W., Heymann, C., Spies, C.: ESPEN (European Society for Parenteral and Enteral Nutrition): ESPEN guidelines on enteral nutrition: intensive care. Clin. Nutr. **25**, 210–223 (2006)
5. Kudsk, K.A., Croce, M.A., Fabian, T.C., Minard, G., Tolley, E.A., Poret, H.A., Brown, R.O.: Enteral versus parenteral feeding. Effects on septic morbidity after blunt and penetrating abdominal trauma. Ann. Surg. **215**(5), 503 (1992)
6. Gramlich, L., Kichian, K., Pinilla, J., Rodych, N.J., Dhaliwal, R., Heyland, D.K.: Does enteral nutrition compared to parenteral nutrition result in better outcomes in critically ill adult patients? A systematic review of the literature. Nutrition **20**(10), 843–848 (2004)
7. Anderson, C.F., Macburney, M.M.: Application of ASPEN clinical guidelines: parenteral nutrition use at a university hospital and development of a practice guideline algorithm. Nutr. Clin. Pract. **11**(2), 53–58 (1996)
8. Doig, G.S.: Parenteral versus enteral nutrition in the critically ill patient: additional sensitivity analysis supports benefit of early parenteral compared to delayed enteral nutrition. Intensive Care Med. **39**(5), 981 (2013)
9. Casaer, M.P., Hermans, G., Wilmer, A., Van den Berghe, G.: Impact of early parenteral nutrition completing enteral nutrition in adult critically ill patients (EPaNIC trial): a study protocol and statistical analysis plan for a randomized controlled trial. Trials **12** (1), 21 (2011)
10. Chase, J.G., Le Compte, A.J., Suhaimi, F., Shaw, G.M., Lynn, A., Lin, J., Pretty, C.G., Razak, N., Parente, J.D., Hann, C.E., Preiser, J.C.: Tight glycemic control in critical care–the leading role of insulin sensitivity and patient variability: a review and model-based analysis. Comput. Methods Programs Biomed. **102** (2), 156–171 (2011)
11. Dickson, J.L., Stewart, K.W., Pretty, C.G., Flechet, M., Desaive, T., Penning, S., Lambermont, B.C., Benyo, B., Shaw, G. M., Chase, G.: Generalisability of a virtual trials method for glycaemic control in intensive care. IEEE Trans. Biomed. Eng. (2017)
12. Chase, J.G., Suhaimi, F., Penning, S., Preiser, J.C., Le Compte, A. J., Lin, J., Pretty, C.G., Shaw, G.M., Moorhead, K.T., Desaive, T.: Validation of a model-based virtual trials method for tight glycemic control in intensive care. Biomed. Eng. **9**(1), 84 (2010)
13. Griesdale, D.E., de Souza, R.J., van Dam, R.M., Heyland, D.K., Cook, D.J., Malhotra, A., Talmor, D.: Intensive insulin therapy and mortality among critically ill patients: a meta-analysis including NICE-SUGAR study data. Can. Med. Assoc. J. **180** (8), 821–827 (2009)
14. Lin, J., Razak, N.N., Pretty, C.G., Le Compte, A., Docherty, P., Parente, J.D., Chase, J.G.: A physiological Intensive Control Insulin-Nutrition-Glucose (ICING) model validated in critically ill patients. Comput. Methods Programs Biomed. **102**(2), 192–205 (2011)
15. Evans, A., Shaw, G.M., Le Compte, A., Tan, C.S., Ward, L., Steel, J., Pretty, C.G., Pfeifer, L., Penning, S., Suhaimi, F., Signal, M.: Pilot proof of concept clinical trials of Stochastic Targeted (STAR) glycemic control. Ann. Intensive Care **1**(1), 38 (2011)
16. Stewart, K.W., Pretty, C.G., Tomlinson, H., Thomas, F.L., Homlok, J., Noémi, S.N., Illyés, A., Shaw, G.M., Chase, J.G.: Safety, efficacy and clinical generalization of the STAR protocol: a retrospective analysis. Ann. Intensive Care **6**(1), 24 (2016)
17. Hann, C.E., Chase, J.G., Lin, J., Lotz, T., Doran, C.V., Shaw, G. M.: Integral-based parameter identification for long-term dynamic verification of a glucose–insulin system model. Comput. Methods Programs Biomed. **77**(3), 259–270 (2005)
18. Fisk, L.M., Le Compte, A.J., Shaw, G.M., Penning, S., Desaive, T., Chase, J.G.: STAR development and protocol comparison. IEEE Trans. Biomed. Eng. **59**(12), 3357–3364 (2012)
19. Gb, C., Rios Benitez, M.R., Irwin, R.S., Jeejeebhoy, K., Kats, D. P., Pingleton, S.K., Pomposelli, J., Rombeau, J.L., Shronts, E., Wolfe, R.R., Zalosa, G.P.: Applied nutrition in ICU patients. A consensus statement of the American College of Chest Physician. Chest **111**, 769–778 (1997)

Investigation of Glucose-Insulin Model Efficacy for Diabetes Patient in the ICU

Athirah Abdul Razak, Asma Abu-Samah, Normy N. Razak, Nurhamim Ahamad, Fatanah M. Suhaimi, Ummu K. Jamaludin, Azrina Md Ralib, and M. B. Mat-Nor

Abstract

This paper investigates the inter-relationship of blood glucose (BG) model fitting error percentage for both diabetic and non-diabetic patient's status with ICING glucose-insulin model efficacy. Data were taken from 132 critically ill patients admitted in ICU of Hospital Tunku Ampuan Afzan (HTAA), Kuantan, Malaysia. The BG fitting error results are important to determine the efficacy and safety of chosen glucose-insulin model-based. Analysis results of per-patient and by-cohort were used to suggest whether diabetes is an influential factor for model BG fitting error efficacy. For diabetic per-patient and by-cohort, median model fit error are 0.29 and 0.31% respectively Meanwhile, for non-diabetic per-patient and by-cohort model fit error are 0.27 and 0.29% respectively. In conclusion, results demonstrated no significant difference between diabetic and non-diabetic patient in model fitting error. Thus, ICING model may be used to deliver real time model-based insulin therapy whether patient is diabetic or non-diabetic.

Keywords

Fitting error • Diabetes • Model-Based • Malaysian patient • Blood glucose

1 Introduction

In the Intensive Care Unit (ICU), critically ill patients are commonly diagnosed with stress induced hyperglycaemia when their blood glucose (BG) is higher than 10 or 11.1 mmol/L [1]. This condition easily occurs even if patient has no prior history of diabetes and it complicates when patients are diabetic, leading to higher risk of other complications and at worst, mortality. A move towards better diabetes control management is crucial as the number of critically ill patients in ICU keep rising yearly. Statistically in 2014, 347 million people in the world have diabetes, either of type 1 or type 2 [2]. The diabetic or non-diabetic critically ill patients whose BG condition evolves regularly provide challenge towards normal glycaemic control. Thus, it is important to gain more understanding towards the diabetic factor while performing glycaemic control to improve patient outcome.

Intensive Insulin Therapy (IIT) as the common BG control practice has been used extensively around the globe, but optimality in maintaining normoglycaemic are still in question for diabetic patients [3]. Glucose-insulin models used for BG control like Minimal Model, Bergman model from previous studies [4] have limitations in accuracy and

A. Abdul Razak (✉) · A. Abu-Samah · N. N. Razak (✉) · N. Ahamad
Universiti Tenaga Nasional, Kajang, Malaysia
e-mail: athirahrazak@gmail.com

N. N. Razak
e-mail: Normy@uniten.edu.my

A. Abu-Samah
e-mail: Asma@uniten.edu.my

F. M. Suhaimi
Universiti Sains Malaysia, Penang, Malaysia

U. K. Jamaludin
Universiti Malaysia Pahang, Pekan, Malaysia

A. Md Ralib · M. B. Mat-Nor
International Islamic University Medical, Pahang, Malaysia

F. Ibrahim et al. (eds.), *2nd International Conference for Innovation in Biomedical Engineering and Life Sciences*, IFMBE Proceedings 67, https://doi.org/10.1007/978-981-10-7554-4_31

physiological interpretation which only used single compartment of glucose and insulin kinetics [5]. In this paper, a clinically validated model-based approach namely Intensive Control Insulin Nutrition Glucose (ICING) that utilizes a mathematical model of glucose-insulin system has been adapted to analyse the concern above. This physiological model has been used in several tight glycaemic control protocols in New Zealand [6], Hungary [7] and latest in Malaysia [6, 7], in the form of a pilot study. Model-based insulin sensitivity (S_I) is identified per-patient and by-cohort by fitting blood glucose level with estimated endogenous insulin secretion and nutrition using this model. However, no works have reported on the performance of its fitting error analysis with diabetic vs. non-diabetic status in questions.

This paper presents the inter-relationship status between diabetic patients fitting error analysis with the efficacy of integrated model-based approach known as ICING. This approach provides better estimation in glucose prediction control when median model fitting error is lower or <1% [8, 9]. This paper analyzes the preliminary result of model BG fit error for diabetic and non-diabetic patients before further assessment simulation of BG management can be done. The preliminary result percentage of BG model fit error is assessed to assure better BG prediction in future simulation output. Accuracy of model fitting is an important step to evaluate the relationship between per-patient and by-cohort diabetic patients BG and insulin sensitivity (S_I) using glucose-insulin model-based. The result of this step will benefit in glycaemic control. Model fit analysis will not only brings safety to the patient as it is done in virtual trial, and prediction of insulin sensitivity captures hourly patients varying metabolic rate, as real clinical data are used to do this fitting analysis. This paper determines whether diabetic patient status affects the analyzed result while using ICING model. This analysis is an important step of clarification to determine whether ICING model is clinically efficient for real-time glycaemic control.

2 Methodology

210 retrospective data from Intensive Care Unit patients in HTAA are extracted and 132 patients are selected and categorized equally as diabetic (66 patients) and non-diabetic (66 patients), to be used for virtual fitting analysis. All non-diabetic patients were diagnosed with critical illness, either with sepsis, kidney failure or hypertension. These patients were selected from the pool of patients who underwent sliding scale protocol IIT from Hospital Tengku Ampuan Afzan (HTAA). The data was obtained with ethics (NMR-13-1592-18706) from Malaysian National Medical Research, National Institute of Health and International Islamic University Medical Research Ethics Committee under a collaborative research between UNITEN,[1] IIUM,[2] UMP,[3] USM[4] and UC.[5] HTAA data are fitted using integral fitting method [10] with glucose-insulin-nutrition ICING model [9]. Intensive Control Insulin Nutrition Glucose (ICING) model is described from Eqs. (1) to (5) and all descriptions can be referred in Table 1 [9],

$$\dot{G} = p_G \cdot G - S_I \cdot G \frac{Q}{1+\alpha_G Q} + \frac{\min(d_2 P_2, P_{max}) + EGP_b - CNS}{V_G} \quad (1)$$

$$\dot{I} = \frac{n_L I}{1+\alpha_I I} - n_K I - (I-Q) n_I + \frac{u_{ex}(t)}{V_I} + (1-x_L)\frac{u_{en}(G)}{VI} \quad (2)$$

$$\dot{Q} = (I-Q) n_I - n_c \frac{Q}{1+\alpha_G Q} \quad (3)$$

$$\dot{P}_1 = -d_1 P_1 + P(t) \quad (4)$$

$$\dot{P}_2 = -\min(d_2 P_2, P_{max}) + d_1 P_1 \quad (5)$$

From fitted data, parameter insulin sensitivity (S_I) profiles are represented in time-varying metabolic status per-patient. Percentage model fit error assessment, is the error between real measurement BG and model fit BG. The efficacy of ICING model is measured from the accuracy of fitting error percentage. Once these simulation profiles are assessed, new output can be carried out for new interventions. For results analysis, Kolmogorov-Smirnov test is used to compare the distribution in two data factor [11]. In this paper, the significant value of Cumulative Distribution Frequency (CDF) per-patient and by-cohort in model fit error is tested. In general, hypothesis is not significant if both vectors come from the same distribution, but since these data are from different patients, the hypothesis for this test is accepted. The result of this test will be 1 when hypothesis is significant at 5% of the significant level or 0 otherwise [11].

These patients data are virtually fitted using HTAA Protocol [12, 13]. Intensive Insulin Infusion adopted in HTAA targeted BG band is between 5.1 and 8.0 mmol/L. The frequency of nurses monitoring patient's BG is hourly once insulin are administered. In addition, BG monitoring will be updated every 2 h when there is no requirement of insulin rate changes. Once the patient is stable, then frequency of monitoring is lessened.

[1]UNITEN is the Universiti Tenaga Nasional, Malaysia.
[2]IIUM is the International University University Medical, Malaysia.
[3]UMP is the Universiti Malaysia Pahang, Malaysia.
[4]USM is the Universiti Sains Malaysia, Malaysia.
[5]UC is the University of Canterbury, New Zealand.

Table 1 Descriptions of symbols from ICING Model

Symbols (units)	Descriptions	Symbols (units)	Descriptions
G (mmol/L)	Total plasma glucose	EGP_b (mmol/min)	Basal endogenous glucose production
I (mU/L)	Total insulin	CNS (mmol/min)	Glucose uptake by central nervous system
Q (mu/L)	Interstitial insulin	V_G (L)	Volume of glucose distribution
S_I (L/mU.min)	Insulin sensitivity	V_I (L)	Volume of insulin distribution
n_L (min^{-1})	Insulin clearance from plasma via renal	α_G (L/mU)	Saturation of plasma insulin clearance by liver
n_K (min^{-1})	Insulin clearance from plasma via hepatic	α_I (L/mU)	Insulin dependent glucose clearance
n_I (min^{-1})	The rate transport between plasma and interstitial insulin compartments	u_{ex} (mU/min)	Exogenous input insulin
n_c (min^{-1})	Cellular insulin clearance rate from interstitium	u_{en} (mU/min)	Endogenous insulin production
x_L (min^{-1})	1st pass hepatic clearance	p_G (min^{-1})	Endogenous glucose clearnce
P_1 (mmol)	Glucose in stomach	P_2 (mmol)	Glucose in gut
d_1 (min^{-1})	Rate trasnsfer between stomach to gut	d_2 (min^{-1})	Rate transfer from gut to bloodstream
P(t) (mmol/min)	Enteral glucose input	P_{max} (mmol/min)	Maximum disposal rate from the gut

Table 2 Median and IQR of model fit error

Patient type	Median (IQR) BG fit error [%]	
	Per-patient	By-cohort
Diabetic	0.29 (0.21–0.39)	0.31 (0.11–1.34)
Non-diabetic	0.27 (0.21–0.36)	0.29 (0.12–0.74)
Kolmogorov-Smirnov test	0	0

3 Results

Table 2 shows median and interquartile range (IQR) of model fit error for per-patient and by-cohort in percentage. Both median and IQR were fitted for diabetic and non-diabetic patients with Kolmogorov-Smirnov test. For diabetic patients, both results of median and interquartile range of per-patient and by-cohort are 0.29 (0.21–0.39)% and 0.31 (0.11–1.34)%. Meanwhile for non-diabetic patients, the result of median and (IQR) of per-patient and by-cohort are 0.27 (0.21–0.36)% and 0.29 (0.12–0.74)%. The result shows that non-diabetic case has lower median model fit error.

Figure 1a shows the CDF for all patients in diabetic and non-diabetic cohort, while Fig. 1b illustrates the overall CDFs. Figure 2 shows an example of diabetic patient's profile. The first panel is measured BG (cross-stitch) and fitted model BG (blue line), the second panel show the total plasma insulin (I), and the interstitial insulin, (Q) in dotted lines. The third panel shows the varying insulin sensitivity

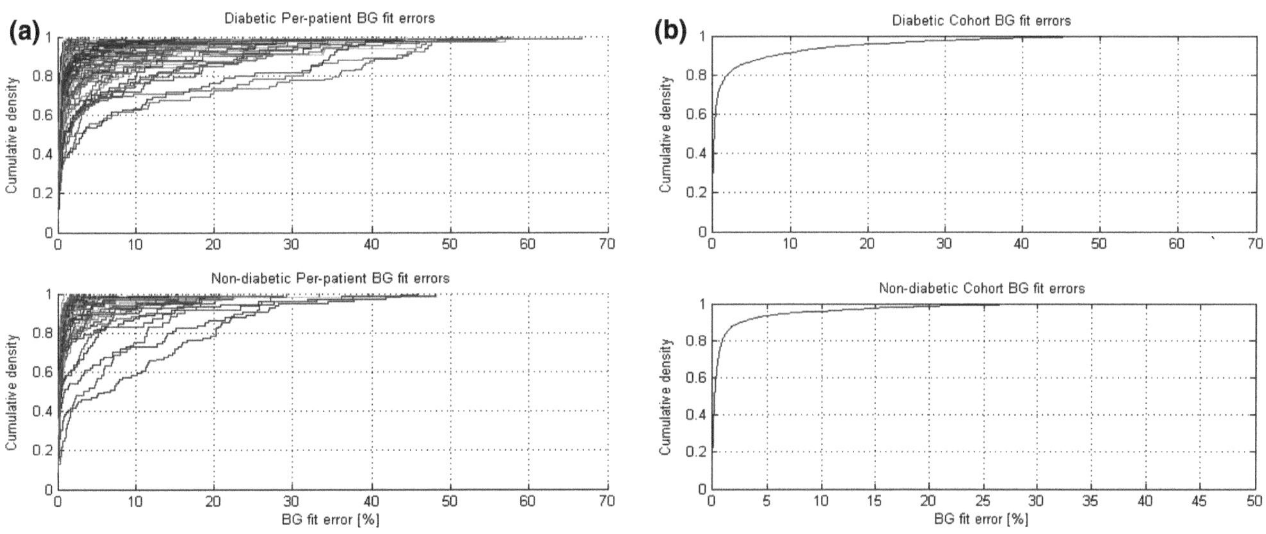

Fig. 1 Cumulative distribution frequency for diabetic and non-diabetic for **a** per-patient and **b** by-cohort analysis

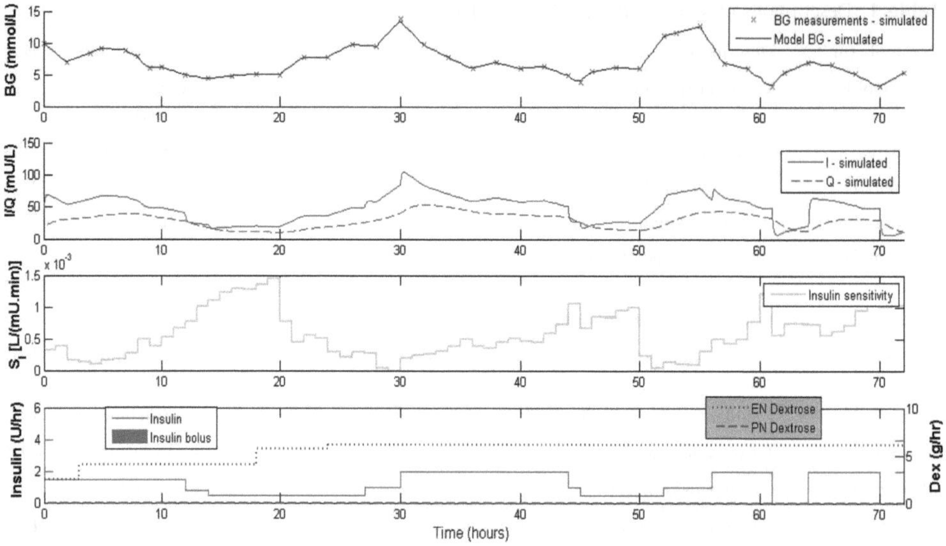

Fig. 2 Patient's profile for one diabetic patient

(S_I) and the last panel shows the insulin and nutrition (dotted line) given to the patient for hourly as shown in Fig. 2.

4 Discussion

Gaining a reliable model fit error in ICU patients is an important step for glycaemic control using automated model-based approach to replace and/or complement the sliding scale protocol. From the results in Table 2, Fig. 1a, b, non-diabetic case shows lower model fit error as compared to diabetic case. However, this result is not significant nor does it indicate diabetes as the cause for slightly higher or lower model fit error. In addition, the patients from the two cohorts are not significantly different as tested using Kolmogorov-Smirnov test. Results from Table 1 indicates that 0 fail to reject the null hypothesis which, if we consider the *p*-value test, the result of model fit error does not show any significance when *p*-value <0.05.

Figure 2 shows a patient's profile from diabetic patient. This patient length of stay was ±70 h and has good model BG fit which is shown in the blue line of first panel. The real BG measured from the glucometer (cross-stitch) is nicely fitted to the model BG (blue line) as in the first panel. The median fitting error is less than 1% for this patient. When fitted BG error is low, the model is efficient to determine prediction accuracy [9, 14]. ICING model shows that this model-based is able to capture highly dynamic patient's profile where in the third panel of Fig. 2 illustrates this patient hourly insulin sensitivity. Insulin sensitivity was predicted in every hour when the current and previous insulin and nutrition of this patient were identified. Insulin and nutrition (known as dextrose) that were measured in the hospital is automated into ICING model as showed in panel four from Fig. 2. As for panel 2, the total plasma insulin (I) and interstitial insulin (Q) are illustrated from the Eqs. (2) and (3).

In summary, these results showed that, ICING is a suitable model to deliver different simulation output despite this patient was diagnosed with diabetic. Diabetes was not related to model fitting error efficacy and ICING model is capable to capture real time model intensive insulin therapy. However, different parameters can be adjusted for more optimized output if glycaemic variation is considered in future. The study developed in this paper conduct analysis by taking into consideration of diabetic patients in general, however a more robust analysis may differ the output when specific type of diabetes is known. If different type of diabetes is known, BG fit error can be adjusted accordingly using ICING model and different output of model fitted parameter can be produced.

5 Conclusion

In conclusion, both per-patient and by-cohort analyses from HTAA case study demonstrates low model fitted error which is less than 1%, suggesting, ICING model is efficient and perform well. Diabetes is not a factor involved in model fitting error efficacy for these patient data and further future analysis can be done to investigate on patient glycaemic management control and glycaemic variabilities.

Acknowledgements We are thankful to UNITEN for the BOLD grant to enable this research. Thank you to IIUM, University of Canterbury, UMP and USM for the collaboration. Ethics were granted by National Institute of Health, Malaysia.

References

1. Hui, M., Kumar, A., Adams, G.G.: Protocol-directed insulin infusion sliding scales improve perioperative hyperglycaemia in critical care. Perioper. Med. **1**, 7 (2012)
2. Wu, C.H.: A patient-centered self-care support system for diabetics. In: Proceedings of 11th IEEE International Conference on E-bus Engineering (ICEBE 2014), Including 10th Workshop on Service Oriented Applications Integration and Collaboration (SOAIC 2014), 1st Workhop on E-Commerce Engineering (ECE 2014), pp. 298–302 (2014)
3. Schlussel, A.T., Holt, D.B., Crawley, E., Lustik, M.B., Wade, C. E., Uyehara, C.F.: Effect of diabetes mellitus on outcomes of hyperglycemia in a mixed medical surgical intensive care unit. J. Diabetes Sci. Technol. **5**(3), 731–740 (2011)
4. Yamamoto, C., Furutani, E., Sumi, S.: Mathematical model of postprandial glucose-insulin metabolism in insulin dependent diabetics with subcutaneous and intravenous insulin administration routes. In: SICE Annual Conference 2011, pp. 2648–2653 (2011)
5. Marmarelis, V., Mitsis, G., Daskalaki, E., Diem, P., Mougiakakou, S.: Data-Driven Modeling for Diabetes Diagnosis and Treatment, pp. 91–116. Springer, Berlin
6. Evans, A., Shaw, G.M., Le Compte, A., Tan, C.-S., Ward, L., Steel, J., Pretty, C.G., Pfeifer, L., Penning, S., Suhaimi, F., Signal, M., Desaive, T., Chase, J.G.: "Pilot proof of concept clinical trials of Stochastic Targeted (STAR) glycemic control. Ann. Intensive Care **1**(1), 38 (2011)
7. Benyo, B., Illyes, A., Nemedi, N.S., Le Compte, A.J., Havas, A., Kovacs, L., Fisk, L., Shaw, G.M., Chase, J.G.: Pilot study of the SPRINT glycemic control protocol in a Hungarian medical intensive care unit. J. Diabetes Sci. Technol. **6**(6), 1464–1477 (2012)
8. Lin, J., Chase, J.G., Shaw, G.M., Lotz, T.F., Hann, C.E., Doran, C. V., Lee, D.S.: Long Term Verification of Glucose-Insulin Regulatory System Model Dynamics, pp. 758–761 (2004)
9. Lin, J., Razak, N.N., Pretty, C.G., Le Compte, A., Docherty, P., Parente, J.D., Shaw, G.M., Hann, C.E., Geoffrey Chase, J.: A physiological intensive control insulin-nutrition-glucose (ICING) model validated in critically ill patients". Comput. Methods Programs Biomed. **102**(2), 192–205 (2011)
10. Hann, C.E., Chase, J.G., Lin, J., Lotz, T., Doran, C.V., Shaw, G. M.: Integral-based parameter identification for long-term dynamic verification of a glucose-insulin system model. Comput. Methods Programs Biomed. **77**(3), 259–270 (2005)
11. Huang, C.H., Chen, T.H., Ng, K.L.: Graph theory and stability analysis of protein complex interaction networks. IET Syst. Biol. **10**(2), 64–75 (2016)
12. Ahamad, N., Razak, N., Pretty, C., Chase, G.: Efficacy and safety of SPRINT and STAR protocol on malaysian critically-ill patients. In: IEEE EMBS Conference on Biomedical Engineering and Science, pp. 370–375 (2016)
13. Jamaludin, U.K., Dzaharudin, F., Razak, N.N.A.: Performance of STAR virtual trials for diabetic and non-diabetic in HTAA intensive care unit. In: IEEE EMBS Conference on Biomedical Engineering and Science, pp. 193–198 (2016)
14. Razak, A.A., Razak, N.N., Suhaimi, F.M., Jamaluddin, U.: Efficacy of Glucose-Insulin Model for Sepsis Patients, pp. 1–5 (2014)

Inflammatory Response of Bioscaffolds Decellularized by Sonication Treatment

Aqilah Hazwani, Munirah Sha'ban, and Azran Azhim

Abstract

Decellularization of sonication system efficiency is dictated by evaluation of host immune response on scaffolds. This aim of this study is to examine the inflammatory response after implantation of bioscaffolds decellularized by sonication treatment at 35 days post-surgery in rats. In this study, aortic tissues decellularized by sonication treatment in 0.1 and 2% of Sodium Dodecyl Sulfate (SDS) detergents. Samples were washed for 5 days in Phosphate Buffer Saline (PBS) following decellularization. Samples were implanted in on the lower right thoracic cavity of rats for 35 days. To examine the inflammatory response, the implanted samples were explanted and evaluated histologically by Hematoxylin Eosin (H-E) staining. From the histological analysis, bioscaffolds decellularized by sonication treatment in response to macrophages shows the minimal inflammatory response. Minimal inflammatory response elicits by bioscaffolds suggests the safety of bioscaffolds prepared by sonication treatment as biomedical implants.

Keywords

Decellularization • Sonication • Bioscaffolds • Aorta • Macrophages

1 Introduction

The decellularization removes whole cellular contents, while maintaining the extracellular matrix components. Complete removal of cellular material is essential as it can evoke immune response. The process of host response to bioscaffolds involved the transformation of macrophages to phagocyte and increase inflammatory reaction as well as stimulate the immune system. Besides, it also plays an important anti-inflammatory role and decrease the immune reaction by expressing high level of cytokines [1].

The sonication decellularization system developed by Azhim et al. [2] which applies the combination of sonication and ionic detergent is able to completely decellularize aorta tissues with uniform cell removal. However, host immune response to decellularized tissue by sonication treatment needs to be evaluated to indicate the success or failure of bioscaffolds in vivo [3].

The aim of this study is to examine the inflammatory response of bioscaffolds decellularized by sonication treatment following implantation at 35 days post-surgery in rats.

2 Methodology

2.1 Sonication Decellularization Treatment

Samples were sonicated at 10 mm from the ultrasonic generator horn from the luminal side. Samples were processed for 10 h in the 0.1 and 2% of circulating Sodium Dodecyl Sulfate (SDS) detergent, oscillating continuously. The power was set at 15 W and ultrasound frequency of 170 kHz. After

A. Hazwani · A. Azhim (✉)
Department of Biotechnology, Kulliyyah of Sciences, IIUM, Kuantan Campus, Malaysia
e-mail: azranazhim@iium.edu.my

A. Hazwani · M. Sha'ban · A. Azhim
Kulliyyah of Allied Health Sciences, IIUM, Kuantan Campus, Malaysia

F. Ibrahim et al. (eds.), *2nd International Conference for Innovation in Biomedical Engineering and Life Sciences*,
IFMBE Proceedings 67, https://doi.org/10.1007/978-981-10-7554-4_32

treatment, samples were washed for 5 days under constant shaking in Phosphate Buffer Saline (PBS) solution.

2.2 Histological Analysis

Samples were fixed in a 4% paraformaldehyde for 24 h at 4 °C. Samples were embedded in paraffin three times for 1 h each at 60 °C after dehydrating with a sequential alcohol gradient of 70, 80, 90 and 100% for 1 h of each gradient and rinsed the samples three times with xylene for 1 h each and xylene/paraffin (50:50% in volume) overnight. Then, the samples were sectioned 8-µm thick using microtome and stained with Hematoxylin-Eosin (H-E) staining.

2.3 Animal Experiment

Native samples, bioscaffolds decellularized by sonication treatment in 0.1 and 2% SDS were implanted in the subcutaneous lower right thoracic cavity of male Wistar rats for 35 days. For histological examination the samples were explanted, followed by fixation and staining with Hematoxylin Eosin (H-E). The inflammatory response was measured by calculating the number of macrophages using ImageJ (National Institute of Health, USA).

3 Results

The population of macrophages surrounds the implanted samples was illustrated in Fig. 1. The more number of macrophages were observed at the interface of implanted native samples (Fig. 1a) followed by bioscaffolds decellularized by sonication treatment with 0.1% (Fig. 1b) and 2% SDS (Fig. 1c). The number of macrophages for bioscaffolds decellularized by sonication treatment in 0.1 and 2% SDS illustrated in Fig. 2 shows significant differences from native samples ($p < 0.05$). The macrophage number of native samples was found to be three times higher than bioscaffolds decellularized by sonication treatment.

4 Discussion

Bioscaffolds decellularized by sonication treatment exhibit minimal inflammatory response after implantation in rats for 35 days. The inflammatory response is stimulated by cellular remnants such as double-stranded DNA (dsDNA), mitochondria, and phospholipids which are not completely removed from the bioscaffolds. The minimal criteria of cellular remnants to avoid adverse effects on cells and host response as recommended by previous study [4]. In this study, the cellular components are completely removed

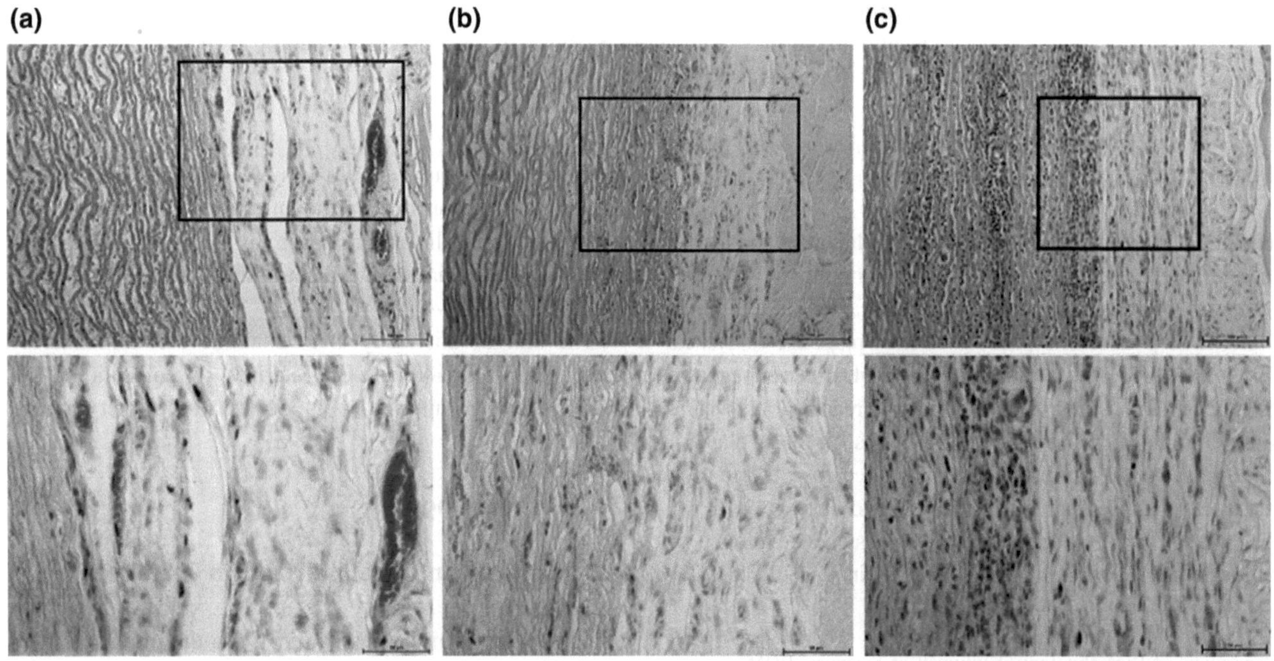

Fig. 1 Inflammatory response at 35 days post-surgery for **a** native samples, **b** bioscaffolds decellularized by sonication treatment with 0.1% SDS, and **c** bioscaffolds decellularized by sonication treatment with 2% SDS. Image with higher magnification represent the area within the black box in the lower magnification image

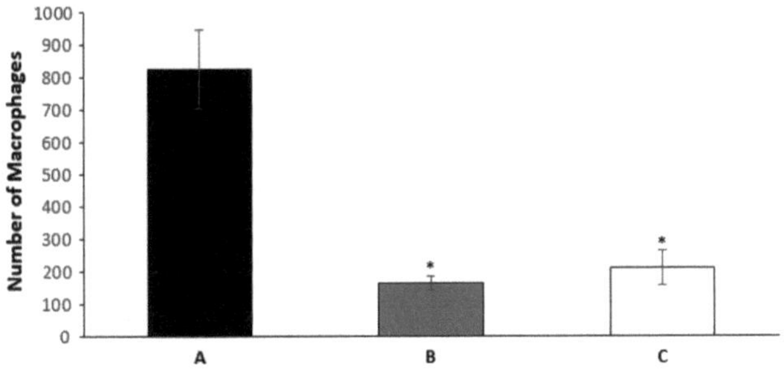

Fig. 2 The number of macrophages at 35 days post-surgery for **a** native samples, **b** bioscaffolds decellularized by sonication treatment with 2% SDS and **c** 0.1% SDS. *Significant difference between native samples and bioscaffolds ($p < 0.05$)

based on H-E staining evaluation and DNA quantification (not published). This result is consistent with our previous study [2].

Macrophages presence is associated with both immune response and tissue regeneration. The continuous present of macrophages in high amount might reduce the healing process and lead to an undesirable inflammatory response. This study has shown that less macrophages were presented within the bioscaffolds and more fibroblasts were infiltrated into the bioscaffolds. This might indicate that the bioscaffolds caused no aggression to the tissues and support faster penetration which characterized by better cell migration. This result is supported by previous study, which used crosslink agents in decellularized esophageal tissues and showed that less macrophages present and fibroblast infiltration, improve the integration of the bioscaffolds surround the tissues [5]. Likewise, bioscaffolds decellularized by sonication treatment able to attenuate the inflammatory response.

5 Conclusion

We conclude that bioscaffolds decellularized by sonication treatment exhibit minimal inflammatory response while providing a suitable environment for repopulation of host cells. More investigation needs to be done on the biocompatibility to affirm the safety of bioscaffolds decellularized by sonication treatments.

Acknowledgements The authors are grateful to the Ministry of Higher Education for financial support through Fundamental Research Grant Scheme (FRGS15-204-0445), Prototype Research Grant Scheme (PRGS16-002-0033) and Transdiciplinary Research Grant Scheme (TRGS16-02-001-0001).

References

1. Sheikh, Z., Brooks, P.J., Barzilay, O., Fine, N., Glogauer, M.: Macrophages, foreign body giant cells and their response to implantable biomaterials. Materials **8**(9), 5671–5701 (2015)
2. Syazwani, N., Azhim, A., Morimoto, Y., Furukawa, K.S., Ushida, T.: Decellularization of aorta tissue using sonication treatment as potential scaffold for vascular tissue engineering. J. Med. Biol. Eng. **35**(2), 258–269 (2015)
3. Smith, M.J., Smith, D.C., White, K.L., Bowlin, G.L.: Immune response testing of electrospun polymers: an important consideration in the evaluation of biomaterials. J. Eng. Fibers Fabr. **2**(2), 41–47 (2007)
4. Crapo, P.M., Gilbert, T.W., Badylak, S.F.: An overview of tissue and whole organ decellularization processes. Biomaterials **32**(12), 3233–3243 (2011)
5. Koch, H., Graneist, C., Emmrich, F., Till, H., Metzger, R., Aupperle, H., Schierle, K., Sack, U., & Boldt, A.: Xenogenic esophagus scaffolds fixed with several agents : Comparative in vivo study of rejection and inflammation. J. Biomed. Biotechnol. **2012**, 1–11 (2012)

Part IV

Medical Device and Clinical Healthcare

Model-Based Insulin-Nutrition Administration for Glycemic Control in Malaysian Critical Care: First Pilot Trial

A. Abu-Samah, N. H. Ahamad, N. N. Razak, F. M. Suhaimi, U. K. Jamaluddin, A. M. Ralib, M. B. Mat-Nor, C. G. Pretty, J. L. Dickson, and G. Chase

Abstract

Stress-induced hyperglycemia is prevalent in critical care, even in patients with no history of diabetes. Control of blood glucose level with tight insulin therapy has been shown to reduce incidences of hyperglycemia leading to reduced mortality and improved clinical outcomes. STAR is a tablet-based glucose control protocol with a specialized user interface into which insulin and nutrition information can be entered and predicted. This research describes the first clinical pilot trial of STAR approach in International Islamic University Hospital, Kuantan, Malaysia. The clinically specified target for blood glucose level is between 4.4 and 8.0 mmol/L. Seven episodes (of 359 h) were recruited based on the need for glucose control. Overall, 43.93% of measurement are in the range of 4.4–8.0 mmol/L band. The blood glucose median is 8.30 [6.32–10.00] mmol/L with only 1 patient having below than 2.22 mmol/L which is the guaranteed minimum risk level. This pilot study shows that STAR protocol is a patient specific approach that provides a good glycemic control in critically ill patients. Nevertheless, its implementation in Malaysian intensive care environments requires modifications and improvements in certain areas.

Keywords

Glycemic control • Model-based approach • Hyperglycemia • Critical care • Clinical trial

A. Abu-Samah (✉) · N. H. Ahamad · N. N. Razak
Department of Electronics & Communication Engineering, College of Engineering, Universiti Tenaga Nasional Malaysia, Kajang, Malaysia
e-mail: Asma@uniten.edu.my

F. M. Suhaimi
Advanced Medical and Dental Institute, Universiti Sains Malaysia, Penang, Malaysia

U. K. Jamaluddin
Faculty of Mechanical, Universiti Malaysia Pahang, Pekan, Malaysia

A. M. Ralib · M. B. Mat-Nor
Kulliyah of Medicine, International Islamic University Malaysia, Gombak, Malaysia

C. G. Pretty · J. L. Dickson · G. Chase
Department of Mechanical Engineering, University of Canterbury, Christ Church, New Zealand

1 Introduction

Stress-induced hyperglycemia is common in critically ill patients, even among those without diabetes mellitus [1–4]. The metabolic response to stress is characterized by major changes in glucose metabolism. Hyperglycemia is not only a marker for severity of illness, it also worsens outcomes, leading to an increased risk of further complications, such as severe infections [5], myocardial infarctions [6], polyneuropathy and multiple-organ failure [3]. Glucose Control (GC) has been shown to reduce intensive care unit (ICU) patient mortality up to 45% [3, 7–9]. However, other GC studies have shown the difficulty in reproducing the

F. Ibrahim et al. (eds.), *2nd International Conference for Innovation in Biomedical Engineering and Life Sciences*,
IFMBE Proceedings 67, https://doi.org/10.1007/978-981-10-7554-4_33

reduced risk of death in patients [10–12]. A recent study showing glycemic outcome is a function of GC given and not of patient condition [13] has reopened debate on GC and how to apply it safely and effectively.

The Stochastic TARgeted (STAR) protocol is based on a clinically validated model [14, 15] to provide specific recommendations for insulin, as well as nutrition, while controlling hyperglycemia and ensuring a maximum of 5% of the risk of light hypoglycemia [16, 17]. STAR is an adaptable model-based control approach that empowers versatile, patient-specific GC. It is applied over a range of clinical conditions and is used for real-time bedside care. Since 2010, two hospitals have been using STAR protocol as standard of care for GC in their ICUs namely Christchurch hospital, New Zealand and Kálmán Pándy Hospital, Hungary.

This paper presents a STAR protocol modulating only insulin infusions toward a targeted glycemic range of 4.4–8.0 mmol/L. The initial pilot trial results of the protocol are compared with the simulations using the same patients' data. This pilot trial tests the ability to adapt the model-based STAR GC framework from its development environment at Christchurch Hospital in New Zealand to the International Islamic University Malaysia (IIUM) Hospital ICU which has an ethnically different cohort and culturally different clinical practice.

2 Method

2.1 STAR Protocol

STAR uses a time varying insulin sensitivity to capture changes in patient condition and insulin-nutrition metabolism over time. Starting criteria for STAR is two blood glucose (BG) measurements over 8 mmol/L within a 4-hour period. The benchmarked BG target range of STAR is set at 4.4–8.0 mmol/L based on reduced risks with increased BG in intermediate bands. STAR specifically captures developing physiological patient condition and inter- and intra-patient fluctuation by identifying insulin sensitivity (SI) and its future variability to optimize safety and GC performance. Its elements can be modified according to clinically determined glycemic targets, control approaches (e.g. insulin only, insulin and nutrition, etc.) and clinical resources (e.g. measurement frequency).

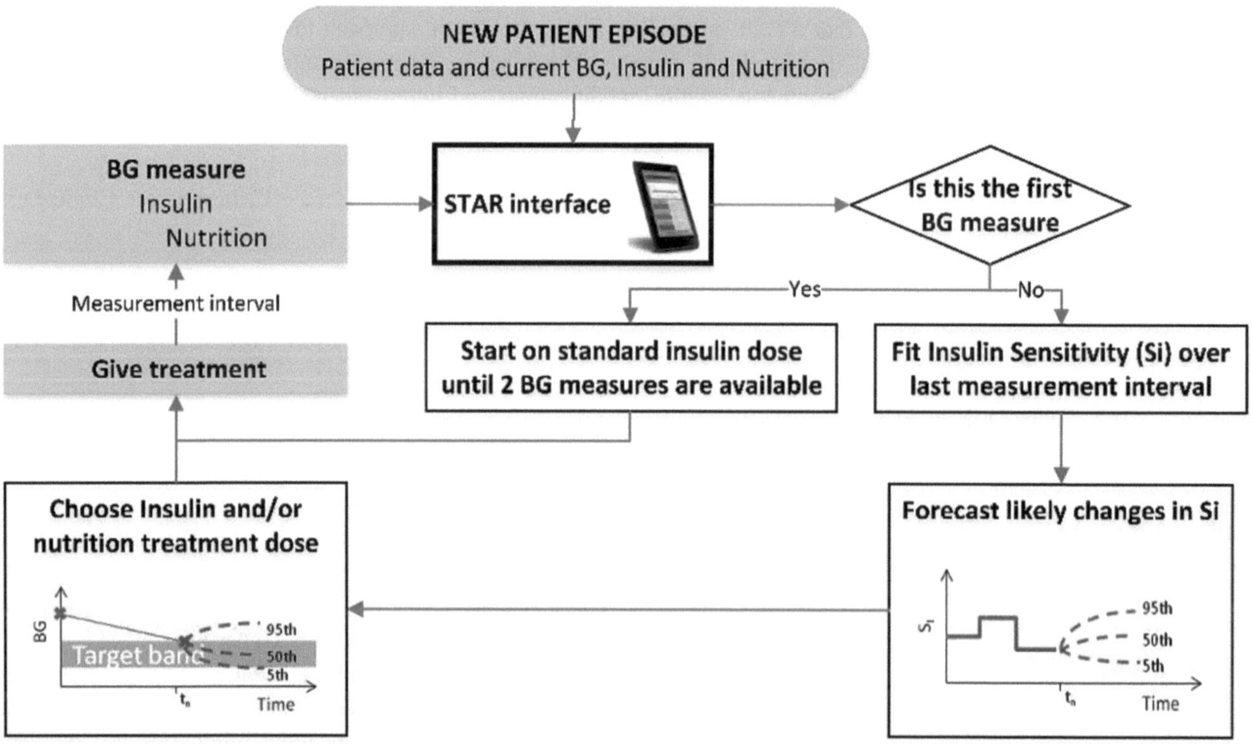

Fig. 1 Summary of the STAR functions. SI is the insulin sensitivity and BG is blood glucose

Table 1 Baseline clinical data of all four patients and corresponding to seven episodes

Patient	Age	Sex	Hours	Initial BG	Diagnosis
1a	56	F	58	14.4	Congestive heart failure, lung fibrosis mixed connective tissue disease, acute pulmonary oedema, bilateral parenchymal disease
1b			46	10.2	
2a	49	M	71	11.0	Motor vehicle accident with polytrauma, hypertension, diabetes mellitus
2b			28	11.6	
2c			85	11.7	
3	70	M	33	17.4	Early glottic CA, left vocal cord, heart disease, diabetes mellitus
4	66	F	38	11.8	Hospital required pneumonia, renal cell carcinoma, with brain metastasis

Table 2 Cohort blood glucose results based on clinical data and the virtual trial

Whole cohort statistics	Clinical results	Virtual trial
Number of episodes	7	7
Total hours	359	359
Number of BG measurements	239	203
BG median [IQR] (mmol/L)	7.86	6.64
BG standard deviation (geometric) (mmol/L)	1.39	1.35
% BG within 4.0–6.1 mmol/L	23.85	38.91
% BG within 4.4–7.0 mmol/L	29.71	47.78
% BG within 4.4–8.0 mmol/L	43.93	66.50
% BG within 8.0–10.0 mmol/L	28.03	18.22
% BG > 10.0 mmol/L	23.85	7.88
% BG < 4.4 mmol/L	5.02	8.37
% BG < 4.0 mmol/L	1.67	5.42
% BG < 2.2 mmol/L	0.42	0
Number of patients < 2.22 mmol/L	1	0
Median insulin rate [IQR] (U/h)	2.0 [1.0–3.8]	3.0 [1.5–6.0]
Median glucose rate [IQR] (g/h)	6.0 [4.9–8.3]	5.5 [3.9–6.0]

STAR is implemented in tablet computer to provide specialized user interface into which nutrition and insulin information can be entered. Once BG measurements are taken, an insulin sensitivity profile can be computed. Insulin sensitivity varies hourly, and stochastic based forecasting is used to determine likely outcomes in insulin sensitivity for any possible insulin and/or nutrition input. This forecasting allows changes in future insulin sensitivity to be determined.

The insulin-nutrition combination that best overlaps the resulting BG range with the target clinical band is selected by placing the 5th percentile BG out come on the lower edge of the clinically specified target range. STAR seeks to maximize nutrition delivery while dosing insulin in this context. If, for maximum nutritional and insulin treatments, the BG outcome range does not sufficiently overlap the clinical target range, then STAR will recommend a drop in nutrition to maintain GC and reduce risk [17]. The nutrition will be raised back to its goal feed value as soon as possible. In cases where the feed must be clinically determined or switched off, STAR can be set to give insulin recommendations only. Figure 1 summarizes how STAR functions.

2.2 STAR Protocol

In this first pilot trial for Malaysian critically ill patients, seven episodes (of 359 h) from four non-diabetic patients were controlled using STAR. These were obtained during their ICU stay in the first semester of 2017. Patient 1 and 2 had more than one episodes as they were taken off GC using STAR tablet due to judgement by medical staffs who were new to its use. The selection criteria includes GC for a minimum of 24 h. Table 1 shows the baseline for each episode (age, sex, hours, initial BG and diagnosis). Written

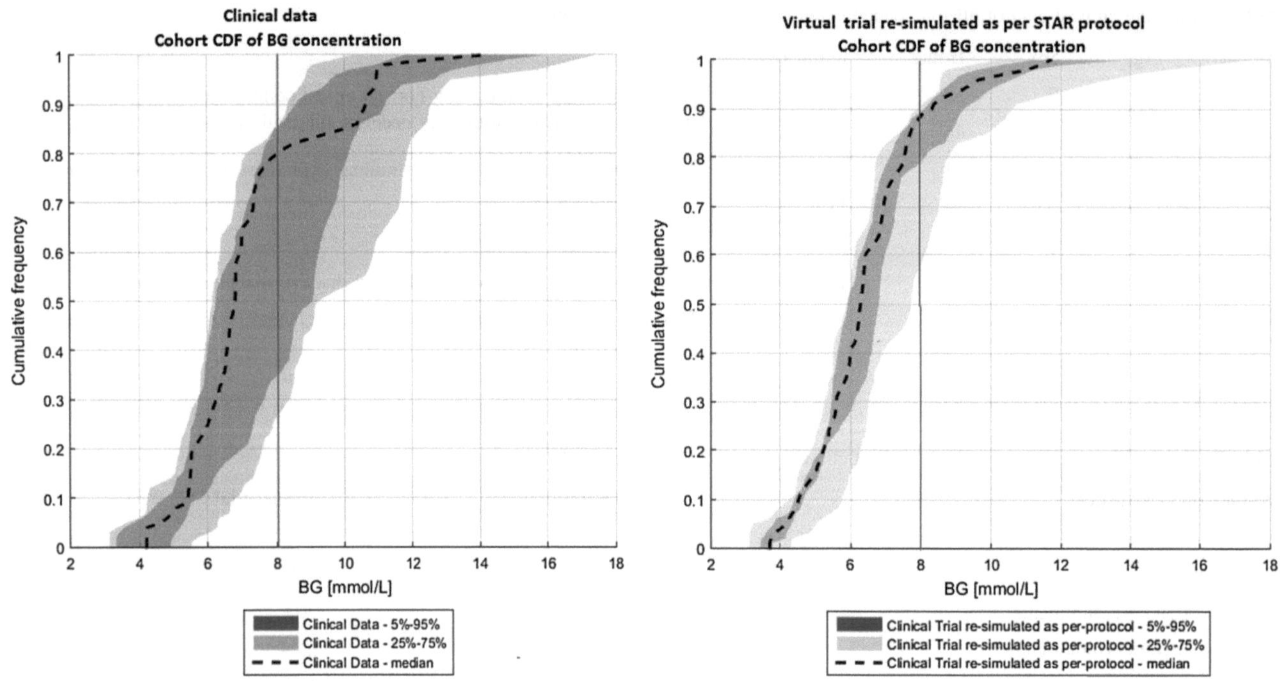

Fig. 2 Median, 25–75% and 5–95% intervals of per-patient BG Cumulative Distribution Functions (CDFs) defines on whole cohort

informed consent was obtained for all patients, and approval (IREC 657) was granted for this study by IIUM Research Ethics Committee and National Institute of Health (NIH).

For each patient, the trial started with a BG measurement made by nursing staff. BG measurements were made using B-Braun glucometers. The approach then identifies a new insulin and nutrition infusion rate, which was then given by the nurse. The time interval until the next BG measurement is also selected by the medical staffs based on 1–3 hourly treatment options offered.

2.3 Virtual Trial

To analyze the compliance of STAR as clinically applied GC protocol, a virtual trial was conducted using the clinical data. This trial has two phases: (i) Fitting and (ii) Simulation. During the fitting, clinical data were used to identify hourly SI values which served as a virtual patient profile. This profile reflects the glycemic response to insulin and nutrition inputs and can be used to simulate responses to different interventions with good accuracy [15, 18]. In the second phase, the profile was used to simulate glycemic response to STAR as simulated, where differences indicates non-compliance to STAR.

3 Results

Clinical results are summarized by whole cohort statistics in Table 2. There were 239 BG measurements taken during 359 h of control. BG median value, 8.3 mmol/L for whole cohort is higher than BG target of 8.0 mmol/L. BG levels are relatively widely distributed, as evidenced by the IQR range (25th–75th per-centile value) of 4.68 mmol/L for the cohort, and the 25–75% confidence interval across patients in Fig. 2. Table 2 shows that 43.93% of BG measurements are between 4.4 and 8.0 mmol/L. The control is tight in this band, as illustrated by the steep slope of BG Cumulative Distributive Function (CDF) for the whole cohort in Fig. 2 and similar per-patient CDFs. There was only one patient with severe hypoglycemic (measurement BG < 2.22 mmol/L) which was patient 2b.

Virtual trial results are also presented in Table 2. The total number of BG measurements is significantly reduced by 36, from an average 16 per-day to the 13 per-day that matches STAR in use elsewhere [19]. Overall statistics for BG measurements recorded better numbers than clinical results. 66.50% patients are between target range of 4.4–8.0 mmol/L and only 7.88% of patients with BG >10.0 mmol/L. There is no episode of severe hypoglycemia.

Table 3 Clinical results for each patient

Patient	BG Median [IQR] (mmol/L)	BG mean (mmol/l)	BG St Dev (mmol/l)	% BG within 4.4–6.1 mmol/L	% BG within 4.4–7.0 mmol/L	% BG within 8.0-10.0 mmol/L	% BG > 10.0 mmol/L	% BG < 4.4 mmol/L	% BG < 2.22 mmol/L	Median Insulin rate [IQR] (U/hr)	Max insulin rate (U/hr)	Median Dextrose rate [IQR] (g/hr)
1a	6.3	6.52	1.33	41.18	61.76	14.71	5.88	5.88	0	1	3	3.8
	[5.4–7.4]									[0.6–1.1]		[3.8–5.7]
2a	8.8	8.52	1.27	11.76	17.65	43.14	29.41	1.96	0	2	8.6	8.3
	[7.9–10.4]									[1.2-3.0]		[8.3-11.4]
2b	9.9	8.92	1.59	10.71	14.29	17.86	46.43	7.14	3.57	2.9	8	6.3
	[7.3–12.2]									[2.0–6.3]		[5.7–8.9]
3	7	7.34	1.49	29.17	37.5	41.67	25.00	16.67	0	2.7	10	2.1
	[5.8–9.4]									[1.1–4.2]		[0.7–4.9]
1b	6.7	6.46	1.35	29.63	48.15	18.52	3.7	7.41	0	1	4	5.1
	[5.8–7.8]									[0.2–1.4]		[5.1–5.1]
2c	9.5	9.05	1.25	7.02	8.77	47.37	31.58	0	0	4.3	10	8.3
	[8.5–10.2]									[2.5–6.0]		[6.4–8.3]
4	6.4	6.93	1.29	38.89	55.56	11.11	11.11	0	0	1.5	4	4.9
	[5.7–7.8]									[1.0–2.0]		[4.9–4.9]

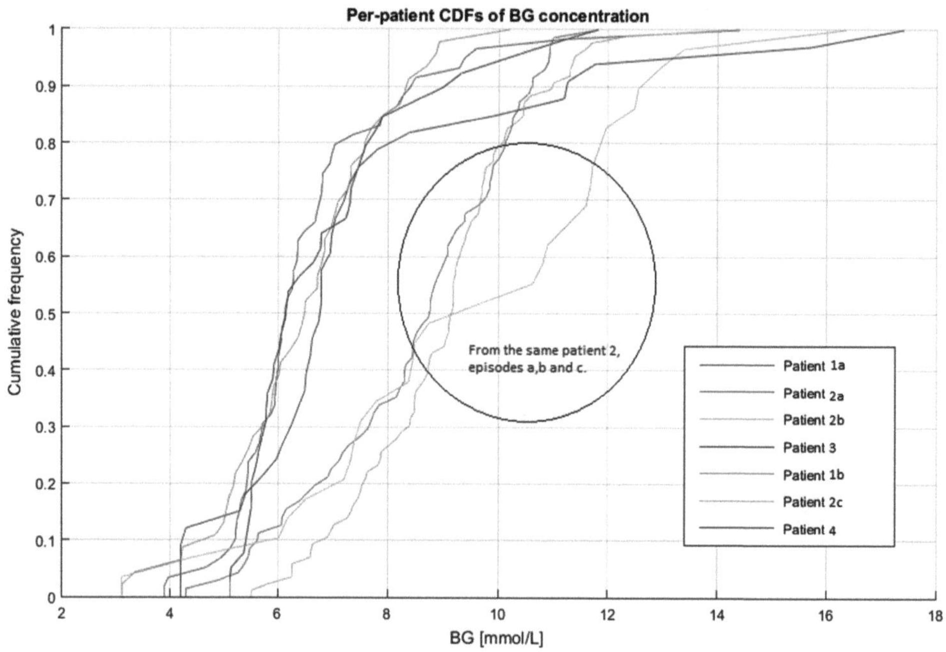

Fig. 3 Per-patient CDFs of BG concentration

Clinical results are provided for all individual episodes in Table 3 and Fig. 3. Overall, these results indicate that these specific patients were particularly insulin resistant. In addition, the pilot trial length was not sufficient to achieve consistently high percentages of BG levels in a tight band around the target where resistance and variability decline significantly after 48–72 h. In this trial, only one episode exceeds this lengths (episode 2c).

4 Discussion

The objective of this pilot trial is to look at the initial assessment of performance, safety and compliance of STAR in a Malaysian ICU. This trial was started off using a benchmarked BG target range of 4.4–8.0 mmol/L. Overall results suggest that the protocol can be considered to be used in Malaysian intensive care. However, the clinical performance assessment in Table 2 where a significant 23.85% of BG level above 10.0 mmol/L and 28.03% of BG level within 8.0–0 mmol/L raises the question of whether to put a different target range for Malaysian patients. Results from virtual trial shows percentage of light hypoglycaemia is above 5% (8.37% for BG < 4.4 mmol/L and 5.42% for < 4.0 mmol/L) also suggest the same proposition.

This pilot clinical trial is the first attempt to use STAR protocol in a Malaysian ICU. Another important result is one patient episode with severe hypoglycemia (BG < 2.22 mmol/L). Figure 4a shows episode 2b profile with hypoglycemia. The results for this episode 2b were re-simulated and the new outcome is shown in Fig. 4b. It demonstrates no hypoglycemia. In fact, this patient has an overall low insulin sensitivity. He was admitted with polytrauma secondary to motor vehicle accident. One main difference comes from the number of BG measurement (26 in clinical trial vs. 23 in simulation) which modifies iterative forecasting of insulin and nutrition input towards patient. As a result, clinical patient has higher BG measurement even though still in target range and a rise in insulin sensitivity is recorded before the hypoglycemia is detected. More differences can be seen on the proposition for insulin and nutrition input. These differences question the nurses' compliance during clinical implementation.

5 Conclusion

This pilot study shows the results of our initial attempt of using Stochastic TARgeted (STAR) protocol in the intensive care unit, IIUM hospital. We demonstrated that this approach was able to provide a good blood glucose control among critically ill Malaysian patients and was adaptable according to patients' clinical condition. We have identified specific issues that warrant modification of the protocol in

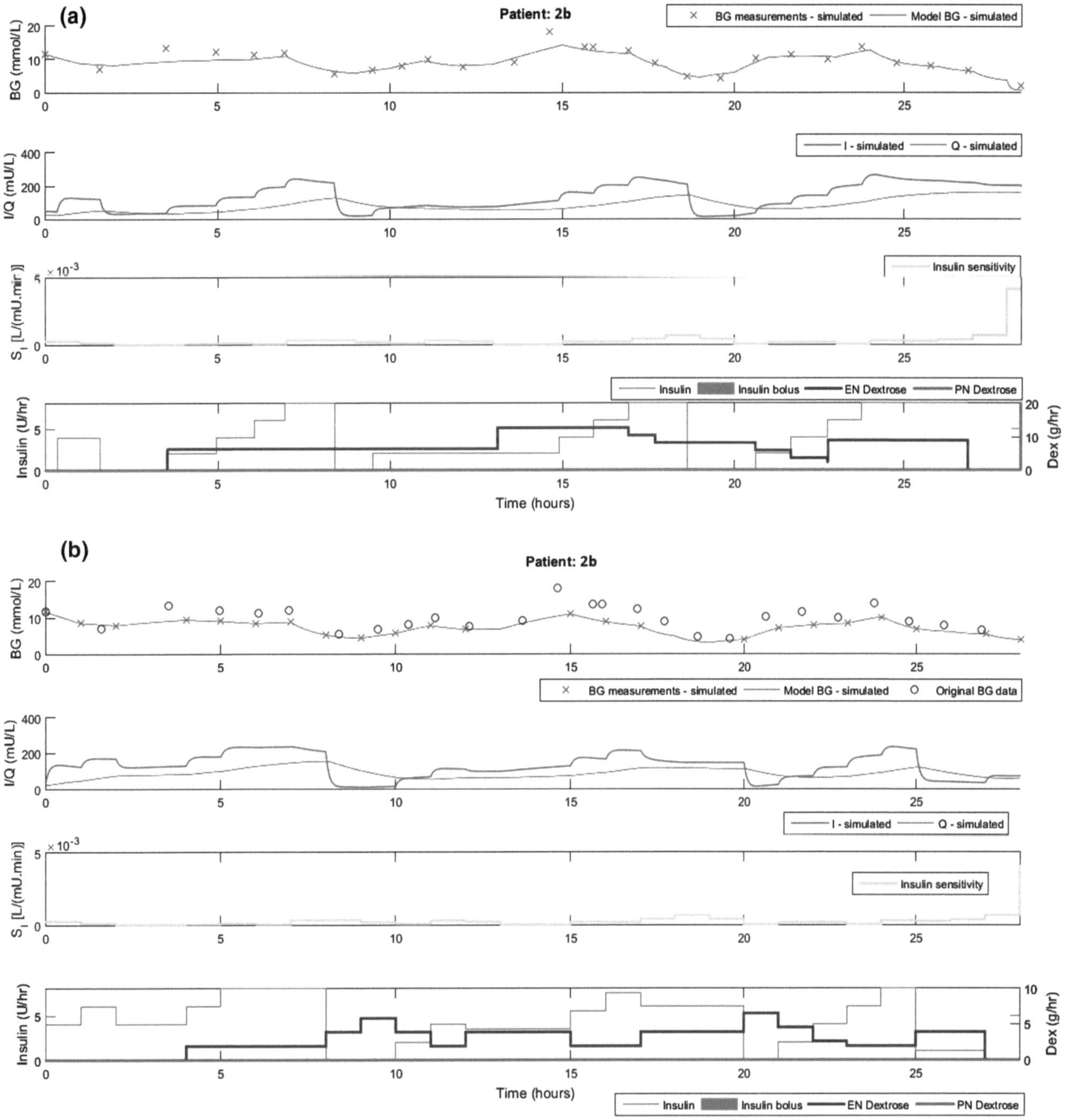

Fig. 4 Patient 2b profiles; **a** is the clinical results and **b** is the clinical data re-simulated with STAR protocol

order to improve its performance, if it is to be used in future trials. This is to ensure its safety and efficiency in achieving good glycemic control in Malaysian ICU population.

Acknowledgements The authors thankfully acknowledge IIUM for their commitment and provision of data for the pilot study. The authors also acknowledge Ministry of Higher Education (MOHE), BOLD Scheme and UNITEN for their support on the ongoing research.

References

1. Marik, P.E., Raghavan, M.: Stress-hyperglycemia, insulin and immunomodulation in sepsis. Intensive Care Med. **30**(5), 748–756 (2004)
2. McCowen, K.C., Malhotra, A., Bistrian, B.R.: Stress-induced hyperglycemia. Crit. Care Clin. **17**(1), 107–124 (2001)
3. Ellahham, S.: Insulin therapy in critically ill patients. Vascular health and risk management **6**, 1089 (2010)

4. Dungan, K.M., Braithwaite, S.S., Preiser, J.C.: Stress hyperglycaemia. Lancet **373**(9677), 1798–1807 (2009)
5. Bistrian, B.R.: Hyperglycemia and infection: which is the chicken and which is the egg? J. Parenter. Enteral Nutr. **25**(4), 180–181 (2001)
6. Capes, S.E., Hunt, D., Malmberg, K., Gerstein, H.C.: Stress hyperglycaemia and increased risk of death after myocardial infarction in patients with and without diabetes: a systematic overview. Lancet **355**(9206), 773–778 (2000)
7. Van den Berghe, G., Wouters, P.J., Bouillon, R., Weekers, F., Verwaest, C., Schetz, M., Vlasselaers, D., Ferdinande, P., Lauwers, P.: Outcome benefit of intensive insulin therapy in the critically ill: insulin dose versus glycemic control. Crit. Care Med. **31**(2), 359–366 (2003)
8. Krinsley, J.S.: Glycemic variability: a strong independent predictor of mortality in critically ill patients. Crit. Care Med. **36**(11), 3008–3013 (2008)
9. Chase, J., Shaw, G., Le Compte, A., Lonergan, T., Willacy, M., Wong, X.W., Lin, J., Lotz, T., Lee, D., Hann, C.: Implementation and evaluation of the SPRINT protocol for tight glycaemic control in critically ill patients: a clinical practice change. Crit. Care **12**(2), R49 (2008)
10. Brunkhorst, F.M., Engel, C., Bloos, F., Meier-Hellmann, A., Ragaller, M., Weiler, N., Moerer, O., Gruendling, M., Oppert, M., Grond, S., Olthoff, D.: Intensive insulin therapy and pentastarch resuscitation in severe sepsis. N. Engl. J. Med. **358**(2), 125–139 (2008)
11. Nice-Sugar Study Investigators: Intensive versus conventional glucose control in critically ill patients. N. Engl. J. Med. **2009** (360), 1283–1297 (2009)
12. Marik, P.E., Preiser, J.C.: Toward understanding tight glycemic control in the ICU: a systematic review and metaanalysis. CHEST J. **137**(3), 544–551 (2010)
13. Uyttendaele, V., Dickson, J.L., Shaw, G.M., Desaive, T., Chase, J. G.: Untangling glycaemia and mortality in critical care. Crit. Care **21**(1), 152 (2017)
14. Lin, J., Razak, N.N., Pretty, C.G., Le Compte, A., Docherty, P., Parente, J.D., Chase, J.G.: A physiological intensive control insulin-nutrition-glucose (ICING) model validated in critialy ill patients. Comput. Methods Programs Biomed. **102**(2), 192–205 (2011)
15. Dickson, J.L., Stewart, K.W., Pretty, C.G., Flechet, M., Desaive, T., Penning, S., Lambermont, B.C., Benyo, B., Shaw, G.M., Chase, G.: Generalisability of a virtual trials method for glycaemic control in intensive care. IEEE Trans. Biomed. Eng. (2017)
16. Fisk, L.M., Le Compte, A.J., Shaw, G.M., Penning, S., Desaive, T., Chase, J.G.: STAR development and protocol comparison. IEEE Trans. Biomed. Eng. **59**(12), 3357–3364 (2012)
17. Evans, A., Shaw, G.M., Le Compte, A., Tan, C.S., Ward, L., Steel, J., Pretty, C.G., Pfeifer, L., Penning, S., Suhaimi, F., Signal, M.: Pilot proof of concept clinical trials of Stochastic Targeted (STAR) glycemic control. Ann. Intensive Care **1**(1), 38 (2011)
18. Chase, J.G., Suhaimi, F., Penning, S., Preiser, J.C., Le Compte, J., Lin, J., Pretty, C.G., Shaw, G.M., Moorhead, K.T., Desaive, T.: Validation of a model-based virtual trials method for tight glycemic control in intensive care. Biomed. Eng. **9**(1), 84 (2010)
19. Stewart, K.W., Pretty, C.G., Tomlinson, H., Thomas, F.L., Homlok, J., No′emi, S.N., Illy′es, A., Shaw, G.M., Chase, J.G.: Safety, efficacy and clinical generalization of the STAR protocol: a retrospective analysis. Ann. Intensive Care. **6**(1), 24 (2016)

Multi-channel Fabric Based Pressure Mapping Data Acquisition System

Normazlianita M. Alias, Zakiran Abd Razak, Mohd Yazed Ahmad, and Nur Azah Hamzaid

Abstract

A smart fabric based multichannel pressure mapping Data Acquisition System (DAS) was specially designed to read and visualize real time pressure data from an array of piezoresistive fabric pressure sensor. This DAS is important to enable rapid monitoring of pressure distribution for biomedical engineering applications. A customized pressure mapping circuit using off-the-shelf components has been designed and fabricated. In addition, a pressure mapping algorithm which runs on Arduino platform and MATLAB was developed to continuously read and visualize pressure profile from the fabric pressure sensor. To ensure low component count and simple hardware, the concept of multiplexing has been employed in the hardware architecture and the firmware was written to support this architecture. This approach allows the system to perform even with single processor and single ADC. The reliability of the system was tested with array of fabric pressure sensor using a special portable load cell (Advance Force Gauge by MECMESIN). Pressure profile for each sensor unit matches the sensor resistive load characteristic, repeatability and accuracy of a commercial data acquisition system. The system is suitable for reading slow changing analog signal. If one desired to measure fast changing signal, the same architecture can still be used but specifications of the components must be higher. In conclusion, a smart multichannel pressure mapping Data Acquisition System (DAS) was designed, fabricated and tested. The unit capable of acquiring and visualizing pressure data from the developed sensor array. The speed of the system can be increased by using higher speed components while maintaining the same architecture.

Keywords

Multichannel data acquisition system • Pressure mapping • Smart sensor array • Pressure array

1 Introduction

In recent years, diverse types of pressure sensors to measure surface pressure have been introduced for biomedical engineering applications [1]. This is due to increasing interest in wearable technologies especially in the field of gait and biomechanics investigations, including in the field of prosthetic and orthotic devices [2]. Multi array pressure mapping system is an important technology in biomedical devices. The mapping can be used to avoid the occurrence of pressure ulcer on the skin surface of bed bound patients as well as to measure the stump pressure distribution of prosthetic limb user [3]. In gait and biomechanics pressure mapping can be used in sport gait and performance investigations [4]. Apart from these, pressure mapping systems are now being considered in wearable devices such as data gloves and heart rate breathing jackets [5, 6]. The high demand on pressure mapping system triggers diverse

N. M. Alias · Z. Abd Razak · M. Y. Ahmad (✉) · N. A. Hamzaid
Department of Biomedical Engineering, Faculty of Engineering, University of Malaya, Kuala Lumpur, Malaysia
e-mail: myaz@um.edu.my

F. Ibrahim et al. (eds.), *2nd International Conference for Innovation in Biomedical Engineering and Life Sciences*, IFMBE Proceedings 67, https://doi.org/10.1007/978-981-10-7554-4_34

investigations to further improve the system. Most developed sensors have a common problem which is the inadaptability towards human body curvature during application of sensors [7]. Most of surface pressure sensors were rigid and have limited flexibility. For example, due to rigid sensor, in lower limb prosthetic biomechanics investigations, users are unable to walk as natural as they can as the sensor rigidity affects the comfort of the user [8]. Often, scientists used multiple pressure sensors to map the whole targeted area such as a lower limb stump [9]. Lack of flexibility, scalability, difficulty interpreting mass collected data, direct acquisition system therefore requires large space and is energy consuming, which cause inconveniences to system users. Therefore, a flexible and comfortable pressure sensor array along with an effective and scalable data acquisition system needs to be developed.

Over time, textile industry has developed conductive materials into fabric which can be manipulated into various electronic and wearable applications [10]. The fabric can be manipulated to become a simple sensing element by adjusting the layer between conductive materials and the degree of conductivity on its surface. In this paper we proposed a novel multichannel fabric based pressure sensor along with the sensor data accusation system capable of mapping the collected pressure data. In this study, we utilized of the-shelf materials to form multi array pressure sensor, the acquisition circuit along with smart algorithms to read the sensor data.

2 Sensor Design

2.1 The Sensor and Electronic Hardware

A multi array fabric based pressure sensors was designed based on a single layer sensor that was constructed on an earlier experiment [11]. The constructed fabric was woven from manipulations of layers from three basic materials; neoprene, conductive fabric and velostat. Neoprene is a non-conductive fabric consisting of synthetic rubber material designed with high flexibility and can maintain its properties even in the presents of high humidity. In this sensor design conductive fabric column acted as wefts and warps of the sensor [12]. The conductive fabric functions by transporting current in and out of the sensor. The material is stretchable and made up from 78% Nylon, 22% elastomer and 99% pure silver. The middle layer is constructed from a single sheet of piezoresistive plastic we named velostat placed in between the two electrodes thus forming piezoresistive effects. Velostat is carbon thin approximately 4 mm infused plastic (polyolefin) that has high resistivity in the range of 500 Ω per cm. The resistance of the velostat changed when pressure is applied.

To read the signal from the fabric based sensor, two multiplexors and Atmega 328 were used in this system. Multiplexor works as an analog input signal collector and forward the selected input into a single line for data transmission into Atmega 328. The microcontroller was programmed using software from Arduino IDE. Serial monitor display analog signal in the form of digital representation. Serial plotter function prints out real-time sensor value data from one output channel at a time The Arduino IDE did not have subplot program which can perform multiple plot at different graph space in real-time. Hence, data representation need to be viewed in the form of pressure mapping. Figure 1 described the entire system starting from sensor to the user interface.

2.2 Data Acquisition System and Software

MATLAB program was used to interface with Atmega 328 by visualization of sensor data using colour-coded mesh plot. This plot collates with the amount of force applied at each individual sensor. The scale of colour ranged from dark blue, light blue, green, yellow, orange and red. The colourmap change would respond in accordance to the amount of force applied to each sensor. When a greater amount of force was applied, the acquisition system would detect the decrease in voltage occurred at the sensor cell. The cell on pressure mapping screen that represent the location of applied force in turn displayed changes in color according to the magnitude of force applied on the sensor cell. At free loading condition, the sensor value displayed on pressure mapping was dark blue. When a slight force was applied, pressure mapping cell will change color to light blue, green, yellow, orange and red in order of increasing force application. Figure 2 illustrates the color changes of the system according to the amount of the force design. Here we can see the pressure was mapped based on the force applied on its surface.

2.3 The Sensor Load Test

Compressive force was applied to the sensor using Advanced Force Gauge (AFG) by Mecmesin to each cell and real-time data was collected to plot force vs resistance profile of the fabric sensor. AFG Mecmesin is a manual device, where the user just need to apply force by their hand as described in Fig. 2a. In this experiment, a steady force was applied to the sensor until the AFG measurement 1 N. Now, the force was maintained at constant magnitude. At the pressure mapping screen, voltage data displayed at the location of applied force were collected. Then, the voltage value was converted into resistance by using Ohm's Law

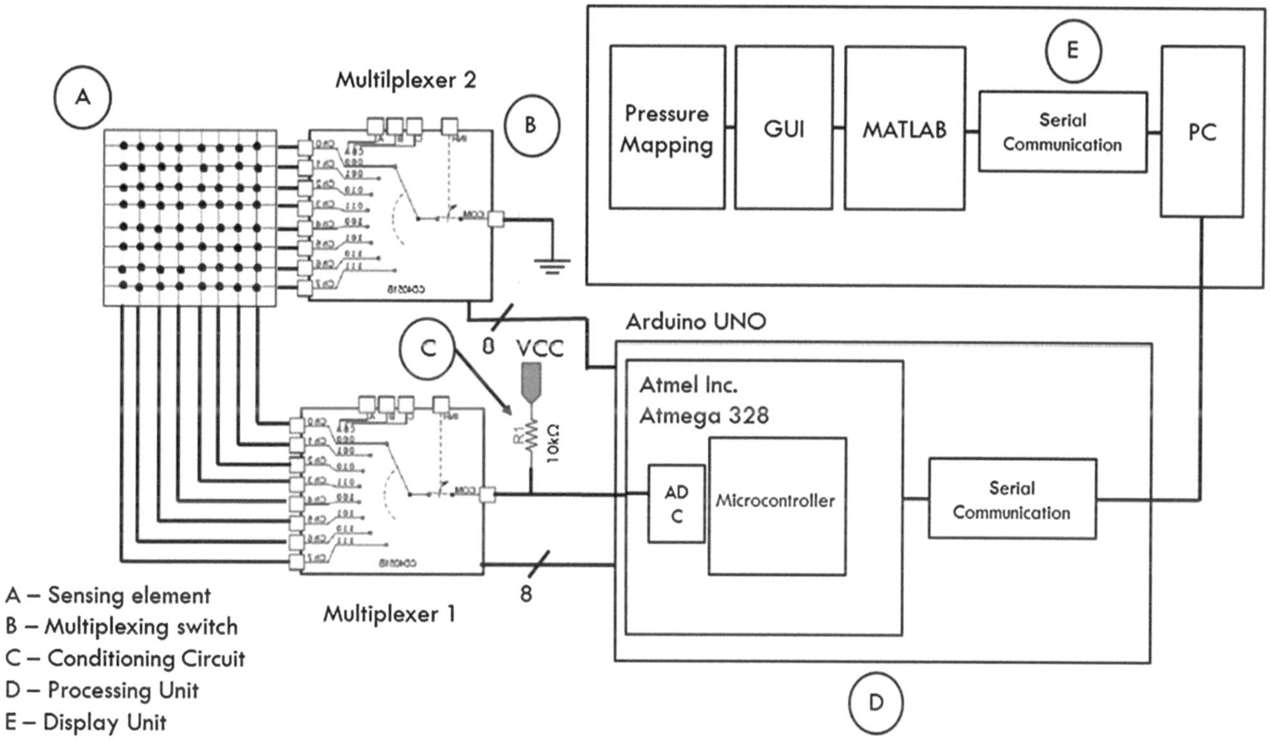

Fig. 1 The architecture of the whole multi array fabric based pressure sensor system. **a** Woven fabric based pressure sensor. **b** Multiplexer used to transmit data from the sensor into microcontroller. **c** Microcontroller that read the input data and convert it into digital data. **d** User interface output mapping on PC screen

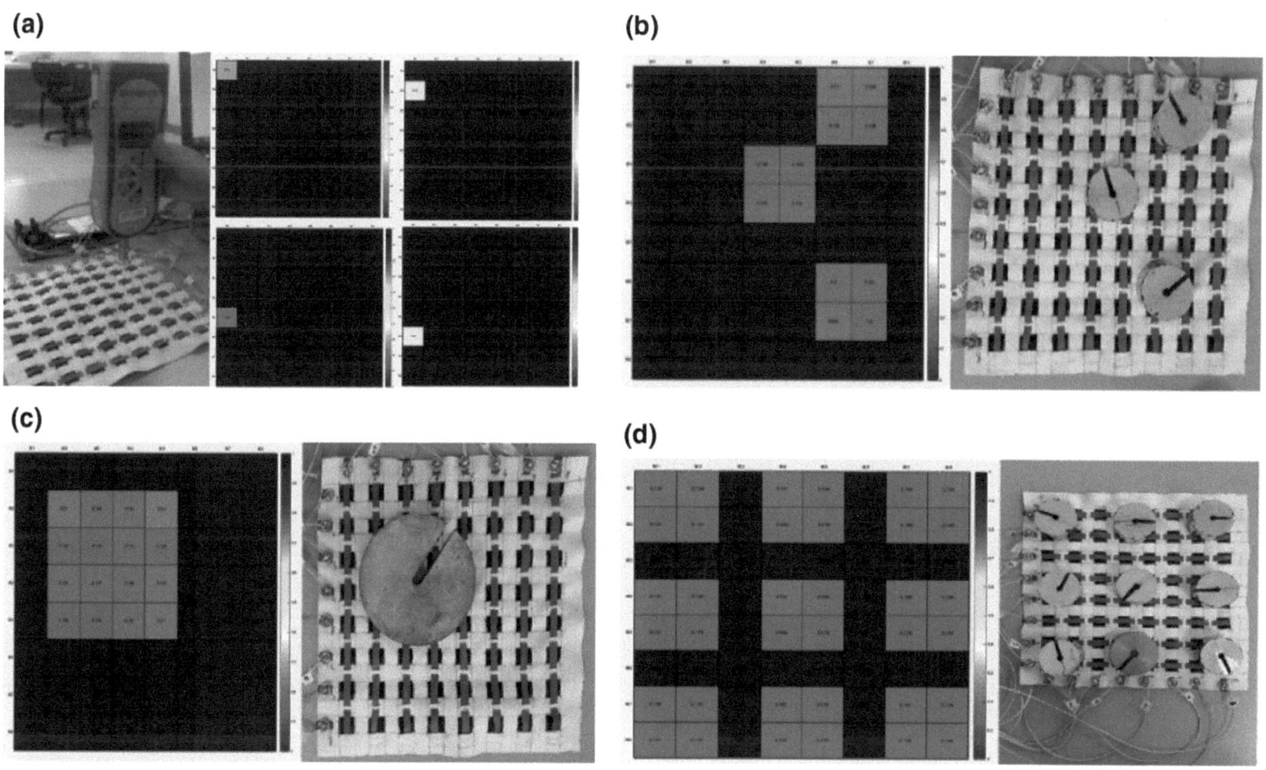

Fig. 2 Static loading of fabric based multi array system and read using data acquisition software. **a** Using Mac Machine, each individual array was tested. **b** Three-point static loading using metal load 1 N each. **c** One-point 5 N metal load tested on the array. **d** Multi array point tested using 1 N metal each

and the resistance values were tabulated. The procedure was repeated until 20 N of force was applied to the sensor. A graph of sensor resistance (Ω) versus magnitude of force application (N) was plotted and a relationship was deduced. The sensor force versus resistance profile was then compared with the characteristics profile of the fabric pressure sensor from Fig. 3. To ensure the system was able to map on different region and different amount of force, circular metal load was used as shown in Fig. 2b–d.

3 Results and Discussion

3.1 Results

The proposed multichannel fabric based pressure sensor mapping DAS aims to analytically visualize force distribution applied on the sensing pad array. Its acquisition unit was designed using a low-cost microcontroller. The system displayed array of pressure values through GUI using an

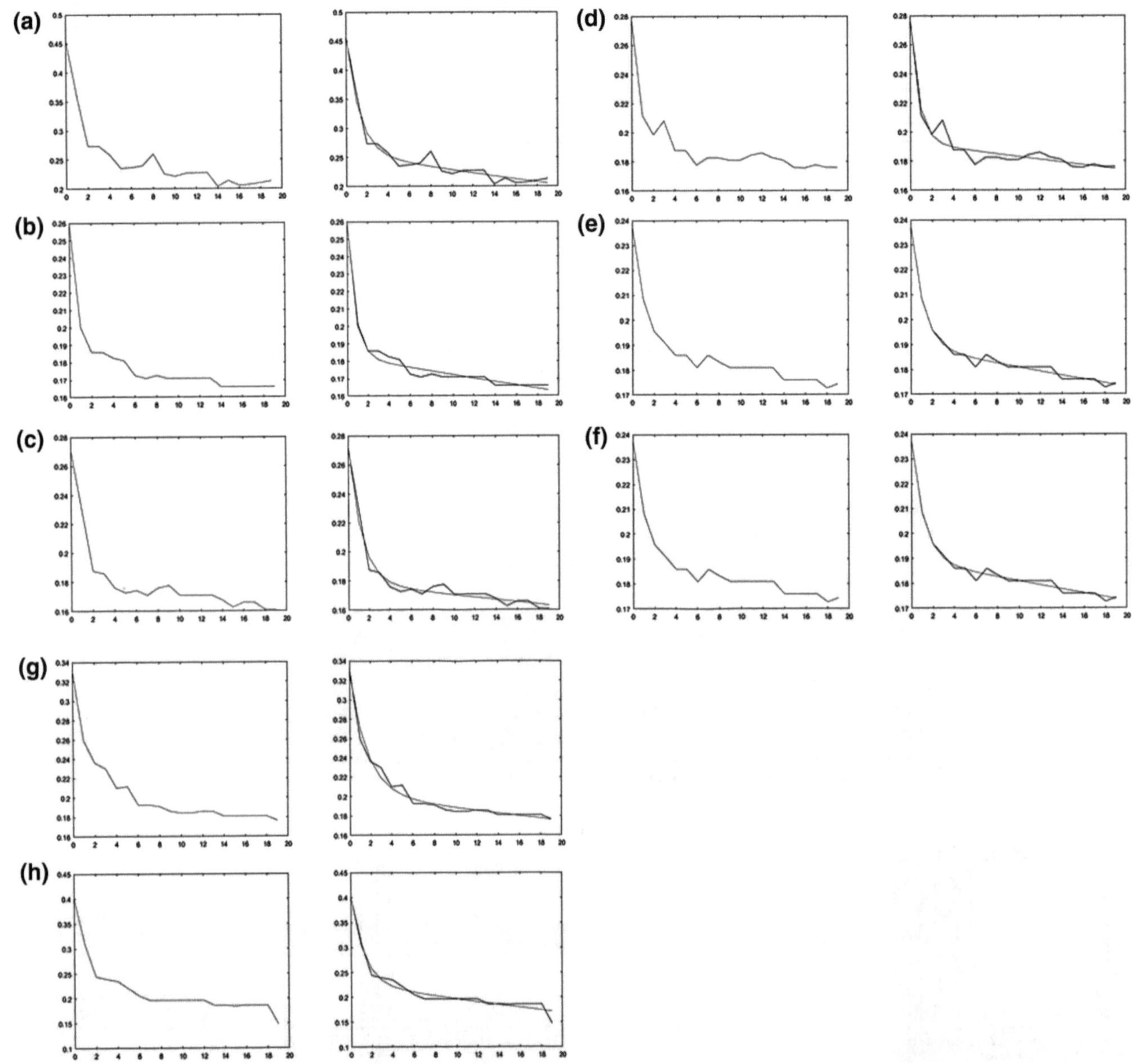

Fig. 3 Data plot for voltage (V) vs force data (N) for sensor 1 to 8 respectively (left side). The exponential model fit curve plot generated using MATLAB fit function (right side). Graphs A to H plots data sensors 1 to 8 respectively

Table 1 Characteristic of DAS

Characteristics	Results	Characteristics
Repeatability	88.9%	Repeatability
Accuracy error	0.25%	Accuracy error
Resolution	4.88 mV/bit	Resolution
Sampling frequency	10 kHz	Sampling frequency
Characteristics	Results	Characteristics

algorithm that was computed based on the voltages receives which corresponded to the pressure applied on the sensing pad array. This was achieved using multichannel data acquisition system design. Table 1 illustrates the characteristic of the DAS based on the tests conducted.

Test results, showed good integrity of data measurement that was suitable to be used in biomedical surface force measurement application. Consistency error in this system was to be expected due to parasitic capacitance. However, the measurement accuracy was good for a system that used a basic front-end interface. The system read data in sequence and visualized the data in the form of pressure mapping GUI. The pressure mapping DAS was able to locate area of static loading as shown in Fig. 3. The sensor voltage was inversely proportional to the load applied on sensor cell. An experiment was conducted to match whether DAS would confirm the sensor characteristic upon load application. As shown in Fig. 3, the relationship between sensor voltage and load application on sensor cell matched the fabric pressure sensor characteristic. DAS can read potential difference in each sensor cells. This voltage data was converted into force value by using the experimental data which correlated with relationship between voltage and force applied on each sensor cell. A Data conversation algorithm is introduced to display force distribution values applied on each sensor cell in real-time. For this system, the exponential model is suitable to approximate the force and voltage curve characteristic of sensor data. A function is generated by determining the four variables using exponential curve fit function command in MATLAB.

3.2 Discussion

From the exponential model function equation shown in Eq. [1], the necessary variable was algebraically derived to express the value of force in terms of voltage reading in each sensor. The coefficients of the model denoted by a, b, c and d were determined by substituting the voltage and force values from the experiment conducted on individual sensor using **Force Formula**:

$$\mathrm{y_n} = \mathrm{ae^{bx_n}} + \mathrm{ce^{dx_n}} \tag{1}$$

$$x_n = \frac{\ln y_n}{ab + cd} \tag{2}$$

where:

y_n	nth number of sensor voltage value
x_n	nth number of sensor force magnitude
a, b, c & d	coefficient of characteristic curve
n	Sensor labelling = 1,2,3…

The graph obtained from 8 sensors using DAS and approximation curve using exponential curve fit the function as shown Fig. 3. The values of coefficient are determined and illustrated in Table 1. This multichannel DAS can currently read data from 8 × 8 array sensor or lower which consist of 64 different time varying signals. The system can be utilized to read more than 64 signals by replacing the off-shelf components into components with higher specifications. The goal of this research was to design a simple yet cost effective multichannel DAS. This system has high potential in several different applications such as prosthetic socket casting measurement system and biomechanics gait analysis. This DAS did not test for residual limb loading and was only tested on static loading on a flat surface. Although the front-end interface design is small and portable, the hardware lack suitable platform that can connect with prosthetic socket complex surface. The circuit development can be further simplified by implementing Printed Circuit Board (PCB) design or wireless module. This approach can be used to reduce the complexity of circuit connection to sensor. However, an algorithm for indicating damaging level of force application on sensor array need to be implemented in this system. This is important to guide clinician on how to identify location of excess force during casting.

4 Conclusion and Future Works

Based on this study, a smart multichannel pressure mapping system was designed, fabricated and tested. A system capable of acquiring signal from pressure sensor array and capable of visualizing pressure data from the sensor can be highly useful for biomedical pressure mapping application such as prosthetic socket fabrication. Performance of the system can be further improved by utilizing sub-components with higher specifications, incorporating noise filter, and integrated with auto calibration.

For future work, the system would need to have an algorithm to produce automatic alarm when a potential damaging force is applied on the sensor. A wireless module can be integrated to allow for wireless communication thus making it more portable and user friendly. More importantly, this could minimize the intertwined wire effects which normally degrade measurement accuracy.

Acknowledgements The authors gratefully acknowledge the funding from Postgraduate Research Grant (PPP) PG167-2015B and University of Malaya Research Grant (UMRG Program)—AET (Innovative Technology (ITRC)) RP009D-13AET from University of Malaya.

References

1. Li, J., Xu, B.: Novel highly sensitive and wearable pressure sensors from conductive three- dimensional fabric structures. Smart Mater. Struct. **24**(12), 125022 (2015)
2. Ahn, Y., Song, S., Yun, K.: Woven flexible textile structure for wearable power-generating tactile sensor array. Smart Mater. Struct. **24**(7), 75002 (2015)
3. Hanson, D., Thompson, P., Langemo, D., Hunter, S., Anderson, J.: Pressure mapping: A new path to pressure ulcer prevention. Wound Care Advis. **1**(1), 15–19 (2012)
4. Shu, L., Hua, T., Wang, Y., Qiao Li, Q., Feng, D.D., Tao, X.: In-shoe plantar pressure measurement and analysis system based on fabric pressure sensing array. IEEE Trans. Inf. Technol. Biomed. **14**(3), 767–775 (2010)
5. Tarchanidis, K.N., Lygouras, J.N.: Data glove with a force sensor. IEEE Trans. Instrum. Meas. **52**(3), 984–989 (2003)
6. Rovira, C., Coyle, S., Corcoran, B., Diamond, D., Stroiescu, F., Daly, K.: Integration of textile-based sensors and Shimmer for breathing rate and volume measurement. In: International Conference Pervasive Computer Technology Healthcare, vol. m, pp. 238–241 (2011)
7. Huang, C.T., Shen, C.L., Tang, C.F., Chang, S.H.: A wearable yarn-based piezo-resistive sensor. Sens. Actuators A Phys. **141**(2), 396–403 (2008)
8. Abdul Razak, A.H., Zayegh, A., Begg, R.K., Wahab, Y.: Foot plantar pressure measurement system: A review. Sensors (Switzerland) **12**(7), 9884–9912 (2012)
9. Ali, S., Osman, N.A.A., Mortaza, N., Eshraghi, A., Gholizadeh, H., Bin Wan Abas, W.A.B.: Clinical investigation of the interface pressure in the trans-tibial socket with Dermo and Seal-In X5 liner during walking and their effect on patient satisfaction. Clin. Biomech. **27**(9), 943–948 (2012)
10. Stoppa, M., Chiolerio, A.: Wearable electronics and smart textiles: A critical review. Sensors (Switzerland) **14**(7), 11957–11992 (2014)
11. Ahmad, M.Y., Alias, N.M., Hamzaid, N.A.: Fabric- based sensor for applications in biomechanical pressure measurement. In: Mohe 2017, pp. 1–4 (2016)
12. Tirosh, O., Begg, R., Passmore, E., Knopp-Steinberg, N., Wearable textile sensor sock for gait analysis. In: Proceedings of International Conference Sensing Technology ICST, pp. 618–622 (2013)

Development of Computer Aids ASPECTS System for Acute Ischemic Stroke Patient: A Preliminary Study

Jenn-Lung Su, Lung Chan, and S. Y. Huang

Abstract

In recent years, stroke ranked within the top ten leading causes of death and the incidence is still rising. As a result of clinical interpretation of Alberta Stroke Program Early CT Score (*ASPECTS*), the relevant personnel to define stroke area and score range are not consistent and cause difficulty to make treatment decision. This study was to develop a computer-aided scoring system for ischemic stroke patient to help doctors effectively determine the severity of ischemic stroke. Image processing technology was used to develop the system. First, an adaptive median filter was used to filter noise in computed tomography (*CT*) image, and then bi-level and regional growth methods were used to obtain effective image information. After texture parameters selection through t-test and support vector machine (*SVM*), regions of interesting (*ROI*) were automatically selected. Finally, the ischemic severity were obtained based on calculated ASPECTS score (by compared the left and right sides of the brain image). The CT images of 80 sets (40 training sets and 40 test sets) were used to evaluate the system by comparing with corresponding DWI-MRI. The results showed that the area under the ROC curve of the training sets and the test sets were 0.952 and 0.938, respectively, when four parameters (*autocorrelation, variance, maximum probability, and homogeneity*) were chosen. Accuracy was 0.90, sensitivity was 0.76, specificity was 1, and Kappa value was 0.52 for test data respectively, and the performance was superior to the physician group.

Keywords

ASPECTS • Acute ischemic stroke • Texture parameters • CT images

1 Introduction

According to statistics Ministry of Health and Welfare of Taiwan cause of mortality statistics, cerebrovascular diseases ranked fourth, and the total number of deaths (11,846) compared with previous year increased by 6.1% [1]. From 2005 to 2015, high percentage of death cause by ischemic stroke occurred within 1 h. In clinical practice, usually the incidence of Middle cerebral artery (MCA) ischemic stroke is higher than that posterior cerebral artery (PCA). In clinical practice, the doctor usually interprets the stroke thru ASPECTS [2] of x-ray CT and/or MRI images (see Fig. 1). The MCA territory is divided into 10 regions of interest (ROI) on two CT axial slices for ASPECTS calculation, and these 10 regions (Caudate, Insula, Lenticular Nucleus, Internal Capsule, and six other Cortical regions of MCA territory designated M1-M6) is shown in Fig. 1. The score is calculated by a total of 10 subtracting by the number of area that appears hypodense in brain CT image. Thus, "10" is normal and "0" for a completely infarcted MCA territory. A sharp increase in dependence and death with ASPECTS score of 7 or less. Thus, the region selection in image is quite

J.-L. Su (✉) · S. Y. Huang
Department of Biomedical Engineering, Chung Yuan Christian University, Taoyuan, Taiwan
e-mail: jlsu@cycu.edu.tw

L. Chan
Department of Neurology, Shuang Ho Hospital, Taipei, Taiwan

F. Ibrahim et al. (eds.), *2nd International Conference for Innovation in Biomedical Engineering and Life Sciences*,
IFMBE Proceedings 67, https://doi.org/10.1007/978-981-10-7554-4_35

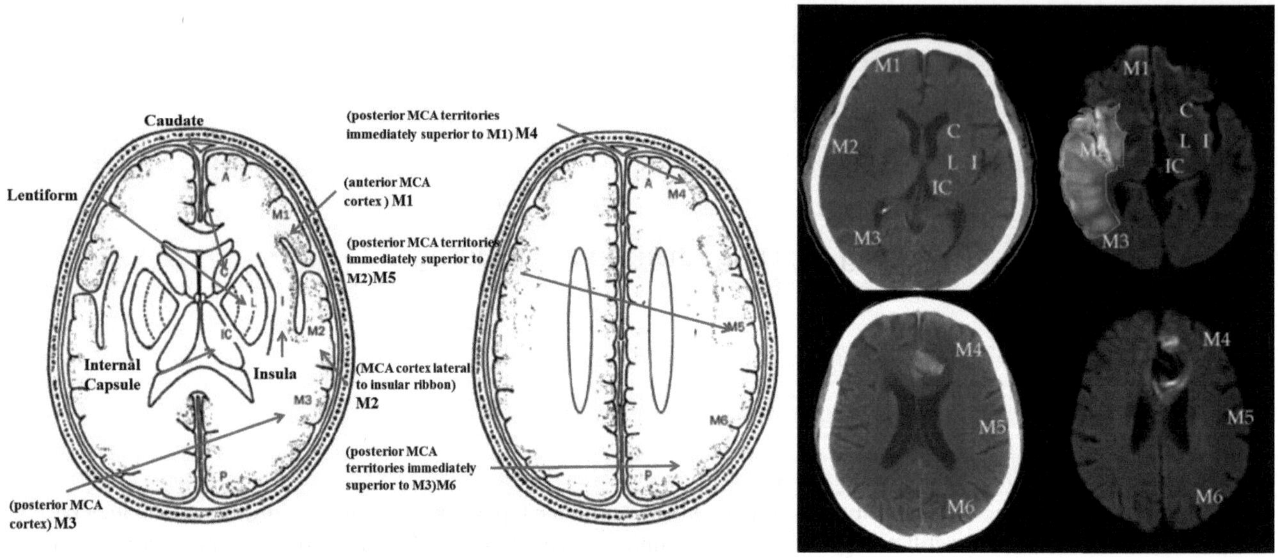

Fig. 1 Typical ASPECTS definition for Atlas, CT image, and MRI (from left to right)

important. Compared with MRI images, CT images are less obvious in showing the boundary of region, and difficult to define ASPECTS, but CT imaging is faster and easier to obtain [2].

In recent years many related research have been done [3–7]. Yongbum Lee group proposed the use of adaptive edge median filter rate of noise for image interpretation, and proposed ROC curve value to increase from 0.876 to 0.926 [3]. The same group also proposed the use of Z-score mapping to combine with ASPECTS score, and the ROC curve was improved from 82.6 to 86.6% [4]. N. Hema Rajini and R. Bhavani use the left and right brain symmetry axis to detect and analyze the image of texture parameters. The use of SVM (*Support Vector Machines*) for statistical analysis also improved specificity to almost 100%, accuracy and sensitivity of 98% [5]; Yeu Sheng Tyan group used three-dimensional curve contrast enhancement method for image enhancement, and used both canny edge detection algorithm and unsupervised regional growth algorithm for segmentation, resulting in increased specificity from 31 to 83% [6]. Berend C. Stoel et al. proposed the use of Brain CT image density Quantitative analysis of ASPECTS blocks, for the left and right sides of the brain density histogram comparison, and this identification system achieved 73% accuracy [7]. In our previous studies, we used NGLDM (*Neighboring Gray-Level Dependence Matrix*) combined with GLCM (*grayscale covariance matrix*) to extract texture feature quantization analysis and classified it with SVM, to develop acute ischemic stroke detection system. Preliminary results reported accuracy, sensitivity, specificity, kappa value of 1 for ischemic stroke detection in mid-brain slice image [8].

Based on those studies, several different image processing methods were used to develop a computer aids ASPECTS system in MCA CT image for acute ischemic stroke patients in this study. We described those methods in more detail in the next section.

2 Materials and Methods

2.1 Image Source and Software Develop Tool

Images in this retrospective study were from Department of Neurology, Shuang Ho Hospital, New Taipei City. Totally CT images and corresponding DWI-MRI of 80 sets were used (40 training sets and 40 test sets). There were 7 lesions and 33 normal cases in training sets, 17 lesions and 23 normal cases in test sets, respectively. There were 8 slices image in one image set and the thickness of each slice of image was 5 mm.

Microsoft Windows 7 was used as PC operating system, Microsoft Visual Studio C # 2010 was used for system programming development, ImageJ severed as validation tool, MATLAB 2013a is used for parametric statistical analysis.

2.2 Preprocessing

The flowchart of this developed system is showed in Fig. 2. Image pre-processing was done for original CT images to eliminate unwanted image information and avoid system execution noise disturbance, and unwanted text on the

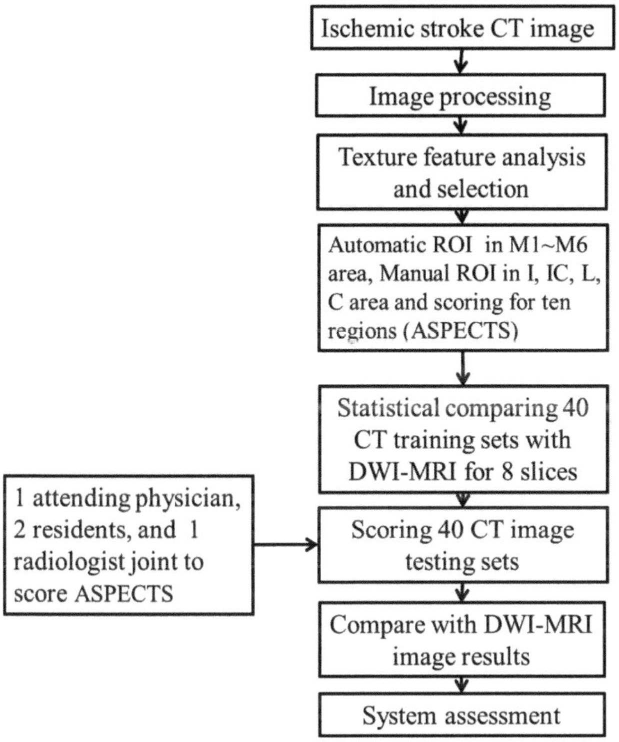

Fig. 2 Flowchart of developed system

image, or skull. In order to remove the noise in the non-enhanced CT image, an adaptive median filter [3] was used followed by region growing method to obtain the effective brain image edge information with optimal threshold [9].

2.3 Texture Feature Parameter Selection

Ten regions in image correspond to ASPECTS system, M1–M6 areas were split automatically and I, IC, L, and C areas were split manually. After selecting ROI (region of interest) region in the left and right sides of the brain tissue image, a 3 by 3mask was applied to ROI, and eight texture parameters (*autocorrelation, contrast, variance, energy, entropy, maximum probability, cluster prominence, homogeneity*) were created for image analysis [5, 7]. T-text was used to train the organizational parameters and the ROC (*Receiver Operating Characteristic*) curve to analysis performance of system. Based on the results, the number of texture parameters can be reduced. The texture parameters for SVM analysis was used to determine whether the region was ischemic. Each region was scored as 0 or 1. After the interpretation of the ischemic region, the individual scores of 10 regions were obtained, and the scores of the eight image slices were selected to obtain the ASPECTS value of the set of images.

2.4 System Assessment

In clinical treatment procedure, the score of ASPECT grade (<7 or ≥ 7) would determine the need for rt-PA injection, thus the system we designed could interpret a case as positive (ASPECT grade < 7, not recommend to treat with rt-PA injection) or negative case, respectively. Because MRI is more sensitive to find acute lesion(s) in acute stroke [10], the score obtained from DWI-MRI is used as standard in this study. By compared the counting score between system and DWI-MRI, the decision matrix of this system was obtained. And then the accuracy, sensitivity, specificity, and kappa [8, 11] of this system were calculated respectively. In order to evaluate the practicality of system, we also compared the scores counted from four physicians (attending physician, radiologists, first year hospital Physician (*R1*), and second year resident (*R2*)) with scores counted from DWI-MRI.

3 Results and Discussion

In this study, texture parameters with P value less than 0.05 were selected to increase system sensitivity in effectively selecting ischemic brain tissue by using only 6 parameters (*autocorrelation, variance, energy, entropy, maximum probability, and homogeneity*). After analyzing the left and right texture parameters through SVM analysis and using ROC curve analysis of its texture parameters of the identification efficiency, we found that the AUC (*area under curve*) value using four parameters (*autocorrelation, variance, maximum probability, and Homogeneity*) was 0.952, which was greater than 0.942 (AUC value) for using the original six texture parameters for the training set image (Fig. 3). We found similar findings for test data set. In Fig. 3, blue line represents 4 parameters results and red line represents 6 parameters results.

By assuming the score from DWI-MRI as a golden standard, systematic interpretation grade (<7 or ≥ 7) of 40 training sets and 40 test sets with MRI corresponding image are shown in Table 1. The comparison of the grade obtained between DWI-MRI and attending physician, radiologist, first year resident, second year resident for test set image, are also shown in Table 1. According to the results of decision matrix for training data (*system training 1 in the table*) and testing data (*system test 1*), the accuracy of the system were 0.925, 0.90, sensitivity were 0.714, 0.76, specificity were 0.97, 1, and Kappa value were 0.728, 0.52, respectively.

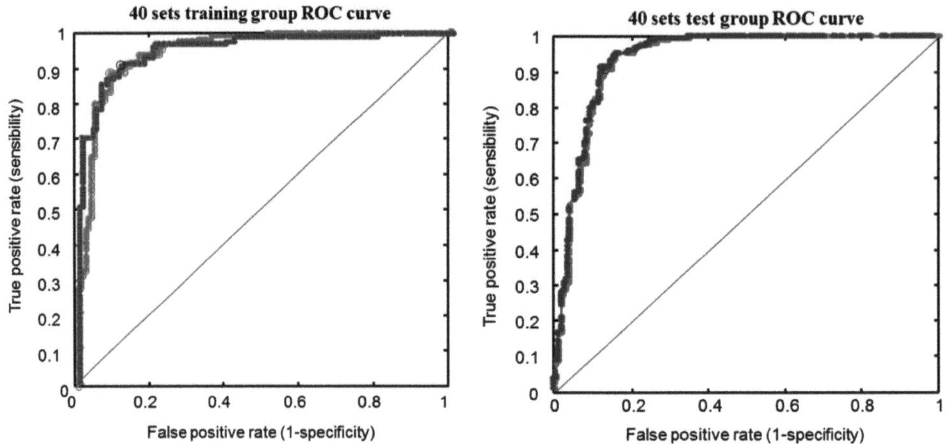

Fig. 3 ROC curves of texture parameters with chosen 4 (in blue) and 6 (in red) were used for 40 sets of training data (left), and 40 sets of test data(right)

Table 1 System performance and compared with physician group

DWI-MRI[a]		P	N	Accuracy	Sensitivity	Specificity	Kappa
System training 1	P	5	1	0.9250	0.7143	0.9697	0.7248
	N	2	32				
System training 2	P	5	0	1.0000	1.0000	1.0000	1.0000
	N	0	33				
System test 1	P	13	0	0.9000	0.7647	1.0000	0.5263
	N	4	23				
System test 2	P	13	0	0.9730	0.9286	1.0000	0.9417
	N	1	23				
Neurology Attending physician	P	15	9	0.7250	0.8824	0.6087	0.4660
	N	2	14				
Radiologist	P	8	2	0.7250	0.4706	0.9130	0.4054
	N	9	21				
2 years resident	P	5	3	0.6500	0.2941	0.9130	0.2244
	N	12	20				
1 years resident	P	5	2	0.6250	0.2941	0.8696	0.0179
	N	12	21				

[a]DWI-MRI score severed as a golden standard, class P and N is determined by ASPECTS (<7 or ≥ 7). i.e. *P* ASPECTS < 7, *N* ASPECTS ≥ 7 and rt-PA treatment is recommended

During test, we found that one type of misjudged case was resulting the old lesion, and another type of misjudged case was resulting both sides have ischemic lesion. Both cases could be excluded by physician with prescreening the image or patient record. With prescreening step applied in this study, the system performance improved in accuracy was 0.973, sensitivity was 0.929, specificity was 1.0, and Kappa value was 0.942 respectively (*system test 2 in Table* 1). Moreover, for the performance test of this system is seem superior to the physician group. These results confirmed the practicality of this developed system.

4 Conclusion

In this preliminary study, we found that the use of autocorrelation, variance, maximum probability, homogeneity of the texture parameter analysis was better at identifying the presence of ischemic blocks from brain CT images. From the result of decision matrix, we also found that with or without prescreening, the performance of system was better than conventional methods. In this stage, only MCA brain CT image was used. PCA CT image will be added in future study with bigger sample size. Moreover, in order to improve this system, the prescreening stage is taken into account as a standard screening step.

Acknowledgements This work was supported the National Science Council, R.O.C. under Grant MOST 105-2221-E-033 -048.

References

1. Taiwan Ministry of Health and Welfare, http://www.mohw.gov.tw/cp-16-33598-1.html 2017/7/28
2. https://teddybrain.wordpress.com/2013/03/16/alberta-stroke-program-early-ct-score-aspects-in-ischemic-stroke/ & http://www.hubstroke.com/en/aspects/ accessed on 2017/7/28
3. Lee, Y., Takahashi, N., Tsai, D.-Y.: Adaptive partial median filter for early CT signs of acute cerebral infarction. Int. J. Comput. Assist. Radiol. Surg. 105–115 (2007)
4. Lee, Y., Takahashi, N., Tsai, D.-Y.: Z-score mapping method for extracting hypoattenuation areas of hyperacute stroke in unenhanced CT. Acad. Radiol. 84–92 (2009)
5. Rajini, N.H., Bhavani, R.: Computer aided detection for ischemic stroke using segmentation and texture features. Measurement 1865–1874 (2013)
6. Tyan, Y.S., Wu, M.C., Chin, C.L., Kuo, Y.L., Lee, M.S., Chang, H.Y.: Ischemic stroke detection system with a computer-aided diagnostic ability using an unsupervised feature perception enhancement method. Int. J. Biomed. Imaging ID947539. 1–12 (2014)
7. Stoel, B.C., Marquering, H.A., Stating, M., Beenen, L.F., Slump, C.H., Roos, Y.B., Majoie, C.B.: Automated brain computed tomographic densitometry of early ischemic change in acute stroke. J. Med. Imaging **2**(1), 014004–014004 (2015)
8. Hu, S.-C.: Texture analysis for aided diagnosis of hemorrhage transformation of acute middle ischemic stroke in CT images, Master Thesis, Chung Yuan Christian University, Chungli (2012)
9. Gonzalez, R.C., Woods, R.E.: Digital Image Processing 3/e, Pearson education Taiwan and Gau Lin Book Co. (2009)
10. Lassalle, Louis: ASPECTS (Alberta Stroke Program Early CT Score) assessment of the perfusion–diffusion mismatch. Stroke **47** (10), 2548–2554 (2016)
11. Hebel, J.R., McCarter, R.J.: Epidemiology and biostatistics, Jones & Bartlett Learning, Burlington (2011)

Testing the Effect of H-field Using Biological Phantom

N. Z. Syahir, Mas S. Mohktar, Sami F. Khalil, M. R. Basar, and M. Y. Ahmad

Abstract

The purpose of this paper is to propose a method to test the biological phantom after an exposure of the electromagnetic field (H-field). The phantom is believed to be influenced by the H-field created by wireless power transfer procedure. In the investigation, the phantom has been exposed for about 150 min. Two parameters were measured from the phantoms which are the impedance and the temperature before and after the exposure of the H-field. From the outcome, we can identify whether there are any changes detected in the assigned frequency. Both parameters change when exposed in the H-field but there are no significant changes after a duration of time exposed in the H-field zone indicates this range of frequency is safe to be used upon the biological phantom. This biological phantom comprises of the cucumber, saline and corn oil. In this test, it is utilizing 20% of corn oil and 80% of saline inside the cucumber as to show the arrangement of fat in typical human body. Since the changes is very low, this WPT setup is safe for testing the biological tissues. Nevertheless, as the complexity of the biological tissues, there are still many aspects to be considered and highlighted as this study play an important role to be referred for the next study about the effect of H-field towards the biological phantom.

Keywords

Phantom • Impedance • Temperature • Electromagnetic field

1 Introduction

Wireless power transmission (WPT) [1] utilizes electromagnetic energy to power the implantable therapeutic devices. WPT system works based on the inductive coupling between primary (transmitting) coil and secondary (receiving) coil. The magnetic field strength, known as H-field produces by the transmitting coil interacts with the receiving coil and induces power in it. The electrical properties of human tissue, to be specific conductivity and permittivity, will be increased by the increasing of the frequency because of human tissue manner that absorbing the electromagnetic waves [2]. H-field is a non-ionizing radiation however it is enough to cause thermal heating. Tissue warming could be happened at over 100 kHz. Thus, the H-field values need to be controlled to avoid any damage on human tissue that being exposed by the WPT. This can be done by controlling the frequency, the amplitude and the time of exposure [3].

Many studies have been conducted to evaluate the safety of WPT exposure on living tissues. One of the study is using computer simulation models [4]. Moreover, most of the studies, measured the specific absorption rate (energy absorption) of the exposure which does not reflecting the wellbeing of the human body itself.

Thus, in this paper, we are proposing a method to study the effect of H-field on composition of biological phantom to

N. Z. Syahir (✉) · M. S. Mohktar · M. Y. Ahmad
Department of Biomedical Engineering, Faculty of Engineering, University of Malaya, Kuala Lumpur, Malaysia
e-mail: syahirzahidin@gmail.com

M. S. Mohktar · S. F. Khalil · M. R. Basar · M. Y. Ahmad
Department of Biomedical Engineering, Faculty of Engineering, Center for Innovation in Medical Engineering, University of Malaya, Kuala Lumpur, Malaysia

F. Ibrahim et al. (eds.), *2nd International Conference for Innovation in Biomedical Engineering and Life Sciences*, IFMBE Proceedings 67, https://doi.org/10.1007/978-981-10-7554-4_36

ensure the safety of the WPT system before it can be applied to either animal or human.

2 Method and Materials

The procedure starts by preparing the phantom and measure two properties (impedance and temperature) before and after exposing it in the H-field for about 150 min. The biological phantom that had been used are composition of cucumber, saline and corn oil [5]. The WPT system was then been set up to a certain frequency. The phantom was placed in the middle of the coil and was exposed in the H-field in a certain period. The temperature and impedance of the phantom were recorded before and after the exposure. Another similar phantom was also prepared as a control phantom for reference (no introduction to the H-field). The surrounding temperature was also controlled in the range between 24 and 26 °C.

2.1 Phantom Preparation

The composition of the phantom is 20% of corn oil and 80% of saline so as mimic human body fat percentage.

2.2 Impedance Measurement

The impedance measurement was taken using AD5933 Evaluation Kit. This is Multi-Frequency Impedance Analyzer.

Four electrodes were connected to the phantom. Two for current infusion (external electrode) and another two for voltage reaction (inward electrodes). The separation of the "external" electrode was ¼ of the length of the depth of the gap at each one end. For "internal" electrode the separation was the same for both terminals. For this experiment, the dimensions of electrodes size were fixed to 2 × 4 cm.

2.3 H-field Generator

The WPT system consists of a cylindrical shape of transmitting coil (TC) with a diameter of 40 cm. A driving/tuning circuit was used with the TC to supply sufficient current and to tune the TC at the desired frequency. The current flows throw the TC produce H-field inside the TC. This experiment demonstrated H-field for two frequencies; 300 and 400 kHz.

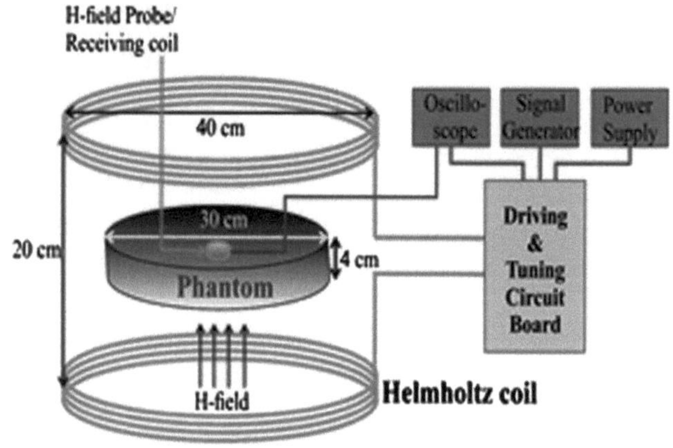

Fig. 1 Setup of the H-field generator with the phantom at the exposure area

2.4 Experimental Setup

The setup for the H-field generator is illustrated in Fig. 1 with the phantom is replaced with biological phantom used in this experiment. The impedance of the phantom was measured using AD5933 assessment board software. Then, the phantom was exposed to the H-field for around 150 min.

The temperature was taken using the infra-red thermometer. The distance of the thermometer to the phantom was fixed at 10 cm.

Two temperatures were measured, the phantom wall and the liquid inside the cucumber temperatures. At the point when taking out the phantom from the H-field to gauge the temperature, it must be taken promptly so that the encompassing temperature did not influence the phantom temperature.

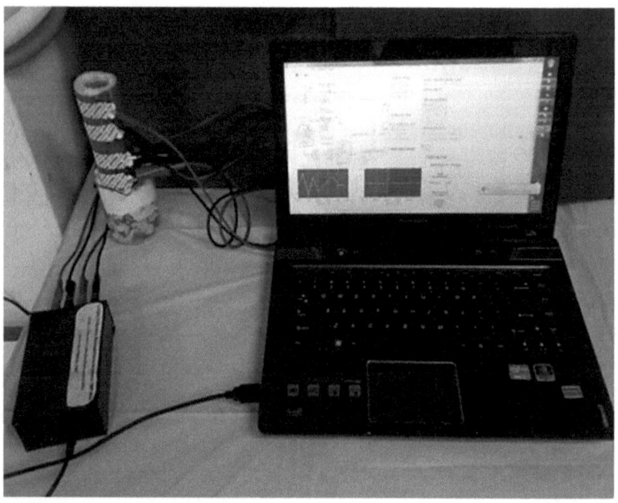

Fig. 2 Setup of the BIA

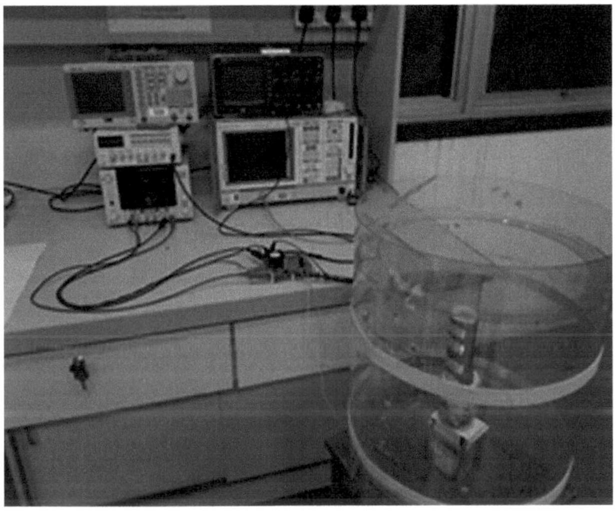

Fig. 3 Setup of the experiment. The phantom will be exposed in the coil fix 150 min

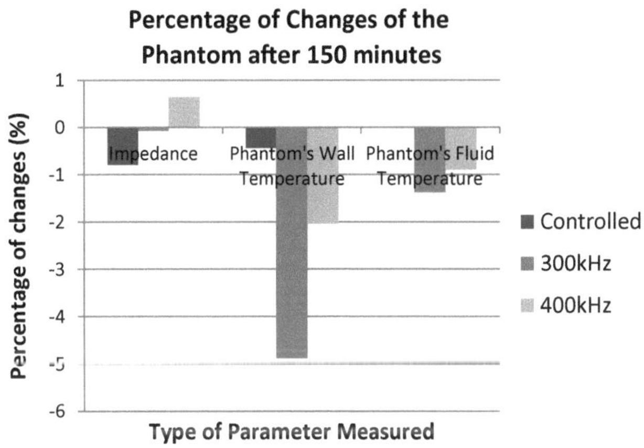

Fig. 4 Graph of the percentage of changes for impedance, phantom's wall temperature and phantom's fluid temperature

Finally, the impedance of the phantom was measured again. The actual experimental set-up is shown in Figs. 2 and 3.

3 Results and Discussion

The changes for both impedance (ohm) and the temperature (°C) were recorded. The negative result indicates that the temperature or the impedance decreased from the initial value of the beginning of the exposure. Usual capsule endoscopy work for about 8 h [6] and more advance capsule need more power and sometimes the capsule needs more than 8 h unless the image and condition in the GI tract after 8 h cannot be examined and analyzed accurately. For that reason, the biological phantom is exposed for 150 min which equivalent 2 h and 30 min as the starting and as the precaution to not to damage the phantom. In the future, it will be exposed for longer time until it can stand until at least 8 h and above.

From the graph in Fig. 4, it can be observed that the temperature using 400 kHz has less changes in wall temperature than the 300 kHz and the control phantom. A study by Xin [1] mentioned that increase in frequency can cause the increase in the temperature of the phantom. Because of that concern, the temperature is chosen as one of the study in this experiment. But it is depend on the purpose of the study, some of other researchers [7] and others used high frequency up till Mega Hertz and above. The frequency in this study, was decided to match the WPT system that requires the frequency to be inside the range of the value of beneath 400 kHz.

As we see the graph in Fig. 4, the changes of the impedance value are not more than 5% for the 3-tested phantoms. The changes of the impedance of the phantom is not more than 1% which is very low. The exposure of frequency with 300 kHz does not cause the impedance to change so much compare to when the exposure in the field using of 400 kHz. Even that, it is still show less than 1% changes. Since the controlled phantom also did not show much changes, it can be deduced that both selected frequency does not affect the phantom's impedance.

The electrical properties (dielectric constant and conductivity) which is not being studied in this experiment must be closely imitate the properties of human tissues over the range of the frequency. That is very important so that they can imitate the interactions of the electromagnetic waves with the biological tissue [8]. This biological phantom is the homogenous and straightforward phantom. There are more things to consider to precisely emulate heterogeneous organs to become accessible economically. We need the phantom to be as simple and cheap as we can so that more test can be done to study more about the biological phantom. The cost for other phantom can go from hundreds to a thousand dollars. However ordinarily there are normally intended for expensive markets and specific and detail applications, and are not adjustable [9]. Those are the motivation behind why this kind of biological phantom set up is chosen to be studied.

Nonetheless, this phantom arises the question in term of which body section is referred to. Since this sort of phantom is based on the human body, more detail of body portion must be made which is the gastrointestinal tract within the area of abdomen of human body to test the device and WPT system.

3.1 Conclusion

The study proven that there was very small change in temperature and impedance of the biological phantom when exposed to low frequencies radio signal. The structure utilized as phantom consisted of 20% corn oil and 80% saline to mimic human body. The frequencies were 300 and 400 kHz. This was based on the WPT system and wireless endoscope-capsule that utilized the range of frequency between 200 and 400 kHz. These range of frequencies were considered safe for human because they did not significantly alter the impedance and temperature of the biological phantom. Our findings may not represent well for human body that consisted of multiple tissue density and thickness. Future study on other tissue properties is recommended. This study is relevant in measuring the effect of using a biological phantom when exposed to H-field. For the future study, we will fabricate more intricate phantom and better materials such as synthetic phantom like gelatin based phantom or fresh meat and setup by increase the H-level or the frequency that may be utilized to test the WPT system.

Acknowledgements Postgraduate Research Grant (PPP), University of Malaya, Kuala Lumpur, Malaysia.

References

1. Xin, W., Yan, G., Wang, W.: Study of a wireless power transmission system for an active capsule endoscope. The Inter. J. Med. Robot. Comput. Assist. Surg. **6**, 113–122 (2010)
2. Chirwa, L.C., Hammond, P.A., Roy, S., Cumming, D.R.: Electromagnetic radiation from ingested sources in the human intestine between 150 MHz and 1.2 GHz. IEEE Trans. Biomed. Eng. **50**, 484–492 (2003)
3. Peterson, D.M.: Tissue Equivalent Phantom Development for Biomedical Applications. University of Florida, Gainesville, FL (2009)
4. Yuan, Q., Chen, Q., Li, L., Sawaya, K.: Numerical analysis on transmission efficiency of evanescent resonant coupling wireless power transfer system. IEEE Trans. Antennas Propag. **58**, 1751–1758 (2010)
5. He, Q., Wang, J., Engelson, E.S., Kotler, D.P.: Detection of segmental internal fat by bioelectrical impedance analysis in a biological phantom. Nutrition **19**, 541–544 (2003)
6. Pan, G., Wang, L.: Swallowable wireless capsule endoscopy: progress and technical challenges. Gastroenterol. res. pract. **2012** (2011)
7. Christ, A., Douglas, M., Nadakuduti, J., Kuster, N.: Assessing human exposure to electromagnetic fields from wireless power transmission systems. Proc. IEEE **101**, 1482–1493 (2013)
8. Mohammed, B., Abbosh, A., Henin, B., Sharpe, P.: Head phantom for testing microwave systems for head imaging. In: 2012 Cairo International Biomedical Engineering Conference (CIBEC), pp. 191–193 (2012)
9. Culjat, M.O., Goldenberg, D., Tewari, P., Singh, R.S.: A review of tissue substitutes for ultrasound imaging. Ultrasound Med. Biol. **36**, 861–873 (2010)

Design Performance of Taylor Spatial Frame in Comparison with Other Orthopaedic Fixation Plate

Ilham Priadythama, Lobes Herdiman, Durkes Herlina Apriani, and I Dewa Nyoman Suci Anindya Murdiyantara

Abstract

Treatment options for limb deformities were delevoped through time and Taylor Spatial Frame was indicated as a Stewart Platform based external fixation system that superior in terms of correction and ease of use. New external fixator (i.e. Ross's external fixator) was created as solution from Taylor Spatial Frame's problems. Several changes on Taylor Spatial Frame design were made to create Ross' external fixator. However, Ross' frame may have stability issue as consequence from Taylor's design changes. Comparation of these frames is needed to develop limb deformities treatment. This paper is focused on the design performance of external fixator's rings based on Stewart Platform. Taylor Spatial Frame's and Ross Frame's ring are compared and evaluated in design form. Both ring models are presented using Autodesk Inventor 2016. Numerical Modelling (i.e. Finite Element Analysis) used for compare both rings design performance. The comparation results will be useful for further research in ring fixation product development.

Keywords

Taylor spatial frame • Ross frame • Finite element analysis

1 Introduction

Limb deformities are not only most frequently caused by abnormal bone growth but also fracture malunion. There are many conventional treatment options for limb deformities but none of these methods give predictable results [1]. In early 20th century Lambotte created the first external fixation for limb deformities [2]. Early type external fixator was developed through time until Professor Gavril Ilizarov created ring fixators [3], a stainless steel rings connected with threaded rods that can be configured in various ways to manage different indications [4]. These external fixator construct and method of fixation provides favorable mechanical and biological environment for bone healing [5, 6]. More recent development of these fixator, Taylor Spatial Frame was created as a Stewart Platform based external fixation system that can produce rotation and translation mechanism to treat simple and complex deformities [7].

There are many data proving that Taylor Spatial Frame is a superior instrument in terms of correction and ease of use [8–13]. Taylor Spatial Frame consists of two external fixation rings attached to bone segments by wires and connected together by six telescopic struts-that may be lenghtened and shortened-with multi planar hinges located at both ends of strut. Several problems found when using Taylor Spatial Frame, include wire positioning, strut length adjustment and time consuming exchange of the struts [14]. In 2013, John D. Ross developed a new external fixator using ball joint instead of universal joint to solve problems that found in Taylor Spatial Frame [14]. However with these change, this external fixator has to re-design Taylor Spatial Frame in struts as well as rings. Ross external fixator may be the new

I. Priadythama (✉) · L. Herdiman · D. H. Apriani
Laboratory of Product Planning and Design, Industrial Engineering Department, Universitas Sebelas Maret, Surakarta, Indonesia
e-mail: priadythama@gmail.com

I. D. N. S. A. Murdiyantara
Orthopedic Surgery and Traumatology Department, Dr. Soeradji Tirtonegoro Hospital, Klaten, Indonesia

F. Ibrahim et al. (eds.), *2nd International Conference for Innovation in Biomedical Engineering and Life Sciences*, IFMBE Proceedings 67, https://doi.org/10.1007/978-981-10-7554-4_37

way to give a better limb correction in orthophaedic surgery. However, Ross' ring has additional side-holes as a consequence of more flexible sturts end mounting instead of holes for key wire installation. Therefore, there will be a possibility of stability issue which caused by ring's strength decrement (Fig. 1).

There are many data comparing Ilizarov and Taylor Spatial Frame in biomechanical, mechanical and real cases of correction using these fixators [4, 8, 9, 16] but there are no data that comparing Taylor Spatial Frame and Ross Frame. This paper report on design performance of Taylor Spatial Frame compared with Ross frame. However, this paper is focused on the ring fixators because frame stability is greatly impacted by ring properties [17]. Numerical modelling used as method to compare fixators that are still in a design form [18]. The comparation results will be useful for further research in ring fixation product development.

2 Methods

Literature review on Taylor's and Ross's frame were done. Designs of Taylor Spatial Frame and Ross Frame were created and tested using Autodesk Inventor 2016.

a. Select inner ring diameter for Taylor Spatial Frame comparable Ross' ring design with the selected ring size.
b. Create details design of both rings using Autodesk Inventor 2016.
c. Select similar material for both rings.
d. Select similar key wire attachment position for both rings.
e. Create static model for determining external load which is exerted to the ring (Inventor's environment features).
f. Input material data, constraints and loads for stress analysis.
g. Set mesh parameters and apply it for the stress analysis.
h. Run simulation.

3 Results

3.1 Ring Fixator Models

Ring fixator models were created using Autodesk Inventor 2016. Created designs were full ring-type with internal diameter 155 mm as a moderate ring internal diameter. Aluminum 6061 T6 was used as main material, as Taylor Spatial Frame used the same material for its rings [4], and Ross Frame had no concern with its material [14].

Taylor's ring had through all-simple holes for key wire installation concentric to inner ring diameter, and through all-simple holes for struts end mounting concentric to outer ring diameter. Unlike Taylor, Ross' strut end holes were tapped in certain distance that was perpendicular to key wire holes.

Ross' ring was slightly modified in this research for manufacturing consideration. We cut off Ross design in such a way that the design is as illustrated in Fig. 2. This modification did not give significant differences to Ross ring.

3.2 Numerical Modelling with Finite Element Method

As for numerical modeling, we selected key wire and strut end attachment positioning. Key wire attachment positioning

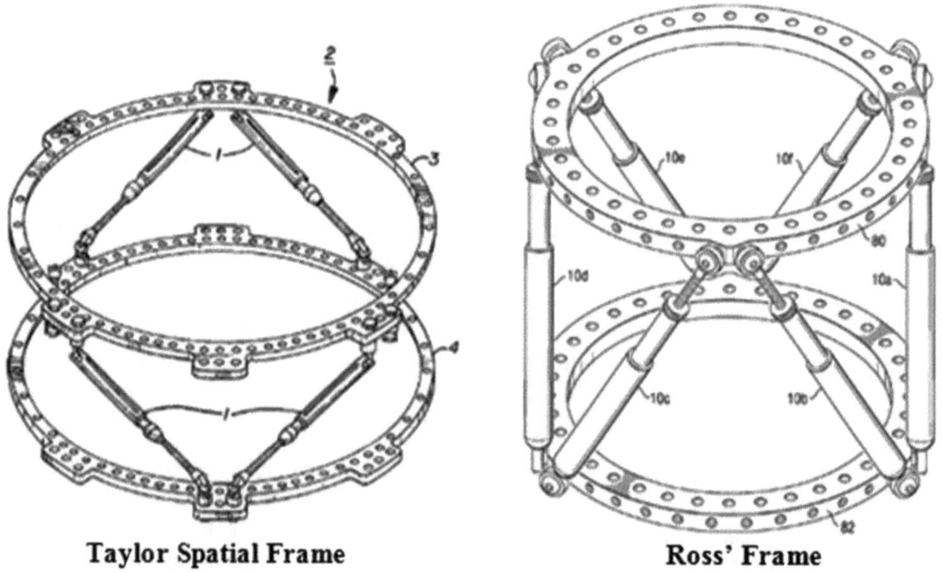

Fig. 1 Stewart platform based external fixator [14, 15]

Fig. 2 Ring fixator models

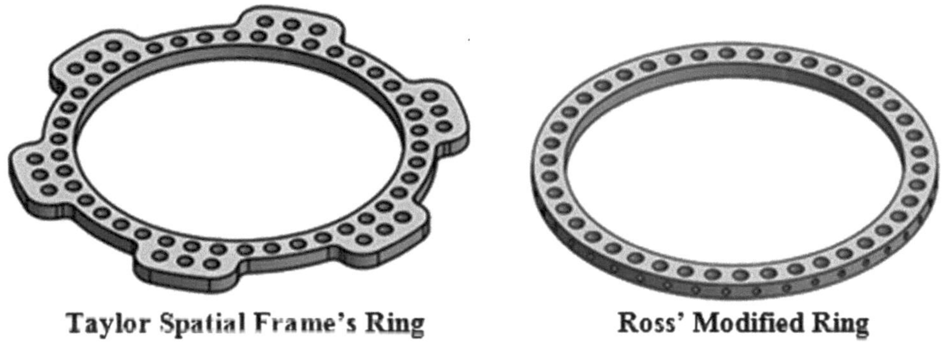

was indicated by wire fixation bolts mounted on inner ring holes, while strut end attachment positioning was indicated by shoulder bolts mounted on outer ring holes for Taylor's ring and perpendicular tapped holes for Ross's ring. Key wire positioning and strut end attachment for each ring placed in the most approaching position form for more reliable results (Fig. 3).

External load was determined from wire tensioning and weight given from normal human body. The reference force of the wire strain in the ring is 900–1100 N [19] and axial load applied to 700 N as maximum weight load for average human weight [20]. In this study we used maximum wire strain and average human weight. Wire load was placed in line with key wire attachment position while weight load was placed in wire fixation bolts and perpendicular with wire load.

Strut end attachment positioning was set to be constraint to ring. Following that, ring models were imported for stress analysis using Finite Element Analysis. In this analysis, meshes were created. Sample of stress analysis results can be seen in Fig. 4. Results of stress analysis comparison is presented in Table 1.

Von mises stress was used to check if fixator ring's design could withstand a given load. Both rings were considered safe because their maximum von mises stresses were below their material strength (Aluminum 6061 T6 yield strength = 386 Mpa [21]). Ross' ring had larger von mises stress value than Taylor's ring, which meant that Taylor's ring was more capable at withstanding higher amount of load.

The 1st principal stress helps you understand the maximum tensile stress induced in the part due to loading conditions while the 3rd principal stress helps you to understand the maximum compressive stress induced in the part due to loading conditions. Both rings maximum 1st principal stress and 3rd principal stress were still below their material strength and therefore both rings were considered safe. Ross' ring had larger tensile and compressive stress value than Taylor's ring, which meant that Taylor's ring was more capable at withstanding higher amount of load.

Stability is the main criteria to generate an accurate limb reconstruction. Less displacement of an object represents higher stability. Table 1 shows that Ross' ring had bigger displacement with 0.0001 mm deviation. This small measurement difference however, may cause a long-term effect to limb reconstruction surgery. Taylor's frame may give better accuracy than Ross.

Fixator rings must be designed to withstand a "design overload" larger than the normally expected load. Safety

Fig. 3 Key wire and strut end attachment positioning

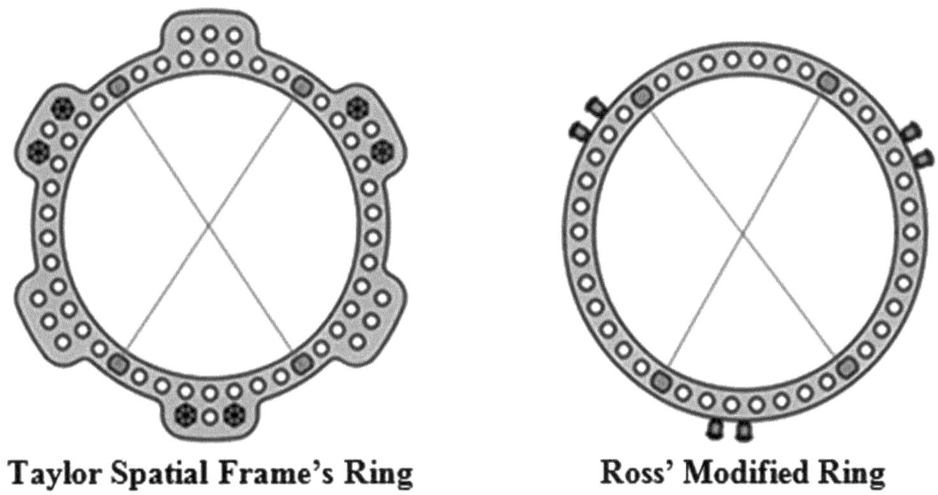

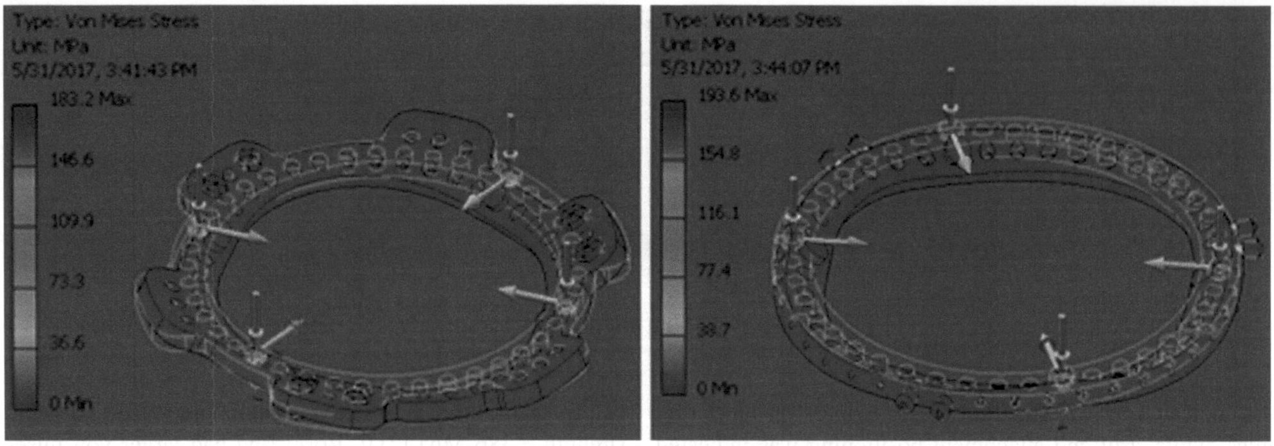

Fig. 4 FE modelling for both ring in term of Von Mises Stress

Table 1 Results of stress analysis—Taylor's ring versus Ross' ring

Attributes	Unit	Taylor's ring	Ross' ring
Max Von Mises stress	Mpa	183.2	193.6
Max 1st principal stress	Mpa	179.8	218.3
Max 3rd principal stress	Mpa	61	64
Max displacement	Mm	0.09712	0.09713
Min safety factor	Ul	1.71	1.62

factor allows you to evaluate the design ability to withstand amount of load applied to that design [22]. Modern engineering design gives a rational safety factor within the range of 1.25–4 [23]. Based on that, Taylor's and Ross' rings were still able to withstand normal load, although Ross' ring had lower safety factor than Taylor's frame.

Taylor's and Ross' rings had some advantages and disadvantages. Ross' frame may provide the solutions on Taylor's frame wire positioning, strut length adjustment and time-consuming exchange of struts [14]. But Taylor's ring provides better design performance, especially in providing stability. Stability is important in limb reconstruction because it has major effect on bone correction success.

Using ball joint as a replacement for universal joint is an interesting idea to solve Taylor's frame problems. However, there is a deep concern on stability issues especially on ring development. This study finding may be useful for further research in ring fixation product development.

4 Conclusion

Both frames have their own advantages and disadvantages. In term of flexibility of use, Ross' frame may offer better flexibility, but in term of frame stability, Taylor's frame is better. However, both frames still provide enough safety for normal loading.

References

1. Marcellin-Little, D.J.: Treating bone deformities with circular external skeletal fixation. Compendium **21**(6) (1999)
2. Karunakar, M.A., Bosse, M.J.: Principles of external fixation. In: Bucholz, R.W., Heckman, J.D. (eds.) Rockwood and Green's Fractures in Adults, 5th edn, pp. 231–244. Lippincott Williams & Wilkins, Philadelphia, PA (2001)
3. Ilizarov, G.A.: New Principles of Osteosynthesis by Means of Crossing Pins and Rings. Internal Publication Book, Kurgan, Russia (1954)
4. Tan, B.B., et al.: A biomechanical comparison between Taylor's spatial frame and Ilizarov external fixator. Malays. Ortho. J. **8**(2), 35–39 (2014)
5. Matthews, S.J., Nikolaou, V.S., Giannoudis, P.V.: Innovations in osteosynthesis and fracture care. Injury **39**(8), 827–838 (2008)
6. Fleming, B., Paley, D., Kristiansen, T.: A biomechanical analysis of the Ilizarov external fixator. Clin. Orthop. Relat. Res. **241**, 95–105 (1989)
7. Taylor, J.C.: Correction of general deformity with the Taylor Spatial Frame fixator. http://www.jcharlestaylor.com/. Accessed 12 June 2016
8. Kristiansen, L.P., Steen, H., Reikeras, O.: No difference in tibial lenghtening index by use of Taylor Spatial Frame or Ilizarov external fixator. Acta Orthop. **77**(5), 772–777 (2006)
9. Manner, H.M., et al.: Accuracy of complex lower-limb deformity correction with external fixation: a comparison of the Taylor Spatial Frame with the Ilizarov ring fixator. J. Child. Orthop. **1**(1), 55–61 (2007)
10. Naqui, S.Z., et al.: Correction of simple and complex pediatric deformities using the Taylor-spatial frame. J. Pediatric Orthopaedics **28**(6), 640–647 (2008)

11. Rozbruch, R., et al.: Does the Taylor Spatial Frame accurately correct tibial deformities? Clin. Orthop. Relat. Res. **468**(5), 1352–1361 (2010)
12. Sluga, M., et al.: Lower limb deformities in children: two-stage correction using the Taylor Spatial Frame. J. Pediatr. Orthop. Part B **12**(2), 123–128 (2003)
13. Taylor, J.C.: Six-axis deformity analysis and correction. In: Paley, D. (ed.) Principles of Deformity Correction. Springer, Berlin (2002)
14. Ross, J.D., et al.: External fixation strut. US Patent 8,439,914 B2 (2013)
15. Taylor, H.S., Taylor, J.C.: Six axis external fixator strut. US Patent 6,030,386 (2000)
16. Bas, A., et al.: Biomechanical comparison of spatial frames with different strut models and Ilizarov type external fixators. http://www.llrs.org/. Accessed 3 June 2016
17. Fragomen, A.T., Rozbruch, S.R.: The mechanics of external fixation. J. Hosp. Spec. Surg. **3**(1), 13–29 (2007)
18. Frydrýšek, K., et al.: Design of external fixators used in traumatology and orthopaedics—treatment of fractures of pelvis and its acetabulum. Proced. Eng. **48**, 164–173 (2012)
19. Solomin, L.N.: The Basic Principles of External Skeletal Fixation Using the Ilizarov and Other Devices, 2nd edn. Springer, Italia (2010)
20. Gessman, J., et al.: The influence of a weight-bearing platform on the mechanical behavior of two Ilizarov ring fixators: tensioned wires vs. half-pins. J. Orthop. Surg. Res. **6**(61) (2011)
21. Aluminum 6061-T6; 6061-T651. ASM homepage. http://asm.matweb.com/search/SpecificMaterial.asp?bassnum=ma6061t6. Accessed 31 May 2017
22. Ullman, J.D.: Compilers, Principles, Technique, and Tools. Addison-Wesley Publishing Company, Boston, Ma (1986)
23. Juvinall, R.C., Marshek, K.M.: Fundamentals of Machine Component Design, 6th edn. Wiley, Hoboken, NJ (2012)

The Effects of Oral Glutathione Supplementation on Human Volunteers: An Observation on Metabolic Syndrome Status

Nur Rasyidah Hasan Basri, Mas Sahidayana Mohktar, Wan Safwani Wan Kamarul Zaman, and Hasif Ilyasa Mohd Yusof

Abstract

Oral glutathione supplementation helps in improving metabolic syndrome status of human volunteers. Glutathione is a cluster of amino acid containing one molecule of L-glutamic acid, L-cysteine, and glycine each. It act as a master antioxidant that combat free radicals in cells and help in detoxification of chemical from our body. This study aim to observe the effects of glutathione on metabolic syndrome symptoms after 12 weeks of supplementation. Reflection of the effects were carried on fasting glucose, blood pressure, cholesterols, waist to hip ratio, body fat percentage and basic measurement such as body mass index and blood pressure. Volunteers were divided into two groups; Group 1: One dosage of 5 g supplement, Group 2: Two dosage of 5 g supplement. A total of 23 volunteers' have been recruited for the study. Based on the metabolic syndrome diagnosis we studied, out of 23, six subjects were found to have metabolic syndrome. After 12 week of supplementation, based on the parameter studied, there seems to be an improvement in the level of HDL cholesterol, body mass index (BMI), and waist to hip ratio for both group volunteers.

Keywords

Glutathione • Metabolic syndrome • Human volunteers

1 Introduction

1.1 Glutathione Supplementation

Glutathione is a cluster of amino acid containing one molecule of L-glutamic acid, L-cysteine, and glycine each. The molecule acts as an antioxidant where it is found in the food supply and in the human body. Supplementation of glutathione is thought to act as catalyst for the enzymes that synthesize glutathione within the cell to exerts antioxidant effects and thus maintain the efficacy of the entire glutathione system.

Glutathione supplementation is normally consumed for its antioxidant power that can help to combat free radicals within the body that can damage the cells. Besides, it also helps in the detoxification of chemicals, including some that that is created naturally in the body, as well as pollutants and drugs. In previous study, a novel GSH demonstrated a highly effective in reducing oxidative stress and supported the potential of GSH precursors as therapeutic intervention for prevention or treatment of oxidative stress induce medical conditions [1]. However, studies on the benefit and effects of glutathione supplementation on human metabolic status are not well established and warrant for further investigation into it.

N. R. Hasan Basri (✉) · M. S. Mohktar · W. S. Wan Kamarul Zaman
Department of Biomedical Engineering, Faculty of Engineering, University of Malaya, Kuala Lumpur, Malaysia
e-mail: shyda.hasan@gmail.com

M. S. Mohktar · W. S. Wan Kamarul Zaman
Faculty of Engineering, Center for Innovation in Medical Engineering, University of Malaya, Kuala Lumpur, Malaysia

H. I. Mohd Yusof
Chakra We Care Resources Sdn Bhd, Kuala Lumpur, Malaysia

F. Ibrahim et al. (eds.), *2nd International Conference for Innovation in Biomedical Engineering and Life Sciences*, IFMBE Proceedings 67, https://doi.org/10.1007/978-981-10-7554-4_38

1.2 Metabolic Syndrome

Metabolic syndrome is a group of risk factors that can affect health which are high blood pressure, high glucose level, and unhealthy cholesterol level and unbalance abdominal fat. Usually having one of the risk factor is considered unhealthy, however, when all factors combined, they will pose serious risk of having chronic diseases. Heart disease, stroke and diabetes are examples of chronic metabolic diseases.

In this 21st century, lifestyle varies with modernization, whereby the numbers of people that are suffering metabolic syndrome worldwide are increasing. This increases the correlation to global epidemic of obesity and diabetes [2]. While some people are aware of their condition, most are still in the dark about their health. Some 23% of adults are at higher risk of cardiovascular disease, diabetes, stroke and diseases related to fatty build-ups in artery walls as the effect of metabolic syndrome [2].

When a person has three or more symptoms with the range as stated in Table 1, a person may be at risk for several metabolic diseases [2].

As body composition is an indicator of healthy lifestyle, many researches has been carried out on healthy subjects as well as subjects that suffered from metabolic diseases. Over the years, people with obesity in Malaysia has been increasing at an alarming rate. According to the 2015 National Health and Morbidity survey, it has reveals that nearly half of the Malaysian population aged 18 and above with 47.7% were overweight or obese [3]. Obesity is one of the factor that cause metabolic syndrome in human. Several studies shows that obesity leads to higher oxidative stress thus depletes glutathione levels in the body. In one study, glutathione peroxidase activity were found decreasing in obese diabetics as compared to non-obese diabetics, indicating higher levels of oxidative stress occur in obese diabetics [4].

As people usually takes glutathione for its antioxidant and detoxification function, the study aim to observe how its function helps in improving metabolic status of these overweight volunteers. Previous study has investigated by comparing the use of the novel sublingual form of GSH with two commonly used dietary supplements, N-acetylcysteine (NAC) and oral GSH, to determine their respective interest for raising GSH levels and acting on the GSH ratio. However, there are no significant outcome regarding to the volunteer metabolic status [5].

2 Methodology

This observation was conducted for about 12 weeks after baseline measurement. The study were adjusted to the supplementation period (3 periods for each subject corresponding to the successive intake of the 3 supplementation). This adjustment allows to take into consideration a potential effect of time on the primary outcome [6]. At the end of 12 weeks, the effect of supplementation on metabolic status of subjects should be observe. The outcomes measured including the fasting glucose, blood pressure, cholesterols, waist to hip ratio, body fat percentage and basic measurement such as body mass index and blood pressure. Volunteers were randomized into groups; Group 1 (one sachet of 5 g daily), Group 2 (two sachet of 5 g daily). The glutathione supplement are in powder form and should be mix with 120 ml of water and consume immediately. Advisable to take the supplement before meals. Group 1 volunteers were instructed to take the supplement before breakfast whereas group 2 were instructed to take one dose before breakfast and dinner. The study ethics were applied and approved; study ethical number: UM. TNC2/RC/H&E/UMREC—152.

2.1 Data Collection

Volunteer subjects were recruited based on criteria in Table 2 with the exclusion criteria which are subject with chronic medical condition or pregnant/lactating mothers. Figure 1 shows the flow chart of study protocol during data collection. The protocol were done with the help of nurses.

Table 1 Metabolic syndrome symptoms diagnosis

Metabolic syndrome symptom	Men	Women
Waist hip ratio	0.9	0.85
HDL cholesterol	<0.9 mmol/L	<1.0 mmol/L
Glucose	$\geq$ 5.6 mmol/L	
Triglyceride	$\geq$ 1.7 mmol/L	
Blood pressure	>140/90 mm Hg	
Body mass index (BMI)	>30	

Table 2 Eligibility criteria of volunteer subjects

Age	Body mass index (BMI)	Waist circumference (WC)		Waist hip ratio (WHR)	
18–65 years old	≥ 25 kg/m^2	Male	Female	Male	Female
		>102 cm	>88 cm	≥ 0.90	≥ 0.85

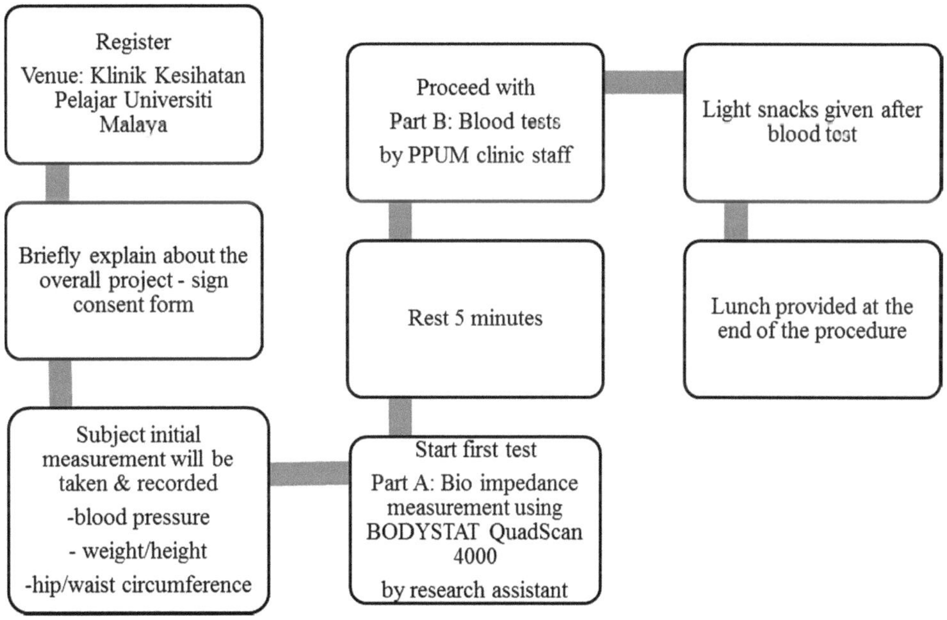

Fig. 1 Flow chart of data collection protocol from volunteer subjects

2.2 Measurements

All measurements were taken in the morning, as subjects were instructed to fast (8–12 h) prior the visits. Usage of any medication also need to be avoided, however subjects with long-term medication were taken noted on their medication intake. They completed consent and questionnaire concerning the use of medication and dietary supplement.

Basic measurement such as blood pressure, height, weight were recorded after the registration. Subjects were then required to withdraw blood to investigate on blood biochemistry (Glucose level), and lipid profile (Triglyceride, HDL cholesterol). The study team also conduct a body composition measurement by using BODYSTAT Quad Scan 4000. The measurement observed were body mass index (BMI), and waist to hip ratio. Subjects were also encouraged to maintain their usual diet and exercise habits during the observation.

3 Results and Discussion

About 23 volunteers have completed the 12 week phases of randomized study and Table 2 shows overall volunteers' age percentage and gender (Figure 2).

Based on the metabolic syndrome diagnosis from Table 1, out of 23, six volunteer subjects were found to suffer from metabolic syndrome according to their baseline measurement. As the subject were randomized into groups, three of them are in Group 1 and another three in Group 2. Out of six, two are male and others are female. 50% of them are in their twenties while other 50% are in late forties to early fifties. Table 3 shows the observation on metabolic syndrome symptoms reading at baseline and week 12 after the supplementation for both group.

Fasting glucose level of group 1 remains unchanged after week 12 while group 2 shows increment in the level. Based on Fig. 3, there seems to be no significant changes in the glucose level after 12 week of supplementation. On another note, the levels of glutathione and glucose may be inversely proportional to each other as it may indicate the severity of diabetes as most of complications from diabetes are linked to oxidative damage that is due to the elevated blood sugar levels. In relation to this, it has been reported that people with diabetes usually have low level of blood glutathione [7]. Therefore, the supplementation should be able to assist in improving one's glucose level. However, in the current study, taking glutathione orally does not appear to increase its level in the body. Some investigators have suggested that GSH is poorly absorbed by oral route mainly due to the

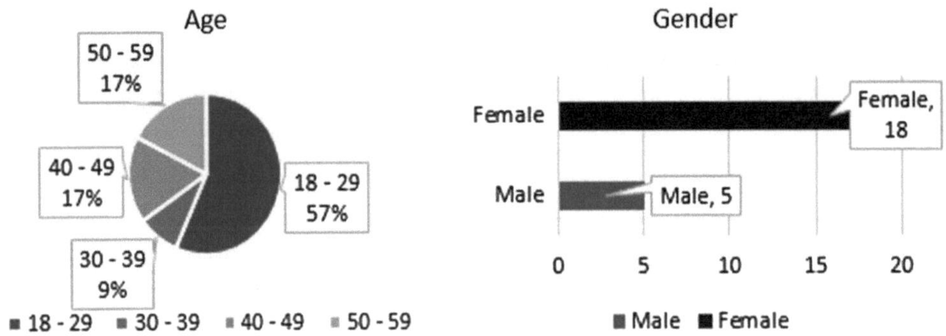

Fig. 2 Volunteers' age and gender

Table 3 Metabolic syndrome subjects' measurement reading during baseline and week 12

	Group 1 (n = 3)		Group 2 (n = 3)	
	Baseline	Week 12	Baseline	Week 12
Glucose	4.87 ± 0.67	4.87 ± 0.21	9.67 ± 1.59	10.77 ± 3.38
Triglyceride cholesterol	1.87 ± 0.55	1.13 ± 0.15	2.17 ± 0.1.44	2.70 ± 1.51
HDL	1.07 ± 0.26	1.10 ± 0.26	1.10 ± 0.18	1.12 ± 0.10
Systolic	126.00 ± 12.53	120.67 ± 14.36	125.33 ± 6.43	135.00 ± 3.46
Diastolic	77.67 ± 0.58	75.33 ± 5.13	79.33 ± 6.66	80.67 ± 5.51
BMI	30.63 ± 5.16	30.23 ± 5.71	37.07 ± 8.98	37.13 ± 8.30
WH ratio	0.9 ± 0.01	0.89 ± 0.05	0.9 ± 0.10	0.89 ± 0.10
Body fat.%	35.43 ± 4.83	34 ± 6.46	42.67 ± 5.03	42.97 ± 3.01

Legend data are in mean ± SD; *HDL* high-density lipoprotein; *WH ratio* waist to hip ratio

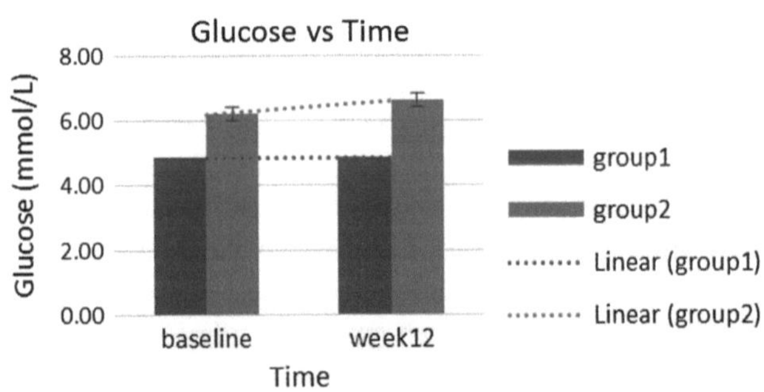

Fig. 3 Fasting glucose of metabolic syndrome subjects from baseline to week 12. Values are mean ± S.E.M

action of an intestinal enzyme, the γ-glutamyl trans-peptidase (GGT) which degrades glutathione [8].

Further investigation on their daily routine should be done to correlate with the glucose result. The aforementioned outcomes from previous studies and the current study may indicate that the daily dietary intake and the dosage time interval of the supplementation may influence the levels of GSH, which require further investigation. Apart from glucose levels, other metabolic parameters used to investigate the effects of GSH are triglyceride and HDL cholesterols, blood pressure, body mass index, waist to hip ratio and body fat percentage.

From Fig. 4, the triglyceride and HDL cholesterol level of both group were observed after 12 weeks. Group 1 shows improvement in decreasing triglyceride cholesterols value while group 2 increase in triglyceride level after 12 weeks. However, both group shows positive increment in HDL value after supplementation.

From Fig. 5, after 12 weeks observation, blood pressure reading for both systolic and diastolic shows improvement toward normal range for group 1 however for group 2 the level seems to increase. However, the range of blood pressure reading for group 2 after 12 weeks still not at an alarming level. Nevertheless, it has been reported that glutathione helped to lower blood pressure by having antioxidant mechanism to fight oxidative stress that are one of the main cause of high blood pressure [5].

Several studies have reported that obesity (BMI > 30) can led to higher oxidative stress that depleted glutathione levels in the body [9], in which glutathione supplementation may be able to neutralize the oxidative stress and improve metabolism. However, based on Fig. 6, the BMI for group 1 seem to show slight reduction but group 2 showed unremarkable difference in the BMI.

Based on Fig. 7, there is an improvement on the waist to hip ratio for both groups after consuming the supplement. According to a previous study of on GSH supplementation

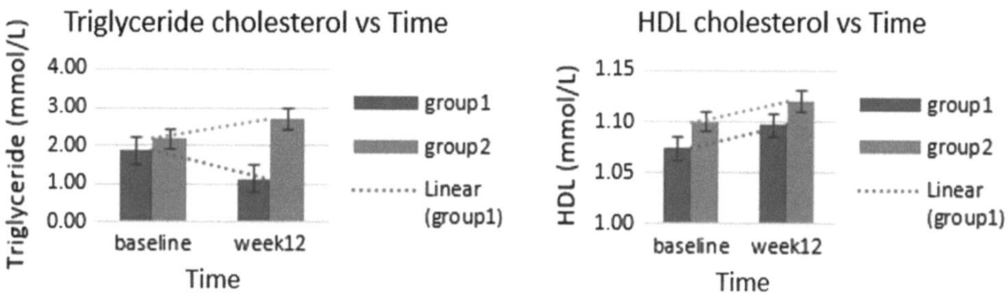

Fig. 4 Triglyceride and HDL cholesterols of metabolic syndrome subjects from baseline to week 12. Values are mean ± S.E.M

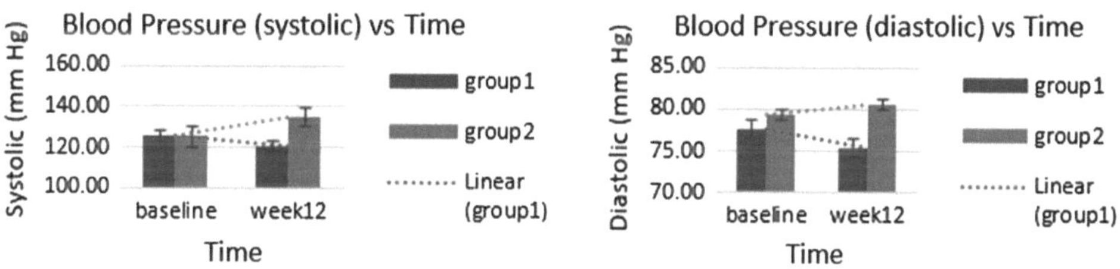

Fig. 5 Blood pressure readings of metabolic syndrome subjects from baseline to week 12. Values are mean ± S.E.M

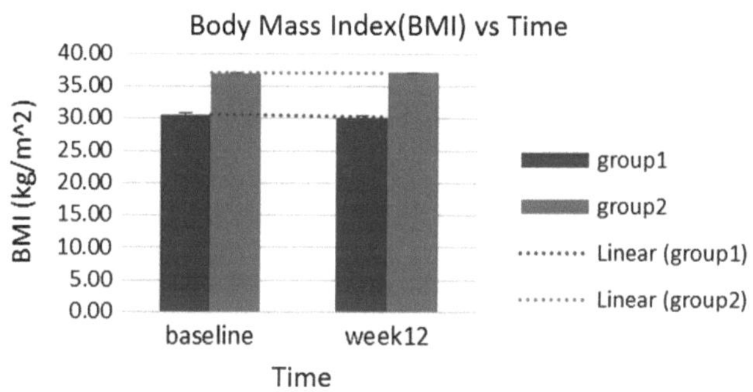

Fig. 6 Body mass index of metabolic syndrome subjects from baseline to week 12. Values are mean ± S.E.M

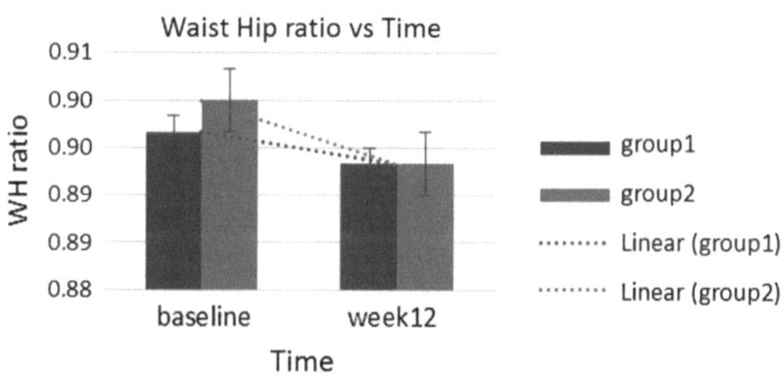

Fig. 7 Waist to hip ratio of metabolic syndrome subjects from baseline to week 12. Values are mean ± S.E.M

carried out on HIV patients, GSH improvement was associated with a modest but significant reduction in body weight. However, there was a remarkable decrease in waist circumference, and waist to hip ratios without any change in the hip circumference, which suggests that the fat loss may

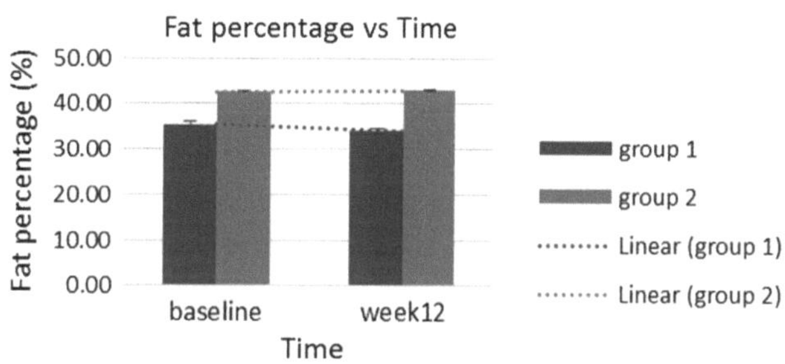

Fig. 8 Fat percentage of metabolic syndrome subjects from baseline to week 12. Values are mean ± S.E.M

have preferentially occurred in the central abdominal adipose tissue compartment [10].

Figure 8 shows the percentage of body fat of subjects during baseline and after 12 weeks of supplementation. There is slight reduction of body fat shown by subjects from group 1, while subjects in group 2 show a slight increase in body fat after 12 weeks. This correlate with the BMI outcome for both group after 12 weeks as shown in Fig. 7.

4 Conclusion and Future Works

After 12 week of supplementation, based on the parameter studied, there seems to be an improvement in the level of HDL cholesterol, BMI, and waist to hip ratio for both group volunteers. However, subjects in group 1 did show more positive effects as compared to group 2, after consuming the supplement. Endogenously, the turnover rate of glutathione is 65% per day, which means most of it will be completely utilized in only 36 h [11]. In relation to this, the subjects in group 2's supplement last intake were taken before dinner of the night before the assessment was carried the following day. This may indicate that the glutathione supplement has not completely taken its effect in the body thus the changes in the levels of most of the metabolic parameters studied were not remarkable and in fact they were similar to the baseline levels. This may be able to explain the reason for the non-observable improvement in the metabolic syndrome in group 2.

In the current study, our finding showed that the improvement of metabolic status can be seen in group 1 where subjects with metabolic syndrome seems to show positive effects while group 2 remain relatively unchanged. From this observation, glutathione supplementation may benefits its consumer on metabolism however more subjects should be recruited and observed for further investigation to gauge the optimum effect of glutathione supplementation. With regards to the effects observed in group 2, dosage time interval of the GSH supplement should be reviewed and modified to see the optimum effects of GSH.

Acknowledgements This research is supported and funded by Chakra We Care Resources Sdn Bhd, Project No. PV004-2017.

References

1. Sinha-Hikim, I., Shen, R., Lee, W.N.N.P., Crum, A., Vaziri, N.D., Norris, K.C.: Effects of a novel cystine-based glutathione precursor on oxidative stress in vascular smooth muscle cells. Am. J. Physiol. Cell Physiol. **299**(3), C638–C642 (2010)
2. Eckel, R.H., et al.: The metabolic syndrome. The Lancet **365** (9468), 1415–1428 (2005)
3. New Straits Times homepage: BERNAMA. Malaysia's the fattest country in Asia, so why aren't we spending on our health? https://www.nst.com.my/news. Accessed 12 Oct 2016
4. Goyal, R., Singhai, M., Faizy, A.F.: Glutathione peroxidase activity in obese and nonobese diabetic patients and role of hyperglycemia in oxidative stress. J. Mid-Life Health **2**(2), 72 (2011)
5. Schmitt, B., Vicenzi, M., Garrel, C., Denis, F.M.: Effects of N-acetylcysteine, oral glutathione (GSH) and a novel sublingual form of GSH on oxidative stress markers: a comparative crossover study. Redox Biol. **6**, 198–205 (2015)
6. Schulz, K.F., Altman, D.G., Moher, D.: CONSORT 2010 statement: updated guidelines for reporting parallel group randomised trials. BMC Med. **8**(1), 18 (2010)
7. Rosen, P., Hans, J.T., Glenn, A.K., Angelo, A.: Antioxidants in Diabetes Management (Oxidative Stress and Disease), 1st edn. CRC Press, California (2000)
8. Zhang, H., Forman, H.J., Choi, J.: Gamma-glutamyl transpeptidase in glutathione biosynthesis. Methods Enzymol. **401**, 468–483 (2005)
9. Furukawa, S., Fujita, T., Shimabukuro, M., Iwaki, M., Yamada, Y., Nakajima, Y., Shimomura, I.: Increased oxidative stress in obesity and its impact on metabolic syndrome. J. Clin. invest. **114** (12), 1752 (2004)
10. Nguyen, D., Hsu, J.W., Jahoor, F., Sekhar, R.V.: Effect of increasing glutathione with cysteine and glycine supplementation on mitochondrial fuel oxidation, insulin sensitivity, and body composition in older HIV-infected patients. The J. Clin. Endocrinol. Metab. **99**(1), 169–177 (2013)
11. Lyons, J., Rauh-Pfeiffer, A., Yu, Y.M., Lu, X.M., Zurakowski, D., Tompkins, R.G., Castillo, L.: Blood glutathione synthesis rates in healthy adults receiving a sulfur amino acid-free diet. Proc. Natl. Acad. Sci. U. S. A. **97**(10), 5071–6 (2000)

Robotic Upper Limb Rehabilitation Using Armeo®Spring for Chronic Stroke Patients at University Malaya Medical Centre (UMMC)

N. Hamzah, N. I. Giban, and M. Mazlan

Abstract

This is a retrospective study of patients with chronic partial arm paresis post stroke who attended neurorehabilitation at University Malaya Medical Centre, Malaysia. In this study we aimed to analyze the clinical and practical outcome of robotic-assisted upper limb rehabilitation. Specifically, we analyzed the impact of therapy on motor and function of chronic stroke arm paresis through structured therapy protocol. We extended our analysis towards user acceptance in robotic-assisted rehabilitation. We applied our Armeo®Spring Therapy Protocol on stroke patients with unilateral partial upper limb paresis of more than six months duration. The outcome measures were muscle strength, spasticity and hand dexterity. Thirty three patients who fulfilled the criteria of treatment protocol attended outpatient therapy session. Fourteen patients completed the treatment protocol in which ten participants were stroke patients. This study reported statistically significant improvement in multiple joint range of motions following therapy. Although there was non progressing arm spasticity, and improved paretic hand dexterity, both latter outcomes were not statistically significant at the end of therapy.

Keywords

Robotic-assisted therapy • Chronic arm paresis • Neurorehabilitation

1 Introduction

Stroke is the most common cause of partial or complete loss of upper limb function with variable motor recovery over time [1–3]. Several studies had reported on the effect of rehabilitation aim at improving upper limb functional outcome at early and chronic stroke [1, 3, 4]. In chronic stroke survivors, upper limb motor and functional recovery are limited [1, 2]. Several neurological and musculoskeletal complications can further influence the recovery of post stroke arm paresis. Spasticity, neurological and mechanical induced pain, joint inflammation and disused further diminish the potential return of upper limb function [1].

The application of robotic-assisted therapy provides physical rehabilitation that can contribute to motor and functional recovery following stroke [2, 4, 5] in part, as it provides the necessary rehabilitation input that relates to favourable outcome [4, 6]. Robotic technology has the potential to deliver high intensity reproducible therapy incorporating task-specific motor practice with high repetitive variable movements [3–6]. There are currently various accepted protocols for upper limb robotic rehabilitation after stroke. The treatment programs vary in frequency, duration, intensity and robotic technology at different stages of stroke onset [3–5].

The aim of this intervention is to incorporate the use of robotic technology for functional arm training in patients with chronic partial paretic arm as a form of treatment that is accepted, safe and effective. In particular, we applied and

N. Hamzah (✉) · M. Mazlan
Department of Rehabilitation Medicine, Faculty of Medicine, UM, Kuala Lumpur, Malaysia
e-mail: norhamizan@ummc.edu.my

N. Hamzah · N. I. Giban · M. Mazlan
Department of Rehabilitation Medicine, UMMC, Kuala Lumpur, Malaysia

F. Ibrahim et al. (eds.), *2nd International Conference for Innovation in Biomedical Engineering and Life Sciences*, IFMBE Proceedings 67, https://doi.org/10.1007/978-981-10-7554-4_39

analyzed the impact of a structured therapy protocol on chronic upper limb paresis and extended our analysis towards user acceptance of robotic assisted rehabilitation.

2 Methodology

This was a retrospective study of patients with chronic stroke partial arm paresis who were registered for outpatient neurorehabilitation therapy, at University Malaya Medical Centre (UMMC), Malaysia from August 2013 until December 2014. This study was approved by Medical Research Ethics Committee UMMC (MECID No 201411-730).

All chronic stroke patients with unilateral partial arm paresis of more than six months duration, registered for neurorehabilitation outpatient treatment UMMC, who fulfilled our inclusion criteria were recruited from August 2013 until December 2014.

Inclusion criteria included intact cognition, good compliance to Armeo®Spring Therapy Protocol, upper limb spasticity grading of Modified Ashworth Scale (MAS) not more than two and Medical Research Council (MRC) muscle power grading of shoulder and elbow not more than two. Patients with uncontrolled pre-existing pain involving the upper limb, severe postural instability, visual impairment, abnormal motor movements which included ataxia, dyskinesia, myoclonic jerks; open skin ulcer and paretic upper limb joint contracture(s) were excluded from the study.

The delivery of therapy was only by a trained and certified Occupational Therapist in Armeo®Spring Therapy Protocol, UMMC. We measured the upper limb spasticity, MRC muscle power grading, range of movement (ROM) of each joint, grip strength (dynamometer in kilogram) at baseline and upon completion of treatment protocol. We chose Box and Block Test (BBT) as measurement of manual hand dexterity. It is highly correlated with the level of independence in elderly population and a significant predictor of physical health [7].

2.1 Armeo®Spring Therapy Protocol

Armeo®Spring is a robotic unilateral exoskeleton with a computer interface Armeocontrol software that provides game-like functional limb movements and exercises supported by an interactive two dimensional virtual reality environment (via computer screen). The exoskeleton is put on the affected upper limb and movements are related to reach and retrieval function, pronation, supination, wrist flexion and extension and grasps and release. These movements involved the shoulder, elbow, wrist and finger joints within a specified workspace.

Armeo®Spring Therapy Protocol incorporated both functional exercise and assessment exercise within the determinate workspace [8]. The functional exercise covered a large range of easy, one-dimensional to complex, three-dimensional exercises involving all the movements listed earlier. All functional exercises were scaled into calibrated workspace. The assessment exercise involved assessing motor ability and coordination. Built-in sensors recorded active arm movements at each joint during all therapy sessions. All the relevant joint angles and the hand position were captured and stored in the equipment database. Based on these data, movement quality and coordination pattern were analyzed.

The Armeocontrol software supplies accurate assessments through the options of A-MOVE (for active reaching distance, reaction time and movement velocity), A-GOAL (for precise, goal oriented movements), A-COORD (for inter-limb coordination during active movements) and A-ROM (for the range of motion during active and passive movements [8].

All participants adhered to Armeo®Spring Therapy Protocol as illustrated in Table 1.

The progression of motor and functional status were reviewed by the same certified therapist weekly. Complexity of exercise was increased according to assessment findings. The final assessment was repeated within one week after completion of 10 sessions Armeo®Spring Therapy Protocol. Data was recorded using a standardized form. The flow of treatment protocol is illustrated in Flowchart 1.

2.2 Patients and Therapists Acceptance

Seven therapists and 10 chronic stroke patients were involved in this part of the study. All therapists were certified Armeo®Spring trained and had designed the treatment application protocol at UMMC. The selected patients received upper arm therapy using Armeo®Spring as outpatient sessions and had completed the treatment protocol.

Acceptance to the Armeo®Spring therapy was evaluated using questionnaires adapted from a previous arm robotic study [9]. Patients section consisted of 5 questionnaires: 4 closed-ended questions and 1 question for patients to provide additional comment. The patients were evaluated twice, after the first therapy session and upon completion of Armeo Spring therapy. Questionnaires for therapists consisted of 28 questions (open and closed-ended) with 6 questions specifically catered to Armeo®Spring use. The Questionnaires were divided into three main groups—safety, system utility and therapist acceptance.

Table 1 Armeo®Spring therapy protocol

Session frequency	10 sessions within 10 weeks
Therapy setting	Outpatient setting
Treatment intensity	Low (1–2 times treatment per week)
Duration	60 min (included set up time)

Flowchart 1 The treatment protocol work flow

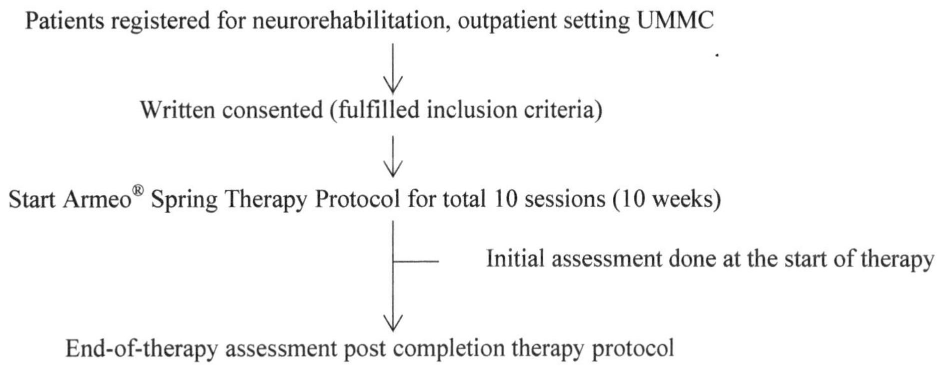

3 Results

3.1 Armeo®Spring Therapy

A total of 33 patients who fulfilled the criteria of treatment protocol attended the outpatient therapy. Fourteen patients completed the treatment protocol, 6 patients were in the ongoing treatment group and 13 patients defaulted treatment (Flowchart 2). Table 2 is the demographic and clinical characteristics of participants who completed the Armeo®-Spring treatment protocol.

One patient recorded improvement of strength of the whole arm. However the mean differences for shoulder, elbow and wrist strengths at baseline and post treatment showed no statistically significant difference. Active ROM of shoulder flexion, elbow flexion and forearm supination, the analysis reported statistically significant improvement in the mean differences.

Twenty percent of patients who improved in MAS spasticity grading were by one grade lower at therapy completion than at baseline, for elbow and wrist flexion spasticity. Although the spasticity grading for the remaining patient did not show improvement, the spasticity however, did not worsen.

Gross manual dexterity assessment via BBT for both upper limbs had scored lower at initial assessment and at end of treatment against age- and gender-specific normal values. The findings also reported slightly higher BBT score mean difference for the affected upper limb. Handgrip strength was not statistically significant for both sides of the upper limbs (Tables 3 and 4).

3.2 Patients and Therapist Acceptance

Eighty percent of therapists and 70% of patients responded to the questionnaires. Due to small number of participants the analysis of result is qualitative in nature. Results were divided into 4 categories.

Patient acceptance. All patients reported positively to the therapy method after completing the first session with

Flowchart 2 Patient distribution receiving the Armeo®Spring treatment during the neurorehabilitation outpatient therapy

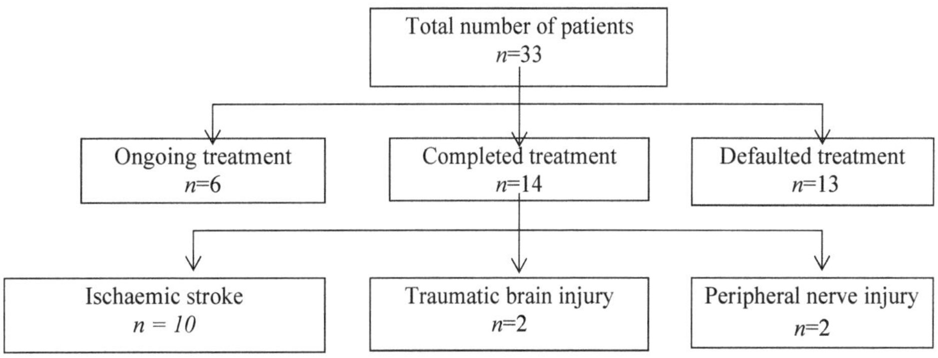

Table 2 Demographic distribution of the chronic stroke patients (n = 10)

Demographic characteristics	Value
Mean age (years)	59 (48–71)
Gender ratio (men/women)	8/2
Side of lesion: cortical/subcortical	7/3
Dominant hand paresis/non dominant hand paresis	5/5

Table 3 Measurement outcome of paretic arm range of motions for patients following completed treatment protocol

aROM (°)	At baseline (mean °) (n = 10)	Completed treatment (mean °) (n = 10)	Mean difference (°)	*P* value
Shoulder flexion	91.5 ± 62.5	102.1 ± 61.43	10.6 ± 21.43	**0.05**
Elbow flexion	95.7 ± 41.1	101.8 ± 47.2	6.1 ± 3.77	**0.01**
Forearm:				
Pronate	33.4 ± 31.5	41 ± 35.4	7.6 ± 12.36	0.06
Supinate	40 ± 34	50.2 ± 38.6	10.2 ± 14.6	**0.03**

Note aROM—Active ROM improvement, is measured as the difference in the degree at baseline to the degree at end of treatment protocol for shoulder, elbow, forearm and wrist joints
°—degree

Table 4 Measurement outcomes for hand dexterity and hand grip strength for patients following completed treatment protocol

Level of measurement	At baseline (mean)	Completed treatment (mean) (n = 10)	Mean difference	*P* value
BBT (number of blocks)				
Affected side	15 ± 20	25 ± 28	10 ± 34	0.14
Unaffected side	49 ± 21	56 ± 15	8 ± 31	**0.02**
Hand grip strength (kg)				
Affected side	5.23 ± 6.65	6.17 ± 6.48	0.94 ± 0.21	0.07
Unaffected side	28.47 ± 6.29	29.93 ± 6.55	1.46 ± 0.90	0.16

Table 5 Patient feedback: acceptance—first experience

Question	Answer	
	Yes	No
1. Did you like this treatment?	7	0
2. Do you feel it was helpful?	7	0
3. Was this boring?	2	5
4. Was it confusing to use?	1	6

improved outlook at the end of therapy protocol (Tables 5 and 6).

Safety. The majority of therapists reflected no risk of injury to patients during therapy and one therapist specified that possible risk of soft tissue injury could occur during rapid arm movements. All therapists had no qualm of letting patient work alone following therapist set up at first-time use. The current protocol however required a therapist to be present throughout each session.

Therapist Acceptance. Ninety percent of therapists were confident in applying Armeo®Spring as a therapy approach at first-time use and their confidence level grew with regular use. The overall response to the technology was positive. However there were variable reviews pertaining to a question on the need of Armeo®Spring equipment in the current service setting. The response received included request for system development that could cater for patients with higher motor power grading. Another recommendation involved incorporating relearning of fine motor hand function exercises beyond finger grip strength (Table 7).

System utility. All therapists stated that the support training was adequate for their needs. They suggested that

Table 6 Patient feedback: acceptance—after completed 10 sessions of therapy protocol

Question	Answer	
	Yes	No
1. Did you like this treatment?	7	0
2. Do you feel it was helpful?	7	0
3. Was this boring?	1	6
4. Was it confusing to use?	1	6

Table 7 The pros and cons of using the Armeo®Spring as reported by occupational therapists

Pros	Frequency
1. Robot provides good exercise/patient work harder and reach further	3
2. Appropriate for high and low stages of motor recovery; easily graded	2
3. Provides an addition to pre-existing treatment	1
4. New and interesting treatment concept	5
5. Patient feels it's provides recovery of the arm	4
6. Robot gives patient's motivation in doing exercise	1
7. Patient able to learn new therapy method using robotic technology and patients feels its easy to use	1
8. Data can be easily retrieved	1
Cons	**Frequency**
1. Robot exercises are boring	2
2. Cannot use system for patients with weak distal arm/hand	1
3. Not challenging for high motor power level patients	1
4. Limited use on other patient population—may not be cost efficient	1
5. Duration of the therapy, patient gets bored	1
6. Inadequate training provided to handle Armeo Spring	1

maximum time set-up should be no more than 10 min (Table 8). The longer duration was predominantly because of patient handling and response to first-time instructions. The subsequent sessions of same patient required less time due to the system software ability to save data for subsequent use of the same patient. More over both patients and therapists were adapting well with the system over time. All therapists also stated that the system was a valuable treatment modality for motor relearning because it provided high repetition and consistency motor training. One recommendation was made to include touch screen function for easier data entry and data retrieval. All therapists agreed on the application of the same technology for other patient population including brachial plexus injury, traumatic brain injury, multiple sclerosis, spinal cord injury and pediatric patients who require upper limb rehabilitation.

4 Discussion

This study focused on chronic stroke patients with established paretic upper limb therapy approach via robotic-assisted neurorehabilitation of the upper limb. One advantage of this approach was controlled manipulation of multiple upper limb joints and simultaneous stretching exercise whilst minimising soft tissue injuries [5]. Although this study's sample size was small, it was able to demonstrate several positive outcomes. These included improved joint active ROM and non progressing spasticity.

Functionally, there was a modest improvement of gross manual hand dexterity measured with BBT. The BBT scores improved for both upper limbs. Bilateral arm improvement was in correlation with better functional outcome [7]. In the elderly, integration of meaningful activities involving both arms may enhance overall levels of function [7].

Table 8 Estimated set-up time by therapists

Estimated set-up time needed (minutes)	Satisfactory		
	Yes	No	Total
10	3	-	3
5–10	1	-	1
5	1	-	1

Robotic-assisted treatment modality also allow for precise measurement and computer-based data storing that makes it a valuable tool for treatment progress monitoring in motor rehabilitation [10].

The duration of therapy program was 10 weeks, which may be a short therapy duration for chronic stroke population. One study that looked at three different sub populations of stroke showed that improvement may occur for chronic stroke patient over a longer period of therapy time for robotic assisted therapy [4]. Patients involved in this study were well beyond the expected spontaneous motor recovery timeline but may further improve with longer treatment duration. The age range was between 48 and 71 years old. This study also took into consideration on the appropriate therapy intensity as to avoid fatigue for the much older patients.

The overall acceptance of therapy and robotic use were both positive. Acceptance of treatment modality is important to encourage compliance of treatment. Although the treatment protocol was implemented to other injuries, we however did not include those injuries in outcome analysis because the sampling size was too small. However, with treatment protocol acceptance and good user feedbacks, we may be able to apply this treatment modality across a wider neurologically impaired population.

5 Conclusion

In chronic stroke patients, robotic-assisted treatment may provide benefit in joint motion improvement and prevention of spasticity progression. Future study should consider longer treatment period and larger sample size.

Acknowledgements This study was funded by Postgraduate Research Grant (PPP) University of Malaya (PO035-2015B) and approved by Medical Research Ethics Committee UMMC (MECID No 201411-730).

References

1. Broeks, J.G., Lankhorst, G.J., Rumping, K., Prevo, A.J.: The long-term outcome of arm function after stroke: results of a follow-up study. Disabil. Rehabil. **21**, 357–364 (1999)
2. Jørgensen, H.S., Nakayama, H., Raaschou, H.O., Vive-Larsen, J., Støier, M., Olsen, T.S.: Outcome and time course of recovery in stroke. II. Time course of recovery: the Copenhagen stroke study. Arch. Phys. Med. Rehabil. **76**, 406–412 (1995)
3. Lo, A.C., Guarino, P.D., Richards, L.G., et al.: Robotic-assisted therapy for long term upper limb impairment in stroke. N Engl. Med. **362**, 19 (2010)
4. Colombo, R., Sterpi, I., Mazzone, A., Delconte, C., Pisano, F.: Robot aided neurorehabilitation in sub-acute and chronic stroke: does spontaneous recovery have limited impact on outcome? NeuroRehabilitation **33**, 621–629 (2013)
5. Abdullah, H.A., Tarry, C., Lambert, C., Barreca, S., Allen, B.O.: Results of clinicians using a therapeutic robotic system in an inpatient stroke rehabilitation unit. J. NeuroEng. Rehabil. **8**, 50 (2011)
6. Levin, M.F., Kleim, J.A., Wolf, S.L.: What do motor "recovery" and "compensation" mean in patients following stroke? Neurorehabil. Neural Repair **23**, 313–319 (2009)
7. Desrosiers, J., Bravo, G., Hébert, R., Dutil, E., Mercier, L.: Validation of the box and block test as a measure of dexterity of elderly people: reliability, validity, and norms studies. Arch. Phys. Med. Rehabil. **75**, 751–755 (1994)
8. Hocoma Armeo®Therapy Concept: http://www.stargen-eu.cz/wp-content/uploads/2015/04/Hocoma_Armeo_BRO_Armeo_Therapy_Concept_120420_en.pdf
9. Dijkers, M.P., deBear, P.C., Erlandson, R.F., Kristy, K., Geer, D. M., Nichols, A.: Patient and staff acceptance of robot technology in occupational therapy: a pilot study. J. Rehabil. Res. Dev. **28**(2), 33–44 (1991)
10. Loureiro, R.C.V., Harwin, W.S., Nagai, K., Johnson, M.: Advances in upper limb stroke rehabilitation: a technology push. Med. Biol. Eng. Comput. **49**, 1103–1111 (2011)

Development of Near-Infrared Diffuse Correlation Spectroscopy for Noninvasive Monitoring of Cerebral Blood Flow

I-Tseng Huang, Kuan-Chen Wu, and Jia-Jin Chen

Abstract

Cerebral blood flow (CBF) is a critical physiological process because it controls the oxygen supply, metabolic consumption, and byproduct clearance in the brain. However, a non-invasive method for long-term CBF monitoring is lacking. In recent years, NIR light has been used to monitor brain activities and cerebral blood flow based on DCS technique because it is able to penetrates human skull. The aim of this study was to develop a near infrared (NIR) Diffuse correlation spectroscopy (DCS) system for CBF monitoring. NIR laser at wavelength of 785 nm with properties of continuous wave and long coherence length (>10 m) was emitted into tissue. Single photons scattered by the red blood cells (RBCs) inside the cerebrovascular, were picked up. The auto-correlation function of the optical signal was calculated by the correlator downstream optical-detector. The developed NIR-DCS was first tested on a phantom in which the particle vibration was changed to simulate the changes in blood flow. Then the system was further tested on rats suffered with hypercapnia, normoxia and hyperoxia to measure the changes in CBF. The rats were connected to the ventilator through two plastic tubes—one for inhalation, the other for exhalation. The content of the exhaled gas was analyzed, and the real-time partial pressure of CO_2 and the current end tidal CO_2 ($EtCO_2$) are measured. The results showed that hyperoxia increased blood flow due to changes in vascular wall tension. The outcome from this study supported DCS as a novel noninvasive method to measure CBF.

Keywords

Diffuse correlation spectroscopy • Near infrared spectroscopy • Cerebral blood flow

1 Introduction

Blood flow is a crucial factor affecting the delivery of oxygen to tissue. Cerebral tissues receive about 20% of the body's oxygen, and the majority of this oxygen is consumed in the process of metabolism via aerobic mechanisms. Many diseases such as stroke, pressure ulcer, and cancer are associated with abnormal blood flow. Hypercapnia is also an effective dilator for cerebral blood. It is known that high CO_2 partial pressure can cause vasodilatation and decrease vascular wall tension (VWT), and it is a simple way to verify the increase or decrease in blood flow between carbon dioxide changes. In a similar vein, the effects of neurovascular coupling on cerebral blood oxygen saturation and blood volume are known to perturb functional near infrared spectroscopy (fNIRS) signals. Indeed, accurate monitoring of CBF has important implications in fundamental neuroscience, clinical modeling, rehabilitation, and for patient management and treatment in the hospital [1–6].

Near-infrared spectroscopy (NIRS) is a spectroscopic method that uses the electromagnetic spectrum (wavelength ranges from 700 to 1000 nm). The molar absorptivity in the NIR region is typically quite small, in this band the main absorber of body light is hemoglobin. The advantage is that

I-T. Huang · K.-C. Wu · J.-J. Chen (✉)
Department of Biomedical Engineering, National Cheng Kung University, Tainan, Taiwan
e-mail: jasonbiolab@gmail.com

F. Ibrahim et al. (eds.), *2nd International Conference for Innovation in Biomedical Engineering and Life Sciences*,
IFMBE Proceedings 67, https://doi.org/10.1007/978-981-10-7554-4_40

NIR can typically penetrate much farther into a sample than mid infrared radiation, such like human skull. The diffusion depth of NIR light is about 1–3 cm below the surface. Diffuse correlation spectroscopy (DCS) is an optical approach to measure the blood flow by using NIR light, which is an extension of single-scattering dynamic light scattering (DLS) to the multiple scattering limit [7–9]. By measuring the speckle fluctuation of photon resulting the scatters from moving red blood cells (RBCs) inside vessels [10]. Blood flow will get by calculating the autocorrelation function with scattered photons. Detailed review of the development for DCS can also be found in Refs. [11–13].

2 Methods and Materials

2.1 Theory

A turbid medium such as biological tissue is characterized by a reduced scattering coefficient μ_s, and an absorption coefficient μ_a, which in general can depend on position and time [14]. There is a positive correlation between the point light source with the unit intensity on the semi-infinite medium and the scattered photon density waveform, unnormalized temporal field autocorrelation defined as [12]:

$$G_1\left(\vec{\rho},\tau\right) = \frac{3\mu_s}{4\pi}\left(\frac{\exp(-k_D r_1)}{r_1} - \frac{\exp(-k_D r_2)}{r_2}\right) \quad (1)$$

where $k_D^2 = 3\mu_s\mu_a + 6\mu_s^2 k_0^2 \alpha D_B \tau$, τ is delay time, αD_B equal the relative change of the cerebral blood flow (CBF), ρ is the distance between the source and detector, n is refractive index of tissue, μ_a is optical absorption coefficient of tissue, μ_s is reduced optical scattering coefficient of tissue, $k_0 = \frac{2\pi}{\lambda}$ (Free-space wavenumber), $k = k_0 n$ (Wavenumber in tissue), $r_1 = \sqrt{\rho^2 + z_0^2}$, $r_2 = \sqrt{\rho^2 + (z_0 + 2z_b)^2}$, $z_o = 1/\mu_s$, $z_b = \frac{2}{3\mu_s}\frac{1+R_{eff}}{1-R_{eff}}$, $R_{eff} = -1.44n^{-2} + 0.71n^{-1} + 0.668 + 0.064n$ (The effective reflection coefficient determined the by the ratio of the reflection).

Measure the time dependence of detected diffuse light intensity, $I\left(\vec{r},t\right)$, and computing the intensity autocorrelation function:

$$g_2\left(\vec{r},\tau\right) \equiv \frac{\left\langle I\left(\vec{r},t\right) I\left(\vec{r},t+\tau\right)\right\rangle}{\langle I\rangle^2} \quad (2)$$

where $\langle\ldots\rangle$ means the time average. $g_2\left(\vec{r},t\right)$ is related to the temporal filed autocorrelation function $g_1(\tau,t)$ through the Siegert relationship [15]:

$$g_2\left(\vec{r},\tau\right) = 1 + \beta\frac{\left|G_1\left(\vec{r},\tau\right)\right|^2}{\langle I\rangle^2} \quad (3)$$

where β is parameter which is depend on the number of speckles detected moreover the coherence length and stability of the laser. Combining the autocorrelation function of the electrical field with the diffusion equation, we are able to calculate an index—αD_B.

2.2 Near-Infrared DCS Technology

The instrument we used in tissue phantom was a suspension of polystyrene microspheres. The CW coherent laser sources at 785 with long-coherence ($\approx$10 m) (DL785-120-S, CrystaLaser, Inc., Nevada). The DCS source emitted NIR light into the tissue via multimode optical fibers with a core diameter of 400 μm. There detector fibers used a single mode fiber with a core diameter of 5 μm (780HP, THORLABS, New Jersey). Photodiodes being utilized in DCS system are fast photon-counting avalanche photodiodes (Excelitas SPCM-AQRH APD modules). APD is capable of generating an electrical TTL pulse for every photon that is detected. The output of APDs was sent to autocorrelator board (Correlator.com, New Jersey). The equipment set up showing the different flow rates is depicted in Fig. 1.

2.3 Experiment

The experiment of phantom was via suspension of polystyrene microspheres by monitoring changes in particle vibration of phantom to simulate the condition in the cerebrovascular. Here, we tested three conditions, normoxia, hypercapnia and hyperoxia. Gas experiment was a very simple way to change blood flow. It is known that high CO_2 partial pressure can cause vasodilatation and decrease vascular wall tension (VWT). In this experiment, we use normoxia, hypercapnia and hyperoxia to control the elevation and decrease of cerebral blood flow, and gas inhaling model on rats in Table 1.

3 Results

The beam of the source of DCS was coupled into an optical fiber of 400 μm for scattering into the tissue. The other fiber of 62 μm was connected to the light detector. After the experiment, the photon signals were arranged into several segments. Every segment corresponded to an analog data point. The segments can be sent to the autocorrelation function written in

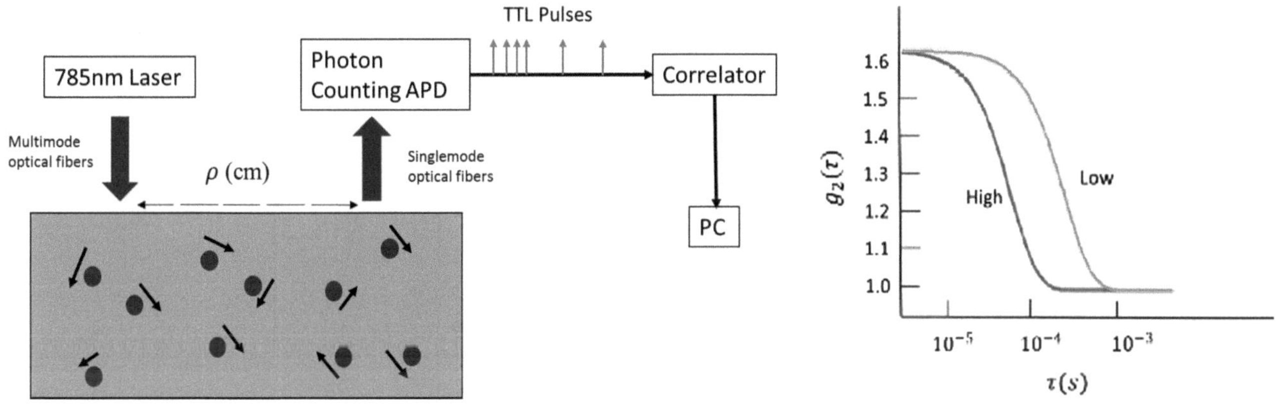

Fig. 1 DCS in the semi-infinite geometry. Laser irradiation on the surface of samples (e.g., blood flow). APD collected scatter signal, and correlator received TTL signal from APD to calculate the DCS correlation function (left). Autocorrelation graph with different flow rates (right)

Table 1 The chart of the gas content, the inspiration pressure, and the respiration rate in four different gas inhalation models

	Room air (L/min)	O_2 (L/min)	CO_2 (L/min)	Δ inspiration pressure (cm-H_2O)	Respiration rate (BPM)
Normoxia	1.2	0.3	0.0	+0	45
Hypercapnia	1.1	0.3	0.1	+0	45
Hyperoxia	0.0	1.5	0.0	+0	45

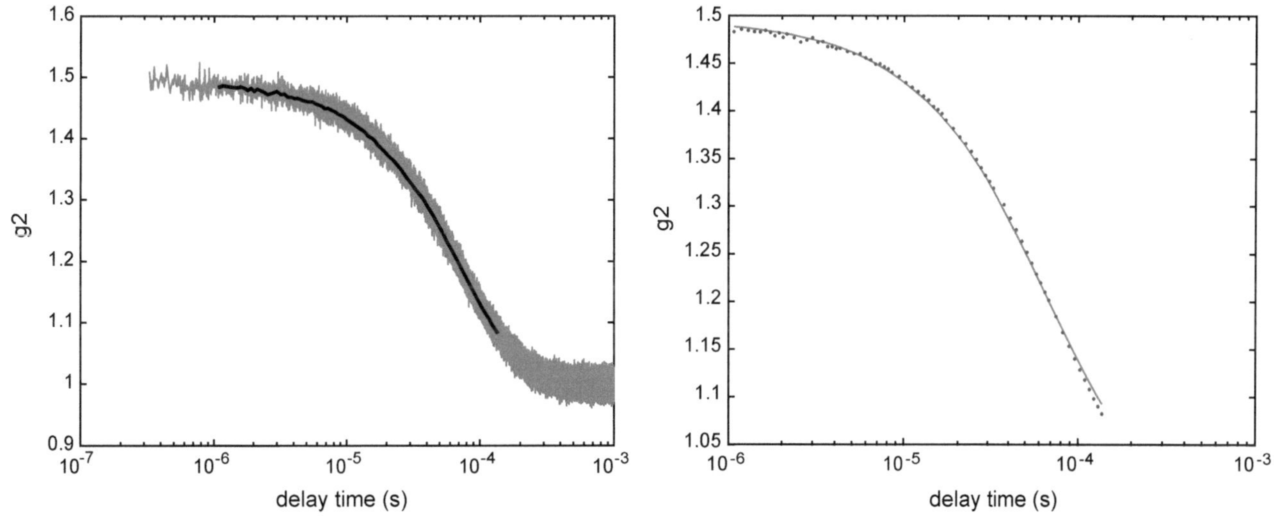

Fig. 2 A plot of the g_2(autocorrelation) measured by the phantom. The g_2 is autocorrelation raw data (blue) and segment g_2 data (black) (left). The g_2 is autocorrelation data (full curve) and fitting with g_2 (dot). Source-detector distance is 2.5 cm, μ_s is 5.2 cm^{-1} and μ_a is 0.042 cm^{-1}. Flow (αD_B) equal to 3.46 × 10^{-9} and β is 0.49 (right)

Matlab. The output of the function was named autocorrelation curves, labeled as g_2 and can be plotted against the corresponding lag. In Diffuse Correlation, the lag of time was called tau represented in a logarithmic scale.

After the phantom defrost, the optical fiber of the light and the fiber of the detector were placed on the surface. Source-detector distance was 2.5 cm, μ_s is 5.2 cm^{-1} and μ_a is 0.042 cm^{-1}. The result of phantom autocorrelation curve as shown in Fig. 2. The surgery started from tracheotomy insertion. A plastic cannula was inserted into the trachea to control respiration and continuous isoflurane inhalation. The other end of the tracheotomy cannula was connected to the ventilator through two plastic tubes—one for inhalation, the other for exhalation. In Normoxia, End tidal CO_2 ($EtCO_2$) was monitored and should be between 35 and 43 mmHg. In this experiment we set it at 38 mmHg. Figure 3 shows the $ETCO_2$ and blood flow for the experiment of hypercapnia, normoxia and hyperoxia.

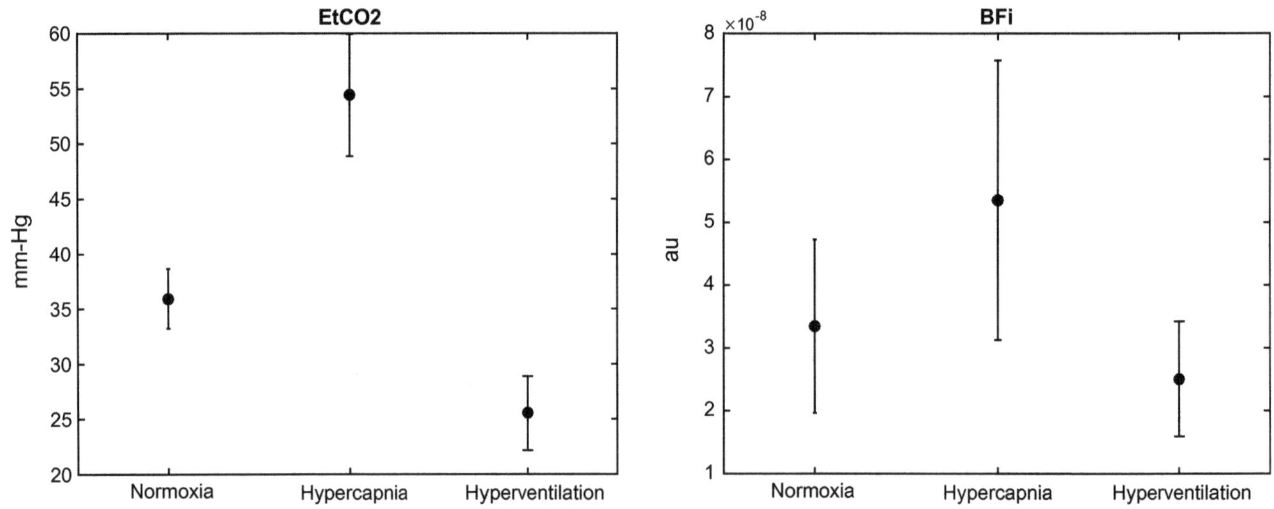

Fig. 3 A plot of end tidal CO_2 with hypercapnia, normoxia and hyperoxia (left). The blood flow (right)

4 Conclusion

We have demonstrated the flow measurement using the DCS on phantom. After fitting the curve with the autocorrelation data, its data looks very reasonable. Our results showed that flow equal to 3.46×10^{-9} and β value of 0.49, similar to human adult measurement. Hypercapnia, normoxia and hyperoxia experiments showed that CO_2 increased blood flow. Our experiment exhibits similar trend.

Acknowledgements This work is supported by the National Cheng Kung University under grant number MoST 104-2314-B-006-007-MY3. We thank J Chen and J Wu for helpful comments and discussions.

References

1. Peltonen, G.L., Harrell, J.W., Aleckson, B.P., LaPlante, K.M., Crain, M.K., Schrage, W.G.: Cerebral blood flow regulation in women across menstrual phase: differential contribution of cyclooxygenase to basal, hypoxic, and hypercapnic vascular tone. Am. J. Physiol. Regul. Integr. Comp. Physiol. (2016) ajpregu. 00106.02016
2. Durduran, T., et al.: Diffuse correlation spectroscopy for non-invasive, microvascular cerebral blood flow measurement. Neuroimage **85**(0 1), 51–63 (2014) (15 Jan 2014)
3. Siesjo, B.K.: Brain energy metabolism. Wiley (1978)
4. Zauner, A., Daugherty, W.P., Bullock, M.R., Warner, D.S.: Brain oxygenation and energy metabolism: part I—biological function and pathophysiology. Neurosurgery **51**(2), 289–302 (2002). [PubMed:12182767]
5. Lin, Y., et al.: Noncontact diffuse correlation spectroscopy for noninvasive deep tissue blood flow measurement. J. Biomed. Opt. **17**(1), 010502 (2012) (03 Feb 2012)
6. Stanimirovic, D., Friedman, A.: Pathophysiology of the neurovascular unit: disease cause or consequence. J. Cereb. Blood Flow Metab. **32**(7), 1207–1221 (2012). [PubMed:22395208]
7. Berne, B.J., Pecora, R.: Dynamic Light Scattering with Applications to Chemistry, Biology, and Physics. Krieger, Malabar (1990)
8. Chu, B.: Laser Light Scattering, Basic Principles and Practice. Academic, New York (1991)
9. Brown, W.: Dynamic Light Scattering: The Method and Some Applications. Clarendon, New York (1993)
10. He, Lian, et al.: Using optical fibers with different modes to improve the signal-to-noise ratio of diffuse correlation spectroscopy flow-oximeter measurements. J. Biomed. Opt. **18**(3), 037001 (2013)
11. Boas, D.A., Campbell, L.E., Yodh, A.G.: Scattering and imaging with diffusing temporal field correlations. Phys. Rev. Lett. **75**(9), 1855–1858 (1995). [PubMed: 10060408]
12. Boas, D.A.: Diffuse photon probes of structural and dynamical properties of turbid media: Theory and biomedical applications. Doctor of Philosophy, Graduate School of Arts and Sciences, University of Pennsylvania (1996)
13. Boas, D.A., Yodh, A.G.: Spatially varying dynamical properties of turbid media probed with diffusing temporal light correlation. J. Opt. Soc. Am. A **14**(1), 192–215 (1997)
14. Cheung, C., et al.: In vivo cerebrovascular measurement combining diffuse near-infrared absorption and correlation spectroscopies. Phys. Med. Biol. **46**, 2053–2065 (2001)
15. Rice, S.: Mathematical analysis of random noise. In: Wax, N. (eds.) Noise and Stochastic Processes, p. 133. Dover, New York (1954)

Positively Charged Silver Nanoparticles as Labels for Paper-Based Colorimetric Detection of Heparin

Sanjay Kumar, Pulak Bhushan, and Shantanu Bhattacharya

Abstract

A novel paper-based colorimetric sensor has been developed for ultrasensitive detection of heparin based on aggregation of positively charged silver nanoparticles. The electrostatic interaction between negatively charged heparin and positively charged silver nanoparticles (AgNPs) resulted in a shift in the surface plasmon resonance (SPR) wavelength leading to a color change from yellow to colorless. The proposed assay can selectively detect heparin with an ultra-low detection limit of 1.898 ng mL^{-1} in solution and 8.68 μg mL^{-1} on paper. In future, the assay can be further employed for detection of heparin in human blood serum samples.

Keywords

Colorimetric detection • Heparin • (+) AgNPs • Paper-based detection

1 Introduction

Heparin, a highly sulfated negatively charged linear acidic polysaccharide, is widely used as an anticoagulant in various clinical practices such as, kidney dialysis, cardiac and vascular surgery, etc. [1, 2]. A heparin concentration beyond the normal levels (i.e. 17–67 μM during cardiovascular surgery and 1.7–10 μM in postoperative and long-term therapy) can induce hemorrhage and thrombocytopenia [3]. Hence, frequent clinical monitoring of heparin levels is imperative to guide antidote treatment. However, as a consequence of chemical heterogeneity, natural polydispersity and poor fluorescence properties of heparin, its quantization is challenging [4]. Currently, laboratory-based traditional methods such as, ion pair high-performance liquid chromatography [5], ion chromatography [6], activated partial thromboplastin time (aPTI) or measurements of the activated clotting time (ACT) [7] have been utilized for quantification of heparin. These techniques are time-consuming, suffer from poor specificity due to potential interferences from other factors and require sophisticated equipment. In the past several years, researchers have attempted to develop reliable and sensitive biosensors based on various detection schemes such as, electrochemical [8], fluorimetry [9], nuclear magnetic resonance [10], light scattering [11], colorimetric [1, 3, 12, 13], etc. Among these, colorimetric sensors are based on observing a color change of the substrate or solution that can be easily visualized by naked eye. They have been widely exploited owing to their simplicity, rapidity, robustness, portability, inexpensiveness and sensitivity. Recently, a wide variety of metal nanocomposites such as gold nanoparticles [12], gold nanorods, gold-graphene oxide composites [13, 14], etc. have been employed for colorimetric detection of heparin, attributed to their distinctive optical, photothermal and electrical properties. Typically, assays employed for heparin detection are based on aggregation of positively charged nanoparticles on interaction with negatively charged

S. Kumar · P. Bhushan · S. Bhattacharya (✉)
Microsystems Fabrication Laboratory, Department of Mechanical Engineering, Indian Institute of Technology Kanpur, Kanpur, India
e-mail: bhattacs@iitk.ac.in

S. Bhattacharya
Design Programme, Indian Institute of Technology Kanpur, Kanpur, India

F. Ibrahim et al. (eds.), *2nd International Conference for Innovation in Biomedical Engineering and Life Sciences*,
IFMBE Proceedings 67, https://doi.org/10.1007/978-981-10-7554-4_41

heparin. However, they are generally complex since they require an additional chemical to induce aggregation, e.g. PEI, protamine, etc. In order to overcome this complexity, very few efforts have been made. Cao et al. reported the direct interaction of heparin with positively charged cysteamine-stabilized gold nanoparticles [1]. Chen et al. proposed chitosan capped, citrate reduced AuNPs-based colorimetric assay for heparin detection [12]. Nonetheless, these detection assays suffer from high detection limits.

In this work, a highly sensitive paper-based colorimetric assay for detection of heparin has been established and developed for the first time. The assay exploits chitosan (a cationic polyelectrolyte) capped positively charged AgNPs for heparin detection. AgNPs have been exploited owing to their unique surface plasmon resonance (SPR), higher extinction coefficient and sharp extinction bands. The electrostatic interaction between the positively charged AgNPs and polyanionic heparin led to the aggregation of (+) AgNPs. This resulted in a color change from yellow to colorless within 5 min (Fig. 1a). The proposed protocol has been further used to develop a point-of-care (POC) paper-based device.

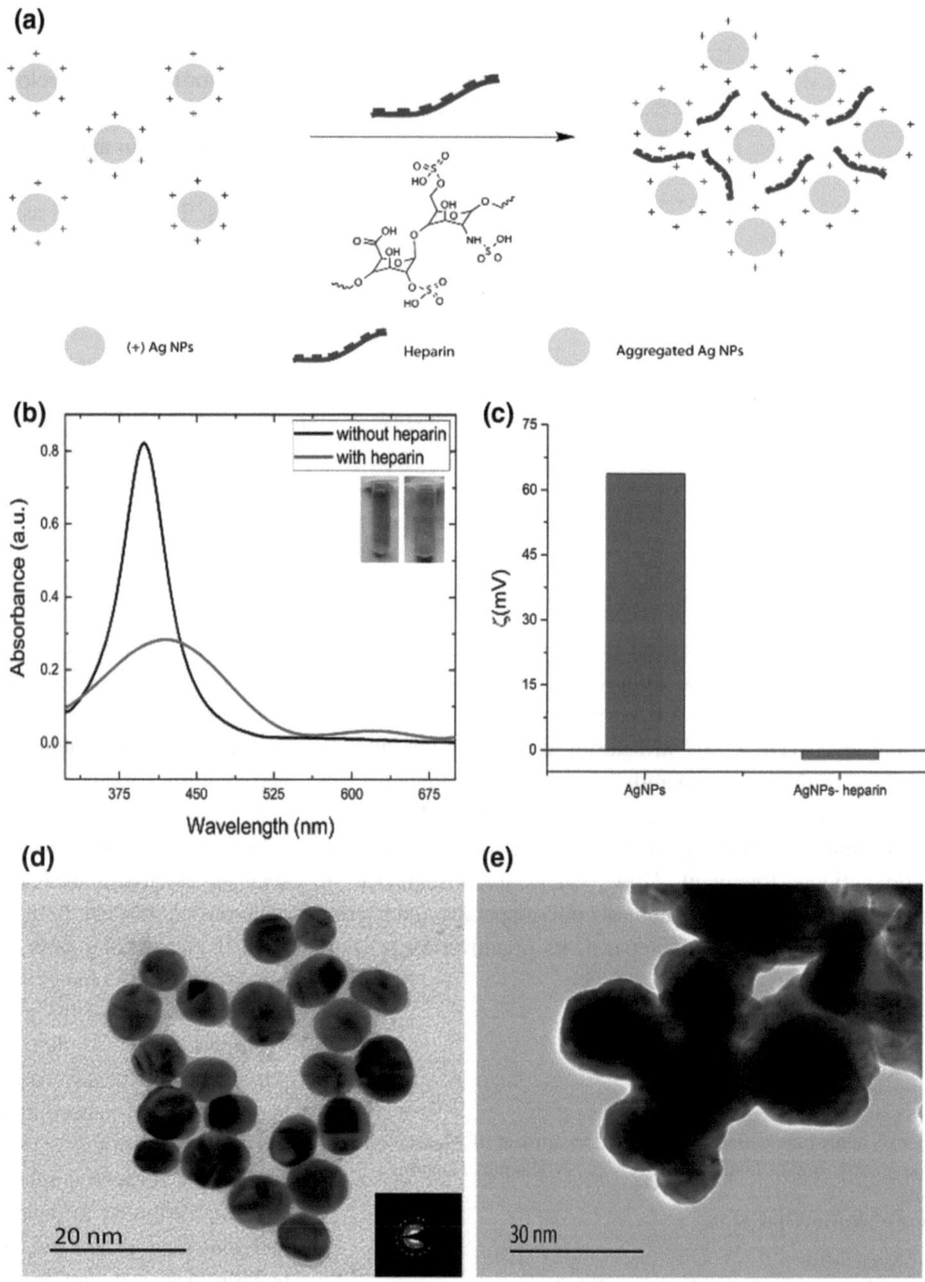

Fig. 1 **a** Schematic diagram of the colorimetric sensor for the visual detection of heparin using (+) AgNPs. Inset shows the photographs of (+) AgNPs aqueous solution without and with heparin (left to right). **b** UV-Vis absorption spectra of (+) AgNPs in the absence (solid blue line) and presence (solid black line) of heparin. Inset shows the corresponding photographic images. **c** Zeta potential measurements of AgNPs and AgNPs-heparin. **d** TEM image of (+) AgNPs and **e** TEM image of heparin induced aggregated AgNPs

2 Experimental

2.1 Materials

Chitosan, silver nitrate, sodium borohydride and heparin were purchased from Sigma-Aldrich, India. All other chemicals and reagents in this study were of analytical grade and used without further purification. All aqueous solutions were made with high performance liquid chromatography (HPLC) water, obtained from Merc Life Science Pvt. Ltd., India.

2.2 Synthesis of Positively Charged Silver Nanoparticles

Silver nanoparticles ((+) AgNPs) were prepared according to a previously reported protocol with a slight modification [15]. Briefly, 0.094 g of chitosan was dissolved in 47 mL of 1% acetic acid under vigorous stirring for 5 h at 60 °C. The solution was then mixed with 2 mL of 0.01 M $AgNO_3$ aqueous solution. After stirring for 30 min at room temperature, 1 mL of freshly prepared $NaBH_4$ (8 mg mL^{-1}) aqueous solution was added rapidly under vigorous stirring for another 90 min at room temperature in dark. The resulting solution was centrifuged at 2000 g for 30 min. The final AgNPs solution was obtained by discarding the supernatant and redispersing the remaining solution in HPLC water. The solution was kept in dark at 4 °C until use.

2.3 Colorimetric Detection of Heparin

The heparin-induced aggregation of (+) AgNPs was monitored by observing the change in UV-vis spectra on addition of heparin to colloidal AgNPs. In a typical procedure, various concentrations of heparin (5, 10, 15, 25, 50 and 100 ng mL^{-1}) were first prepared using 10 mM of Britton–Robinson (BR) buffer solution (pH 3.6). Next, 50 μL of as-prepared heparin was added to 200 μL of (+) AgNPs solution (0.05 mg mL^{-1}) and the mixture was diluted with a buffer solution to 1 mL. After incubation of the above solution for 5 min, the UV-vis spectra were recorded.

3 Results and Discussion

3.1 Characterization

The as-synthesized (+) AgNPs colloid and heparin-induced aggregation of AgNPs were characterized using UV-vis spectroscopy, zeta-potential and transmission electron microscopy (TEM). Figure 1b shows the absorption spectra of the (+) AgNPs in absence (solid blue line) and presence (solid black line) of heparin. The peak at 399 nm confirms the presence of AgNPs. It was observed that aggregation of AgNPs in presence of heparin resulted in a characteristic red-shift and broadening of the surface plasmon band. The positive charge of as-synthesized AgNPs colloid was confirmed through zeta-potential measurements (Fig. 1c). A decrement of the surface charge from +63.78 mV to −2 mV, on addition of heparin confirmed the successful aggregation of AgNPs. TEM images (Fig. 1d) confirmed the monodispersity of the spherical shaped AgNPs having an average size of about 10 nm. The irregular aggregation of AgNPs in presence of heparin was also confirmed through TEM (Fig. 1e).

3.2 Quantitative Determination of Heparin

The color variation of (+) AgNPs solution on addition of heparin with different concentrations was recorded by UV-vis spectrometer. The process of heparin induced aggregation of AgNPs was studied by monitoring the shift in plasmon band position and change in absorption. The color variation of the solution from yellow to colorless was directly observed by the naked eye. It was discerned from Fig. 2a, that on addition of heparin, the absorption of AgNPs at 399 nm gradually decreases with a red-shift in the SPR wavelength. Figure 2b shows the calibration curve of absorbance against the heparin concentration. The curve was linear in a range from 5 to 25 ng mL^{-1}. The linear regression equation for heparin was expressed as $Y = -0.0136 X + 0.8353$ with a correlation coefficient of 0.9991 (Fig. 2b inset), where X is the heparin concentration and Y is the absorbance. The limit of detection for heparin was calculated to be 1.898 ng mL^{-1}, which is lower than previously reported values (Table 1).

Further, to assess the applicability of this assay, a paper-based rapid detection device was developed. The device was essentially circular shaped paper strip made of Whatman no. 1 filter paper. In a typical experiment, the paper strips were soaked into the (+) AgNPs colloidal solution and dried overnight at room temperature. Next, 150 μL of heparin of varying concentrations (0.1, 0.5, 5, 10, 15, 25, 50 and 100 μg mL^{-1}) was pipetted onto the above paper strips. After 5 min of incubation, the yellow color of the paper strips discolored depending on the heparin concentration. The uniform color intensity of the paper strip confirms the even distribution of AgNPs onto the paper (Fig. 2c). For quantitation of heparin, images of the paper strips were captured using a Nikon DSLR camera. The gray level intensities of the images were obtained

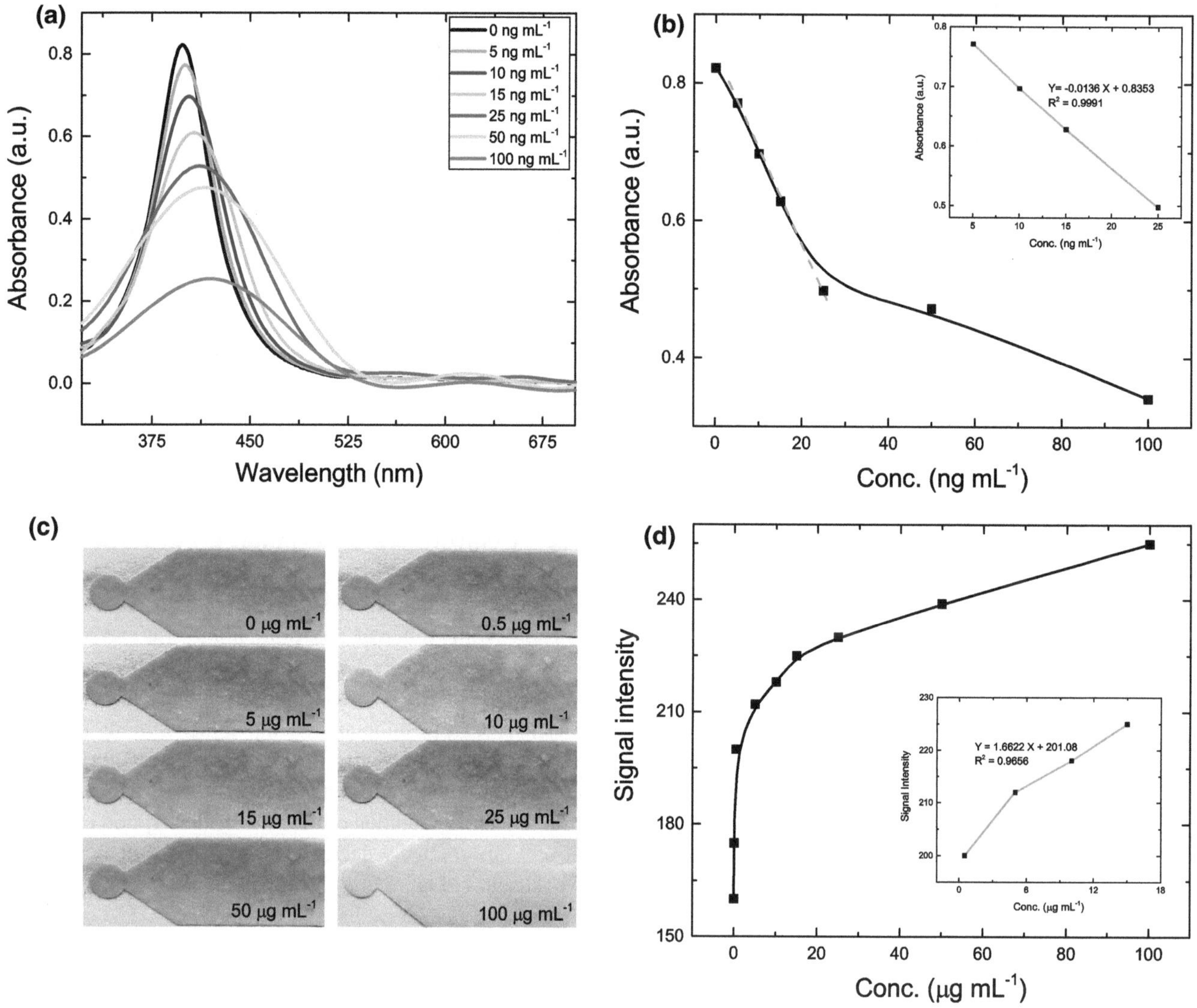

Fig. 2 **a** UV-vis absorption spectra of (+) AgNPs aqueous solution with various concentrations of heparin (0–100 ng mL^{-1}). **b** The dose–response curve for heparin detection in solution phase. Inset: the calibration curve of the (+) AgNPs system for the detection of heparin. **c** Photographs of the (+) AgNPs soaked paper strips after addition of various concentrations of heparin (0–100 μg mL^{-1}). **d** The dose–response curve for heparin detection on paper strip. Inset: the calibration curve of the (+) AgNPs system for the detection of heparin

Table 1 Comparison of the reported colorimetric biosensors for heparin

Probes	Linear range	Correlation coefficient	LOD
Cationic polythiophene [3]	1.7–10 μM	–	1.36 μM
Chitosan capped AuNPs [12]	0.2–60 μM	0.998	0.8 μM
AuNPs (protamine-coated) [16]	1.2–10 μg mL^{-1}	–	0.6 μg mL^{-1}
(+) AuNPs [1]	0.09–3.12 μg mL^{-1}	0.9958	0.03 μg mL^{-1}
CTAB established AuNRs [13]	0.02–0.28 μg mL^{-1}	0.9957	5 ng mL^{-1}
AuNPs/GO/protamine [14]	0.06–0.36 μg mL^{-1}	0.9936	3 ng mL^{-1}
(+) AgNPs [this work]	5–25 ng mL^{-1}	0.9991	1.898 ng mL^{-1}

using Image J and plotted against the heparin concentration (Fig. 2d). The color intensity of the paper strip was linearly dependent on the concentration of added heparin in the range of 0.5–25 μg mL^{-1} (Fig. 2d inset). The detection limit calculated using the developed calibration curve was 8.68 μg mL^{-1}.

3.3 Selectivity

To investigate the selectivity of the above established colorimetric assay, several potential interfering compounds coexisting in human serum, such as, $Na^{+}, Ca^{2+}, k^{+}, Mg^{2+}, Cl^{-}, Co_{3}^{2-}, Po_{4}^{3-}$, glucose, cholesterol, hyaluronic acid, heparin and their mixture were examined by recording the respective absorbance responses. As shown in Fig. 3, the absorbance obtained for the interfering compounds was very low as compared to heparin. Moreover, upon addition of all the coexisting substances with heparin resulted in a minor change in the absorbance value (3%). Therefore, it can be concluded that the electrostatic attraction between (+) AgNPs and heparin was significantly stronger than any other interfering substance, ensuring high selectivity for heparin.

4 Conclusion

This study demonstrates that a paper-based device utilizing positively charged silver nanoparticles can be employed for simple quantitative analysis of heparin. The colorimetric sensing assay takes advantage of the strong electrostatic interaction between (+) AgNPs and negatively charged heparin, thereby successfully attaining an easy colorimetric readout without requiring additional instrumentation. The as-developed paper-based platform for ultrasensitive detection of heparin can be further used for detection of heparin in human serum/plasma samples.

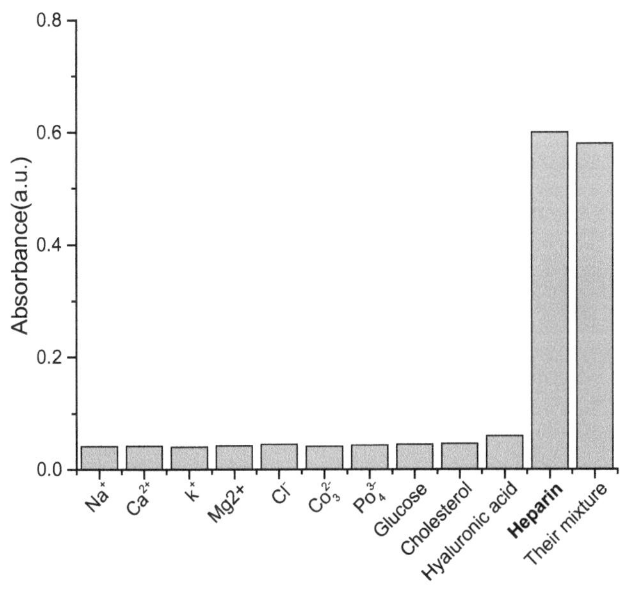

Fig. 3 Relative absorption responses of (+) AgNPs based biosensing system upon addition of heparin (0.1 μg mL^{-1}) and various potential interfering substances. Na^{+} (50 mM), Ca^{2+} (1.5 mM), k^{+} (2 mM), Mg^{2+} (0.5 mM), Cl^{-} (5 mM), Co_{3}^{2-} (1 mM), Po_{4}^{3-} (0.5 mM), glucose (1 μg mL^{-1}), cholesterol (1 μg mL^{-1}) and hyaluronic acid (0.5 μg mL^{-1})

Acknowledgements This work was supported by the Ministry of Human Resource Department (MHRD), India under the Design Innovation Centre (DIC) scheme (Project number-MHRD/DESP/2016142H).

References

1. Cao, R., Li, B.: A simple and sensitive method for visual detection of heparin using positively-charged gold nanoparticles as colorimetric probes. Chem. Commun. **47**(10), 2865–2867 (2011)
2. Rabenstein, D.L.: Heparin and heparan sulfate: structure and function. Nat. Prod. Rep. **19**(3), 312–331 (2002)
3. Zhan, R., Fang, Z., Liu, B.: Naked-eye detection and quantification of heparin in serum with a cationic polythiophene. Anal. Chem. **82** (4), 1326–1333 (2009)
4. Meyerhoff, M.E., Fu, B., Bakker, E., Yun, J.H., Yang, V.C.: Peer reviewed: polyion-sensitive membrane electrodes for biomedical analysis. Anal. Chem. **68**(5), 168A–175A (1996)
5. Patel, R.P., Narkowicz, C., Jacobson, G.A.: Effective reversed-phase ion pair high-performance liquid chromatography method for the separation and characterization of intact low-molecular-weight heparins. Anal. Biochem. **387**(1), 113–121 (2009)
6. Ander, B., Karlsson, A., Öhrlund, Å.: Determination of heparin on intraocular lens surfaces by ion chromatography. J. Chromatogr. A **917**(1), 105–110 (2001)
7. Murray, D.J., Brosnahan, W.J., Pennell, B., Kapalanski, D., Weiler, J.M., Olson, J.: Heparin detection by the activated coagulation time: a comparison of the sensitivity of coagulation tests and heparin assays. J. Cardiothorac. Vasc. Anesth. **11**(1), 24–28 (1997)
8. Gemene, K.L., Meyerhoff, M.E.: Reversible detection of heparin and other polyanions by pulsed chronopotentiometric polymer membrane electrode. Anal. Chem. **82**(5), 1612–1615 (2010)
9. Pu, K.Y., Liu, B.: Conjugated polyelectrolytes as light-up macromolecular probes for heparin sensing. Adv. Funct. Mater. **19**(2), 277–284 (2009)
10. Zhang, Z., Li, B., Suwan, J., Zhang, F., Wang, Z., Liu, H., Mulloy, B., Linhardt, R.J.: Analysis of pharmaceutical heparins and potential contaminants using 1H-NMR and PAGE. J. Pharm. Sci. **98**(11), 4017–4026 (2009)
11. Yan, H., Wang, H.F.: Turn-on room temperature phosphorescence assay of heparin with tunable sensitivity and detection window based on target-induced self-assembly of polyethyleneimine capped Mn-doped ZnS quantum dots. Anal. Chem. **83**(22), 8589–8595 (2011)
12. Chen, Z., Wang, Z., Chen, X., Xu, H., Liu, J.: Chitosan-capped gold nanoparticles for selective and colorimetric sensing of heparin. J. Nanopart. Res. **15**(9), (2013)
13. Fu, X., Chen, L., Li, J., Lin, M., You, H., Wang, W.: Label-free colorimetric sensor for ultrasensitive detection of heparin based on color quenching of gold nanorods by graphene oxide. Biosens. Bioelectron. **34**(1), 227–231 (2012)

14. Fu, X., Chen, L., Li, J.: Ultrasensitive colorimetric detection of heparin based on self-assembly of gold nanoparticles on graphene oxide. Analyst **137**(16), 3653–3658 (2012)
15. Huang, H., Yuan, Q., Yang, X.: Preparation and characterization of metal–chitosan nanocomposites. Colloids Surf. B **39**(1), 31–37 (2004)
16. Jena, B.K., Raj, C.R.: Optical sensing of biomedically important polyionic drugs using nano-sized gold particles. Biosens. Bioelectron. **23**(8), 1285–1290 (2008)

Single Image Super-Resolution for MRI Using a Coarse-to-Fine Network

Jia Liu, Fang Chen, Huabei Shi, and Hongen Liao

Abstract

Single Image Super-Resolution (SISR) which aims to recover a high resolution (HR) image from a low-resolution (LR) image has a wide range of medical applications. In this paper, we present a novel Super-Resolution Coarse-to-Fine Network (SRCFN) that recovers the finer texture details strongly and enables precise high-frequency detail to address this challenging task. First, we apply some residuals units to achieve a coarse Super-Resolution result. Second, we add a fine module using the idea of segmentation networks to combine more high-frequency detail into the coarse results for final Super-Resolution results. In addition, we use a combined loss function of Mean square error loss and SSIM loss. Our proposed method applied to medical MRI outperforms previous methods of accuracy (PSNR and SSIM) and visual improvements.

Keywords

Single image super-resolution • MRI • CNN

1 Introduction

Magnetic resonance scanning is a non-invasive diagnostic technology of the body's anatomy and physiology and is widely used in medical imaging of brain disease. In the clinical and research status, high-resolution and high-contrast Magnetic Resonance Image (MRI) is preferred because it provides significant structural details of a smaller voxel size. Comparatively, low-resolution MRI has fuzzy tissue boundaries of lower contrast. However, currently, HR MRI acquisition equipment is expensive and less available in hospitals and clinical centers because of long scanning time. Therefore, single image super-resolution (SISR) which can restore a high-resolution (HR) MRI from a given low-resolution (LR) one, is significant to reduce scanning time for the image quality assurance. One major difficulty of SISR is the ambiguity of solutions to the underdetermined super-resolution problem. Especially, in high downsampling factors, texture detail in the reconstructed HR images is typically absent so that the ill-posed nature of the SISR problem is obviously appearance.

Despite the difficulty of the above problem, research received growing interest in the computer vision community due to SISR have a wide range of applications including video streaming and medical applications. Many methods and substantial advances have been studied in SISR problem [1, 2]. Early methods mainly focus on interpolation technology such as bicubic interpolation and Lanczos resampling [3]. Interpolation method is very fast, but usually fails to recover the high-frequency image information on an overly smooth solution. Sparse coding [4, 5] methods use a learned compact dictionary based on sparse signal representation to address the task of SISR.

In recent years, due to the development of Deep Learning (DL), especially Convolutional Neural Network (CNN), CNN is widely used to address the ill-posed inverse problem of Super Resolution (SR), improving the accuracy and computational speed. CNN has performed much better than reconstruction-based [6] and other learning methods [7, 8]. As the pioneer CNN model for SR, Super-Resolution

J. Liu · F. Chen · H. Shi · H. Liao (✉)
Tsinghua University, Beijing, 100084, People's Republic of China
e-mail: liao@tsinghua.edu.cn

F. Ibrahim et al. (eds.), *2nd International Conference for Innovation in Biomedical Engineering and Life Sciences*,
IFMBE Proceedings 67, https://doi.org/10.1007/978-981-10-7554-4_42

Convolutional Neural Network (SRCNN) [9] predicts non-linear LR-HR mapping and significantly outperforms classical non-DL methods. However, SRCNN relies on the context of small image regions and training to converge too slowly. Kim et al. [10] also presented the Deeply-Recursive Convolutional Network (DRCN) which presented a high performance in accuracy with a small number of model parameters. Overall, deeper network architectures have been shown to increase performance and "the deeper the better" might be the case of SR. In addition, residual learning is widely used in deeper network architectures.

Based on the achievement of deeper network architectures and residual learning applied in SISR and segmentation task, we propose a novel Super-Resolution Coarse-to-Fine Network (SRCFN). Different from previous works, we add a fine network consists of some full-resolution residual units (FRRU) after the coarse super resolution network and explore a PSNR and SSIM combined loss. Our new network with perceptual losses can jointly compute good high-level features and good low-level features. Experimental results show that our method is superior to other methods, both in quantitative studies and human judgements.

2 Method

2.1 Problem Formulation: Image Super Resolution

The SISR is formulated as a problem that learns the relationship between the LR and the corresponding HR images [9]. The problem formulation is as follows:

$$I_{LR} = \mathrm{DB}I_{HR} + \varepsilon \tag{1}$$

Here, where I_{LR} is the LR image; I_{HR} is the HR image; B is a blurring (low pass filter) operator; D is a decimation operator that discards every other pixel for a given scaling factor r (r = 4 in our method). ε denotes an additive noise, normally i.e. white Gaussian noise.

The solution to this problem, we train a generator network as a feed-forward CNN $\Phi(I_{LR}, \Theta)$ parametrized by Θ. Here $\Theta = \{\mathrm{W1} : \mathrm{L}; \mathrm{b1} : \mathrm{L}\}$ denotes the weights and biases of a L-layer deep network and is obtained by optimizing a SR specific loss function Ψ. For given training images I^n_{HR}, n = 1, …, N with corresponding I^n_{LR}, n = 1, …, N, we solve the minimizes loss function Ψ defined by:

$$\widehat{\Theta} = \underset{\Theta}{\mathrm{argmin}} \frac{1}{N} \sum_{n=1}^{N} \Psi\left(\Phi\left(I^n_{LR}, \Theta\right) - I^n_{HR}\right) \tag{2}$$

In this work, we will specifically design a perceptual loss Ψ as a weighted combination of several loss components that model distinct desirable characteristics of the recovered SR image. The individual loss functions are described in more detail in Sect. 2.3.

2.2 Network Architectures for SISR

In this section, we introduce each main component of our framework. As sketched in Fig. 1, the proposed framework includes two components: Coarse super resolution Networks, Fine super resolution Networks.

Coarse Super Resolution Networks. We design a Non-Deep CNN structure for coarse super resolution networks. The configuration is outlined in Fig. 1. The first layer is the upsampling layer (factor = 4) which interpolates LR image to the desired size as the new input I1. After that, the first convolution layer with kernel size 5 × 5 is designed as a spacious receptive field to capture as much image information as possible, as illustrated in [11]. Then, we add a ResNet for feature extraction, which consists of two convolution filters of the size 3 × 3 × 48, ReLU activation function and an addition operation.

A ResNet is composed of four residual units (RUs). As depicted in Fig. 2, the output x_n of the n-th RU in a ResNet is computed as

$$x_n = x_{n-1} + (Fx_{n-1}; W_n) \tag{3}$$

where $Fx_{n-1}; W_n$ is the residual, which is parametrized by W_n. Thus, instead of computing the output x_n directly, F only computes a residual that is added to the input x_{n-1}.

Fine Super Resolution Networks. Inspired by the Residual Networks and FRRN, we design a new FRRU based feature extraction structure for Fine super resolution Networks. The configuration is outlined in Fig. 2. Fine super resolution Networks is composed of six of full-resolution residual units (FRRUs).

Full-resolution residual units are a modification of residual units. Each FRRU has two inputs (R_{n-1} and P_{n-1}) and two outputs (R_n and P_n), and it simultaneously operates. Figure 2b shows the structure of a FRRU. Let R_{n-1} be the residual input to the n-th FRRU and let P_{n-1} be its pooling input. Then the outputs are computed as:

$$R_n = R_{n-1} + H(P_{n-1}, R_{n-1}; W_n) \tag{4}$$

$$P_n = \mathrm{G}(P_{n-1}, R_{n-1}; W_n) \tag{5}$$

where W_n are the parameters of the pooling function G and residual function H, respectively. If G = 0, then a FRRU corresponds to an RU since it disregards the pooling input y_n, and the network effectively becomes an ordinary ResNet. On the other hand, if H = 0, then the output of a FRRU only depends on its input via the function G. Hence, no residuals

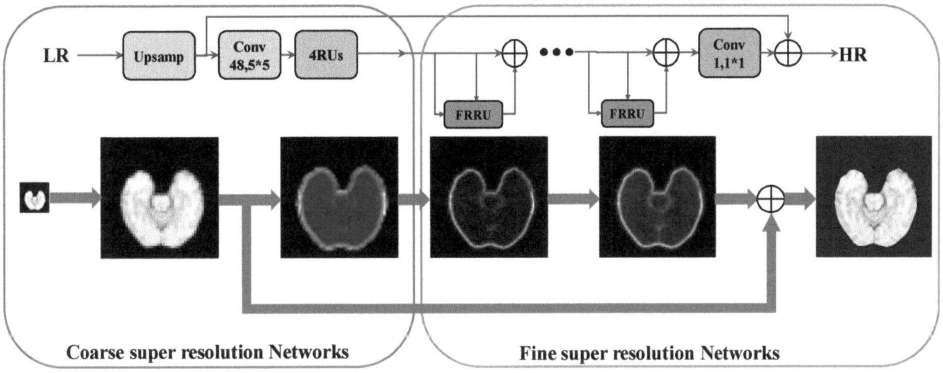

Fig. 1 Illustration of the proposed framework for single image super-resolution. Given an input LR image, our framework first applies one upsampling module to interpolate the image to the desired size and present a coarse super resolution network and a fine super resolution networks. The final HR estimation is obtained by two step super-resolution operations

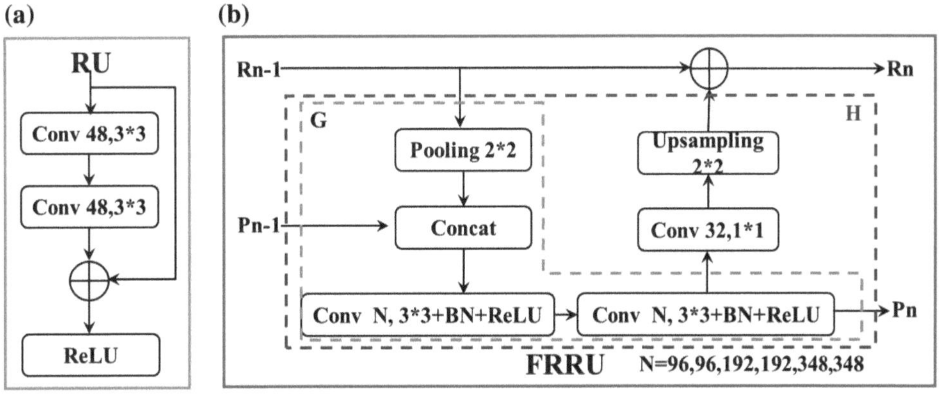

Fig. 2 An illustration of the residual unit (RU) and full-resolution residual unit (FRRU). **FRRU**: Total 6 FRRUs parts in our fine SR Network, the Number of convolution filter for each FRRU: N = 96, 96, 192, 192, 348, 348. BN: Batch normalization layer. Upsampling (2 * 2) is implemented by resize function

are computed and we obtain a traditional feedforward network. By carefully constructing G and H, we can combine the two network principles.

After a sequence of FRRUs (number = 6), some high level feature maps were achieved and added to the input of Fine super resolution Networks to fine the coarse SR image.

2.3 Perceptual Loss Function

With respect to an ideal result image, the Peak Signal to Noise Ratio (PSNR) computes the mean squared error between reconstruction results and the original image. Higher PSNR means more noise removed. However, as least squares result, it is slightly biased towards over smoothed (blurry) results, i.e. an algorithm that removes not only the noise but also a part of the textures will have a good score. Structural Similarity (SSIM) [12] has been developed to have a quality reconstruction metric that also takes into account the similarity of the edges (high frequency content) between the reconstruction image and the original one. Hence, SSIM as the object loss function, will not only remove the noise while also preserving the edges of the objects. Thus, our proposed perceptual loss function has combined PSNR and SSIM.

Content Loss. PSNR is most commonly used to measure the quality of reconstruction results of SISR. PSNR is most easily defined via the mean squared error (MSE). Thus, we choose MSE loss as our content loss. The pixel-wise MSE loss is calculated as:

$$\mathcal{L}_{MSE}^{SR} = \frac{1}{r^2 WH} \sum_{x=1}^{rW} \sum_{y=1}^{rH} \left(I_{x,y}^{HR} - G\left(I_{x,y}^{LR}\right)\right)^2 \tag{6}$$

Here, G(ILR) is a reconstructed image and IHR is the reference image. However, while achieving particularly high PSNR, solutions of MSE optimization problems often lack high-frequency content which results in perceptually unsatisfying, overly smooth solutions.

Structural Similarity Loss. The goal is for SISR network to learn to produce visually pleasing images. It stands to reason that the error function should be perceptually motivated, as is the case with SSIM. SSIM for pixel p is defined as

$$\mathrm{SSIM}(p) = \frac{2\mu_x\mu_y + C_1}{\mu_x^2 + \mu_y^2 + C_1} \cdot \frac{2\sigma_{xy} + C_2}{\sigma_x^2 + \sigma_y^2 + C_2} \tag{7}$$

$$= l(p) \cdot cs(p) \tag{8}$$

Here, μ_x the average of x; μ_y the average of y; σ_x^2 the variance of x; σ_y^2 the variance of y;σ_{xy} the covariance of x and y; $C_1 = (k_1 L)^2, C_2 = (k_2 L)^2$ two variables to stabilize the division with weak denominator; L the dynamic range of the pixel-values; $k_1 = 0.01$ and $k_2 = 0.03$ by default.

Where we omitted the dependence of means and standard deviations on pixel p. Means and standard deviations are computed with a Gaussian filter with standard deviation. The loss function for SSIM can be then written setting:

$$\mathcal{L}_{SSIM}^{SR} = \frac{1}{N} \sum_{p \epsilon P} 1 - \mathrm{SSIM}(p) \tag{9}$$

Finally, our proposed perceptual loss function is defined as follow:

$$\Psi = \mathcal{L}_{MSE}^{SR} + 10 * \mathcal{L}_{SSIM}^{SR} \tag{10}$$

3 Experiments and Results

3.1 Training Details and Parameters

We trained all networks on 2016 MR images from the BRATS 2015 training data [13]. We obtained the LR images (48 * 48) by downsampling the HR images (192 * 192) using bicubic kernel and downsampling factor r = 4. Note that we can apply the generator model to images of arbitrary size as it is fully convolutional. For optimization we use Adam and an auto learning rate. Training a SRCFN with epoch = 30 roughly takes 13.3 h on a GeForce GTX 1080 GPU. For testing, we perform experiments on 480 MRI from the BRATS 2015 training data. For fair quantitative comparison, all reported PSNR [dB] and SSIM measures were calculated.

3.2 Comparison with Other Methods

We evaluation the results of quantitative comparisons with the DL (SRCNN [9]) and non-DL (bicubic) methods. For compared with SRCNN, we train the same training images and testing images with SRCNN network, epoch = 30 and learning rate = 0.0001. Table 1 summarizes quantitative results on the testing sets, by citing the results of prior methods. Our SRCFN model outperforms than bicubic and SRCNN methods in both PSNR and SSIM. Our SRCFN significantly advances the state of the art, with the improvement margin of 4.62 dB/0.10 and 0.61 dB/0.02 on scale factor 4 than bicubic and SRCNN. Qualitative comparisons, among bicubic, SRCNN and our SRCFN are

Table 1 Comparison of methods: bicubic, SRCNN and our SRCFN on 480 test MR images. Highest calculated measures [PSNR (dB), SSIM] in bold

Parameters	PSNR (dB)	SSIM
Method		
Bicubic	23.7448	0.7941
SRCNN	27.7551	0.8793
SRCFN (our)	**28.3695**	**0.8977**

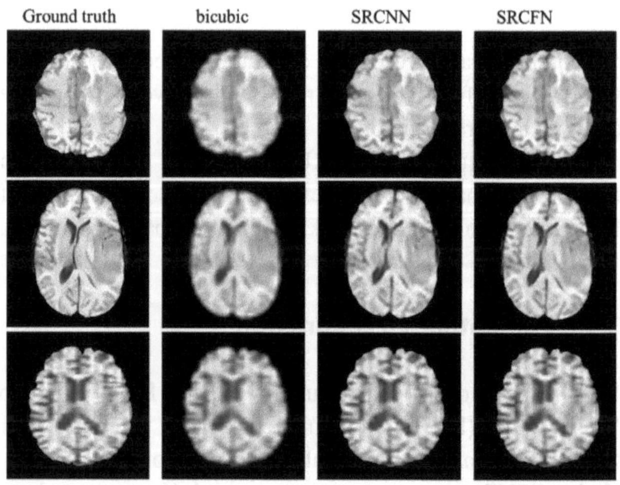

Fig. 3 Qualitative comparison. *1* The first column shows the ground truth MR image. *2* The second column shows SISR results of bicubic method. *3* The third column shows SISR results of SRCNN. *4* The forth column shows SISR results of our SRCFN. SRCFN and SRCNN recover sharp lines, bicubic give blurry results. Row 1: brats_tcia_pat103_0001/60 slice, row2: brats_tcia_pat165_0001/52 slice, row3: brats_2013_pat0027_1/41 slice

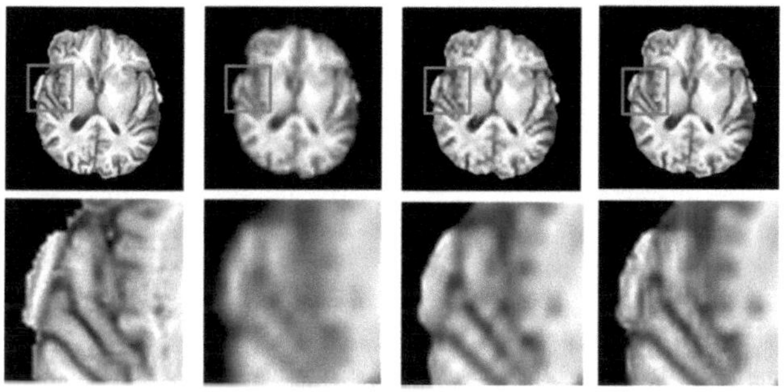

Fig. 4 Qualitative comparison. *1* The first row shows qualitative comparison of three methods in MR image (brats_tcia_pat247_0001/44 slice). *2* The second row shows local detail of results

illustrated in Fig. 3. As we can see, our method produces relatively sharper edges with respect to patterns, while bicubic method may give blurry result and SRCNN may give lower PSNR and SSIM result. DL (SRCNN) achieve good result applied in both nature images and medical images with higher PSNR and SSIM than non-DL (bicubic) method. For the quantitative comparisons of PSNR and SSIM, our method (SRFCN) is better than SRCNN method. For all that, DL based method is better than non-DL (Shown in Table 1 and Fig. 4).

4 Conclusion

We have described a deep Super-Resolution Coarse-to-Fine Network (SRCFN) with perceptual losses, which can jointly compute good high-level features and good low-level features. SRCFN achieved better experimental results for MRI super-resolution than state of the art models. Our method is significant to reduce scanning time under the image quality assurance for clinical application. We have proposed a new perceptual loss for MRI-SR, which combined PSNR and SSIM. We will focus on applied and improved this method in other medical applications in the future, such as SR for diffusion tensor imaging.

Acknowledgements This work was supported in part by the National Natural Science Foundation of China (81771940, 81427803), National Key Research and Development Program of China (2017YFC0108000), Beijing Municipal Science & Technology Commission (Z151100003915079), and Beijing Municipal Natural Science Foundation (7172122).

References

1. Nasrollahi, K., Moeslund, T.B.: Super-resolution: a comprehensive survey. Mach. Vis. Appl. **25**(6), 1423–1468 (2014)
2. Yang, C.Y, Ma, C, Yang, M.H.: Single-image super-resolution: a benchmark. In: European Conference on Computer Vision, pp. 372–386. Springer, Cham (2014)
3. Duchon, C.E.: Lanczos filtering in one and two dimensions. J. Appl. Meteorol. **18**(8), 1016–1022 (1979)
4. Yang, J., Wright, J., Huang, T.S., et al.: Image super-resolution via sparse representation. IEEE Trans. Image Process. **19**(11), 2861–2873 (2010)
5. Zeyde, R., Elad, M., Protter, M.: On single image scale-up using sparse-representations. In: International conference on curves and surfaces, pp. 711–730. Springer, Berlin, Heidelberg (2010)
6. Glasner, D., Bagon, S., Irani, M.: Super-resolution from a single image. In: Computer Vision, 2009 IEEE 12th International Conference on, pp. 349–356. IEEE (2009)
7. Pérez-Pellitero, E., Salvador, J., Ruiz-Hidalgo, J., et al.: PSyCo: manifold span reduction for super resolution. In: Proceedings of the IEEE Conference on Computer Vision and Pattern Recognition, pp. 1837–1845. (2016)
8. Salvador, J., Pérez-Pellitero, E.: Naive bayes super-resolution forest. In: Proceedings of the IEEE International Conference on Computer Vision, 325–333. (2015)
9. Dong, C., Loy, C.C., He, K., et al.: Learning a deep convolutional network for image super-resolution. In: European Conference on Computer Vision, pp. 184–199. Springer, Cham (2014)
10. Kim, J., Lee, J.K., Lee, K.M.: Deeply-recursive convolutional network for image super-resolution, 1637–1645 (2015)
11. He, K., Zhang, X., Ren, S., et al.: Deep residual learning for image recognition. In: Proceedings of the IEEE conference on computer vision and pattern recognition, 770–778. (2016)
12. Wang, Z., Bovik, A.C., Sheikh, H.R., et al.: Image quality assessment: from error visibility to structural similarity. IEEE Trans. Image Process. **13**(4), 600–612 (2004)
13. Menze, B.H., Jakab, A., Bauer, S., et al.: The multimodal brain tumor image segmentation benchmark (BRATS). IEEE Trans. Med. Imaging **34**(10), 1993–2024 (2015)

High-Quality 3D Autostereoscopic Medical Display Method with Optical Enhancement and Image Compensation

Xinran Zhang, Tianqi Huang, and Hongen Liao

Abstract

With the growing demand for intuitive and accurate medical image visualization, autostereoscopic 3D display becomes a new developing trend in image guided diagnosis and therapy. However, the limited resolution and inadequate accuracy of current autostereoscopic techniques are two common problems that restrict the application of 3D medical display. This research introduces an image enhancement and optical compensation method based high quality autostereoscopic medical display system. An image enhancement module is designed to improve the spatial resolution and viewing angle of the same time. The accuracy degradation caused by optical aberration is compensated by the proposed raytracing rendering. Experiments demonstrated the feasibility in the improvement of 3D image quality and accuracy. Anatomic details of volunteer data were well shown by the proposed method. With further improvements in real-time image rendering, the proposed methods will have wide applicability in improving the display performance in the field of AR guidance system, microscope and head-mounted display systems.

Keywords

Autostereoscopic medical display • Image enhancement • Image compensation • Image guided surgery

1 Introduction

With the fast development of medical imaging techniques, the visualization methods of medical images also become hot areas of researches. Since pre-/intra-operative medical images play an increasingly important role in computer assisted diagnosis and therapy, the accuracy and intuitiveness of image visualization are two main important factors influencing the efficiency of operations. Conventional medical image visualization techniques often display two-dimensional (2D) cross-sectional images and three-dimensional (3D) volume rendering or surface rendering models on common 2D panel screens. Although the 3D rendering algorithm can achieve a vivid reconstruction of anatomic structures, doctors still need to rotate or move the 3D model to understand the spatial relationships of anatomic structures [1]. The loss of depth information may have a negative impact on the correct understanding and judgement of doctors especially when the anatomic relation gets complicated.

In recent years, 3D displays and augmented reality (AR) techniques are widely applied in medical applications. 3D information can significantly accelerate the speed of human cognition compared with 2D information [2]. Stereo 3D display technique, which is frequently used in 3D surgical microscope, 3D surgical robot and head-mounted display (HMD), can produce the depth cues by providing two different 2D images separately to each eye of observer [3]. Visual tiredness is easy to occur after long-time watching the stereo 3D images when users must wear special glasses. The more important thing is that the accuracy inconsistency introduced by individual difference restricts the use of stereo

X. Zhang · T. Huang · H. Liao (✉)
Department of Biomedical Engineering, School of Medicine,
Tsinghua University, Beijing, China
e-mail: liao@tsinghua.edu.cn

F. Ibrahim et al. (eds.), *2nd International Conference for Innovation in Biomedical Engineering and Life Sciences*,
IFMBE Proceedings 67, https://doi.org/10.1007/978-981-10-7554-4_43

3D techniques in high-accuracy augmented reality applications. Autostereoscopic 3D display techniques, which can provide the depth cues without any supplemental equipment, often use optical lens or grating to achieve naked-eye 3D imaging. Although autostereoscopic 3D display shows a great potential in medical use, the inadequate accuracy and resolution of autostereoscopic 3D display are two important problems we need to handle [4].

The resolution limitation of real 3D display such as integral photography (IP), lenticular arrays and parallax barrier based techniques are mainly caused by the sampling effect of the optical elements to the elemental image. Due to the high difficulty in fabricating super high resolution 2D display and micro lens array/grating, resolution of real-3D displays cannot satisfy the requirement in medical use if users want to distinguish the detailed structures [5]. The accuracy degradation of optical lens-based autostereoscopic 3D displays is mainly caused by the mismatch between the parameters of rendered 3D image and that of the real optical instruments during computer-generated (CG) 3D rendering [6]. For example, in conventional CG rendering of IP, the convex lens in micro lens array is simplified as an ideal pinhole. Furthermore, in real 3D augmented reality (AR) systems, HMD and floating display systems, more optical lenses or mirrors are used in the display system. How to take the aberration of real optical model into the consideration during CG real 3D medical imaging is quite important for the promotion of display accuracy.

2 Methods

2.1 High Accuracy and Resolution Real 3D Medical Display Method

To solve the difficulty in exact understanding that often occurs when surgeons visualizing complex and detailed 3D anatomic structures based on common 2D medical image, real 3D medical display with high quality and accuracy that will preserve the depth cues of structures are highly necessary [7].

The resolution of real 3D display is determined by the spatial pixel density of image. We design a set of negative optical lenses in front of conventional 3D display to demagnify the 3D image [8]. The compression of the pixel size and their distance means the improvement of spatial resolution. Meanwhile, lights emitted from 3D image passed through the optical system are diverged into a larger angle, resulting in the increase of viewing angle (Fig. 1a). This method breaks the tradeoff between viewing angle and resolution in conventional real 3D displays [9]. To further handle the accuracy loss introduced by aberrations of optical elements in the display system, we add the aberration models of optical lenses in the 3D image rendering pipeline during the image generation procedure instead of using the ideal optical models in traditional approaches (Fig. 1b).

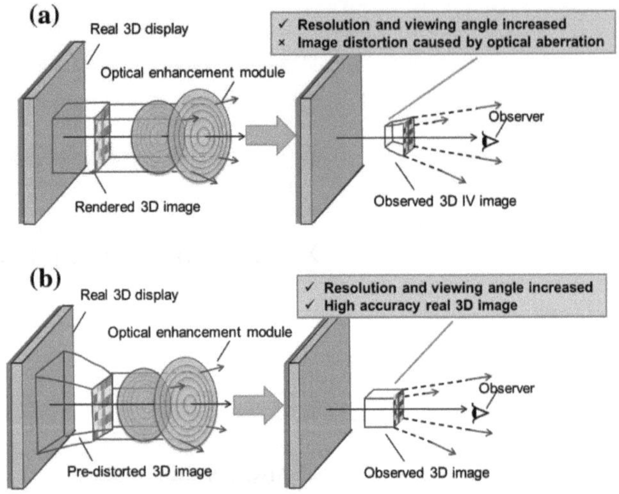

Fig. 1 The principle of the accuracy and resolution enhancement method proposed by this article. **a** A set of negative optical lenses are placed in front of the 3D display as the image enhancement module to increase resolution and viewing angle. **b** Aberration parameters of optical device in the display system are considered during the CG 3D image process

In this way, the proposed method breaks the tradeoff between the viewing angle and resolution in IP based real 3D medical display technique. The image distortion correction method also provides an effective approach to achieve high-accuracy medical image acquisition and visualization.

2.2 Analysis of Resolution and Viewing Angle of Real 3D Medical Display

The optical image enhancement module is composed of a set of lenses, which equals to a concave lens. In this section we use IP as an example to analyze the enhancement effect of resolution and viewing angle to real 3D images. The spatial resolutions in lateral and depth directions are determined the maximum size of a single light point in space. In traditional IP, the tradeoff between spatial resolution, viewing angle and image depth is unavoidable. After we place the optical image enhancement module in front of the conventional IP display, the spatial sizes of the light point both in horizontal and depth directions turn smaller. The enhancement of resolution and viewing angle of images at the same time makes real 3D medical displays are able to show more anatomic details in a more natural way. The spatial size and the viewing angle of the light point can be control by the focal length and position of the adjustable optical enhancement system, making it flexible and easy to satisfied various needs in biomedical applications.

2.3 Aberration-Compensated Real 3D Image Rendering

To reproduce a high-accuracy spatial 3D image that is in consistent with the medical data, we proposed an optical aberration-compensated image rendering method to address the image distortion problem in 3D medical display systems [10].

We assume that one voxel s_i in the medical data emits rays to space. Each ray $\overrightarrow{n_f}$ passes through all the optical lenses in the display system and is finally recorded on the 2D elemental image plane as $\mathrm{E}(a, b)$. Instead of the ideal optical imaging model used in traditional methods to calculate the position of E, we calculate every intersection points P_1 *to* P_n between the ray and all optical surfaces s_1 *to* s_n, as shown in Fig. 2.

The calculation of P_1 *to* P_n requires to know the surface parameters of optical lenses. We can use a 3D triangular mesh to describe the surfaces. When the surface of optical lens can be described using mathematical formula, the calculation of P_1 *to* P_n can be further accelerated. When medical data becomes large, the raytracing rendering will be time-consuming. The tracing of different rays can be calculated in parallel using graphics processing unit. The proposed aberration-compensated real 3D image rendering method solve the image distortion problem in many lens-based autostereoscopic display systems. The method is also suitable in the image correction in surgical microscope, HMD and other AR medical systems.

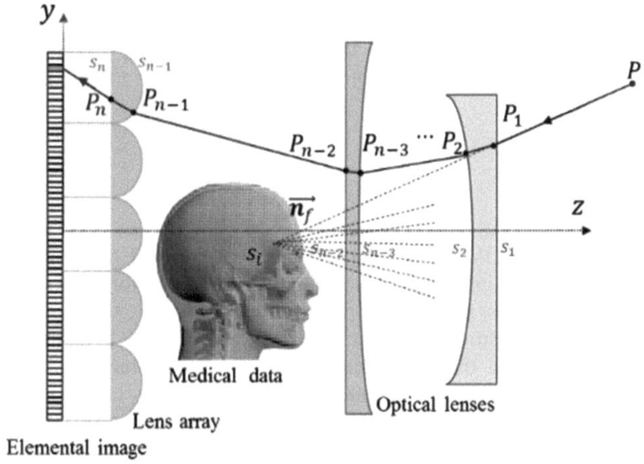

Fig. 2 Optical aberration-compensated real 3D image rendering using a front-to-back raytracing method

3 Experiments and Results

3.1 Materials

To validate the feasibility of the proposed real 3D medical visualization methods, we built up an IP-based autostereoscopic display system. The resolution of the liquid crystal display (LCD) used to show the elemental image is 1536 × 2048 (264 ppi). The lens pitch and focal length of the hexagonal micro lens array are 2.32 and 3.0 mm, respectively. The focal length of the proposed optical image enhancement system is −130 mm.

3.2 Quantitative Accuracy and Resolution Evaluation of the High Quality 3D Display Method

We used the proposed medical display and the traditional IP display to show three different 3D cube patterns to evaluate the image resolution and accuracy.

The comparison of performance with and without the image enhancement and aberration compensation is illustrated in Fig. 3. The edges and apexes were significantly clearer after image enhancement due to the promotion of resolution (Fig. 3b, d). Meanwhile, the aberration compensation method efficiently corrected the distortion both in horizontal and depth directions (Fig. 3c, d). We further used a stereo camera to evaluate the spatial accuracy of three displayed cube patterns. The average deviations of distance of measured value and theoretical value are shown in Fig. 3e. The results showed that the proposed method reduce the display error in distance compared with traditional approach.

3.3 Volunteer Experiments and the Prototype AR Navigation System

We used a 3D brain data of a clinical volunteer acquired by CT to evaluate the performance of the proposed method. Details of small and complex structures can be easily distinguished from the real 3D image. We also applied the display in an AR surgical guidance system to provide an intuitive guidance in minimally invasive diagnosis and therapy (Fig. 4).

4 Discussion and Conclusion

In this research, we proposed a high-accuracy and resolution enhanced 3D autostereoscopic medical display technique using optical enhancement and image compensation

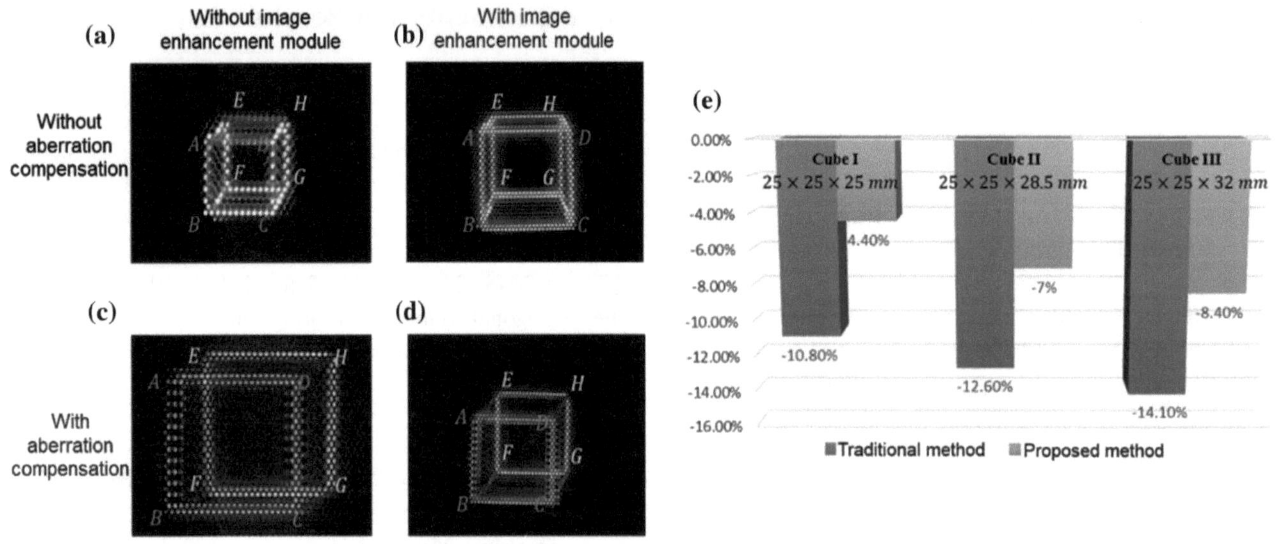

Fig. 3 **a–e** The observed autostereoscopic 3D image generated with and without the proposed image enhancement module and aberration compensation algorithm. **e** Average deviations of distance of measured value and theoretical value in accuracy evaluation

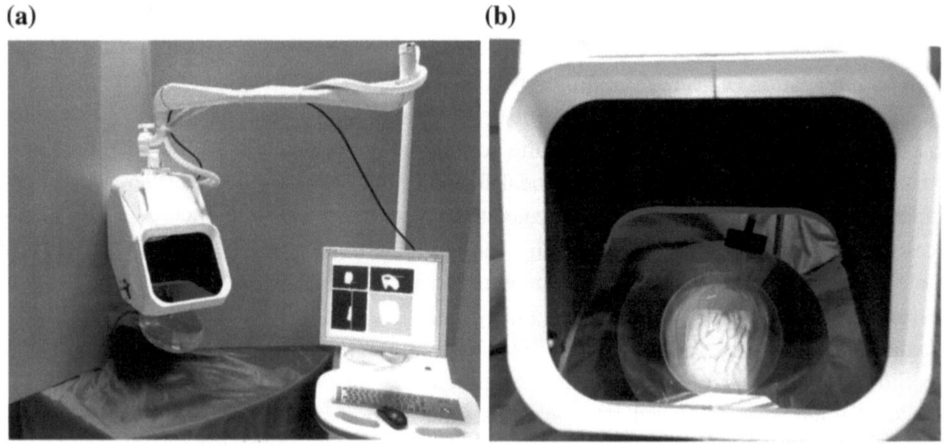

Fig. 4 **a** The prototype surgical navigation system based on real 3D medical display. **b** The AR surgical scene observed through the viewing window. The accurately overlaid images of internal structures provided an intuitive guidance in precise operation

methods. We design an optical image enhancement module and an aberration compensation rendering method to address the accuracy and resolution limitation problem of 3D display in medical use. Experiments demonstrated the feasibility and the potential in medical applications. The display error in distance can be further reduced by the calibration of the related position of optical lenses, the display and the patient. By using real-time imaging techniques such as ultrasound etc., the proposed display can also realize real-time image update and provide a true dynamic picture of the anatomy for surgeons.

The proposed method can be further applied in medical AR, virtual reality systems and floating displays, providing a more accurate and vivid image visualization and minimizing the exploratory procedures in surgical navigation, medical training and telemedicine.

Acknowledgements This work was supported in part by National Key Research and Development Program of China (2017YFC0108000), National Natural Science Foundation of China (81427803, 81771940), Beijing Municipal Science & Technology Commission (Z151100003915079), Beijing Municipal Natural Science Foundation (7172122), and Soochow-Tsinghua Innovation Project (2016SZ0206).

References

1. Zhang, X., Fan, Z., Wang, J., et al.: Augmented reality based orthopaedic interventions, computational radiology for orthopaedic interventions, pp. 71–90. Springer, New York, NY, USA (2015)
2. Orden, K., Broyles, J.: Visuospatial task performance as a function of two- and three-dimensional display presentation techniques. Displays **21**(1), 17–24 (2000)
3. Surman, P.: Stereoscopic and autostereoscopic displays, 3D-TV system with depth-image-based rendering, pp. 375–411. Springer, New York, NY, USA (2013)

4. Chen, G., Zhang, X., Fan, Z., et al.: An innovative calibration based integral photography rendering algorithm for medical application and its evaluation, the 37th annual international conference of the IEEE EMBC, pp. 4226–4229. Milano, Italy (2015)
5. Park, S., Yeom, J., Jeong, Y., et al.: Recent issues on integral imaging and its applications. J. Inf. Disp. **15**(1), 37–46 (2014)
6. Liao, H.: Super long viewing distance light homogeneous emitting three-dimensional display. Sci. Rep. **5**, 9532 (2015)
7. Geng, J.: Three-dimensional display technologies. Adv. Opt. Photonics 5(4), 456–535 (2014)
8. Zhang, X., Chen, G., Liao, H.: In situ real 3D see-through microsurgical guidance system. In: The 12th Asian Conference on Computer Aided Surgery, Korea (2016)
9. Min, S., Kim, J., Lee, B., et al.: New characteristic equation of three- dimensional integral imaging system and its applications. Jpn. J. Appl. Phys. **44**(1L), L71–L74 (2005)
10. Fan, Z., Zhang, S., Weng, Y., et al.: 3D quantitative evaluation system for autostereoscopic display. IEEE/OSA J. Disp. Technol. **12**(10), 1185–1196 (2016)

Application of Sinogram-Based Moving Window Subtraction Technique in OSEM Algorithm to Reduce Ring Artifacts for Cone-Beam Micro-CT Image Quality Assessment

David Shih-Chun Jin, Wen-Hsiang Chou, and Jyh-Cheng Chen

Abstract

Severe ring artifacts in cone-beam micro-CT degrade image quality and usually cause error recognition during image-based-observer study or computed-aid-diagnosis (CAD). It causes wrong values, such as size of the tumor. The aim of this study was to derive a moving window subtraction method before the iterative reconstruction step to reduce ring artifacts. To access acceptable image quality data, we used a uniform water cylinder phantom and followed the performance measurement guide, ISO-15708 and EN 16016, to evaluate the imaging performance. Results show there are 1.23 times improvement in SNR and 5% improvement in uniformity. Most importantly, we can reduce the error estimate of the object size from 6 pixels to less than 2 pixels. Our proposed method can efficiently reduce the error rate caused by ring artifacts for automatic size estimation applications.

Keywords

Ring artifacts • Ordered-subsets expectation-maximization • Image quality assessment

1 Introduction

Cone-beam micro-CT (CBμCT) uses a 2D X-ray detector to acquire projections while the gantry holding an X-ray source and detector pair moves around the center of the object. Since the pixels of many 2D flat-panel detectors are smaller than that of discrete detectors in typical medical CTs, the x-ray flux is shift variant, and there is no consistent linear response between the detector and the x-ray flux. This phenomena causes ring artifacts and makes it very difficult to achieve a perfect uniform sensitivity [1, 2]. Severe ring artifacts in CBμCT degrade image quality and cause wrong measurement in image-based-computed aided diagnosis (CAD).

Many researchers mention that a simple flat-field correction with a dark-current subtraction will not work for multiple projection images taken from CBμCT, even with a long flat-field acquisition time or a frame averaging method. This is why ring artifacts are usually shown on CBμCT images [3].

Some iterative algorithms have been used in CBμCT in recent years to improve image quality by continuously updating reconstructed images between iterations. The initial sinogram data, which was forward projected by a predefined initial image value, will be compared with the real projection data collected by the micro-CT system. An updated sinogram can then be back projected to obtain a corresponding updated image. However, the intrinsic artifacts from the original projection data cannot be removed and will be enhanced by iterations into the reconstructed image. Here we introduce a sinogram based efficient ring reduction technique

D. Shih-Chun Jin · W.-H. Chou · J.-C. Chen (✉)
Department of Biomedical Imaging and Radiological Sciences, National Yang-Ming University, Taipei, Taiwan
e-mail: jcchen@ym.edu.tw

J.-C. Chen
Biophotonics and Molecular Imaging Research Center, National Yang-Ming University, Taipei, Taiwan

J.-C. Chen
Radiology College, Tai-Shan Medical University, Shan-Dong, China

F. Ibrahim et al. (eds.), *2nd International Conference for Innovation in Biomedical Engineering and Life Sciences*,
IFMBE Proceedings 67, https://doi.org/10.1007/978-981-10-7554-4_44

that can be easily implemented with iterative reconstruction methods.

2 Materials and Methods

2.1 Flowchart of Reduction Method

This study applied a moving window subtraction method before an iterative reconstruction kernel to prevent artifact enhancement during ordered-subsets expectation maximization (OSEM) reconstruction, which was derived by [5]. The flowchart of preprocessing, proposed moving window subtraction, OSEM reconstruction and image quality assessment is shown (Fig. 1a). The following steps are involved.

Step.1: We need to do a basic standard procedure of projection image preprocessing, flat-field, dark-field, and exposure corrections [4]. Next, the line pattern in pre-corrected sinogram $\boldsymbol{S}$ is reduced by detecting this pattern from homogeneous row segments in $\boldsymbol{S}$. In the following steps, $\boldsymbol{S}$ is assumed to be of size $r \times \theta$.
Step.2: A window $\boldsymbol{W}_n$ of fixed size $w \times \theta$ is slid over $\boldsymbol{S}$. We move the window ranges from $n = 0$ to $r - w$ for each row $w_{r,m}$ of $\boldsymbol{W}_n$, where $m = 1,\ldots, \theta$. The signal variance $\sigma_{r,m}$ is then computed. If $\sigma_{r,m}$ is greater than three times its mean, $w_{r,m}$ will remove the straight-line artifacts by moving the outliers into an artifact map $\boldsymbol{A}_n$. For each row of $\boldsymbol{A}_n$, the median value is computed. This results in an artifact template vector a_n of size w. Finally, the artifact matrix $\boldsymbol{A}$, the sets of all $a_{n,}$ is subtracted from $\boldsymbol{S}$ to obtain the processed sinogram $\mathbf{S}_{\mathrm{o}}$.
Step.3: To access acceptable image quality data, we set the OSEM algorithm stopping criterion as the root-mean-square error (RMSE) of the sinogram between iteration numbers. The RMSE is calculated by the following formula:

$$RMSE = \sqrt{\frac{(\mathbf{S}_n - \mathbf{S}_o)(\mathbf{S}_n - \mathbf{S}_o)^T}{N}}, \quad (1)$$

where $\mathbf{S}_n$ denotes the *n-th* iterated sinogram, the "T" denotes matrix transposition, and N denotes the total pixel number in one sinogram.

2.2 Experimental Design and Setting

In this study, all data were collected from our handmade CBμCT, which includes an X-ray tube (Hamamatsu L10951) and a flat-panel detector (Dexela1512) (Fig. 1b). All 360-degree projection data were reconstructed by OSEM with 10 subsets. The distance from the X-ray focal spot to the detector is 150.994 mm, and the distance from the rotation axis to the detector is 58.357 mm.

Since excessive iterations cause artifact enhancement, which must be avoided, we calculated the effective diameter from the reconstructed image to estimate the size of a uniform water cylinder phantom (QRM GmbH), which is made of a stable transparent plastic providing a low density (1.0 g/cm^3) in diameter 32.5 mm (equivalent to 706.8 pixels) (Fig. 1c). This phantom can also be used to assess the signal-to-noise ratio (*SNR*) and the uniformity (*U*) measurement as mentioned in ISO-15708 and EN-16016-3 regulation guides [6–8]. The *SNR* and *U* are defined as:

$$SNR = \frac{\mu_{ROI}}{\sigma_{ROI}}, \quad (2)$$

$$U = \frac{|Max_{global} - Min_{global}|}{|Max_{global} + Min_{global}|} \times 100\%, \quad (3)$$

where the notation μ_{ROI} and σ_{ROI} are the mean and standard deviation of the ROI, respectively, and the area of the ROI is suggested to include at least 100 pixels. Here we choose five different ROIs at the central, upper, lower, left and right parts of the uniform region. After choosing the whole uniform region and finding their global extreme values to calculate uniformity, we found that the more uniform the region is, the lower the *U* we obtain. We used the circle detection algorithm [9, 10] to automatically detect the boundary of the water phantom image, calculate the effective diameter (D) of the phantom, and simulate the size estimation by CAD.

3 Results and Discussion

We found that choosing w = 15 is optimal for removing almost all artifacts in this study because most horizontal line patterns were deleted (Fig. 2b). Thus, we chose this parameter as a basic setting before OSEM reconstruction. We found that there are two steep decreases in the RMSE; the RMSE drops to 10^{-4} before 3 iterations, and then to 10^{-6}

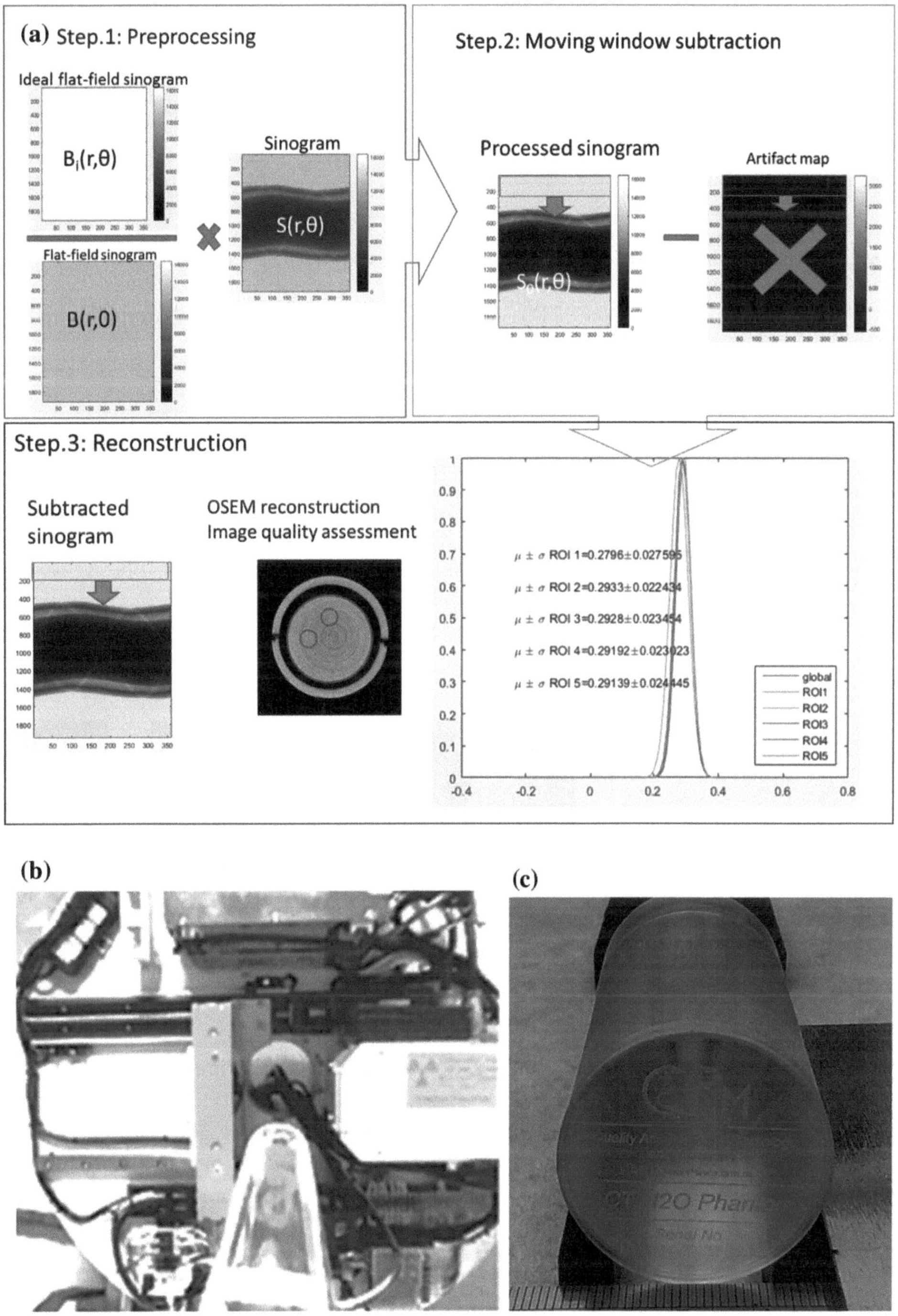

Fig. 1 **a** The flowchart of the sinogram moving window subtraction technique combined with OSEM reconstruction. **b** Handmade CBμCT system. **c** QRM uniform water cylinder phantom

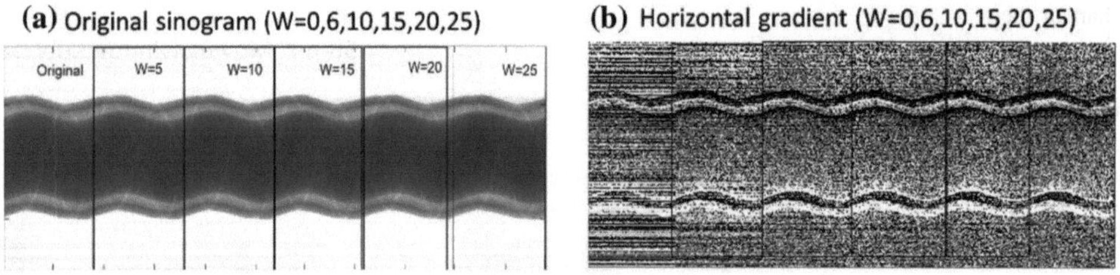

Fig. 2 Different moving window size selection to reduce the artifacts in the (a) original sinogram. And we use (b) horizontal gradient (applied sobel kernel) to observe this phenomena

Fig. 3 (a) Plot and (b) semi-log plotting of the RMSE values from OSEM image to avoid over-iteration. Early stopping point at the RMSE < 10^{-7}. The OSEM images are used to compare (c) OSEM only and (d) OSEM with the proposed ring artifact reduction image (W = 15)

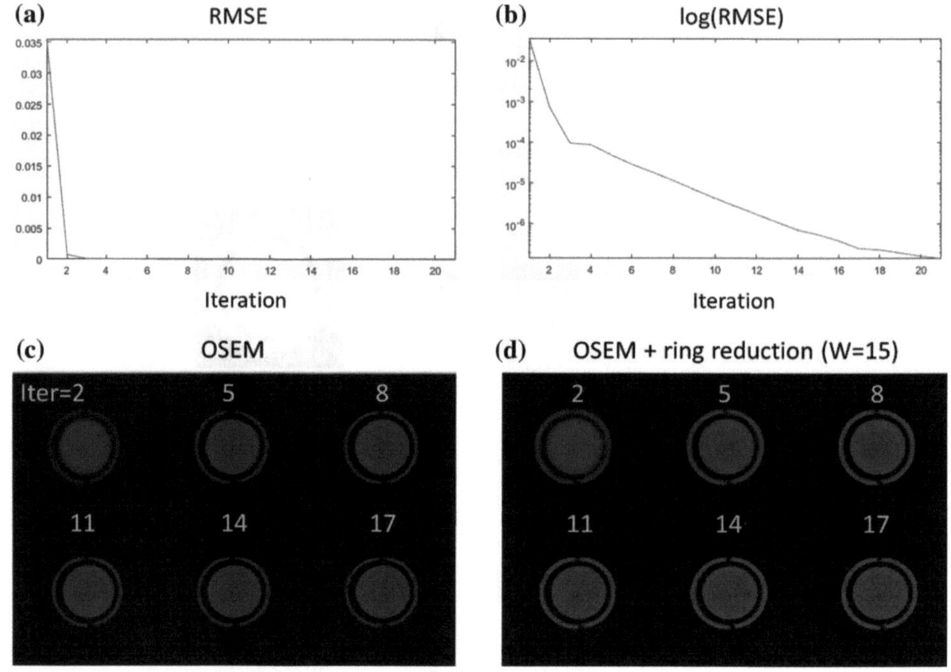

before 14 iterations (Fig. 3a and b). These figures indicate only 2 iterations is enough to reconstruct the disk shape. Sufficient detail enhancement requires at least 17 iterations (Fig. 3c and d). As expected, the ring pattern is also enhanced.

Table 1 records the image quality assessment from OSEM reconstructed images with and without applied reduction algorithm by calculating SNR, U and D (Fig. 3c and d). A simple method is to select a region-of-interest (ROI) that is applied in this procedure. We find that *SNR* tends to monotonically decrease, and the value of *U* tends to increase with increasing iterative number no matter whether the ring reduction algorithm is applied or not. This is because in our OSEM, we used a uniform matrix as our initial guess to start the reconstruction. The more OSEM iterations applied, the more the details, noise, and ring pattern are enhanced. After applying the ring reduction algorithm, a *D* value closer to the true diameter (706.8 pixels) can be estimated. The ring reduction algorithm also improves SNR by 1.23 times and uniformity by 3–5%. This means the accuracy of the CAD depends on whether the source image contains artifacts. Our proposed method can efficiently reduce the ring artifacts and thus avoid error size estimation.

Table 1 Image quality assessment for uniform water phantom measurement

Method	OSEM only				OSEM + ring reduction (W = 15)					
# of Iter.	SNR	U (%)	D (pixel)	Error (pixel)	SNR	Improve SNR	U (%)	Improve U (%)	D (pixel)	Error (pixel)
8	18.5	28.7	700.7	−6.1	22.8	1.23	27.9	0.9	707.6	0.8
11	14.7	33.9	702.0	−4.9	18.0	1.22	30.7	3.3	707.8	1.0
14	12.5	39.3	702.4	−4.4	15.0	1.20	34.3	5.0	708.1	1.3
17	10.9	43.9	702.6	−4.2	13.0	1.18	38.6	5.4	708.4	1.6

**Note* diameter = 706.8 pixels

4 Conclusions

This paper provides a solution for reducing ring artifacts to avoid recognizable errors when we use the CAD method to evaluate the circle diameter. We apply the projection based flat-field pre-correction, the proposed sinogram based moving window subtraction, and the OSEM algorithm. Image quality assessment results show that SNR improves by 1.23 times, and image uniformity improves by 5%. Notably, the proposed method reduced the error size of the diameter to less than 2 pixels. This valuable finding can support our next mission to improve the correct size recognition by fully automatic CAD algorithm.

Acknowledgements The authors thank for research funding supported by project YM-104-C-008. We also acknowledge Meng-Tse Wu for helping us calibrate the CBμCT system and supporting the gantry alignment technique.

References

1. Lee, S.C., Kim, H.K., Chun, I.K., et al.: A flat-panel detector based micro-CT system: performance evaluation for small-animal imaging. Phys. Med. Biol. **48**(24), 4173–4185 (2003)
2. Zuber, M., Hamann, E., Ballabriga, R., et al.: An investigation into the temporal stability of CdTe-based photon counting detectors during spectral micro-CT acquisitions. Biomed. Phys. Eng. Express. **1**(2), 025205 (2015)
3. Seibert, J.A., Boone, J.M., Lindfors, K.: Flat-field correction technique for digital detectors. In: Proceedings SPIE. 3336, Medical Imaging 1998: Physics of Medical Imaging, pp. 348–354 (1998)
4. Patel, V., Chityala, R.N., Hoffmann, K.R., et al.: Self-calibration of a cone-beam micro-CT system. Med. Phys. **36**(1), 48–58 (2009)
5. De Man, B., Nuyts, J., Dupont, P., et al.: An iterative maximum-likelihood polychromatic algorithm for CT. IEEE Trans. Med. Imag. **20**(10), 999–1008 (2001)
6. ISO 15708-1: Non-destructive testing—radiation methods—computed tomography—Part 2: examination practices (2002)
7. ISO 15708-4: Non-destructive testing—radiation methods for computed tomography—Part 4: qualification (2017)
8. BS EN 16016-3: Non-destructive testing. Radiation methods. Computed Tomography. Operation and interpretation (2011)
9. Atherton, T.J., Kerbyson, D.J.: Size invariant circle detection. Imag. Vis. Comp. **17**(11), 795–803 (1999)
10. Yuen, H.K., Princen, J., Illingworth, J., et al.: Comparative study of Hough transform methods for circle finding. Imag. Vis. Comp. **8** (1), 71–77 (1990)

Is There a Relationship Between Bone Mineral Density, Mammographic Breast Density and Serum Estrogen Concentration Level? Study on Malaysian Women

N. M. Zain, K. C. Kanaga, E. S. Latif, and A. Shantini

Abstract

Estrogen is known to affect both mammographic breast density and bone mineral density. Breast density is a well-known risk factor of breast cancer determined by the physiological hormonal changes. Studies shown inconclusive relationship of breast density and bone mineral density as breast cancer risk factors. This study aimed to determine the relationship of bone mineral density; mammographic breast density and serum estrogen concentration level in Malaysian women. A cross sectional study on 108 women aged above 40 years who underwent bone scan and mammography screening was conducted at the General Hospital Kuala Lumpur from 2016 to 2017. Subjects were chosen using convenient sampling technique. With women commenced cancer treatment and women who had mastectomy done being excluded from the study. An independent-t and chi-square statistical test were used to analyze the continuous and categorical data whilst binary logistic regression was used to determine the mentioned relationship. A total of 75.5% of the respondents had a normal bone mineral density in the hip and 76.9% in the spine. Abnormality in bone mineral density were recorded to be higher in postmenopausal women with the bone mineral density in spine (23.2%) region were slightly higher compared to the hip (18.5%). There was no statistical significant relationship between mammographic breast density and serum concentration level with bone mineral density in hip and spine.

Keywords

Bone mineral density • Mammographic breast density • Estrogen

The original version of this chapter was revised: Incorrect figure has been corrected. The erratum to this chapter is available at https://doi.org/10.1007/978-981-10-7554-4_51

N. M. Zain · K. C. Kanaga (✉)
Diagnostic Imaging & Radiotherapy Program, Faculty of Health Sciences, Universiti Kebangsaan Malaysia, Kuala Lumpur, Malaysia
e-mail: kanagakkc@yahoo.com

N. M. Zain
Medical Imaging Program, School of Health Sciences, KPJ Healthcare University College, Nilai, Malaysia

E. S. Latif · A. Shantini
Program of Biomedical Science, Faculty of Health Sciences, Universiti Kebangsaan Malaysia, Kuala Lumpur, Malaysia

1 Introduction

There is high number and proportion of female with breast cancer globally. More than one million women diagnosed with breast cancer every year, and more than 410,000 may result in mortality [1]. With an estimated 1.38 million new cases diagnosed worldwide in women in 2008, it represents about 23% of all cases of cancer [2]. A marked geographical variation was observed, with the highest incidence in northern Europe and North America and lowest in Africa and Asia. In more developed countries, the age-standardized incidences

F. Ibrahim et al. (eds.), *2nd International Conference for Innovation in Biomedical Engineering and Life Sciences*,
IFMBE Proceedings 67, https://doi.org/10.1007/978-981-10-7554-4_45

were 95 per 100,000 compared to 20 per 100,000 in less developed countries [3]. Variation in incidence rates worldwide is thought to be due to differences in reproductive patterns and other hormonal factors as well as early detection rate [4, 5].

In Malaysia, breast cancer is increasingly common [6]. Based on study by Omar [7] a total of 3242 new cases of breast cancer were reported in 2007, given an age-standardized incidence rate of 29.1 per 100,000 women. Malaysian women present at earlier age compared to women in Western countries [8]. In addition, a collaborative study between two tertiary academic hospitals in Malaysia and Singapore found that approximately 50% of women were diagnosed before the age of 50 years [9]. Several risk factors standard measures in detecting breast cancer. In addition, the risk factors for breast cancer are often used as prognostic factors too [10].

Another commonly discussed risk factor is bone mineral density (BMD) [11]. BMD is influenced by multiple factors including calcium and vitamin D intake, physical activity, weight, ethnicity and endogenous and exogenous estrogen [12, 13]. Estrogen has a central role in the maintenance of bone integrity. Its deficiency produces accelerated bone resorption leading to decreased BMD [14]. Thus, BMD may be a biomarker for cumulative estrogen exposure over a woman's lifetime, potentially providing a measure of breast cancer risk [15]. Therefore, this study was designed to determine the relationship between BMD, mammographic breast density (MBD) and serum estrogen concentration level on how these three factors are linked to each other.

2 Methodology

This cross-sectional study was conducted at Department of Radiology, General Hospital Kuala Lumpur for one year from July 2016 to July 2017. This study was approved by Institutional Review Board at National Medical Research Register [NMRR-15-198-24414(IIR)] and University Kebangsaan Malaysia Medical Center (NN-017-2015). Women aged 40 years old and above who underwent mammogram were offered bone scan and blood test. Women who had mastectomy; or undergoing cancer treatment, or consumed calcium supplement were excluded from this study. A convenient sampling technique was used to recruit 130 women. Twenty-two respondents were unable to complete three examinations were excluded from the study. Final recruitment was 108 participants. The overview of the methodology is in Fig. 1.

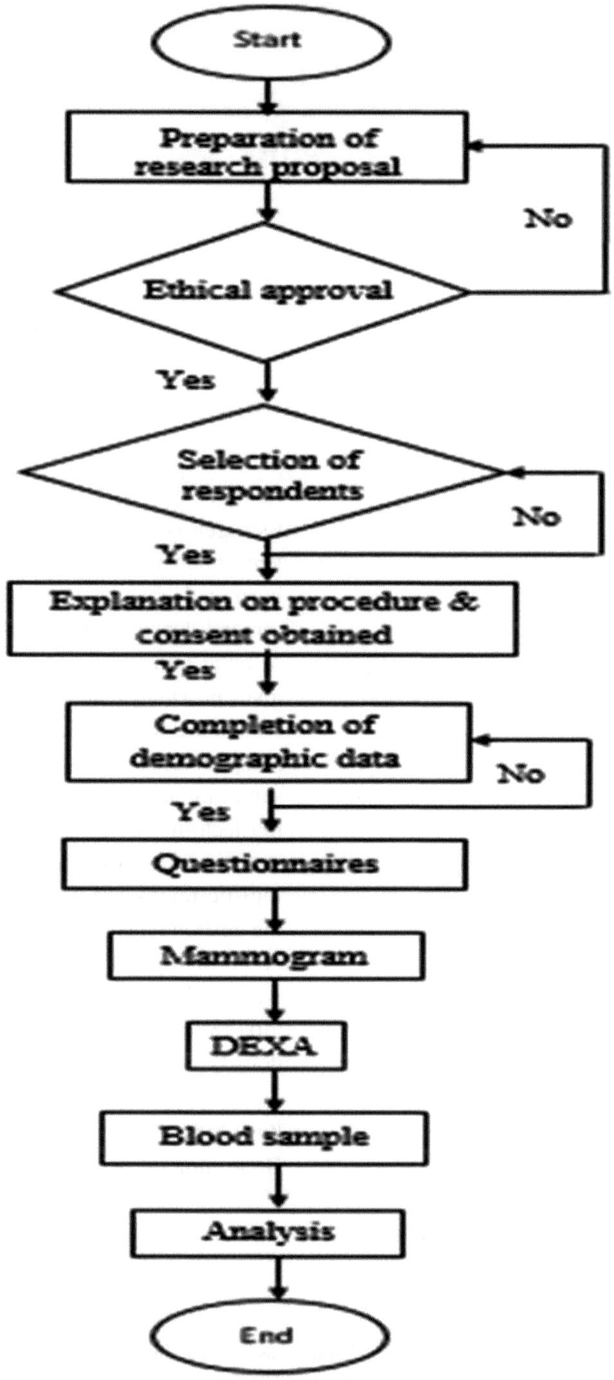

Fig. 1 Research flow chart

2.1 Mammogram

We used digital mammography system (*Senographe Essential, General Electric Company, United States*) to obtain breast tissue images. Breast composition was defined by the visual estimation content of fibroglandular-density tissue within the breasts by an experienced radiologist according to standard Breast Imaging-Reporting and Data System (BI-RADS). The breast density was categorized as follows; BI-RADS a: the breasts are almost entirely fat, BI-RADS b: there are scattered areas of fibroglandular density, BI-RADS c: the breasts are heterogeneously dense, and BI-RADS d: the breasts are extremely dense [16].

2.2 Dual Energy X-ray Absorptiometry (DEXA)

A DEXA unit (*Discovery QDR Series, Hologic Discovery W, Hologic Inc. Bedford, MA*) was used to measure BMD at the femoral neck and lumbar spine in grams per centimeter (g/cm^2). The BMD was also expressed as T-scores [17] with normal is ≥ -1 standard deviation (SD); osteopenia is between −1 and −2.5 SDs and osteoporosis is ≤ -2.5 SDs. In this study, a score <1 SD was considered abnormal. Serum Estrogen Concentration Level.

2.3 Serum Estrogen Concentration Level

We obtained 10 ml of blood via the brachial vein [18] and the samples were kept separately into two different tubes. A plain non-heparinized vacutainer was used to store 4 ml of blood sample taken and the remaining samples were transferred into a heparinized vacutainer. The separation of blood components were done and stored at −80 °C. Estrogen was tested using a Total Estrogens ELISA (*Cat TE192S (96 Tests), Calbiotech Inc. Spring Valley, CA*).

2.4 Statistical Analysis

Descriptive analysis was done for demographic data. Analysis was presented in groups of pre- and postmenopausal women. An independent-t and chi-square tests were used to analyse continuous and categorical data respectively. Significant value was $p < 0.05$. Binary logistic regression was used to determine relationship of BMD, MBD and serum estrogen concentration level. All analysis were performed by using Statistical Packages for Social Science (SPSS) version 22.0 (SPSS Inc. Chicago IL, USA).

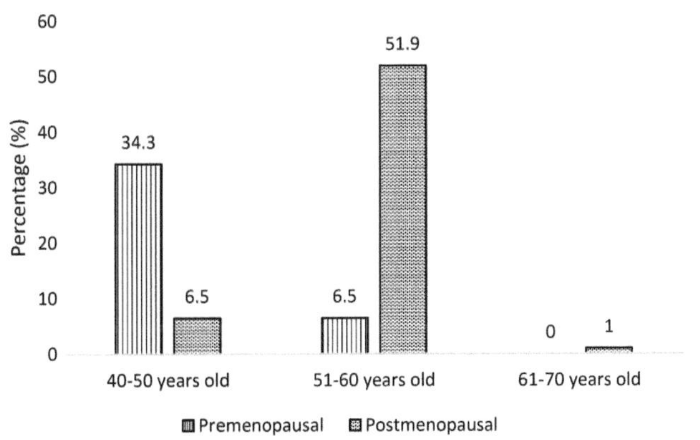

Fig. 2 Age of pre- and postmenopausal Malaysian women

3 Results

Majority of respondents were postmenopausal women aged 51–60 years old and the premenopausal women were within the age range of 40–50 years old (see Fig. 2). A total of 69.7% women had breast density BI-RADS *b* (scattered fibroglandular) followed by 12.2% BI-RADS *c* (heterogeneously dense), 7.1% BI-RADS *d* (extremely dense) and 11.1% women with BI-RADS *a* (fatty tissue). The distribution of breast density type according to menses status is in Fig. 3.

Majority of respondents were shown to have a normal BMD hip (75.5%) and BMD spine (76.9%) T-score. Abnormal BMD was higher in spine (23.2%) and slightly less in hip (18.5%) with postmenopausal women shown to have higher frequency of having abnormal BMD as compared to premenopausal women (see Fig. 4).

Comparison of BMD and MBD hip and spine reported no statistical significance between the two groups of women. Majority of them (13.1%) with scattered fibroglandular breast tissue density had normal BMD. No abnormal BMD was recorded for women with extremely dense breast (Table 1). Mean serum estrogen concentration level was

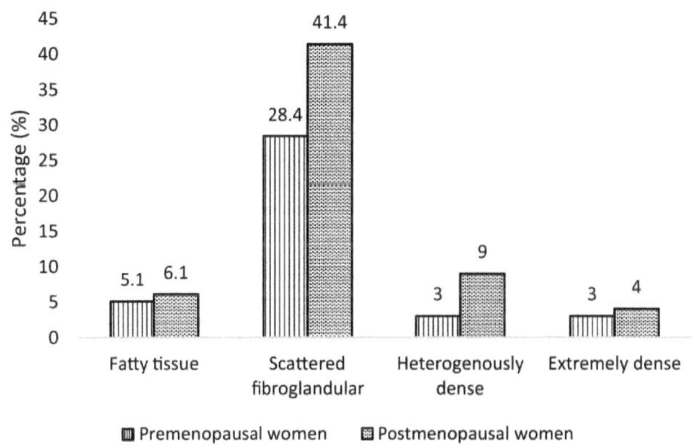

Fig. 3 Distribution of breast density type according to menses status

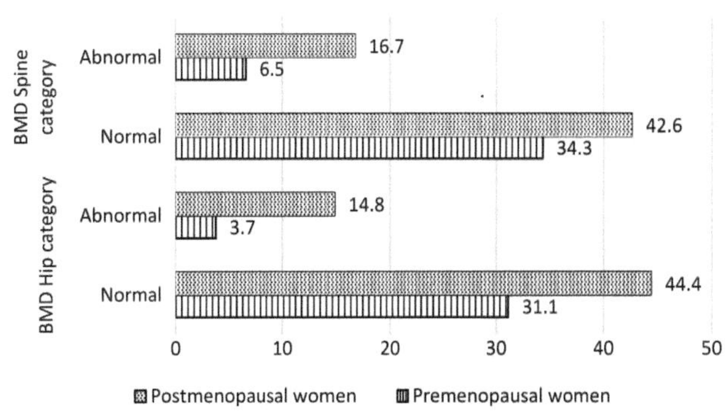

Fig. 4 BMD hip and spine according to menses status

Table 1 Association of BMD and MBD

	BMD hip			BMD spine		
	Normal	Abnormal	*p*-value	Normal	Abnormal	*p*-value
MBD			0.521[a]			0.536[a]
BI-RADS *a* (%)	9.1	2.0		7.1	4.1	
BI-RADS *b* (%)	57.1	13.1		53.1	17.3	
BI-RADS *c* (%)	81.6	4.1		9.1	3.1	
BI-RADS *d* (%)	6.1	0.0		6.1	0.0	

[a]Chi-square analysis

higher in premenopausal women (35.3 pg/ml) as compared to postmenopausal women (12.5 pg/ml).

Binary logistic regression analysis showed no significant relationship of MBD ($p > 0.05$) and serum concentration level towards BMD hip [OR = 0.998 (95% CI = 0.992–1.003)] ($p > 0.05$) and spine [OR = 0.997 (95% CI = 0.992–1.003)] ($p > 0.05$).

4 Discussion

BMD and MBD are likely connected to each other through inherited estrogen in women. Hence, there should be a relationship between BMD and MBD [19]. However, the relationship between MBD and BMD has not been well studied therefore, the relationship is unclear [20]. Nonetheless, in this cross-sectional work, we found contradictory results, which is similar to the results in the study by Chen [21] and Buist [22]; it is likely that unique organ responses obscure the relationship between BMD and MBD as cumulative estrogen exposure [22].

In addition, this opposing relationship can occur possibly because there are many factors that made the BMD level in women's body vary [23]. The variance of MBD appearances and BMD levels could be possibly due to different ethnicities and the reliant of their lifetime exposure to estrogen.

In a systematic analysis done on nine studies related to the association of BMD to MBD as predictors of breast cancer; estrogen also contributes a significant role in affecting and sustaining BMD. Through this study, we investigate the relationship of BMD and MBD. We found that there were no significant outcome as is in line with findings by Zain [24].

Crandall [19] reported that BMD and MBD were positively associated in women who were not recent users hormone therapy (HRT), suggesting a unifying biological mechanism linking BMD and MBD. Author mentioned that HRT may obscure the relationship by having a persistent effect on breast tissue. We found inconsistencies despite the fact that there was a small number of respondents (13%) using HRT.

In this study, we also investigated the relationship of BMD; MBD together with serum estrogen concentration level. Our research involved premenopausal and postmenopausal women of different ethnicity in Malaysia which increases the generalizability of the results obtained from the analysis done as it represent the Malaysian population. However, earlier studies included women from different ethnic groups stated no relationship between BMD and MBD were found [22, 25].

5 Conclusion

There was no significant relationship between BMD, and serum estrogen concentration level in Malaysian women reported in this study. Future research with larger sample size and inclusion of other factors are recommended to determine the complex link of BMD and MBD.

Acknowledgements This research work is supported by the University Kebangsaan Malaysia Research Grant (GUP-2014-063). Authors would also like to express their heartfelt gratitude to the Radiology Department, Kuala Lumpur General Hospital for facilitating the study.

References

1. Coughlin, S.S., Ekwueme, D.U.: Breast cancer as a global health concern. Cancer Epidemiol. **33**, 315–318 (2009)
2. Jemal, A., Bray, F., Center, M.M., et al.: Global cancer statistic. CA Cancer J. Clin. **61**, 69–90 (2011)
3. Yip, C.H., Taib, N.A., Mohamed, I.: Epidemiology of breast cancer in Malaysia. Asian Pac. J. Cancer Prev. **7**, 369–374 (2006)
4. Jemal, A., Center, M.M., De Santis, C., Ward, E.M.: Global patterns of cancer incidence and mortality rates and trends. Cancer Epidemiol. Biomark. Prev. **19**(8), 1893–1907 (2010)
5. Mackay, J., Jemal, A., Lee, N.C., et al.: The Cancer Atlas. The American Cancer Society, Atlanta (2006)

6. Yip, C.H., Emran, N.A., Ibrahim, A.W., et al.: Breast cancer services in Malaysia [Power point slides]. Retrieved from Healthcare Performance Measurement & Reporting System (HPMRS) (2012)
7. Omar, Z.A., Ibrahim Tamin, N.S.: National cancer registry report: Malaysia cancer statistics—data and figure 2007. Ministry of Health Malaysia (2011)
8. Yip, C.H., Pathy, N.B., Teo, S.H.: A review of breast cancer research in Malaysia. Med. J. Malays. **69**(Supplement A August 2014), 8–22 (2014)
9. Pathy, N.B., Yip, C.H., Taib, N.A., Hartman, M., Saxena, N., Iau, P., et al.: Breast cancer in a multi-ethnic Asian setting: results from the Singapore-Malaysia hospital-based breast cancer registry. Breast. **20**(Suppl 2), S75–S80 (2011)
10. Qu, X., Zhang, X., Qin, A., et al.: Bone mineral density and risk of breast cancer in postmenopausal women. Breast Cancer Res. Treat. **138**, 261–271 (2013)
11. Zain, N.M., Seriramulu, V.P., Chelliah, K.C.: Bone mineral density and breast cancer risk factors among premenopausal and postmenopausal women—a systematic review. Asian Pac. J. Cancer Prev. **17**(7) (2016)
12. Cauley, J.A., Gutai, J.P., Kuller, L.H., et al.: Endogenous estrogen levels and calcium intake in postmenopausal women: relationships with cortical bone measures. JAMA **260**(21), 3150–3155 (1988)
13. Felson, D.T., Zhang, Y., Hannan, M.T., et al.: The effect of postmenopausal estrogen therapy on bone density in elderly women. N. Engl. J. Med. **329**(16), 1141–1146 (1993)
14. Fraenkel, M., Novack, V., Liel, Y., et al.: Association between bone mineral density and incidence of breast cancer. PLoS ONE **8**, 70980 (2013)
15. Santen, R.J., Boyd, N.F., Chlebowski, R.T., et al.: Critical assessment of new risk factors for breast cancer: considerations for development of an improved risk prediction model. Endocr. Relat. Cancer **14**(2), 169–187 (2007)
16. American College of Radiology: Breast Imaging and Reporting Data System, 5th edn. American College of Radiology, Reston (2004)
17. World Health Organization: Assessment of fracture risk and its application to screening for postmenopausal osteoporosis. WHO Technical Report series 843. WHO, Geneva (1994)
18. NCBI Homepage. https://www.ncbi.nlm.nih.gov/books/NBK138665/. Last accessed 6 June 2017
19. Crandall, C.J., Zheng, Y., Karlamangla, A., Sternfeld, B., Habel, L.A., Oestreicher, N., Greendale, G.A.: The association between mammographic breast density and bone mineral density in the study of women's health across the nation. Ann. Epidemiol. **17**(8), 575–583 (2007)
20. Gupta, R., Alhajri, F.A., Mohammed, A.M., Chavan, V.N., Gupta, P.: Relationship between mammography breast density and bone mineral density. J. Clin. Densitom. **11**(3), 431–436 (2008)
21. Chen, Z., Arendell, L., Aickin, M., Cauley, J., Lewis, C.E., Chlebowski, R.: Hip bone density predicts breast cancer risk independently of Gail score: results from the women's health initiative. Cancer **113**(5), 907–915 (2008)
22. Buist, D.S.M., Anderson, M.L., Taplin, S.H., La Croix, A.Z.: The relationship between breast density and bone mineral density in postmenopausal women. Am. Cancer Soc. **101**(9), 1968–1976 (2004)
23. Yee, Y.S.S., Zaiton, Y., Chan, Y.M., Norhaizan, M.E.: Association between anthropometric status, dietary intake and physical activity with bone health status among premenopausal Chinese women in the Klang Valley, Malaysia. Malay. J. Nutr. **19**(3), 293–302 (2013)
24. Zain, N.M., Seriramulu, V.P., Kanaga, K.C., Soin, N., Shantini, A.: Association of bone mineral density and mammographic breast density in premenopausal and postmenopausal Malaysian Women: study on women with and without breast cancer. In: Bio-Engineering for Smart Technologies, BioSMART, pp. 1–4. IEEE, Dubai, United Arab Emirates (2016)
25. Kerlokowske, K., Shepherd, J., Creasman, J., Tice, J.A., Ziv, E., Cummings, S.R.: Are breast density and bone mineral density independent risk factors for breast cancer? J. Natl. Cancer Inst. **97**(5), 368–374 (2005)

Blood Glucose and Sepsis Score on Sepsis Patients Requiring Insulin Therapy

Fatanah Mohamad Suhaimi, Ummu Kulthum Jamaludin, Normy Norfiza Abdul Razak, Christopher G. Pretty, Azrina Md. Ralib, Mohd Basri Mat Nor, and Fatimah Dzaharudin

Abstract

Early treatment of sepsis is crucial in improving the patient condition and reduces mortality. The lack of information and methods to diagnose sepsis at an early stage is a significant barrier to early treatment. Treatment is normally based on clinical judgment since blood cultures are negative in the majority of sepsis or septic shock cases as reported in several studies. Even with blood cultures result, a delay in diagnosis may happen while waiting for the results. Therefore, clinical guidelines are still required to provide guidance for the clinician caring for a patient with severe sepsis or septic shock. In this study, a validated glucose-insulin model is used to capture patient-specific insulin sensitivity profile. The validated model is incorporated with a sepsis system to create a glucose-insulin model that describe sepsis occurrence. The relationship between insulin sensitivity, blood glucose and sepsis score is investigated using a retrospective data of 8 patients admitted in the Tengku Ampuan Afzan Hospital, Kuantan. Results indicate that there was a significant relationship between blood glucose level and sepsis score ($p < 0.001$), and insulin sensitivity with a sepsis score ($p < 0.001$). Additionally, blood glucose level was higher in a severe sepsis group compared to the non-sepsis group. Whereas, insulin sensitivity is lower in a severe sepsis group. Insulin sensitivity profile can be incorporated with the sepsis system for monitoring sepsis patients requiring insulin therapy as seen in this study. Thus, the sepsis glucose-insulin model can potentially be used as an indicator or tools for sepsis diagnosis.

Keywords

Blood glucose • Insulin sensitivity • Insulin therapy • Sepsis

F. Mohamad Suhaimi (✉)
Advanced Medical and Dental Institute, Universiti Sains Malaysia, Penang, Malaysia
e-mail: fatanah.suhaimi@usm.my

U. K. Jamaludin · F. Dzaharudin
Faculty of Mechanical Engineering, Universiti Malaysia Pahang, Pahang, Malaysia

N. N. Abdul Razak
Universiti Tenaga Nasional, Selangor, Malaysia

C. G. Pretty
Department of Mechanical Engineering, University of Canterbury, Christchurch, New Zealand

A. Md. Ralib · M. B. Mat Nor
Department of Anaesthesiology & Intensive Care, Kulliyyah of Medicine, International Islamic University Malaysia, Pahang, Malaysia

1 Introduction

Sepsis is described as a clinical syndrome with the presence of infection and a systemic inflammatory response (SIRS). The infection in sepsis is caused by the penetration of either sterile tissue or fluid or body cavity by pathogenic microorganisms [1]. Often the presence of SIRS may or may not indicate the presence of sepsis. This is because SIRS can be triggered by a variety of infectious and noninfectious conditions.

The number of patients with sepsis increases significantly each year as well as the number of sepsis-related deaths [2]. Early diagnosis is critical because early interventions have

F. Ibrahim et al. (eds.), *2nd International Conference for Innovation in Biomedical Engineering and Life Sciences*,
IFMBE Proceedings 67, https://doi.org/10.1007/978-981-10-7554-4_46

been documented to reduce mortality from 46.5 to 30.5% and thus showing significant potential. Inappropriate antibiotic in early goal directed therapy (EGDT) associated with increased mortality [3]. Early goal-directed resuscitation is recommended for patients with sepsis, particularly during the first 6 h after infection recognition and administration of broad-spectrum antimicrobials therapy within 1 h of the recognition of septic shock [4, 5]. Equally, almost 50% of all sepsis that is clinically defined is culture negative [6]. Hence, the ability to recognize infection and diagnose sepsis as early as possible will consequently improve mortality outcome and patient condition.

However, to initiate antimicrobial therapy and antibiotic administration can cause significant delay due to the process of obtaining appropriate cultures. This process requires incubation and standard procedures that result in a delayed outcome of 2–3 days. Additionally, at least two blood cultures recommended to be obtained prior to antibiotics [4, 5]. To date, blood bacteria culture is considered as the standard method for confirming an infection for sepsis. Identifying infection and consequently diagnosing sepsis is a challenging process in critical care.

Significant success in the incidence of sepsis has been observed by implementing a blood glucose control protocol [7, 8] indicating that controlling hyperglycemia may result in controlling the emergence of sepsis. Among the blood glucose control protocol adapted in clinical center, model-based glycemic control is the emerging treatment approach for managing hyperglycemia in critical illness. Model-based glycemic control method able to directly capture patient-specific physiological dynamics of human metabolism [9, 10]. Thus, the patient-specific variability such as insulin sensitivity can potentially be used to represent individual physiological response towards treatment and patient condition.

The aim of this study is to identify the relationship between insulin sensitivity (S_I), blood glucose level and sepsis score in sepsis patients receiving insulin therapy. S_I profiles are captured using a validated glucose-insulin model [11] and analyze with a sepsis system to create a marker that describe sepsis progression. Therefore, the sepsis glucose-insulin model serves as an indicator or tools for sepsis diagnosis.

2 Methods

2.1 Glucose-Insulin Systems

A clinically validated Intensive Control Insulin Nutrition Glucose (ICING) model by Lin et al. [11] is used in this study to determine hourly insulin S_I of the patients. The model equations are defined in Eqs. 1–7.

In Eqs. 1–7, G is defined as blood glucose level, Q is interstitial insulin, and I is plasma insulin. P_1 and P_2 are stomach glucose content and gut glucose content, respectively. P is rate of glucose appearance in plasma. U_{en} is endogenous insulin secretion rate. Other parameter and kinetic values of the ICING model include patient endogenous glucose removal represents by p_G, insulin sensitivity is S_I, the saturation parameter of insulin-mediated glucose removal is α_G, endogenous glucose production rate is EGP, central nervous system glucose uptake is CNS, and plasma glucose distribution volume is V_G.

$$\dot{G}(t) = -p_G G(t) - s_I G(t)\frac{Q(t)}{1+\alpha_G Q(t)} + \frac{P(t) + EGP - CNS}{V_G} \tag{1}$$

$$\dot{Q}(t) = n_I(I(t) - Q(t)) - n_c\frac{Q(t)}{1+\alpha_G Q(t)} \tag{2}$$

$$\dot{I}(t) = -n_k I(t) - n_L\frac{I(t)}{1+\alpha_I I(t)} - n_I(I(t) - Q(t)) + \frac{u_{ex}}{V_I} + \frac{u_{en}G}{V_I}(1 - x_L) \tag{3}$$

$$P_1(t) = -d_1 P_1 + D(t) \tag{4}$$

$$P_2(t) = -\min(d_2 P_2, P\max) + d_1 P_1 \tag{5}$$

$$P(t) = \min(d_2 P_2, P\max) + PN(t) \tag{6}$$

$$u_{en}G = \begin{cases} u_{min}, & u_{min} > k_1 G(t) + k_2 \\ k_1 G(t) + k_2, & u_{min} \le k_1 G(t) + k_2 \le u_{max} \\ u_{max}, & u_{max} < k_1 G(t) + k_2 \end{cases} \tag{7}$$

In Eqs. 2 and 3, n_I, n_c, n_k, and n_L represent plasma interstitium insulin diffusion rate, receptor bound insulin degradation rate, renal insulin clearance rate and hepatic insulin clearance rate, respectively. Saturation parameter for hepatic insulin clearance is denoted by α_I, insulin distribution volume is V_I, exogenous insulin input is u_{ex}, whereas first pass hepatic clearance is denoted by x_L. In Eqs. 4–6, d_1 represents the rate of glucose transport through the enteral, d_2 represents the rate of glucose transport into the bloodstream, and D represents rate of oral glucose input from enteral nutrition. P_{max} and PN represent maximal gut glucose flux and rate of intravenous glucose input from parenteral nutrition.

Other population constant defined in Eq. 7 is u_{min} that represents a minimum pancreatic secretion rate and u_{max} represents a maximum pancreatic secretion rate. k_1 and k_2 are pancreatic insulin secretion glucose-sensitivity and pancreatic insulin secretion offset, respectively. The ICING model parameter and methods used in this analysis have been validated through several studies [12, 13].

2.2 Sepsis Classification

Sepsis classification is determined based on criteria of the American College of Chest Physicians (ACCP)/Society of Critical Care Medicine (SCCM) [1]. The classification of a sepsis score considers two main elements (1) Systemic Inflammatory Response Score (SIRS), and (2) Sepsis-related Organ Failure (SOFA) score. Sepsis score is clinically used to indicate the increased complexity of the disease and similarly represents evaluations and descriptions of the disease.

Sepsis score is calculated based on the clinical data collected during the patient's stay includes SIRS, organ failure, fluid resuscitation and inotropes administration. Blood test results were recorded for every 6 or 12 h. Other clinical parameters such as temperature, blood pressure, and heart rate were recorded hourly and when necessary at the bedside. The control protocol used in this study is designed based on clinical protocol adapted at the Tengku Ampuan Afzan Hospital, Kuantan, Pahang as documented by the Ministry of Health, Malaysia.

3 Results

Table 1 shows the demographic of HTAA sepsis patients selected in this study. There were 8 patients selected for this study, with 25% female and 75% male. Sepsis was confirmed based on blood culture results. Median Apache II score was 22 and the median sepsis score was 1. Median [IQR] of blood glucose level was 10.9 [7.6–15.7] mmol/L.

Figure 1 shows the boxplot of BG level with median and inter quartile range (IQR) of 8 patients in the sepsis cohort. Figure 2 shows the boxplot of S_I of the sepsis cohort. Most of the patients had median BG between 10 and 12 mmol/L. Only 2 patients had slightly lower median BG, which were Patient 4 and Patient 8 with 7.7 and 8.8 mmol/L, respectively. Similarly, interquartile range (IQR) shows that most of the patients had a normal average range of BG, whereas Patient 4 and Patient 8 had lower IQR compared to others. In contrast, Patient 7 had higher IQR compared to other patients.

Median S_I shown in Fig. 2 indicates more variability even though most of the patient lay in the range of 10–20 $\times$ 10^{-5} L min/mU. Interestingly, Patient 4 and Patient 8 have significantly higher median and IQR of S_I. Additionally, the lowest blood glucose profiles seen in Fig. 1 for Patient 4 has the highest insulin sensitivity profiles in Fig. 2.

Table 2 presents median and inter quartile range of blood glucose and insulin sensitivity, according to the sepsis score. At higher sepsis score, the BG was larger compared to the lower sepsis score value. However, S_I was lower at a higher sepsis score value. Additionally, S_I decreases with the increased in BG. There was a significant relationship between BG level and sepsis score with $p < 0.001$, and S_I versus sepsis score with $p < 0.001$ using Mann-Whitney test. Additionally, there was also a significant relationship between BG and S_I on the sepsis score basis, as shown in Table 2.

4 Discussion

This paper presents analysis of glucose-insulin model incorporated with the sepsis system to model the occurrence of sepsis in critical patients requiring intensive insulin therapy. Data was collected from 8 sepsis patients treated in the Hospital Tengku Ampuan Afzan, Kuantan in 2016. In this study, S_I profiles were determined using the glucose-insulin model. These profiles were then analyzed with sepsis system and sepsis-related parameter to determine whether it has any relation with the progression of sepsis. The use of sepsis score not only to help clinicians determine suitable therapy, but also provides more accurate monitoring of patient condition. In particular, sepsis score was interpolated as the hourly sepsis score to match the hourly S_I of a patient.

The relationship between S_I, sepsis score and BG level is analyzed in this paper. Per-patient analysis was also done to determine the relation and variability of the cohort. Most of the patients have very low S_I, as expected, due to their sepsis condition, even though the BG value was more variable. From the results presented in this paper the blood glucose level had a direct relationship with sepsis score. It can be seen that the blood glucose level was higher in a severe sepsis group compared to the non-sepsis group. Nevertheless, S_I was proportional to the sepsis score. As sepsis score increases, S_I was lower. Therefore, BG was proportional to S_I as seen from this cohort. Importantly, S_I represents patient-specific parameter that indicates physiological response, dynamics and patient variability. Whereas, BG represents the patient blood glucose condition and reaction towards insulin treatment.

There are four groups of sepsis score as identified by the SCCM. The score was increased from normal group to the severe sepsis group. In this cohort, sepsis score ranged from

Table 1 Demographic of sepsis patients

Demographic	Median [IQR]
No of patients	8
Female (%)	25
Male (%)	75
Height (cm)	161.5 [157.5–167.5]
Weight (kg)	70.0 [62.5–85.0]
Apache II score	22 [19–26]
Sepsis score	1 [1–2]
Blood glucose level per cohort (mmol/L)	10.9 [7.6–15.7]
Median blood glucose per patient (mmol/L)	11.3 [9.7–14.7]

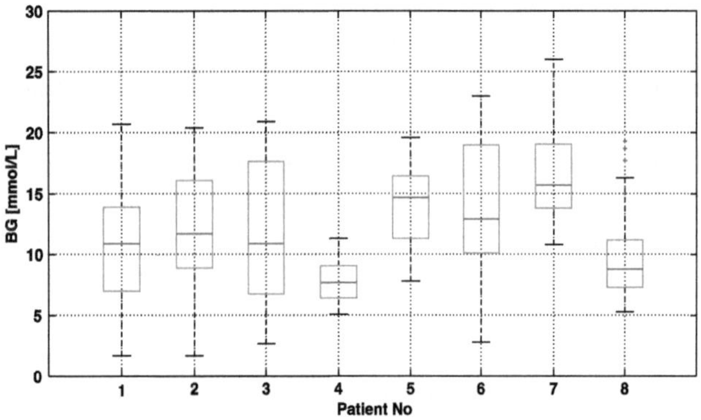

Fig. 1 Median and interquartile range of blood glucose level in sepsis cohort

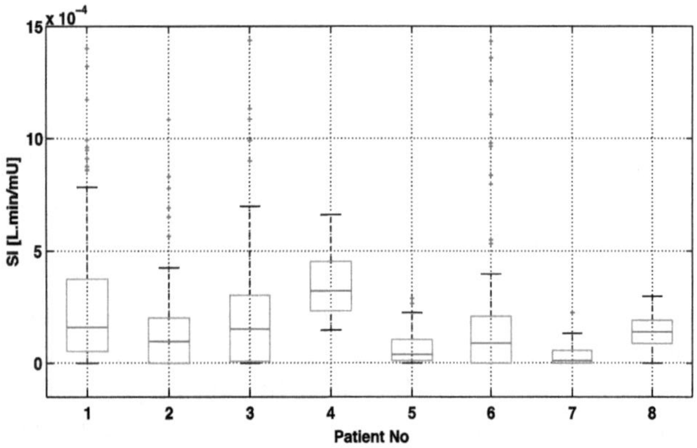

Fig. 2 Median and interquartile range of insulin sensitivity in sepsis cohort

0 to 2, and most of the patients likely scored 1, indicating that this cohort did not experience very severe sepsis episodes. Thus, this cohort did not represent patients with septic shock (sepsis score = 3), and refractory septic shock (sepsis score = 4), and thus the findings of this study do not represent patients in these two categories. Moreover, it should be noted that the number of patients included in this analysis was small even though these patients had a long length of stay. Thus, a larger number of samples are required in the future to establish a strong relation between these parameters. However, in the case where sepsis condition could not be identified earlier, S_I and BG or changes in these two values may potentially be used for aiding in sepsis diagnosis in the ICU.

5 Conclusions

Insulin sensitivity profile identified using the glucose-insulin model can be used to indicate sepsis severity among sepsis patients requiring insulin therapy. More importantly, this profile can be identified hourly and potentially to be used for predicting sepsis occurrence. Additionally, blood glucose level also has a significant relationship with sepsis score. The high blood glucose level indicates the increase severity of sepsis and vice versa. On the other hand, lower S_I corresponds to the increasing severity of sepsis.

Table 2 Blood glucose and insulin sensitivity on a sepsis score basis

Sepsis score	Blood glucose (mmol/L)	Insulin sensitivity ($\times$ 10^{-5} L min/mU)	*P* value
0	8.9 [6.4–12.6]	18.68 [7.50–29.83]	0.000*
1	12.2 [8.7–16.5]	9.39 [0.80–22.07]	0.000*
2	16.3 [15.6–18.3]	0.00 [0.00–3.40]	0.000*

*significant at $p < 0.001$

References

1. Levy, M.M., Fink, M.P., Marshall, J.C., Abraham, E., Angus, D., Cook, D., et al.: 2001 Sccm/Esicm/Accp/Ats/Sis international sepsis definitions conference. Crit. Care Med. **31**, 1250–1256 (2003)
2. Walkey, A.J., Wiener, R.S., Ghobrial, J.M., Curtis, L.H., Benjamin, E.J.: Incident stroke and mortality associated with new-onset atrial fibrillation in patients hospitalized with severe sepsis. JAMA **306**, 2248–2254 (2011)
3. Kalil, A.C., Johnson, D.W., Lisco, S.J., Sun, J.: Early goal-directed therapy for sepsis: a novel solution for discordant survival outcomes in clinical trials. Crit Care Med (2017)
4. Dellinger, R.P., Levy, M.M., Carlet, J.M., Bion, J., Parker, M.M., Jaeschke, R., et al.: Surviving sepsis campaign: international guidelines for management of severe sepsis and septic shock: 2008. Intensiv. Care Med. **34**, 17–60 (2008)
5. Dellinger, R.P., Levy, M.M., Rhodes, A., Annane, D., Gerlach, H., Opal, S.M., et al.: Surviving sepsis campaign: international guidelines for management of severe sepsis and septic shock, 2012. Intensiv. Care Med. **39**, 165–228 (2013)
6. Carrigan, S.D., Scott, G., Tabrizian, M.: Toward resolving the challenges of sepsis diagnosis. Clin. Chem. **50**, 1301–1314 (2004)
7. Hirasawa, H., Oda, S., Nakamura, M.: Blood glucose control in patients with severe sepsis and septic shock. World J. Gastroenterol. **15**, 4132–4136 (2009)
8. Takahashi, G., Matsumoto, N., Shozushima, T., Onodera, C., Kan, S., Akitomi, S., et al.: Retrospective study on the effect of tight glucose control in postoperative sepsis patients using an artificial pancreas. J. Infect. Chemother. **17**, 812–820 (2011)
9. Chase, J.G., Le Compte, A.J., Suhaimi, F., Shaw, G.M., Lynn, A., Lin, J., et al.: Tight glycemic control in critical care—the leading role of insulin sensitivity and patient variability: a review and model-based analysis. Comput. Methods Programs Biomed. **102**, 156–171 (2011)
10. Evans, A., Shaw, G.M., Le Compte, A., Tan, C.S., Ward, L., Steel, J., et al.: Pilot proof of concept clinical trials of stochastic targeted (STAR) glycemic control. Ann. Intensiv. Care **1**, 38 (2011)
11. Lin, J., Razak, N.N., Pretty, C.G., Le Compte, A., Docherty, P., Parente, J.D., et al.: A physiological intensive control insulin-nutrition-glucose (ICING) model validated in critically ill patients. Comput. Methods Programs Biomed. **102**, 192–205 (2011)
12. Chase, J.G., LeCompte, A., Shaw, G.M., Blakemore, A., Wong, J., Lin, J., et al.: A benchmark data set for model-based glycemic control in critical care. J. Diabetes. Sci. Technol. **2**, 584–594 (2008)
13. Hann, C.E., Chase, J.G., Shaw, G.M.: Integral-based identification of patient specific parameters for a minimal cardiac model. Comput. Methods Programs Biomed. **81**, 181–192 (2006)

Contact Pattern of /b/ and /p/ in Malay Language Using Electropalatography

Syatirah Mat Zin, Fatanah Mohamad Suhaimi, Siti Noor Fazliah Mohd Noor, Ahmad Fakrurrozi Mohamad, Nur Fatehah Md Shakur, and Nurulakma Zali

Abstract

This article highlights the use of Electropalatography (EPG) to analyse the pattern of consonants /b/ and /p/ production. These plosive consonants occur when the upper and lower lips are brought together bringing the soft palate close to the posterior wall of the throat. This condition causes the air from the lung to be trapped in the mouth and voice channel. Once air is released, two phenomenon occurs. Firstly, if the air vibrates the vocal cord upon release, plosive consonant (/b/) is produced. Secondly, if the air do not vibrate the vocal cord, voiceless plosive consonant (/p/) is produced. This study aimed to observe the differences between the plosive production of /b/ and /p/ among three Malay adults aged from 24 to 52 years old. The subjects were required to wear the Reading EPG palate with 62 electrodes for monitoring their tongue-palate contact. Data were analysed using Articulate Assist 1.18 software. The results highlighted that the contact patterns for /b/ and /p/ productions were almost similar during the articulation. The contact patterns for the production of /p/ and /b/ occurred at the velar zone. However, significant difference was observed on the contact pattern of S1 due to higher upper arch depth.

Keywords

Bilabial • Consonant plosive • Malay language

1 Introduction

Different speech sounds of different languages such as vowel and consonant exist in phonetic field. The sound of vowel is produced when there are no air restrictions and is determined by the position of tongue and shape of mouth during articulation. The consonant sound is produced when there is restriction of airflow in the channel sound [1]. Consonants are classified into voiceless or voiced, place of the articulation, the way of the air restriction, and the manner of articulation. There are six types of consonant consisting of plosive (/p, b, t, d, k, g/), fricative (/s, h/), affricate (/c, j/), nasal (/m, n, ng/), lateral approximant (/l/), and approximant (/w, r, j/) [2, 3]. Three locations of articulation for plosive existed such as bilabial, alveolar and velar. Bilabial produces voiceless consonant (/p/) and voiced consonant (/b/). Meanwhile, the alveolar consists of consonant /t/ as voiceless consonant and /d/ as voiced consonant. Velar consists of consonant /k/ as a voiceless and /g/ as a voiced consonant.

Various methods are available such as conventional and modern to determine the place of articulation and contact pattern during its production. By using conventional method, researcher should record a sound and replay it using a computer. However, this method has weaknesses and results in inaccurate findings and tend to produce human errors. Magnetic Resonance Imaging (MRI) and Electropalatography (EPG) are one of the alternatives to determine the place of articulation. Masaki et al. (1996) performed research using the MRI-based analysis of consonant /r/ and /l/ in the English language, and compare them with Native American

S. Mat Zin · F. Mohamad Suhaimi (✉) · S. N. F. Mohd Noor · A. F. Mohamad · N. F. Md Shakur · N. Zali
Craniofacial and Biomaterial Sciences Cluster, Advanced Medical and Dental Institute, Universiti Sains Malaysia, Bertam, 13200 Kepala Batas, Penang, Malaysia
e-mail: fatanah.suhaimi@usm.my

F. Ibrahim et al. (eds.), *2nd International Conference for Innovation in Biomedical Engineering and Life Sciences*,
IFMBE Proceedings 67, https://doi.org/10.1007/978-981-10-7554-4_47

English speaker and Japanese speaker. Their results highlighted the advantage of using MRI such as being non-invasive which permits data accumulation from many subjects. Additionally, the result output is in 3D and it is easier to see the image of a place of articulation.

Besides MRI, electropalatography (EPG) is another method used to study the place of articulation. EPG is cheaper than MRI, and provides dynamic real-time visual feedback of the location and timing of tongue contacts with the hard palate [4]. There are many studies involving the use of an EPG as an instrument in phonetic research. Hardcastle et al. (1972) conducted a study about the use of EPG in phonetic research. Their findings showed that the EPG provides real-time information of contact pattern on the tongue and hard palate [5]. Besides, Dagenais et al. (1994) showed that there is significant differences in the contact between tongue and hard palate for the production of /s/ and /z/ [6]. Hence, EPG is an appropriate instrument in determining the contact patterns between the tongue and palate.

In the current study, the contact pattern for bilabial plosive consonants in the production of /b/ and /p/ is determined using EPG, and was analysed using Articulate Assistant ™ version 1.18.

2 Materials and Methods

2.1 Subjects

Three Malay adults with age ranging from 27 to 52 years old (mean age of 37 years) were selected. The subjects were one male speaker and two female speakers. They are medically fit with no history of speech, language or hearing difficulties. This study has been approved by the Human Research Ethics Committee of Universiti Sains Malaysia Human Ethics Committee. During the procedure, the subject was required to wear a customized Reading palate.

2.2 Material

The Reading palate was used to monitor the contact pattern between the tongue and hard palate. The Reading palate is an artificial palate which has 62 silver electrodes embedded on the palate. The silver electrodes were connected to the multiplexer. The software, Articulate Assist 1.18 was used to analyse the data [7].

2.3 Procedure

All subjects consented to take part in the study. Initially, the subject's upper palate impression was imprinted (Fig. 1).

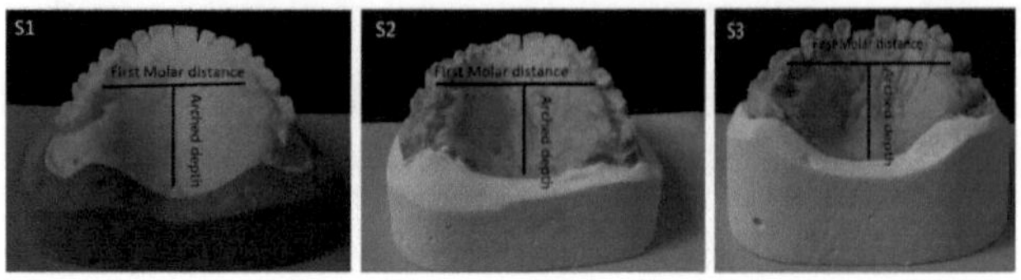

Fig. 1 Arch depth and first molar distance for three subjects

The impression was used as a template to fabricate the EPG palate and training plates. Subjects were asked to wear the training plate at least 4 h a day to allow the subjects to become accustomed to an unfamiliar object in the mouth [4]. The recording was conducted in a soundproof room. During speech production, the artificial palate will transmit the tongue-palate contact information to the computer. Data were recorded as the subjects read out a single consonant which is the sound of consonant /b/ and consonant /p/. Subjects were required to read a single consonants /b/ and /p/ for three times within 0.6 s for each reading, and the average from the three readings was calculated using the formula shown in Eq. 1.

$$\frac{\text{Total electrode contacts during the production} \times 100}{\text{Total electrode contact}} \quad (1)$$

3 Result and Discussion

Table 1 shows the contact patterns between the tongue and hard palate during the production of /b/ and /p/, subject 1 (S1), subject 2 (S2) and subject 3 (S3). In the production of / b/ for S1, there were 20 contacts which represent 32%. There was no contact between tongue and hard palate at the alveolar and post-alveolar zone. There are 14 contacts at the palatal zone and 5 contacts at the velar zone. For the production of consonant /p/, 19 (30%) contacts were detected. Additionally, there was no contact between tongue and hard palate at the alveolar zone but there was one contact at the post-alveolar zone, 14 contacts at the palatal zone and 4 contacts at the velar zone.

For S2, there was 25 (40%) contacts for the production of /b/. No contact at the alveolar zone, 7 contacts at the post-alveolar zone, 12 contacts at the palatal zone and 6 contacts at the velar zone. In the production of /p/, one contact at the alveolar zone, 8 contacts at the post-alveolar zone, 13 contacts at the palatal zone and 6 contacts at the velar zone, resulting into 45% of contacts.

Meanwhile, S3 shows a total of 31 (50%) contact for the production of /b/ and 24 (39%) contacts for the production of /p/. At the alveolar zone, there were 3 contacts in the production of /b/ and 2 contacts in the production of /p/. At the post-alveolar zone, there were 7 contacts in the production of /b/ and 5 contacts in the production of /p/. At the palatal

Table 1 Pattern of the consonant /b/ and /p/

Subject	Gender	Age (year)	Arch depth (cm)	First inter-molar distance (cm)	Consonant /b/	Percentage (%)	Consonant /p/	Percentage (%)
S1	Male	52	2.3	4.1	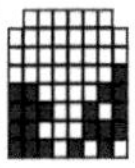	32	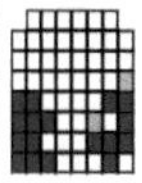	30
S2	Female	32	2.0	4.1	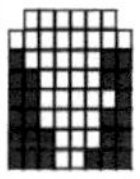	40	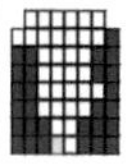	45
S3	Female	27	2.0	4.5	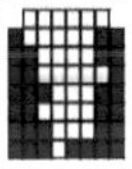	50	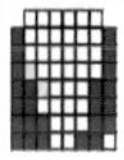	39

zone, there were 14 contacts in the production of /b/ and 13 contacts in the production of /p/. Lastly, at the velar zone there were 7 contacts in the production of /b/ and 5 contacts in the production of /p/.

Additionally, Table 1 shows the arch depth and first inter-molar distance for all subjects. S1 has the highest arch depth, whereas S2 and S3 have similar arch depth. S3 has the highest first inter-molar distance of 4.5 cm followed by S1 and S2 with 4.1 cm.

The consonants /b/ and /p/ are produced when upper and lower lips are in contact with each other. Simultaneously, soft palate is raised up while production of single sound of consonants /b/ and /p/ without alteration of tongue position. This situation explains the results shown in Table 1. All subjects showed a similar contact pattern for the production of consonant /b/ and consonant /p/ at the posterior 1/3 of the palate. Concurrently, it adheres with the characteristics of consonants /b/ and /p/ that have been discussed above. However, S2 and S3 touched the edges of alveolar and post-alveolar zone compared to S1 who only touched the edges of the palatal and the velar zone. This occurred because S2 and S3 have similar arch depth, indicating that arch depth may influenced articulations during speech [8]. Oliver et al. indicates that articulation is influenced by arch depth [8]. In addition, altered palate contour such as in the case of cleft palate due to wider palatal area may provide better tongue position [9]. However, this study shows that the first molar distance does not affect the productions of /b/ and /p/.

4 Conclusion

The productions of consonants /b/ and /p/ have been classified as the plosive type and the place of articulation is bilabial. Besides, tongue position due to different arch depth also contributes towards the results of this study. This study shows that the pattern /b/ and /p/ productions were recorded at the velar zone. However, a larger number of subjects will be more beneficial for comparison.

Acknowledgements This study was conducted at the Craniofacial and Biomaterial Science Cluster, Advanced Medical and Dental Institute, Universiti Sains Malaysia. Authors would like to acknowledge the financial support provided by the Ministry of Science, Technology and Innovation (MOSTI). This study is part of the project granted by Science Fund Grant (Project no: 06-01-05-SF0718).

References

1. Zahid, I.H., Omar, M.S.: Fonetik dan fonologi. Akademia (2006)
2. Chaer, A.: Linguistik umum. Penerbit Rineka Cipta (2007)
3. Zin, S.M., Suhaimi, F.M., Noor, S.N.F.M., Ismail, N.I., Zali, N., Das, K.T., et al.: Analysis of consonant /s/ and syllables in Malay language using electropalatography. In: AIP Conference Proceedings, 2016, p. 020002
4. Hardcastle, W., Jones, W., Knight, C., Trudgeon, A., Calder, G.: New developments in electropalatography: a state-of-the-art report. Clin. Linguist. Phon. **3**, 1–38 (1989)
5. Hardcastle, W.J.: The use of electropalatography in phonetic research. Phonetica **25**, 197–215 (1972)
6. Dagenais, P.A., Lorendo, L.C., McCutcheon, M.J.: A study of voicing and context effects upon consonant linguapalatal contact patterns. J. Phon. **22**, 225–238 (1994)
7. Wrench, A.A.: Advance in EPG palate design. Adv. Speech Lang. Pathol. **9**, 3–12 (2007)
8. Oliver, R., Evans, S.: Tongue size, oral cavity size and speech. The Angle Orthod. **56**, 234–243 (1986)
9. Bhagyalakshmi, G., Renukarya, A., Rajangam, S.: Metric analysis of the hard palate in children with Down syndrome-a comparative study. Down Syndr. Res. Pract. **12**, 55–59 (2007)

Knee Cartilage Ultrasound Image Segmentation Using Locally Statistical Level Set Method

Amir Faisal, Siew-Cheok Ng, Siew-Li Goh, and Khin Wee Lai

Abstract

Boundary delineation of a hypoechoic layers between the surrounding tissues and the bone structure is a necessary step in order to compute the knee cartilage thickness from ultrasound images. Speckle noise and intensity bias often complicates the segmentation task in the ultrasound images. This paper presents knee cartilage boundary segmentation using locally statistical level set method (LSLSM). Comparing to other methods in segmenting the cartilage, LSLSM produces a more satisfying outcome. Application of LSLSM on a set of 80 images illustrates a significant agreement with Cohen's κ coefficient equal to 0.73 for the segmentation quality of the cartilage region rated by two raters. The quantitative evaluation measures of Dice coefficient and Hausdorff distance indicate the overall average values of 0.91 ± 0.01 and 6.21 ± 0.59 pixels, respectively. These good and consistent segmentation performances indicate that the segmented images can be applied for making the thickness computation in the ultrasound images.

Keywords

Cartilage • Level set • Segmentation • Ultrasound

1 Introduction

Knee osteoarthritis is a prevalent disease among elderly [5]. Cartilage degeneration is one of the primary features of this disease [6]. Ultrasound imaging is useful for the evaluation of extra-articular structures [6]. It has been applied to quantify the cartilage thickness and diagnose the cartilage degeneration [1] in patients with osteoarthritis, rheumatoid arthritis, [4] and knee pain [6].

Segmentation is an important task that could significantly affect the accuracy of the thickness measurement [3]. In ultrasound images, the femoral condylar cartilage is shown as a monotonous hypoechoic band between both interfaces of the soft tissue-cartilage and the cartilage-bone as shown in Fig. 1 [6]. Thus, the goal in segmenting the cartilage is to delineate the boundaries between both interfaces. Delineating the cartilage boundary from the adjacent tissues is difficult because the boundary between different tissues is hard to distinguish.

Speckle noise and intensity inhomogeneity occur caused by physical constraint in the ultrasound image acquisition, which often adversely affect the image contrast. If only intensity bias is considered and not by speckle, this problem could be solved similarly to the inhomogeneity correction in magnetic resonance images [10]. The intensity bias correction is often addressed by assuming that intensity inhomogeneity associated with a component of an observed image is

A. Faisal · S.-C. Ng · K. W. Lai (✉)
Department of Biomedical Engineering, Faculty of Engineering, University of Malaya, 50603 Kuala Lumpur, Malaysia
e-mail: lai.khinwee@um.edu.my

A. Faisal
e-mail: amirf415al@yahoo.com

S.-C. Ng
e-mail: siewcng@um.edu.my

S.-L. Goh
Faculty of Medicine, University of Malaya, 50603 Kuala Lumpur, Malaysia
e-mail: gsiewli@um.edu.my

F. Ibrahim et al. (eds.), *2nd International Conference for Innovation in Biomedical Engineering and Life Sciences*, IFMBE Proceedings 67, https://doi.org/10.1007/978-981-10-7554-4_48

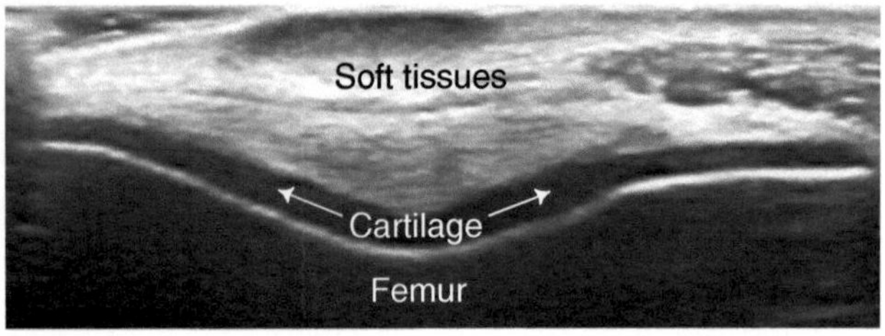

Fig. 1 The knee cartilage shown as a monotonous hypoechoic band between the soft tissue-cartilage and the cartilage-bone interfaces

modelled as the multiplicative noise model. Furthermore, the multiplicative noise model is associated with the classic reflection imaging equation of ultrasound physics of image formation [10]. It is employed retrospectively in the images and usually incorporated with the segmentation algorithm where level set techniques for simultaneous segmentation and intensity inhomogeneity estimation have been presented [7, 11]. While these intensity-based segmentation methods are in general robust to noise, the usage of local intensity and joint intensity inhomogeneity correction could handle the intensity bias.

In this paper, boundary segmentation and thickness computation methods in two dimensional (2-D) knee cartilage ultrasound images are presented. To locate the cartilage boundary corrupted by speckle noise and intensity bias, the locally statistical level set technique is used using the energy derived from Gaussian distributions of local intensity and multiplicative noise model. Segmentation and computational performances of LSLSM are compared to other level set techniques when segmenting the knee cartilage. In addition, the segmentation results of these level set techniques on the total 80 data sets are evaluated qualitatively and quantitatively using Cohen's κ coefficient, Dice similarity coefficient, and Hausdorff distance measures, respectively.

2 Materials and Methods

2.1 Data Acquisition

The Toshiba Aplio MX ultrasound system with a 8–12 MHz, 2-D linear array probe (PLT-805AT) was used to capture axial views of the femoral cartilage revolved the knee. The knee joint was 120° flexed with the subject positioned in the supine posture. The probe was put transversely to the leg and perpendicular to the bone surface above the patella [6, 8]. Total 10 asymptomatic participants (male with age range between 23 and 27 years were registered with the written consent for data collection. The cartilage of both knee joints were acquired four times by repositioning the ultrasound transducer. The image resolution is 0.1316×0.1316 mm stored in DICOM format. Professional sonographer conducted this musculoskeletal sonography. The ethics approval letter of this study was obtained from UMMC Medical Ethics Committee (MECID No. 20147-396).

2.2 Locally Statistical Level Set Method

The two-phase case of the statistical and variational multiphase level set method or referred as the locally statistical level set method (LSLSM) is considered [11]. The energy of LSLSM is obtained from derivation of the Gaussian distributions of local intensity and multiplicative noise model. The energy functions e_i are expressed as

$$e_i(\mathbf{x}) = \int_{\Omega} K(\mathbf{y}-\mathbf{x}) \left(\frac{|I(\mathbf{x}) - b(\mathbf{y})c_i|^2}{2\sigma_i^2} + \frac{\log(2\pi\sigma_i^2)}{2} \right) d\mathbf{y}. \tag{1}$$

The functions e_i are computed by the equivalent expression as follows

$$\begin{aligned} e_i(\mathbf{x}) &= \frac{1}{2\sigma_i^2}(I^2\mathbf{1}_K - 2c_iI(b*K) + c_i^2(b^2*K)) \\ &+ \frac{1}{2}\log(2\pi\sigma_i^2)\mathbf{1}_K, \end{aligned} \tag{2}$$

where b, c_i, and σ_i^2 for $i = 1, 2$ are accordingly the restored bias field, the piecewise constants, and the variances. $*$ is the convolution operation. The function $\mathbf{1}_K$ is defined as $\int K(\mathbf{y}-\mathbf{x})d\mathbf{y}$. The kernel function K chosen in this paper is given by

$$K(\mathbf{z}) = \begin{cases} a & \text{for} \quad |\mathbf{z}| \le \rho \\ 0 & \text{for} \quad |\mathbf{z}| > \rho \end{cases}. \tag{3}$$

where a is a positive constant such that $\int K(\mathbf{z})d\mathbf{z} = 1$ and ρ represents the kernel's radius.

In the attempt of reducing the overlapping image intensity distribution, only intensities $I(\mathbf{x})$ in the neighborhood of $\mathbf{y}$ are considered in the energy functions e_i. The size of the neighborhood depends to the kernel scale. The small neighborhood is able to cope with intensity bias due to the intensities are only involved in the local region [7].

The intensities are estimated by spatially varying means bc_i and variances σ_i^2. To achieve simultaneous segmentation and intensity inhomogeneity estimation, the means are estimated by multiplication between the bias field b that accounts for intensity bias and the piecewise constants c_i estimating the true image signal in each region. The functions e_i represent an image segmentation and a intensity bias correction. To incorporate these functions to the level set formulation, these functions are combined with membership function $M_i(\phi)$. Therefore, the energy functional of LSLSM is defined by

$$E(\phi, c_i, b, \sigma_i) = v \int_{\Omega} |\nabla H_\varepsilon(\phi(\mathbf{x}))| d\mathbf{x} + \int_{\Omega} \sum_{i=1}^{2} e_i(\mathbf{x}) M_i(\phi(\mathbf{x})) d\mathbf{x}, \quad (4)$$

where the first term is the regularization term to compute the arc length of the zero level set, which its relative strength is determined by the parameter v.

The membership functions defined by $M_1(\phi) = H(\phi)$ and $M_2(\phi) = 1 - H(\phi)$ represent both regions Ω_1 and Ω_2, respectively. The regularized Heaviside function $H_\varepsilon(\phi)$ and the smoothed Dirac delta function $\delta_\varepsilon(\phi)$ with $\varepsilon = 1$ [2], are defined by

$$H_\varepsilon(\phi) = \frac{1}{2}\left(1 + \frac{2}{\pi}\arctan\left(\frac{\phi}{\varepsilon}\right)\right), \quad (5)$$

$$\delta_\varepsilon(\phi) = \frac{1}{\pi}\left(\frac{\varepsilon}{\varepsilon^2 + \phi^2}\right). \quad (6)$$

By minimizing the energy function, image partition and bias intensity estimation are accomplished together by approximating the piecewise constants c_i, the restored bias field b, the variances σ_i^2, and the membership functions $M_i(\phi)$. The minimization of the energy functional with respect to each variable ϕ, c_i, b, and σ_i is performed in the iterative process. These variables are obtained from the derivation of the convolution expression of the energy functional. The optimal c_i, b, and σ_i^2 are given by

$$c_i(\mathbf{x}) = \frac{\int_\Omega (b * K) I M_i(\phi) d\mathbf{y}}{\int_\Omega (b^2 * K) M_i(\phi) d\mathbf{y}}. \quad (7)$$

$$b(\mathbf{y}) = \frac{\sum_{i=1}^{2} \frac{c_i}{\sigma_i^2} (I M_i(\phi) * K)}{\sum_{i=1}^{2} \frac{c_i^2}{\sigma_i^2} (M_i(\phi) * K)}. \quad (8)$$

$$\sigma_i^2 = \frac{\int_\Omega \left(I^2 \mathbf{1}_K - 2c_i I (b * K) + c_i^2 (b^2 * K)\right) M_i(\phi) d\mathbf{x}}{\int_\Omega (M_i(\phi) * K) d\mathbf{x}}. \quad (9)$$

Keeping c_i, b, and σ_i fixed, the energy functional $E(\phi, c_i, b, \sigma_i)$ is minimized with respect to ϕ by solving the gradient flow equation $\frac{\partial \phi}{\partial t} = -\frac{\partial E}{\partial \phi}$. The Gâteaux derivative $\frac{\partial E}{\partial \phi}$ can be computed by using calculus of variations. The corresponding gradient flow equation is defined by

$$\frac{\partial \phi}{\partial t} = \delta_\varepsilon(\phi)\left(v \operatorname{div}\left(\frac{\nabla \phi}{|\nabla \phi|}\right) - e_1 + e_2\right). \quad (10)$$

For each iteration of Eq. (10), the level set function is diffused by Eq. (11) to keep the level set evolution stable [12].

$$\phi^{n+1} = \phi^n + \Delta t \cdot \Delta \phi^n, \quad (11)$$

where ϕ^n represents the level set function of the n-th iteration of Eq. (10), Δt is the diffusion strength, and Δ is the Laplacian operator.

3 Results and Discussion

3.1 Comparison with Other Level Set Methods

Several relevant level set techniques in segmenting a real knee cartilage ultrasound image are compared in this subsection. The other two level set techniques without and with multiplicative noise estimation are summarized as follows. First, the local Gaussian distribution fitting (LGDF) model [9] considers a Gaussian distribution with locally varying mean and variance similar to LSLSM. Because LGDF does not approximate bias field, it can be used for segmentation purpose only. Meanwhile, LSLSM can be applied for simultaneous segmentation and bias correction. Second, the locally weighted K-means variational level set (WKVLS) method is considered [7]. WKVLS does not consider the variance component which helps LSLSM to differentiate the boundary from surrounding tissues more satisfactorily. Both WKVLS and LSLSM are essentially designed for simultaneous segmentation and intensity inhomogeneity correction.

In this experiment, all the methods were implemented in MATLAB R2014a in an Intel (R) Xeon (R), 2.00 GHz, 32 GB RAM using the settings as follows. The kernel's scale $\rho = 5$ was set to be small to produce more accurate segmentation result. The parameter v was chosen as small as 0.001×255^2 for images with intensity range in $[0, 255]$ when capturing objects of any size. The time steps for level set evolution Δt_1 and for regularization Δt_2 were set as $\Delta t_1 = 0.01$ for LGDF, $\Delta t_1 = 0.1$ and $\Delta t_2 = 0.1$ for WKVLS, and $\Delta t_1 = 0.01$ and $\Delta t_2 = 0.01$ for LSLSM. The image size is of 420×150 pixels.

Figure 2 depicts segmentation performances of the three different level set methods when employed to the cartilage boundary segmentation. The initial contour is in circle shape with 10 pixels radius and positioned around the middle of the images. In general, these three level set methods were able to delineate the desired object in the image corrupted by speckle noise and intensity inhomogeneity. This is because the local intensity defined in the local neighborhood that reduces the overlapped intensity distribution. With the joint bias field estimation, WKVLS and LSLSM could suppress

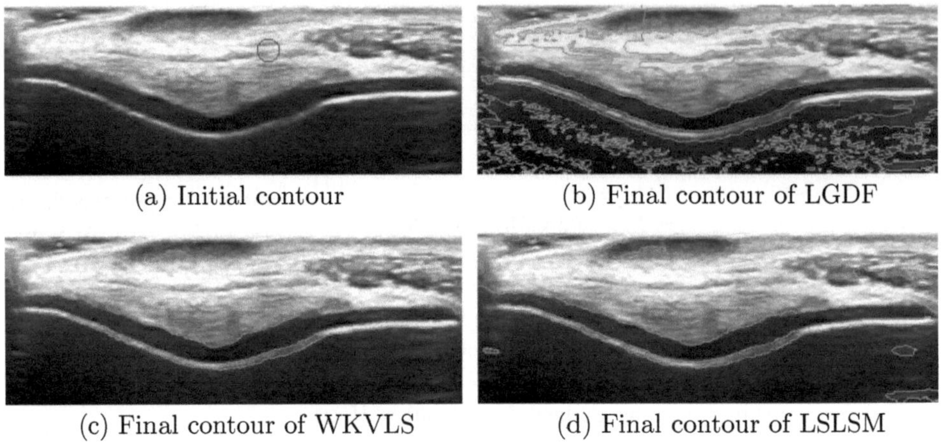

Fig. 2 Segmentation outcomes of three relevant level set techniques in the attempt of segmenting the knee cartilage. The initial contour is depicted by the red circle with 10 pixels radius. The final contours are represented by the green lines

the intensity bias therefore delineate the boundaries between surrounding tissues satisfactorily as depicted in Fig. 2c, d. Without the joint bias field estimation, LGDF produces some misclassified and unnecessary contours inside and around the object as seen in Fig. 2b. Both methods yield satisfactory segmentation outcomes, while LSLSM that takes into account the variance component achieved a more desirable segmentation outcome than WKVLS.

The validation metrics of DSC and HD were computed from the manual outline and the isolated cartilage area as illustrated in Fig. 3. The connected-component labeling was used to extract the cartilage region and remove the other adjacent tissues in the final contours. This is to ensure that the DSC and HD metrics are computed based on the cartilage area only and unaffected by other tissue regions. The first, second, and third rows of the matrices $\begin{bmatrix} 0.9027 \\ 0.9148 \\ 0.9423 \end{bmatrix}$ and $\begin{bmatrix} 6.8557 \\ 7 \\ 6.3246 \end{bmatrix}$ summarized DSC and HD measures for the segmentation outcomes of LGDF, WKVLS, and LSLSM in Fig. 2b–d, respectively. LSLSM obtained DSC value higher than WKVLS and LGDF. Meanwhile, LSLSM obtained HD value smaller than WKVLS and LGDF. Moreover, LGDF, WKVLS, and LSLSM spent the total computational time of 54.82, 13.77, and 12.97 s for 500 iterations, respectively.

3.2 Knee Cartilage Ultrasound Image Segmentation

An application of the three level set techniques in segmenting a set of 80 cartilage images is presented in this subsection. The data sets consist of the real knee cartilage ultrasound images scanned four times each from both knee joints of the ten participants. Figure 4 illustrates a subset of ten segmentation outcomes achieved by LSLSM from both left and right knee cartilages of a subset of five participants. Qualitative and quantitative evaluations are performed to the total 80 segmentation outcomes obtained by LGDF, WKVLS, and LSLSM. While Cohen's κ statistics is employed to validate the segmentation outcomes qualitatively, DSC and HD measures are used to assess the segmentation results quantitatively. The manual segmentation results as gold standard were compared against the isolated cartilage area extracted by the level set methods to be examined qualitatively and quantitatively. The cartilage are delineated manually by the expert from each cartilage ultrasound scan. The connected-component labeling was employed to isolate the cartilage area depicted in Fig. 3 from the adjacent tissues in the final segmentation contours.

The qualitative assessment of the segmentation results was performed by differentiating the boundaries between the both interfaces of the soft tissue-cartilage and the cartilage-bone with the following observations. From the observed agreements of 67 images (83.75% of the observations), 39 images (48.75%) are as grade 1 (excellent), 21 images (26.25%) are as grade 2 (good), 5 images (6.25%) are as grade 3 (poor), 2 images (2.5%) are as grade 4 (bad). The number of agreement due to chance is 32.05 images. Cohen's $\kappa = 0.73$ shows a significant agreement for the overall cartilage segmentation quality rated by two raters.

Figure 5 depicts segmentation results of LGDF, WKVLS, and LSLSM evaluated by DSC and HD measures on a set of 80 cartilage images. DSC values of LGDF, WKVLS, and LSLSM computed from 80 images illustrated in Fig. 5a are ranging from 0.84 to 0.94, 0.29 to 0.95, and 0.82 to 0.95, respectively. A good agreement in size and location of the two comparing contours, which correspond to more accurate segmentation outcomes is indicated by the higher value of DSC. Figure 5b shows HD values of LGDF, WKVLS, and LSLSM fall in the range between 4.47 and 8.83, 5.39 and

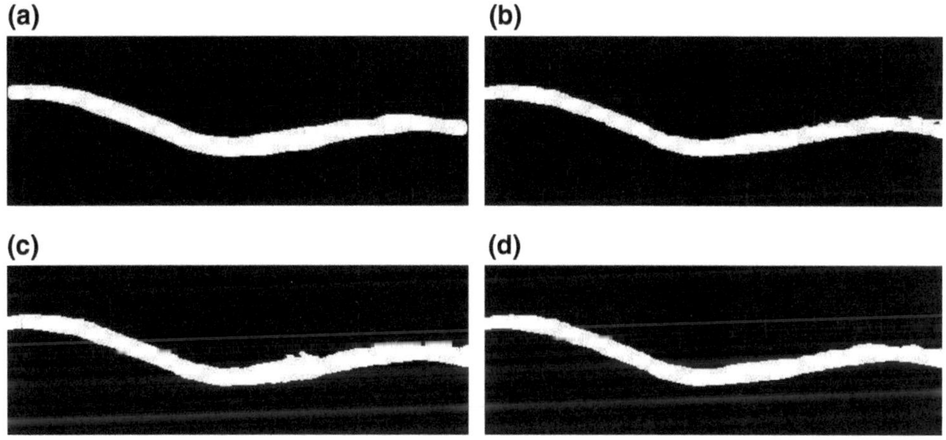

Fig. 3 **a** Manual delineation of the cartilage. Isolated cartilage regions obtained from the segmented images by **b** LGDF, **c** WKVLS, and **d** LSLSM

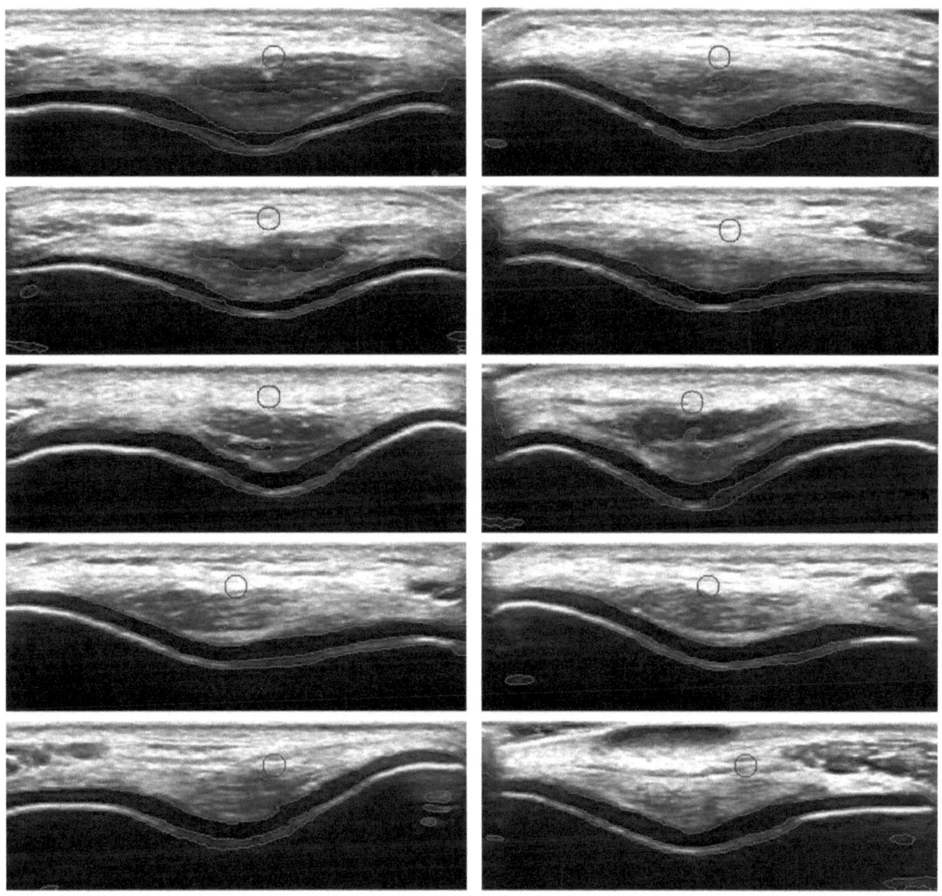

Fig. 4 Left and right columns comprise of the segmentation outcomes achieved by LSLSM from left and right knee cartilages of a subset of five subjects, respectively. The red circles with ten pixels radius put around the middle of the image represent the initial contours. The final contours are depicted by the green lines

19.10, and 4.69 and 8.25 pixels, respectively. The minimal shape difference between the contour pair corresponds to the smaller HD values.

Table 1 summarizes the average values, standard deviations, and p-values for DSC and HD measures of the three methods computed from the total data sets of 80 images. It indicates that the means of DSC values of LSLSM is larger than of LGDF and WKVLS. Moreover, LSLSM obtained smaller means of HD values than LGDF and WKVLS. It can be implied that LSLSM produces an overall satisfying segmentation performance on all set of data illustrated by a good area similarity and the least shape difference of the compared contours. In addition, while the means of LGDF and LSLSM are statistically significant from WKVLS, the means of LGDF is not significantly different from LSLSM.

The overlapping intensity distributions between surrounding tissues caused the segmentation errors. The boundary between the adjacent soft tissue and the bone

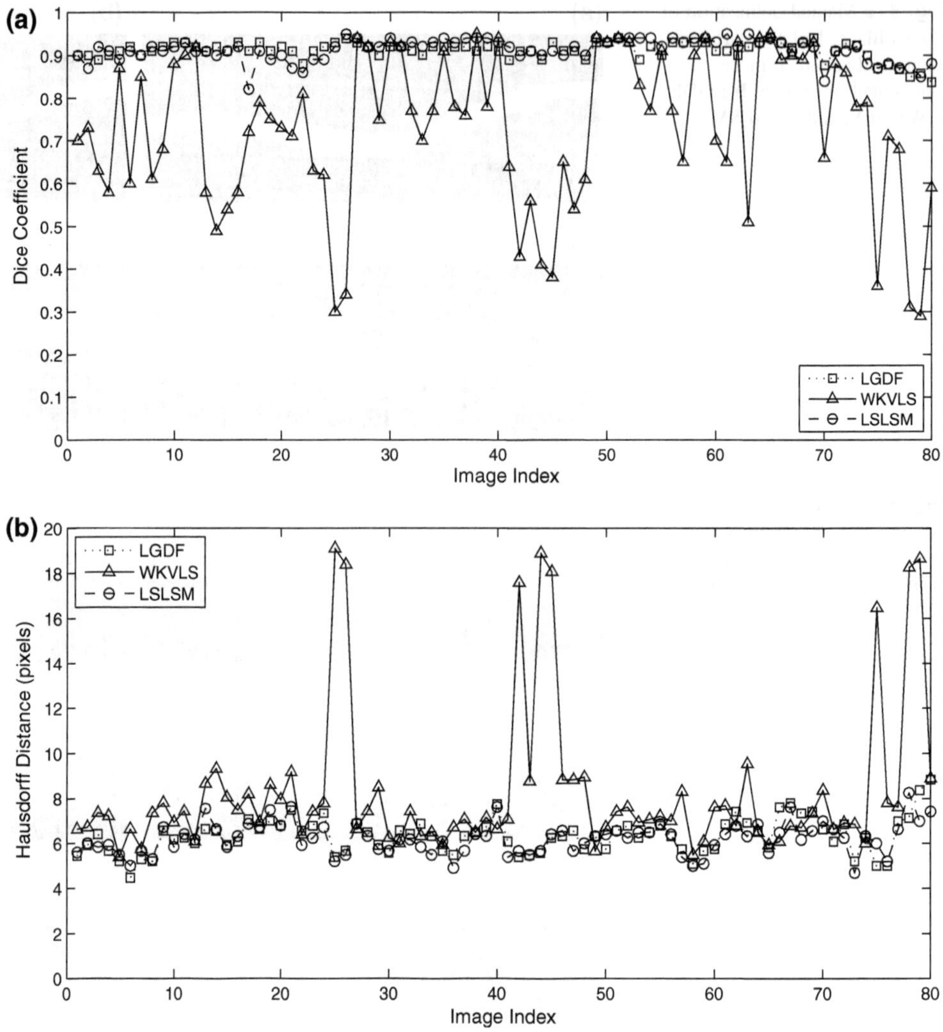

Fig. 5 **a** DSC and **b** HD values obtained by the three level set methods from the total data sets of 80 images

Table 1 Statistics of the validation measures

Methods	DSC		HD (pixels)	
	Mean ± SD	*p*-value	Mean ± SD	*p*-value
LGDF	0.90 ± 0.02	0.69	6.33 ± 0.62	0.27
WKVLS	0.73 ± 0.14	<0.0001	8.32 ± 2.17	<0.0001
LSLSM	0.91 ± 0.01	–	6.21 ± 0.59	–

surface is hard to distinguish. The variance component in the Gaussian distributions considered in LGDF and LSLSM contributes in locating the cartilage boundary more accurately. Although WKVLS take into account the bias field estimation, it has a tendency to misclassify the two interfaces because it does not consider the variance component. DSC measures lower than 0.8 and HD measures higher than 7 pixels in the graph indicates the less satisfactory segmentation result caused by the high variety of intensity bias between the scanned images.

4 Conclusion

The knee cartilage boundary segmentation in the 2-D ultrasound axial view is a challenging task. LSLSM has obtained a more satisfactory outcome than other level set techniques in capturing the cartilage. A significant agreement of the segmentation quality rated by two raters was indicated by Cohen's κ coefficient. A consistent segmentation performance was indicated by DSC and HD measures computed from all available datasets. These segmentation results suggest that the cartilage thickness computation can be made using the segmented cartilage images.

Acknowledgements This work was supported in part by the University of Malaya Research Grant (RP020A-13AET), in part by the International Graduate Research Assistantship Scheme, and in part by the Postgraduate Research Grant (PG003-2014B).

References

1. Aisen, A.M., McCune, W.J., MacGuire, A., Carson, P.L., Silver, T.M., Jafri, S.Z., Martel, W.: Sonographic evaluation of the cartilage of the knee. Radiology **153**(3), 781–784 (1984)
2. Chan, T.F., Vese, L.A.: Active contours without edges. IEEE Trans. Image Process. **10**(2), 266–277 (2001)
3. Fripp, J., Crozier, S., Warfield, S.K., Ourselin, S.: Automatic segmentation and quantitative analysis of the articular cartilages from magnetic resonance images of the knee. IEEE Trans. Med. Imag. **29**(1), 55–63 (2010)
4. Iagnocco, A., Coari, G., Zoppini, A.: Sonographic evaluation of femoral condylar cartilage in osteoarthritis and rheumatoid arthritis. Scand. J. Rheumatol. **21**(4), 201–203 (1992)
5. Jackson, D., Simon, T., Aberman, H.: Symptomatic articular cartilage degeneration: the impact in the new millennium. Clin. Orthop. Relat. Res. **391**, 14–15 (2001)
6. Kazam, J.K., Nazarian, L.N., Miller, T.T., Sofka, C.M., Parker, L., Adler, R.S.: Sonographic evaluation of femoral trochlear cartilage in patients with knee pain. J. Ultrasound Med. **30**(6), 797–802 (2011)
7. Li, C., Huang, R., Ding, Z., Gatenby, C., Metaxas, D.N., Gore, J. C.: A level set method for image segmentation in the presence of intensity inhomogeneities with application to MRI. IEEE Trans. Image Process. **20**(7), 2007–2016 (2011)
8. Naredo, E., Acebes, C., Moller, I., Canillas, F., de Agustin, J.J., de Miguel, E., Filippucci, E., Iagnocco, A., Moragues, C., Tuneu, R., Uson, J., Garrido, J., Delgado-Baeza, E., Saenz-Navarro, I.: Ultrasound validity in the measurement of knee cartilage thickness. Ann. Rheum. Dis. **68**, 1322–1327 (2009)
9. Wang, L., He, L., Mishra, A., Li, C.: Active contours driven by local Gaussian distribution fitting energy. Signal Process. **89**(12), 2435–2447 (2009)
10. Xiao, G., Brady, M., Noble, J.A., Zhang, Y.: Segmentation of ultrasound B-mode images with intensity inhomogeneity correction. IEEE Trans. Med. Imag. **21**(1), 48–57 (2002)
11. Zhang, K., Zhang, L., Lam, K.M., Zhang, D.: A level set approach to image segmentation with intensity inhomogeneity. IEEE Trans. Cybern. **46**(2), 546–557 (2016)
12. Zhang, K., Zhang, L., Song, H., Zhang, D.: Reinitialization-free level set evolution via reaction diffusion. IEEE Trans. Image Process. **22**(1), 258–271 (2013)

Adaptive Network Based Fuzzy Inference System (ANFIS) for an Active Transfemoral Prosthetic Leg by Using In-Socket Sensory System

Nur Hidayah Mohd Yusof, Yan Chai Hum, Nur Azah Hamzaid, and Khin Wee Lai

Abstract

Prosthetic leg is known as one of the solutions to help the amputee to regain back their ambulation ability. However, most of the current existing knee components still lacks in the ability to provide active body propulsion, which in turn. Thus, higher metabolic energy consumption is required by the amputee in doing locomotion movement. Hence, this study proposed the idea of developing both the mechanical structure as well as an ANFIS knowledge-based control system of the active actuated knee joint for transfemoral (TF) prosthetic leg. ANFIS was adopted using Matlab software to analyze human gait phase recognition necessary for cadence and torque control required by the knee joint mechanism while the actuated knee joint was developed using Inventor CAD software. Physical simulation of the controller presented a realistic simulation of the actuated of the knee joint in terms of knee mechanism. The fuzzy system could replicate human gait cycle by categorizing the cycle into seven gait phases.

Keywords

ANFIS • Microprocessor • Prosthetic knee joint • Controller • Transfemoral leg • In-socket sensory system

1 Introduction

Lower limb prosthetics should effectively restore the mobility of an amputee. The evolution of lower limb prosthetics through the years has brought better joint mechanism and structures. Prosthetics have been able to improve lower limb mobility of amputees by offering wide variety of prosthetic components and mechanisms. However, the transfemoral amputees' gait still has noticeable abnormalities, and the prosthetic legs generally have limited ability to perform active body propulsion. In spite of significant development of new technologies during the last decade [1–6], commercial above knee prosthesis are still energetically passive devices. However, many locomotive functions like climbing up stairs and slopes, need significant power in knee and ankle joints. In addition, it has been proved that above knee amputee consumed more metabolic energy during walking compared to their non-amputees counterparts [7–9]. The additional power needed for the previously mentioned activities must be achieved by means of external energy sources that should be integrated in prosthetic components such as the microprocessor knee [2, 3, 10, 11].

In order to compensate for the required robustness of the external power sources, there are a few computer-controlled prosthesis nowadays which adjust the movement of the prosthesis to any speed and compensate any gait deviations and amputee's usage of metabolic energy. The most modern generation of controlled prosthesis [5, 12, 13] with computer controlled phases of standing and swinging are considered to be an important step forward, since they offer the closest proximity to natural walking.

N. H. Mohd Yusof · N. A. Hamzaid · K. W. Lai (✉)
Biomedical Engineering Department, Faculty of Engineering, University of Malaya, Malaya, Malaysia
e-mail: lai.khinwee@um.edu.my

Y. C. Hum
Deparment of Mechatronics and Biomedical Engineering, Lee Kong Chian Faculty of Engineering and Science, Univerisiti Tunku Abdul Rahman, Sungai Long, Malaysia

F. Ibrahim et al. (eds.), *2nd International Conference for Innovation in Biomedical Engineering and Life Sciences*, IFMBE Proceedings 67, https://doi.org/10.1007/978-981-10-7554-4_49

In this study, the design of prosthetic knee should be able to fit all the components supposed to be for a controller to function, i.e.; microproscessor and battery supplied for the system. The primary objective of this study is to develop structural design of microcontroller-based transfemoral prosthesis that is light and small enough to be utilizing within the demographic range and mimic normal leg anthropomorphic Hence, the motivation of this research is to design an active prosthetic knee which is effective and has a modular control/sensing architecture.

2 ANFIS Development Method

2.1 Development of ANFIS for the Control Framework of Transfemoral Prosthetic Leg

ANFIS-based controller was developed to control the actuators movement with respect to the feedback signals. Moreover, it also addressed a new method of human gait phase recognition necessary for cadence control. The control and actions were performed according to a rule-base written in natural language. This information enables the controller to adjust the braking moment automatically, depending on the gait mode, gait velocity and other locomotion requirements. The control action was based on a finite state representation of the gait cycle. A systematic procedure of developing a knowledge based prosthetic controller and the effective way for estimating the gait speed were presented. Particular pattern of the knee angle and moment of the knee joint was used as reference to test the behavior of the knee mechanism. The efficiency of the proposed system was evaluated via simulation and experiment.

Unlike the prior work, the approach proposed herein as shown in Fig. 1 to generate knee torques utilizes a knowledge-based system that considers the uncertainty of inputs and is not dependent of walking cadence dynamic model. Only one PID, as a secondary controller, accompanies the aforementioned system to correct the prosthetic position discrepancies from the desired values due to disturbances like uneven terrain.

2.2 Design of ANFIS for Control Framework

A novel "echo control" has been developed for gait control in which a modified knee trajectory from the sound leg was played back on the contra-lateral side. FIS is used in the proposed control framework to send a correct value of torque to the actuator of the controller-based in-socket sensory system of transfemoral prosthetic knee. The magnitude of torque was associated with the angular position of the tibia and femur of the healthy leg. The rule based system eliminates the need for the mathematical model of the gait cycle during implementation of the control process. Moreover, fuzzy logic is suitable for experiments that utilize low-cost sensors as that is the main purpose of the study. In addition, unity feedback also can be utilized to compare the desired angle of the prosthesis with that of the real position. Then, the errors are sent back to PID controller to adjust the real position of the active knee. In this study, the introduced FIS is a type-1 TSK model [16, 17] which has four inputs the current position of the femur and tibia and their last states and two outputs the command torque and desired angle of the prosthesis. The thigh/leg angle and body-mass-normalized knee torque specifications during one period of the normal gait cycle, are derived from the work of Winter [18]. The repetitive inputs of training data, which is femur and tibia angular position are clustered based on seven phases of the human walking cycle.

A type-1 TSK FIS can be considered as a network that is composed of five layers. Inputs of the FIS come into the first layer where the premise parameters are stored. Membership values of inputs are calculated and sent to the next layer. The second layer computes the firing level of rules based on the defined t-norm. In the third layer, the ratio of each rule's firing level to the sum of all rules's firing levels will be calculated [19–21], and the results will launch the fourth stage. The forth layer gives f1(x) y1 (x) for each rule based on the accumulated consequent parameters in this layer. The fifth layer computes the overall output as a summation of all incoming signals. In order to achieve the desired input-output mapping, the given training data are fed forward to the network, and consequent parameters at layer four are identified by the least square estimate [19]. In the backward pass, while output parameters are fixed, the error rates propagate backward and premise parameters are updated by the gradient descent. Therefore in this technique, although the antecedent parameters are fixed ahead of time, they will be tuned during the training process with the consequent parameters. This iterative method continues until the sum of the squared errors over all the N training data becomes less than a predefined threshold.

3 ANFIS Data Analysis

Simulation of the ANFIS controller proved that the system can evaluated the efficacy of the sensing method that was added on the smart prosthetic leg system and also the accuracy of the mechanical movement based on the controller requirement. Fifteen percent of the data represented by Winter [18] were selected as testing data. Testing data were presented to the trained FIS in order to see how well the FIS model predicts the corresponding

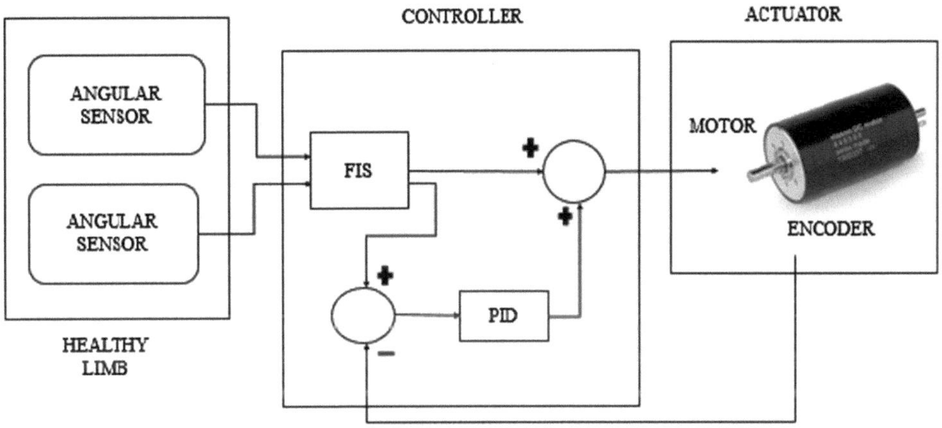

Fig. 1 Control diagram of an active knee prosthesis

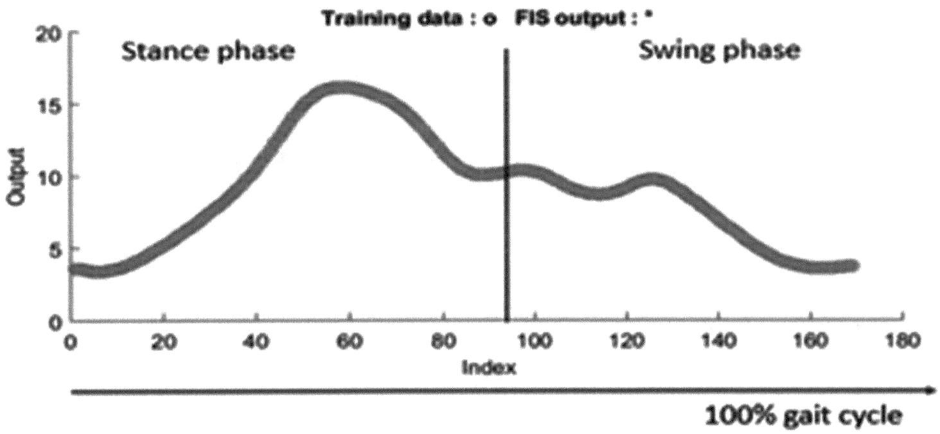

Fig. 2 ANFIS output knee angle versus training data

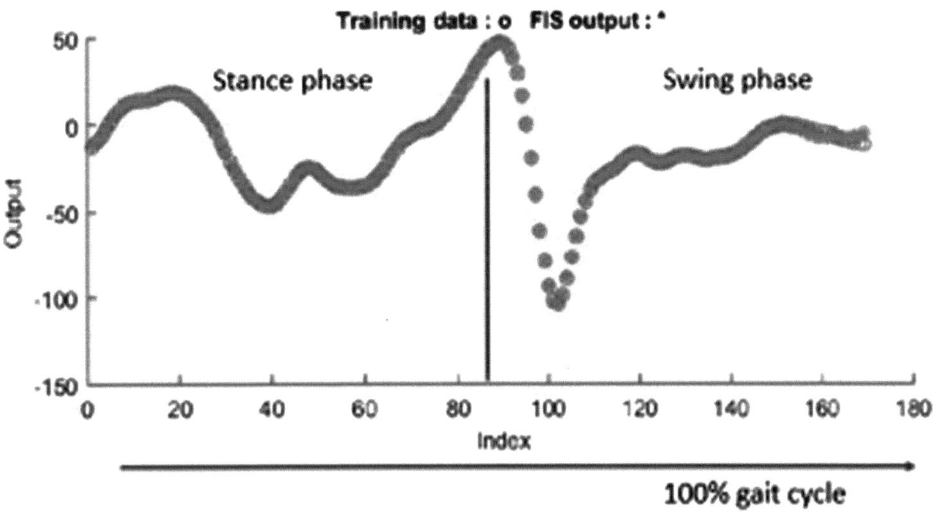

Fig. 3 ANFIS output knee torque versus training data

data set output values. As described before, during training, the proposed FIS was considered as two separated fuzzy systems for each output, (y1 TSK,1 (x) and y2 TSK,1(x)), and they were trained individually even though they have identical input membership functions. After training, both systems considered as one FIS with two outputs and two inputs.

Upon gathering the parameters of the FIS from the training data, testing data utilized new sets of input and outputs. In order to verify the validity, the data was fed into the FIS and was tuned. The FIS outputs for the training data are depicted in Figs. 2 and 3. Meanwhile, Figs. 4 and 5 showed surface viewer for whole ANFIS system using sub-cluster method.

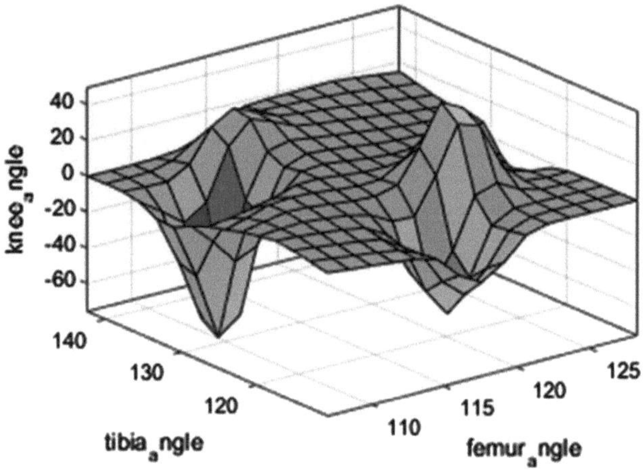

Fig. 4 Surface viewer of ANFIS for knee angle output

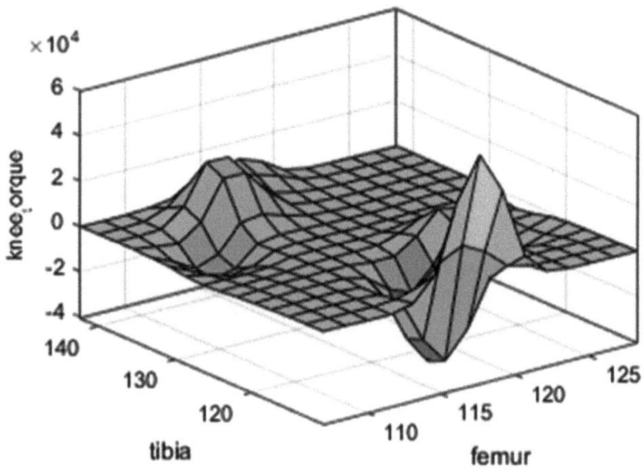

Fig. 5 Surface viewer of ANFIS for knee torque output

In addition, uncertainty in joint angles measurements due to the different patterns of normal walking cycle between individual or even day-to-day activity of same subject (inter-and intra-subject) are considered in designing the knowledge-based system. Moreover, uncertainty of input data corrupted by measurement noise was discussed very briefly. However, due to uncertainty of definition of slow, normal, and fast walking, uncertainty of Gaussian input membership functions must be considered to increase the robustness of the model. This leads to uncertainty about determining the membership value of the entered data (type-2 FIS). Therefore, further study is needed to develop a type-2 FIS for the actuated prosthetic knee controller

4 Conclusion

The mechanical structure was successfully developed and angle of the mechanism was measured at all gait phases during walking. The overall physical dimension of the prosthetic leg allows fitment and user comfort. ANFIS control system, allows high precision and accuracy for the prosthetic legs to replicate human motion ambulation based on simulation the various phases of gait cycle. Further analysis will be carry out later with more patient participation.

Acknowledgements This work was supported in part by the Ministry of Higher Education, Fundamental Research Grant Scheme FRGS (FP047-2014B), and in part by the University Malaya Postgraduate Research Grant (PG125-2015B).

References

1. Zlatnik, D.: Intelligently Controlled Above Knee (A/K) Prosthesis. Institute of Robotics, ETH-Zurich, Switzerland (1998)
2. Aeyels, B., Van Petegem, W., Sloten, J.V., Van Der Perre, G., Peeraer, L.: An EMG-based finite state approach for a microcomputer-controlled above-knee prosthesis. In: Proceedings of 17th International Conference on Engineering in Medicine and Biology Society, vol. 2, pp. 1315–1316 (1995)
3. Zhong, J., et al.: Design and simulation of PD, PID and fuzzy logic controller for industrial application. IEEE Trans. Syst. Man. Cybern. **3**(8), 1–5 (2014)
4. Zahedi, S., Sykes, A., Lang, S., Cullington, I., Zahedi, S.: Adaptive prosthesis—a new concept in prosthetic knee control. Robotica **23**(23), 337–344 (2005)
5. Arieta, A.H., Katoh, R., Yokoi, H., Wenwei, Y.: Development of a multi-DOF electromyography prosthetic system using the adaptive joint mechanism. Appl. Bionics Biomech. **3**(2), 101–112 (2006)
6. Segal, A.D., et al.: Kinematic and kinetic comparisons of transfemoral amputee gait using C-Leg and Mauch SNS prosthetic knees. J. Rehabil. Res. Dev. **43**(7), 857–870 (2006)
7. Huang, C.T., et al.: Amputation: energy cost of ambulation. Arch. Phys. Med. Rehabil. **60**(1), 18–24 (1979)
8. Inman, V.T.: Conservation of energy in ambulation. Arch. Phys. Med. Rehabil. **48**(9), 484–488 (1967)
9. Fisher, S.V., Gullickson, G.: Energy cost of ambulation in health and disability: a literature review. Arch. Phys. Med. Rehabil. **59** (3), 124–133 (1978)
10. Thiele, J., Westebbe, B., Bellmann, M., Kraft, M.: Designs and performance of microprocessorcontrolled knee joints. Biomed. Tech. **59**(1), 65–77 (2014)
11. Kusagur, A., Kodad, S.F., Ram, S.: Modelling & simulation of an ANFIS controller for an AC drive. World J. Model. Simul. **8**(1), 36–49 (2012)

12. Herr, H., Wilkenfeld, A.: User-adaptive control of a magnetorheological prosthetic knee. Ind. Rob. **30**(1), 42–55 (2003)
13. Dyck, W.R., Onyshko, S., Hobson, D., Winter, D., Quanbury, O.: A voluntarily controlled electrohydraulic above-knee prosthesis. Bull. Prosthet. Res. 169–186 (1975)
14. Winter, D.A.: 12 Appendix A Kinematic, Kinetic, and Energy Data (2009)
15. Borjian, R., Khamesee, M.B., Melek, W.: Feasibility study on echo control of a prosthetic knee: sensors and wireless communication. Microsyst. Technol. **16**(1–2), 257–265 (2010)
16. Sugeno, M., Kang, G.T.: Structure identification of fuzzy model. Fuzzy Sets Syst. **28**(1), 15–33 (1988)
17. Feng, G.: A survey on analysis and design of model-based fuzzy control systems. IEEE Trans. Fuzzy Syst. **14**(5), 676–697 (2006)
18. Winter, D.A.: The biomechanics and motor control of human gait: normal, elderly, and pathological (1991)
19. Hong, T.-P., Lee, C.-Y.: Induction of fuzzy rules and membership functions from training examples. Fuzzy Sets Syst. **84**(1), 33–47 (1996)
20. Bai, Y., Wang, D.: Fundamentals of fuzzy logic control—fuzzy sets, fuzzy rules and defuzzifications. Adv. Fuzzy Log. Technol. Ind. Appl. 334–351 (2006)
21. Zadeh, L.: Fuzzy sets. Inf. Control **8**(3), 338–353 (1965)

Vibroarthrography Difference Between Left and Right Knee for Osteoarthritis Detection

Farshad Golshan, Yan Chai Hum, Belinda Pingguan-Murphy, and Khin Wee Lai

Abstract

Osteoarthritis (OA) is the most common type of Arthritis; mostly occur on knees and hip joint. Although majority of elderly suffer from OA, all the existing diagnosis methods have major drawbacks. Past studies have discovered the relationship between knee alignment and weight distribution which are directly related to the OA in the elderly and can be diagnosed years before occurrence of the OA. In this study, specially made Vibroarthrography (VAG) data logging tool was used to collect the knee vibrations of 20 subjects, where 10 of the subjects were confirmed OA patients with grade 2 and higher severity. The test was completed by placing the sensor mounted on knee band of both left and right knees. The goal of this research is to compare the left and right knee of both healthy and OA subjects using Vibroarthrography method. Current results obtained shows significant difference between left and right knee VAG signals for both Healthy and OA knees

Keywords

Vibroarthrography • Leg dominance • STFT

1 Introduction

Osteoarthritis (OA) is one of the most common degenerative knee disease occur on the cartilage between the human joints [3]. Due to wear and tear, the lubricating surfaces between the two cartilage surfaces are removed [4]. Degeneration of the cartilage surface will put direct friction between two knees joint. To overcome the damage, our body will respond to produce spurs on the tip of the joints. Large amount of spurs would block the regeneration of the cartilage and worsen the severity of the OA [5]. OA has become a major issue in our current society due to its painful experiences and could lead to total immobility. However, the major issue with OA is irreversible damage to the joint. Prediction of OA and treatment before its occurrence is the best current solution [14].

Even though exact cause of the OA has not been identified to this day, the indirect causes related to the OA are weight, DNA, gender and age. Recent study explored the weight distribution between the left and right knees and its relation to OA. The study has discovered the subjects with knee bent more than 5° have worse OA progress compared to subjects with less than 5° [12]. Other research studied the athletes and OA relationship associate to the non-dominant knees, result shows that athletes are more susceptible to OA on non-dominant leg rather than dominant leg [6].

Vibroarthrography (VAG) is a primary non-invasive OA detection method currently. However, no major findings have been demonstrated to show the reliability of this method for commercial use [8–10]. VAG was inspired from Vibrography technique where it rely on the detection and visualization of vibration and acoustic emission produced from mechanical structure. It has been used for years in industry to detect faulty in the complex machineries and

F. Golshan · B. Pingguan-Murphy · K. W. Lai (✉)
Biomedical Engineering Department, Faculty of Engineering, University of Malaya, Kuala Lumpur, Malaysia
e-mail: lai.khinwee@um.edu.my

Y. C. Hum
Department of Mechatronics and Biomedical Engineering, Lee Kong Chian Faculty of Engineering and Science, Univerisiti Tunku Abdul Rahman, Sungai Long, Malaysia

F. Ibrahim et al. (eds.), *2nd International Conference for Innovation in Biomedical Engineering and Life Sciences*, IFMBE Proceedings 67, https://doi.org/10.1007/978-981-10-7554-4_50

motors [1, 2]. By placing the sensors on the body of machinery, the vibration in specific area can be detected and hence troubleshooting can be performed [13]. Human knee joint too is a complex structure which can perform complex mechanical motion in its own range of motion (ROM). While knee bones are exposed, abnormal vibration and acoustic emission are produced and distinct to the vibration and acoustic emission produced by normal knee. This difference stem from the extra friction produced by the bones uneven surface [15]. The uneven surface of the patella, tibia and femur motivates the use of vibrograph techniques in detection and comparison of OA knees [7]. The principle of VAG signal lies in detection of the vibration and acoustic emission produced by knee's internal structures using highly sensitive sensors and comparing the results with the ideal signal; the signal's frequency and amplitude difference are the main focus points in VAG signal feature extraction.

This research will focus on the unexplored aspect of the VAG signal analysis by monitoring knee vibration signature on both the healthy and OA knee. This study will monitor the VAG signals and to find the relationship between the VAG signatures signals by using accelerometer. We have created a new data set using noise less capacitive accelerometer and performed the tests on the subjects under load (sit-stand-sit) test. Sit-stand-sit test was previously used to monitor the knee posture indicating any issue with spine as well as joint issues [12]. Based on the previous studies, non-stationary characteristic of the VAG signals promote the use of time-frequency signal analysis method. Longer duration of a certain frequency will result in higher magnitude (dB). However, for OA patient's cartilage, the surface is uneven. The uneven surface will produce higher frequencies, and that in turn translates to higher magnitude (dB) at higher frequencies and lower magnitude (dB) at lower frequencies, if it is compared to Healthy knee.

2 Data Acquisition Method

2.1 Data Acquisition

Data acquisition was taken in collaboration with University Malaya Medical Centre (UMMC), Department of Sports Medicine. Medical Ethics was granted prior to perform the data acquisition on OA patients and subjects with healthy knee (Medical Ethics Number: 20165-19). Total 10 OA patients age between 45 and 75 years old and 10 healthy subjects age between 18 and 30 years old participated in this study. The requirements for subjects include the ability to stand up and sit down without the need of any support, and the OA subjects should have K-L grading of 2 and above approved by designated physician using previously captured knee X-ray.

The sensor band was placed on top of the mid patella, similar to the previous experiments performed as it can minimize the noise produced by the muscles [9]. In this experiment, the subject will sit on a rigid chair and to perform the sit-stand-sit test at specific time interval of 7 s.

2.2 VAG Data Logger

The selected accelerometer model to capture vibrograph signal is MPU6050, together with a tri-axis accelerometer equipped with sampling rate of 1480 samples per second. MPU6050 is a capacitive microcontroller with built-in 12-bit analog to digital converter (ADC) and digital motion processing chip (DMP). The selected sensor dedicated to sensitive measurements in the applications of motion tracking, sport science as well as robotics movement. In comparison to the piezo electric accelerometer used in the previous studies, MPU6050 could provide higher accuracy, noise less data at lower frequencies, and provide all three axis vibration information (unlike mono axis piezo electric accelerometers). This sensor was placed on a sensor pad specially designed by 3D-printer technology using rubber material. A 0.8 mm rubber material was placed between the sensor and skin. The layer must be sufficient thin to capture knee vibration and to provide comfort ability for the subject while performing the STS procedure. Figure 1 illustrates the developed sensor band used to hold the accelerometer. Figure 2 shows the 3D printed prototype mounted under the knee band.

Figure 3 illustrates the basic block diagram of the data logger circuit design. The circuit collects the row vibration data and stores it in the SD card. The use of SD card is essential because of the data transferring rate using Wire serial communication to the computer lead to the data loss problem. Wire serial communication has lower transfer speed (lower bit rate) as compared to the data capturing speed from accelerometer. To obtain the relationship between the right and the left knee, two sensors and data loggers were placed on each knee. The two data loggers were connected together by using a Serial connection. The

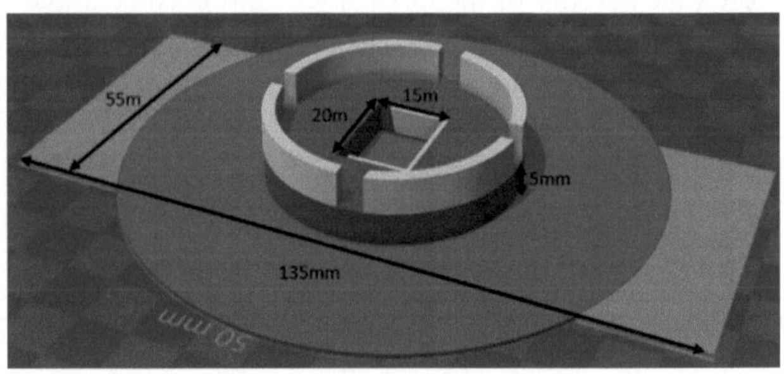

Fig. 1 Knee sensor band where the accelerometer will be placed in

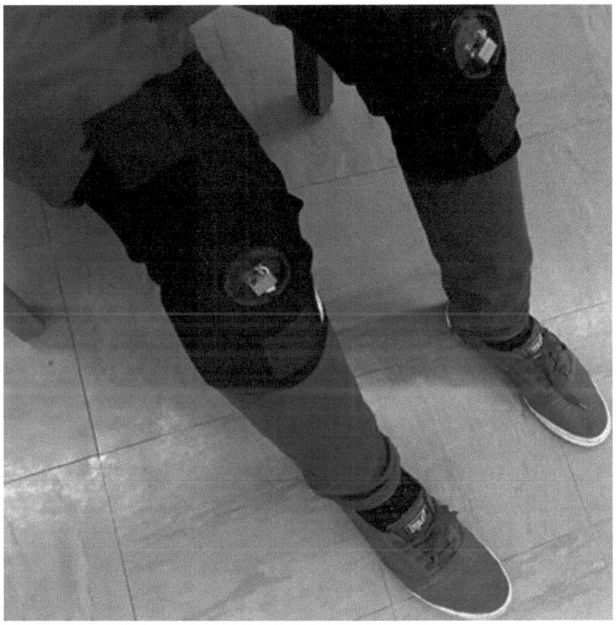

Fig. 2 3D printed sensor band and its placement on the knee

data acquired from the accelerometer was stored on the (class-10) SD card. The sensor and SD card are connected to the AT-MEGA 2560 microcontroller. AT-MEGA 2560 is an 8-bit Atmel microcontroller which can operate at 16 MHz clock speed.

3 Data Analysis

Figure 4 shows the Healthy knees VAG signal after normalization processes, and Fig. 5 shows the stage three OA VAG signal.

Analysis of the signals in frequency spectrum tend to provide more information on the signal properties [11]. Equation 1 presents the per sample root mean square (RMS) of the all three axis signals:

$$RMS_{xyz}(n) = \sqrt{\frac{x(n)^2 + y(n)^2 + z(n)^2}{3}}, \quad 0 \leq n \leq 10{,}000 \tag{1}$$

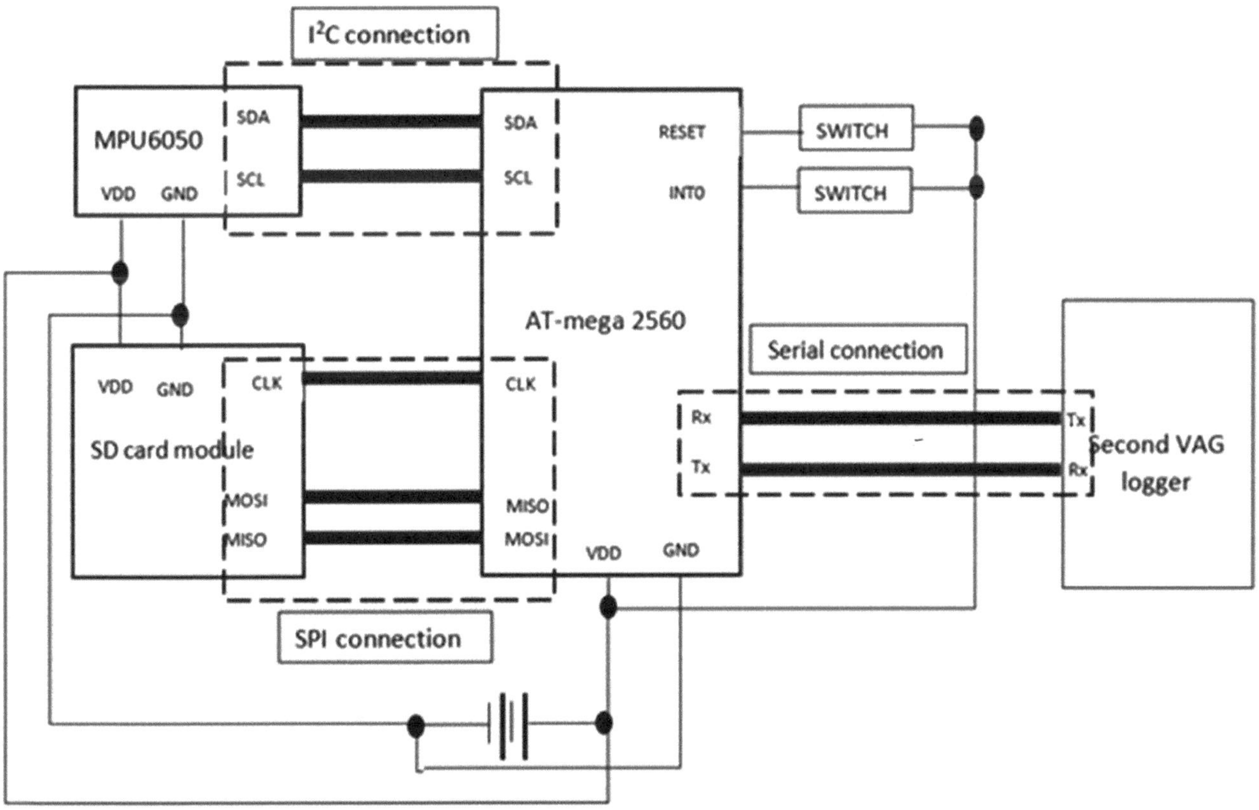

Fig. 3 Simplified block diagram of the developed VAG data loggers

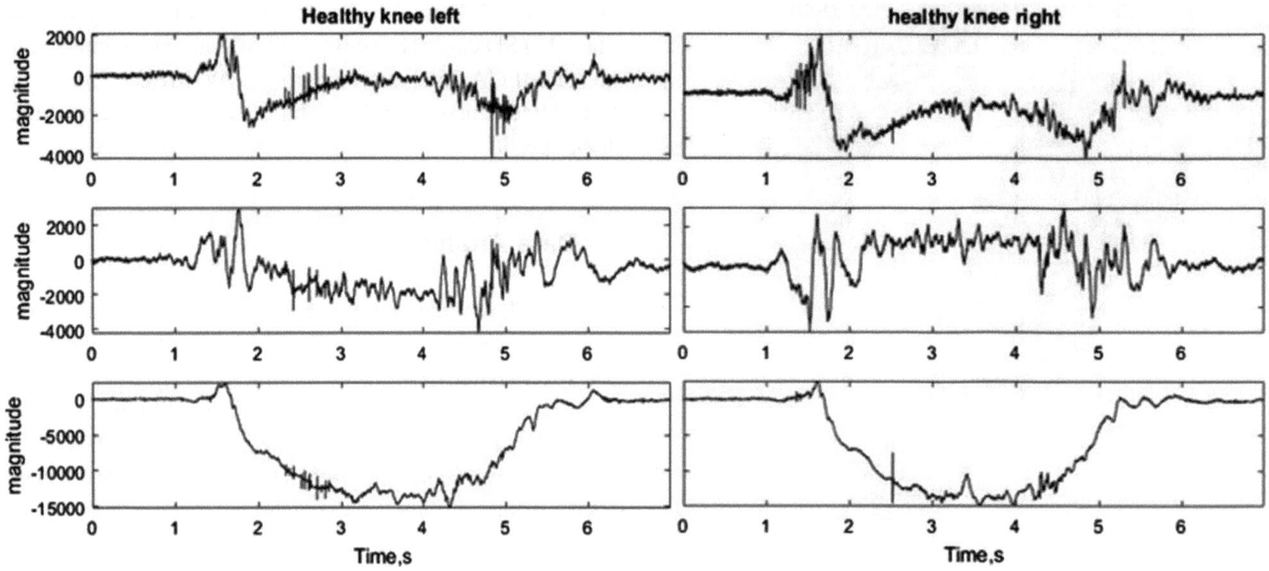

Fig. 4 Healthy knee VAG signal captures from top, x, y, z axis

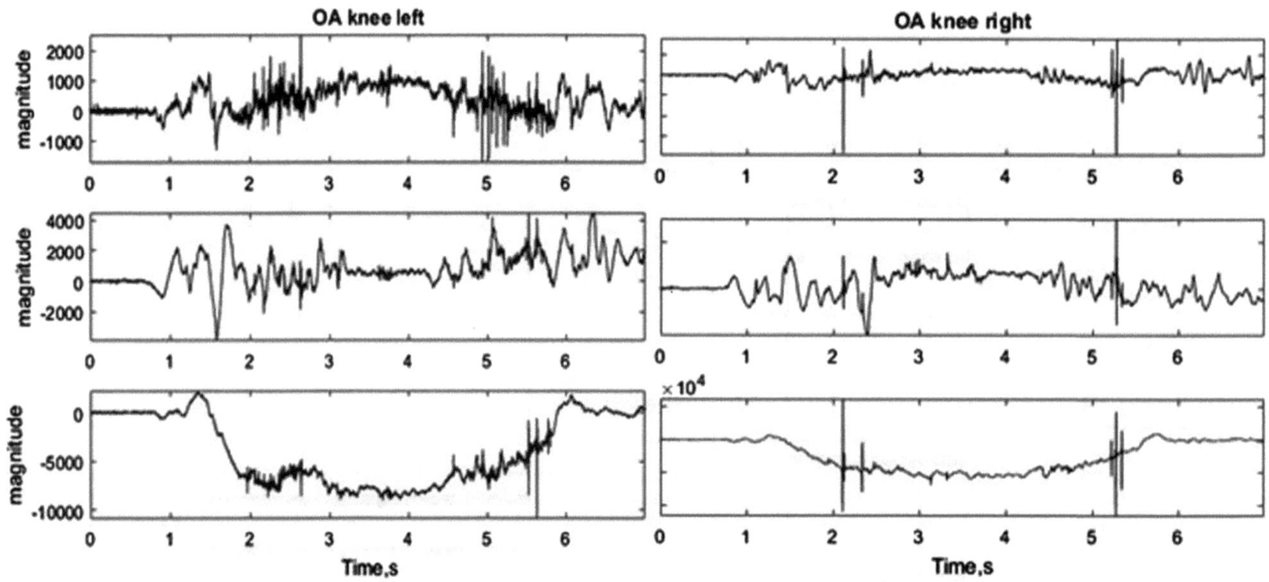

Fig. 5 OA knee VAG signal captures from top, x, y, z axis

RMS of the 3-axis retain the features of all three signals in the frequency domain. Considering that VAG signals are non-stationary, the use of fast Fourier transform may not provide detailed indication on properties and features of the signals. Time-frequency signal analysis of vibrograph and VAG signals can provide details on signal frequency and magnitude at each stage of standing up and sitting down. STFT equation used for this study is as follows:

$$STFT(t,f) = \int_{0}^{7} \left([x(t) \cdot \omega'(t - t')] \cdot e^{2\pi ft} \right) dt \quad (2)$$

The selected humming window is 150 samples; the window size for each Fourier transform is 11.66 ms. Figure 6 shows the spectrogram of left and right healthy knees, and Fig. 7 shows the spectrogram of OA knees.

For better perspective of the findings (Figs. 6 and 7), the maximum magnitude at each sample has been taken and the comparison plot of the left and right knee is shown in Fig. 8 (both OA and healthy knee).

As shown in Fig. 7, the left knee VAG signals on both OA and healthy knees have much lower frequency magnitude compared to right knee. This phenomenon may be related to the knee alignment and weight distribution of the

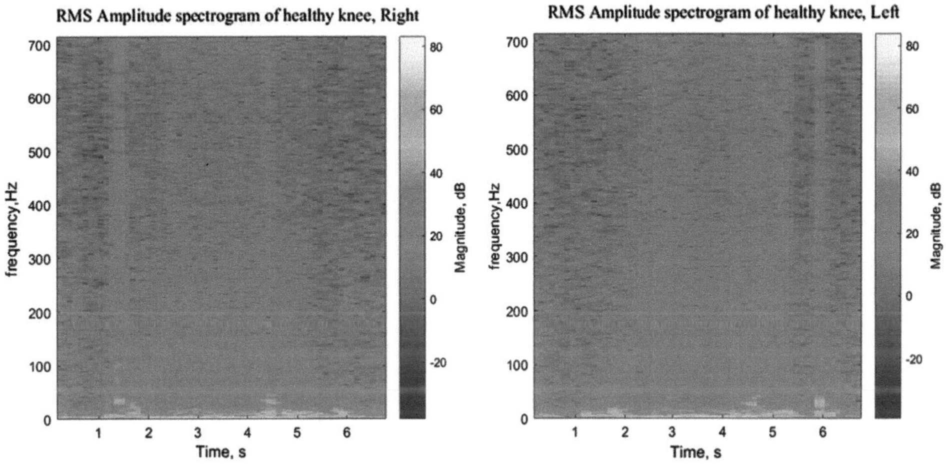

Fig. 6 Spectrogram of STFT RMS of Healthy knee, left and right

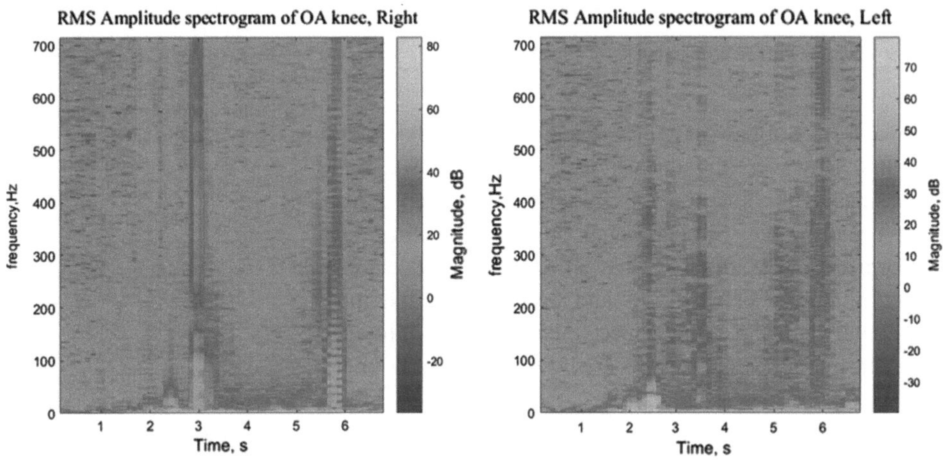

Fig. 7 Spectrogram of STFT RMS of OA knee, left and right

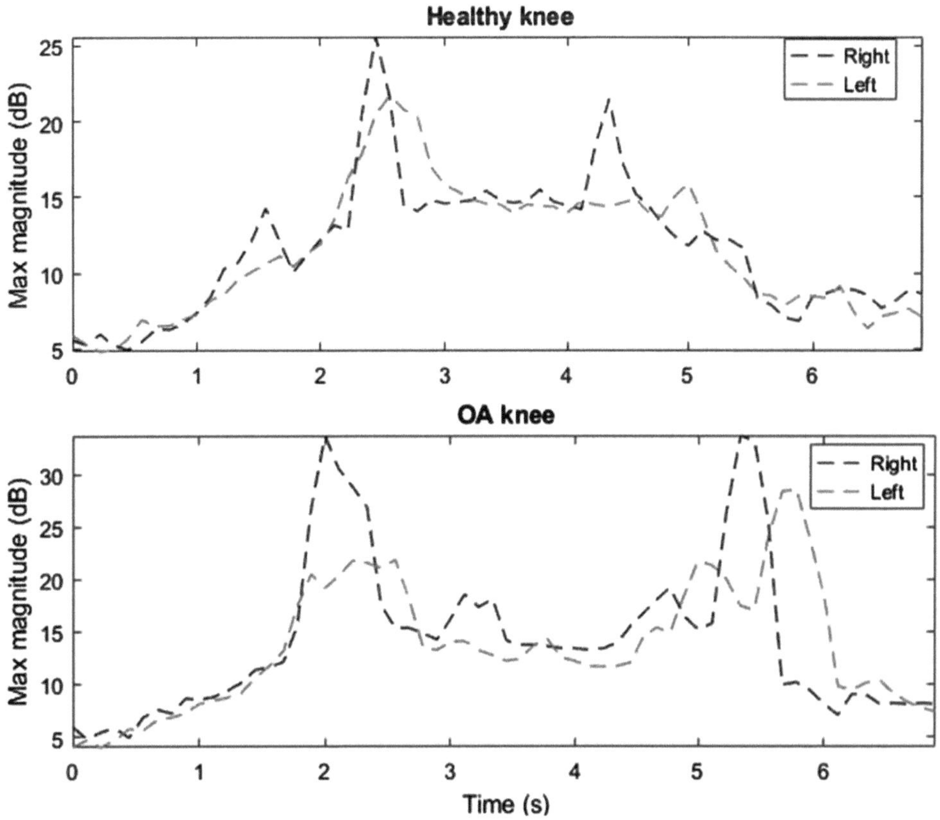

Fig. 8 Left and Right knee comparison of the OA and healthy knee

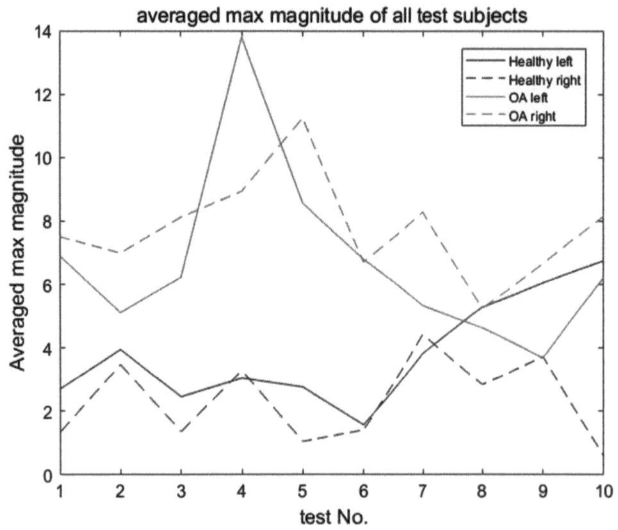

Fig. 9 Overview of max magnitude of all subjects

knee and show a flaw in human posture which could lead to arthritis. However, the fluctuation difference between the both sides of the knee on the OA knee is relatively higher than healthy knee. Similar signal processing method was performed on all 20 subjects as shown in Fig. 9, there is major difference between OA and healthy knee but most importantly, on the same subject there is major difference between left and right knee for both OA and healthy knee.

4 Conclusion

Prediction and detection of the main cause of OA would be the best outcome due to current limited treatment methods. This study introduces a new insight in prediction arthritis long before the first signs of the disease. We compared the left and right knee VAG signals from OA and Healthy knees. Depending on the leg dominance, the fluctuation and deformity on the cartilage may differ due to body weight distribution and knee alignment.

Acknowledgements This work was supported in part by the University of Malaya Research University Grant (RU018 N-2016), in part by the International Graduate Research Assistantship Scheme, and in part by the Postgraduate Research Grant (PG202-2016A).

References

1. Baydar, N., Ball, A.: A Comparative Study of acoustic and vibration signals in detection of gear failures using Wigner–Ville distribution. Mech. Syst. Signal Process. **15**, 1091–1107 (2001)
2. Baydar, N., Ball, A.: Detection of gear failures via vibration and acoustic signals using wavelet transform. Mech. Syst. Signal Process. **17**, 787–804 (2003)
3. Callaghan, J.: The Adult Knee. Lippincott Williams & Wilkins, Philadelphia (2003)
4. Davis, J, Eaton, C.B., Lo, G.H., et al.: Knee symptoms among adults at risk for accelerated knee osteoarthritis: data from the Osteoarthritis Initiative. Clin. Rheumatol. **36**, 1083–1089 (2017)
5. Hinterwimmer, S., Krammer, M., Krotz, M., et al.: Cartilage atrophy in the knees of patients after seven weeks of partial load bearing. Arthritis Rheum. **50**, 2516–2520 (2004)
6. Krajnc, Z., Vogrin, M., Recnik, G., et al.: Increased risk of knee injuries and osteoarthritis in the non-dominant leg of former professional football players. Wiener klinische Wochenschrift **122** (Suppl 2), 40–43 (2010)
7. Krishnan, S., Rangayyan, R.M., Bell, G.D., et al.: Adaptive time-frequency analysis of knee joint vibroarthrographic signals for noninvasive screening of articular cartilage pathology. IEEE Trans. Bio-medical Eng. **47**, 773–783 (2000)
8. Park, H.-J., Kim, S.S., Lee, S.-Y., et al.: A practical MRI grading system for osteoarthritis of the knee: Association with Kellgren–Lawrence radiographic scores. Eur. J. Radiol. **82**, 112–117 (2012)
9. Rangayyan, R.M., Krishnan, S., Bell, G.D., et al.: Parametric representation and screening of knee joint vibroarthrographic signals. IEEE Trans. Biomed. Eng. **44**, 1068–1074 (1997)
10. Rangayyan, R.M., Yunfeng, W.: Screening of knee-joint vibroarthrographic signals using parameters of activity and radial-basis functions. In: 2008 Canadian Conference on Electrical and Computer Engineering, pp. 000057–000060 (2008)
11. Scruby, C.B.: An introduction to acoustic emission. J. Phys. E: Sci. Instrum. **20**, 946 (1987)
12. Sharma, L., Song, J., Felson, D.T., et al.: The role of knee alignment in disease progression and functional decline in knee osteoarthritis. JAMA **286**, 188–195 (2001)
13. Tan, C.K., Irving, P., Mba, D.: A comparative experimental study on the diagnostic and prognostic capabilities of acoustics emission, vibration and spectrometric oil analysis for spur gears. Mech. Syst. Signal Process. **21**, 208–233 (2007)
14. Veerapen, K., Wigley, R.D., Valkenburg, H.: Musculoskeletal pain in Malaysia: a COPCORD survey. J. Rheumatol. **34**, 207–213 (2007)
15. Wierzcholski, K.: Friction forces for human hip joint lubrication at a naturally permeable cartilage. Appl. Mech. Eng. **11**, 515–527 (2006)

Erratum to: Is There a Relationship Between Bone Mineral Density, Mammographic Breast Density and Serum Estrogen Concentration Level? Study on Malaysian Women

N. M. Zain, K. C. Kanaga, E. S. Latif, and A. Shantini

Erratum to:
Chapter "Is There a Relationship Between Bone Mineral Density, Mammographic Breast Density and Serum Estrogen Concentration Level? Study on Malaysian Women" in: F. Ibrahim et al. (eds.), 2nd International Conference for Innovation in Biomedical Engineering and Life Sciences, IFMBE Proceedings 67, https://doi.org/10.1007/978-981-10-7554-4_45

The original version of the book was inadvertently published with same figure for both Figures 2 and 3 in Chapter "Is There a Relationship Between Bone Mineral Density, Mammographic Breast Density and Serum Estrogen Concentration Level? Study on Malaysian Women" in which Figure 3 has been corrected now.

The updated online version of this chapter can be found at
https://doi.org/10.1007/978-981-10-7554-4_45

F. Ibrahim et al. (eds.), *2nd International Conference for Innovation in Biomedical Engineering and Life Sciences*,
IFMBE Proceedings 67, https://doi.org/10.1007/978-981-10-7554-4_51

Author Index

F. Ibrahim et al. (eds.), *2nd International Conference for Innovation in Biomedical Engineering and Life Sciences*,
IFMBE Proceedings 67, https://doi.org/10.1007/978-981-10-7554-4

MIX
Papier aus verantwortungsvollen Quellen
Paper from responsible sources
FSC® C105338

If you have any concerns about our products,
you can contact us on
ProductSafety@springernature.com

In case Publisher is established outside the EU,
the EU authorized representative is:
Springer Nature Customer Service Center GmbH
Europaplatz 3, 69115 Heidelberg, Germany

Printed by Libri Plureos GmbH
in Hamburg, Germany